Kawasaki 400, 500 & 550 Fours Owners Workshop Manual

by Jeremy Churchill

Models covered

Z400 J1. 399cc. UK 1980 to 1981
Z400 J2. 399cc. UK 1981 to 1982
Z400 J3. 399cc. UK 1981 to 1983
ZR400 A1. 399cc. UK 1983
ZR400 B1. 399cc. UK 1984 to 1985
ZX400 C2. 399cc. UK 1984 to 1987
Z500 B1. 497cc. UK 1979
Z500 B2. 497cc. UK 1980
KZ/Z550 A1. 553cc. UK 1980 to 1981, US 1980
KZ/Z550 A2. 553cc. UK and US 1981
KZ/Z550 A3. 553cc. UK 1981 to 1983, US 1982
ZR550 A1. 553cc. UK 1982 to 1983
KZ550 A4. 553cc. US 1983
ZR550 A2. 553cc. UK 1984 to 1987
KZ/Z550 C1. 553cc. UK 1980 to 1981, US 1980
KZ/Z550 C2. 553cc. UK 1981 to 1983, US 1981
KZ550 C3. 553cc. US 1982
KZ550 C4. 553cc. US 1983
KZ/Z550 D1. 553cc. UK 1981 to 1982, US 1981
KZ/Z550 H1. 553cc. UK and US 1982
KZ/Z550 H2. 553cc. UK 1982 to 1983, US 1983
ZX550 A1/A1L. 553cc. UK 1983 to 1985, US 1984
ZX550 A2/A2L. 553cc. UK 1985 to 1987, US 1985
ZX550 A3. 553cc. UK 1986 to 1987
ZX550 A4. 553cc. UK 1988 on
Z550 G1. 553cc. UK 1983 to 1984
Z550 G2. 553cc. UK 1984 to 1986
Z550 G3. 553cc. UK 1986 to 1987
Z550 G4. 553cc. UK 1987 on
KZ550 F1. 553cc. US 1983
KZ550 F2/F2L. 553cc. US 1984
KZ550 M1. 553cc. US 1983

ISBN 978 1 85010 486 5

(910-12AS8)

J H Haynes & Co. Ltd.
Haynes North America, Inc

www.haynes.com

Disclaimer

There are risks associated with automotive repairs. The ability to make repairs depends on the individual's skill, experience and proper tools. Individuals should act with due care and acknowledge and assume the risk of performing automotive repairs.

The purpose of this manual is to provide comprehensive, useful and accessible automotive repair information, to help you get the best value from your vehicle. However, this manual is not a substitute for a professional certified technician or mechanic.

This repair manual is produced by a third party and is not associated with an individual vehicle manufacturer. If there is any doubt or discrepancy between this manual and the owner's manual or the factory service manual, please refer to the factory service manual or seek assistance from a professional certified technician or mechanic.

Even though we have prepared this manual with extreme care and every attempt is made to ensure that the information in this manual is correct, neither the publisher nor the author can accept responsibility for loss, damage or injury caused by any errors in, or omissions from, the information given.

British Library Cataloguing in Publication Data

Churchill, Jeremy, 1954-
Kawasaki 400, 500 & 550 fours owners workshop manual
1. Motorcycles. Maintenance & repair. Amateurs' manuals
I. Title II. Series
629.28'775
ISBN 1-85010-486-7

Library of Congress Catalog Card Number

88-81607

Acknowledgements

Our thanks are due to Kawasaki Motors (UK) Ltd who provided the ZX550 A1 model featured throughout this Manual and the bulk of the service literature; also to the staff of that company who willingly helped clarify various points. We would also like to thank E.A. Taylor and Son of Misterton, Crewkerne, Somerset who gave valuable assistance, John Rose and Ian Barnes who allowed their machines, a Z550 H2 and Z550 A3 respectively, to be examined on several occasions, and CW Motorcycles of Dorchester who supplied the ZX550 featured on the front cover.

The Avon Rubber Company supplied information on tyre care and fitting, and NGK Spark Plugs (UK) Ltd provided information on spark plug maintenance and electrode conditions.

About this manual

The purpose of this manual is to present the owner with a concise and graphic guide which will enable him to tackle any operation from basic routine maintenance to a major overhaul. It has been assumed that any work would be undertaken without the luxury of a well-equipped workshop and a range of manufacturer's service tools.

To this end, the machine featured in the manual was stripped and rebuilt in our own workshop, by a team comprising a mechanic, a photographer and the author. The resulting photographic sequence depicts events as they took place, the hands shown being those of the author and the mechanic.

The use of specialised, and expensive, service tools was avoided unless their use was considered to be essential due to risk of breakage or injury. There is usually some way of improvising a method of removing a stubborn component, providing that a suitable degree of care is exercised.

The author learnt his motorcycle mechanics over a number of years, faced with the same difficulties and using similar facilities to those encountered by most owners. It is hoped that this practical experience can be passed on through the pages of this manual.

Where possible, a well-used example of the machine is chosen for the workshop project, as this highlights any areas which might be particularly prone to giving rise to problems. In this way, any such difficulties are encountered and resolved before the text is written, and the techniques used to deal with them can be incorporated in the relevant section. Armed with a working knowledge of the machine, the author undertakes a considerable amount of research in order that the maximum amount of data can be included in the manual.

A comprehensive section, preceding the main part of the manual, describes procedures for carrying out the routine maintenance of the machine at intervals of time and mileage. This section is included particularly for those owners who wish to ensure the efficient day-to-day running of their motorcycle, but who choose not to undertake overhaul or renovation work.

Each Chapter is divided into numbered sections. Within these sections are numbered paragraphs. Cross reference throughout the manual is quite straightforward and logical. When reference is made 'See Section 6.10' it means Section 6, paragraph 10 in the same Chapter. If another Chapter were intended, the reference would read, for example, 'See Chapter 2, Section 6.10'. All the photographs are captioned with a section/paragraph number to which they refer and are relevant to the Chapter text adjacent.

Figures (usually line illustrations) appear in a logical but numerical order, within a given Chapter. Fig. 1.1 therefore refers to the first figure in Chapter 1.

Left-hand and right-hand descriptions of the machines and their components refer to the left and right of a given machine when the rider is seated normally.

Motorcycle manufacturers continually make changes to specifications and recommendations, and these, when notified, are incorporated into our manuals at the earliest opportunity.

We take great pride in the accuracy of information given in this manual, but motorcycle manufacturers make alterations and design changes during the production run of a particular motorcycle of which they do not inform us. No liability can be accepted by the authors or publishers for loss, damage or injury caused by any errors in, or omissions from, the information given.

Contents

Right-hand view of the Z400 J1

Right-hand view of the Z500 B2

Left-hand view of the Z550 A1

Right-hand view of the ZR550 A1

Left-hand view of the Z550 C1

Right-hand view of the ZX550 A1

Left-hand view of the Z550 G1

Right-hand view of the KZ550 F2

Introduction to the Kawasaki 400, 500 and 550 Fours

This manual covers the Kawasaki four-stroke four-cylinder models of 400, 500 and 550cc capacity produced from 1979 onwards. The information given in this introduction is sufficient to identify each model exactly; it is essential that there is no confusion when working on a machine or when ordering replacement parts. Note that models are referred to at all times in this Manual by their full Kawasaki code, and not by their popular name, whether the latter is written on the machine or not. Also note that US models are prefixed KZ, whereas UK models are prefixed Z. Where any information given applies only to the UK or US version of a model the relevant prefix is used, but where information applies to both UK and US versions, the full identification is written thus: KZ/Z.

Given below, grouped by engine capacity and model type, are the major identifying features of each model, with details of paintwork, followed by the production year, full identification code, popular name (where applicable), frame and engine numbers. Where only one frame number is given for any model, that is the number with which production commenced. Note that the dates refer to the year of production by Kawasaki; this is not necessarily the same as the date of sale, or registration.

Z400 J1, J2, J3, ZR400 A1, B1

These models were sold only in the UK, their styling and specification being that of a standard sports/touring motorcycle. The Z400 J1 was very similar to the Z500 model, except that it had a rear drum brake instead of a disc, and was finished in Grand Prix Silver. The J2 version was fitted with air-assisted front forks, adjustable damping rear shock absorbers, tubeless tyres, a passenger grab rail and different fuel tank/seat unit striping, and was finished in Moondust Silver or Candy Royal Blue. The J3 version is distinguished by its megaphone-shaped silencers and chromed top collar on the rear shock absorbers, and was finished in Luminous Gun Blue or Passion Red. The machine was heavily modified in 1983, this resulting in the ZR400. This model is easily distinguished by its completely new styling, similar to the KZ/Z550 H1 and H2 but without the fairing, and is fitted with the 'Unitrak' rear suspension system of those models. The ZR400 A1 was finished in Polar White or Candy Cobalt Blue; although the ZR400 B1 is fitted with a much-modified engine this is not obvious from the outside and the two can be distinguished only by the wider stripes on the fuel tank, side panels and seat unit of the B1, which is finished in Polar White or Luminous Nautical Blue.
Engine and frame numbers are as follows:

Year	Model code and name	Frame number	Engine number
1980	Z400 J1	KZ400J-000001	KZ400EE011501
1981	Z400 J2	KZ400J-003101	KZ400EE028001
1982	Z400 J3	KZ400J-008501-008508, 008701	KZ400EE048001
1983	ZR400 A1 (Z400 F)	ZR400A-000001	KZ400EE065001-065004, 072001
1984	ZR400 B1 (Z400 F)	ZR400B-000001	ZX400AE000001

Z500 B1, B2

The models on which all others in this Manual are based, these were sold only in the UK in the form of a standard sports/touring motorcycle. The B1 version was finished in Firecracker Red or Metallic Crystal Silver, while the B2 version was finished in Firecracker Red or Ebony. The two can be distinguished only by the fact that the B2 has stripes on its fuel tank and seat unit. Engine and frame numbers are as follows:

Year	Model code	Frame number	Engine number
1979	Z500 B1	KZ500B-000001	KZ500AE000001
1980	Z500 B2	KZ500B-004001	KZ500AE000101

KZ/Z550 A1, A2, A3, KZ550 A4, ZR550 A1, A2

These are the standard sports/touring motorcycles which differed from the 500 largely in having a 3 mm larger cylinder bore and a drum rear brake instead of a disc, also US models were fitted with single disc front brakes, as opposed to the twin disc brake of all UK models. The Z550 A1 was finished in Grand Prix Silver, while the KZ550 A1 was finished in Ebony and fitted with a chromed front fender. The A2 version is distinguished by its air-assisted front forks, adjustable damping rear shock absorbers, passenger grabrail, tubeless tyres, megaphone-shaped silencers and by the different striping on fuel tank and seat unit; it was finished in Luminous Passion Red, the UK version also being available in Sundance Blue. The A3 version can be distinguished only by the chromed top collars of its rear shock absorbers and, on the UK model only, by its rearset footrests and gearchange linkage. Paintwork was finished in Galaxy Silver or Ebony on the US model and Galaxy Silver or Luminous Passion Red on the UK model. In 1983 the KZ550 A4 was produced for the US only, being identified by the adoption of constant-depression carburettors and finished in Ebony, while the ZR550 appeared in the UK. The ZR550 is identical in appearance and specification to the ZR400 described above; again only the wider stripes on the fuel tank, sidepanels and seat unit of the A2 distinguish the 1983 and 1984 models. The ZR550 A1 was finished in Candy Wine Red or Candy Silver, while the A2 is finished in Candy Cardinal Silver or Ebony. Engine/frame numbers are:

Year	Model code and name	Frame number	Engine number
1980	Z550 A1	KZ550A-000001	KZ550AE000001
1980	KZ550 A1	KZ550A-000001-003792	KZ550AE000001-011013
1981	Z550 A2	KZ550A-004101	KZ550AE011601
1981	KZ550 A2	KZ550A-004101-005735	KZ550AE011601-035400
1981	KZ550 A2-New VIN No.	JKAKZFA1* BA005736-010900	KZ550AE011601-035400
1982	Z550 A3	KZ550A-010901	KZ550AE035401-036087
1982	Z550 A3	KZ550A-010901	KZ550AE036093
1982	KZ550 A3-made in Japan	JKAKZFA1* CA010901-012000	KZ550AE035401-055500
1982	KZ550 A3-made in US	JKAKZFA1* CB500001-502605	KZ550AE035401-055500
1983	KZ550 A4	JKAKZFA1* DA012001-013451	KZ550AE055501-061053
1983	ZR550 A1 (Z550F)	ZR550A-000001	KZ550DE027501
1984	ZR550 A2 (Z550F)	ZR550A-002501	KZ550DE041801

KZ/Z550 C1, C2, KZ550 C3, C4 – the Ltd or Custom models

Produced in the style of factory custom machines, these are all known as 550 LTD models. Differing from the 500 model in the same way as the 550 A models, these machines were fitted from the start with air-assisted front forks, adjustable damping rear shock absorbers and tubeless tyres, in addition to the stepped seats, teardrop fuel tanks, pullback handlebars and other styling features common to this type of machine. There are no external features by which individual models can be distinguished, as only minor modifications were made throughout their life. The C1 version was finished in Luminous Dark Red, with the UK model also being available in Ebony, while the C2 version was finished in Luminous Midnight Red, again with the UK model also being available in Ebony. The KZ550 C3 was finished in Ebony or in Sundance Blue, while the C4 version was available in Ebony only. All models had a chromed front mudguard. Engine and frame numbers are as follows:

Year	Model code	Frame number	Engine number
1980	Z550 C1	KZ550C-000001	KZ550AE000001
1980	KZ550 C1	KZ550C-000001-007910	KZ550AE000001-010942
1981	Z550 C2	KZ550C-008101	KZ550AE011601
1981	KZ550 C2	KZ550C-008101-011608	KZ550AE011601-035400
1981	KZ550 C2-New VIN No.	JKAKZFC1* BA011609-021700	KZ550AE011601-035400
1982	KZ550 C3-Made in Japan	JKAKZFC1* CA021701-023700	KZ550AE035401-055500
1982	KZ550 C3-Made in US	JKAKZFC1* CB500001-513905	KZ550AE035401-055500
1983	KZ550 C4	JKAKZFC1* DB514000-518102	KZ550AE055501-061103

KZ/Z550 D1, H1, H2, ZX550 A1, A1L – the GPz models

Based on the KZ/Z550 A2, but with up-rated engine, brakes and suspension, and finished in the characteristic Firecracker Red paintwork with black-painted engine and exhausts, these models have been the 550cc member of Kawasaki's family of super sports machines, the 'GPz' range. The KZ/Z550 D1 is easiest to distinguish as it is the only model not fitted with Unitrak rear suspension. The H1 and H2 models can be distinguished only by their graphics: the H2 fairing has stripes (the H1 item is plain) the H2 fuel tank/tail unit stripes are wider than on the H1 and while the H1 sidepanels have thin stripes along their upper edge, those on the H2 only have a large flash in the centre. The ZX550 A1 is a heavily restyled and modified version of the H1/H2 models, with a still more sophisticated specification. The ZX550 A1L is sold in California only, and differs in being fitted with a sealed fuel tank and carburettors, these being vented to atmosphere only via a charcoal filter to remove all traces of petrol vapour. Note that the Z550 H1 was also available in a finish of Metallic Sonic Gold, while ZX550 A1 models are also available in Galaxy Silver. Engine and frame numbers are as follows:

Year	Model code	Frame number	Engine number
1981	Z550 D1	KZ550D-000001	KZ550DE000001
1981	KZ550 D1	JKAKZFD1* BA000001	KZ550DE000001
1982	Z550 H1	KZ550H-000001	KZ550DE011301
1982	KZ550 H1	JKAKZFH1* CA000001-015000	KZ550DE011301-027500
1983	Z550 H2	KZ550H-015001	KZ550DE027501
1983	KZ550 H2-Made in Japan	JKAKZFH1* DA015001-025612	KZ550DE027501-041258
1983	KZ550 H2-Made in US	JKAKZFH1* DB500001-501900	KZ550DE027501-041258
1984	ZX550 A1-UK model	ZX550A-000001	KZ550DE041801
1984	ZX550 A1, A1L-US model	JKAZXFA1* EA000001	KZ550DE041801

Z550 G1, G2 – the GT550 models

Sold in the UK only these machines are fitted with shaft drive, built in luggage carriers, large capacity fuel tanks and fully adjustable air suspension at front and rear to equip them for the role of touring motorcycles. Although both are finished in Candy Wine Red the G2 model can be distinguished by the stripes on its fuel tank (against the G1's plain tank), by its black wheels and bevel drive cover (as opposed to the G1's gold components), by the chromed exhaust system and by the polished highlights and side cover on its engine/gearbox unit (as opposed to the all-black finish of the G1). Engine/frame numbers are as follows:

Year	Model code	Frame number	Engine number
1983	Z550 G1	KZ550G-000001	KZ550FE000001
1984	Z550 G2	KZ550G-002001	KZ550FE006101

KZ550 M1, F1, F2, F2L

Sold in the US only these machines are all factory-custom styled. The KZ550 M1 is a shaft-driven version of the KZ550 C models, with a similar finish, and was known as the 550 (Shaft) Ltd. The KZ550 F1 was similar, but was fitted with fully adjustable air suspension at front and rear and finished in Ebony/Sonic Gold or Ebony/Firecracker Red. With its striking gold finish on the engine/gearbox unit, this model was known as the Spectre 550. In 1984 a degree of rationalisation was achieved, the M-model being deleted and the F2 losing its gold finished engine but retaining its air suspension to become the 1984 Ltd model. The KZ550 F2 is finished in Ebony or in Luminous Crimson Red and is fitted with a chromed front fender. The KZ550 F2L is sold in California alone, and differs only in having the evaporative emission control system similar to that of the ZX550 A1L. Engine and frame numbers are:

Year	Model code and name	Frame number	Engine number
1983	KZ550 M1 (Ltd)	JKAKZFM1* DA000001	KZ550FE000001
1983	KZ550 F1 (Spectre)	JKAKZFF1* DA000001	KZ550FE000001
1984	KZ550 F2, F2L (Ltd)	JKAKZFF1* EA002401	KZ550FE006101

Note: *The digit indicated by the asterisk (*) in the new VIN numbers of US models varies from machine to machine.*

Model dimensions and weights

	Z400 J1,J2,Z500 B1,B2,Z550 A1,A2	Z400 J3,Z550 A3	ZR400 A1,B1 ZR550 A1,A2	KZ550 A1,A2 A3,A4	KZ/Z550 C1, C2, KZ550 C3,C4
Overall length	2150 mm (84.6 in)	2150 mm (84.6 in)	2235 mm (88.0 in)	2100 mm (82.7 in)	UK – 2190 mm (86.2 in) US – 2160 mm (85.0 in)
Overall width	740 mm (29.1 in)	745 mm (29.3 in)	750 mm (29.5 in)	A1, A2 – 785 mm (30.9 in) A3, A4 – 795 mm (31.3 in)	850 mm (33.5 in)
Overall height	1095 mm (43.1 in)	1095 mm (43.1 in)	1105 mm (43.5 in)	1125 mm (44.3 in)	1200 mm (47.2 in)
Wheelbase	1395 mm (54.9 in)	1395 mm (54.9 in)	1450 mm (57.1 in)	1395 mm (54.9 in)	1420 mm (55.9 in)
Seat height	805 mm (31.7 in)	805 mm (31.7 in)	795 mm (31.3 in)	805 mm (31.7 in)	770 mm (30.3 in)
Ground clearance	145 mm (5.7 in)	150 mm (5.9 in)	160 mm (6.3 in)	A1, A2 – 145 mm (5.7 in) A3, A4 – 150 mm (5.9 in)	140 mm (5.5 in)
Dry weight	Z400 J1 – 194.0 kg (428 lb) Z500 B1, B2 Z550 A1 – 192.0 kg (423 lb) Z400 J2,Z550 A2 191.0 kg (421 lb)	191.0 kg (421 lb)	400 – 180.0 kg (397 lb) 550 – 184.0 kg (405 lb)	A1 – 189.0 kg (417 lb) A2, A3, A4 – 188.0 kg (415 lb)	UK – 198.0 kg (436 lb) US – 192.0 kg (423 lb)

	KZ/Z550 D1	KZ/Z550 H1,H2	ZX550 A1, A1L	Z550 G1,G2	KZ550 F1,F2,F2L,M1
Overall length	UK – 2150 mm (84.6 in) US – 2100 mm (82.7 in)	UK – 2240 mm (88.2 in) US – 2180 mm (85.8 in)	UK – 2205 mm (86.8 in) US – 2155 mm (84.8 in)	2230 mm (87.8 in)	2185 mm (86.0 in)
Overall width	740 mm (29.1 in)	750 mm (29.5 in)	720 mm (28.3 in)	755 mm (29.7 in)	855 mm (33.7 in)
Overall height	1185 mm (46.7 in)	1180 mm (46.5 in)	1245 mm (49.0 in)	1100 mm (43.3 in)	1190 mm (46.9 in)
Wheelbase	1400 mm (55.1 in)	1450 mm (57.1 in)	1445 mm (56.9 in)	1475 mm (58.1 in)	1485 mm (58.5 in)
Seat height	805 mm (31.7 in)	780 mm (30.7 in)	780 mm (30.7 in)	800 mm (31.5 in)	F1 – 775 mm (30.5 in) F2, F2L – 765 mm (30.1 in) M1 – 755 mm (29.7 in)
Ground clearance	145 mm (5.7 in)	160 mm (6.3 in)	160 mm (6.3 in)	155 mm (6.1 in)	155 mm (6.1 in)
Dry weight	199.5 kg (440 lb)	UK – 193.0 kg (425 lb) US – 192.0 kg (423 lb)	A1, UK and US – 191.0 kg (421 lb) A1L – 191.5 kg (422 lb)	201.0 kg (443 lb)	F1,F2 – 197.6 kg (436 lb) F2L – 198.0 kg (437 lb) M1 – 196.7 kg (434 lb)

Ordering spare parts

When ordering spare parts for the models described in this Manual it is recommended that the owner should deal with an official Kawasaki dealer. Be wary of using 'pattern' parts. Whilst in many cases these may be of acceptable quality, there is always risk of failure, perhaps causing extensive damage to the machine or even the rider. Note also that the use of pattern spares will invalidate the warranty.

Retain any worn or broken parts until the correct replacements have been obtained; they may prove essential to identify the correct replacement where detail design changes have taken place. When ordering, always quote the engine and frame numbers in full. These are located on the crankcase and steering head respectively.

Some of the more expendable parts such as spark plugs, bulbs, tyres, oils and greases etc., can be obtained from accessory shops and motor factors, who have convenient opening hours, charge lower prices and can often be found not far from home. It is possible to obtain parts on Mail Order basis from a number of specialists who advertise regularly in the motor cycle magazines.

Engine number is stamped on raised boss near oil filler cap

Frame number is stamped on right-hand side of steering head

Safety first!

Professional motor mechanics are trained in safe working procedures. However enthusiastic you may be about getting on with the job in hand, do take the time to ensure that your safety is not put at risk. A moment's lack of attention can result in an accident, as can failure to observe certain elementary precautions.

There will always be new ways of having accidents, and the following points do not pretend to be a comprehensive list of all dangers; they are intended rather to make you aware of the risks and to encourage a safety-conscious approach to all work you carry out on your vehicle.

Essential DOs and DON'Ts

DON'T start the engine without first ascertaining that the transmission is in neutral.
DON'T suddenly remove the filler cap from a hot cooling system – cover it with a cloth and release the pressure gradually first, or you may get scalded by escaping coolant.
DON'T attempt to drain oil until you are sure it has cooled sufficiently to avoid scalding you.
DON'T grasp any part of the engine, exhaust or silencer without first ascertaining that it is sufficiently cool to avoid burning you.
DON'T allow brake fluid or antifreeze to contact the machine's paintwork or plastic components.
DON'T syphon toxic liquids such as fuel, brake fluid or antifreeze by mouth, or allow them to remain on your skin.
DON'T inhale dust – it may be injurious to health (see *Asbestos* heading).
DON'T allow any spilt oil or grease to remain on the floor – wipe it up straight away, before someone slips on it.
DON'T use ill-fitting spanners or other tools which may slip and cause injury.
DON'T attempt to lift a heavy component which may be beyond your capability – get assistance.
DON'T rush to finish a job, or take unverified short cuts.
DON'T allow children or animals in or around an unattended vehicle.
DON'T inflate a tyre to a pressure above the recommended maximum. Apart from overstressing the carcase and wheel rim, in extreme cases the tyre may blow off forcibly.
DO ensure that the machine is supported securely at all times. This is especially important when the machine is blocked up to aid wheel or fork removal.
DO take care when attempting to slacken a stubborn nut or bolt. It is generally better to pull on a spanner, rather than push, so that if slippage occurs you fall away from the machine rather than on to it.
DO wear eye protection when using power tools such as drill, sander, bench grinder etc.
DO use a barrier cream on your hands prior to undertaking dirty jobs – it will protect your skin from infection as well as making the dirt easier to remove afterwards; but make sure your hands aren't left slippery. Note that long-term contact with used engine oil can be a health hazard.
DO keep loose clothing (cuffs, tie etc) and long hair well out of the way of moving mechanical parts.
DO remove rings, wristwatch etc, before working on the vehicle – especially the electrical system.
DO keep your work area tidy – it is only too easy to fall over articles left lying around.
DO exercise caution when compressing springs for removal or installation. Ensure that the tension is applied and released in a controlled manner, using suitable tools which preclude the possibility of the spring escaping violently.
DO ensure that any lifting tackle used has a safe working load rating adequate for the job.
DO get someone to check periodically that all is well, when working alone on the vehicle.
DO carry out work in a logical sequence and check that everything is correctly assembled and tightened afterwards.
DO remember that your vehicle's safety affects that of yourself and others. If in doubt on any point, get specialist advice.
IF, in spite of following these precautions, you are unfortunate enough to injure yourself, seek medical attention as soon as possible.

Asbestos

Certain friction, insulating, sealing, and other products – such as brake linings, clutch linings, gaskets, etc – contain asbestos. *Extreme care must be taken to avoid inhalation of dust from such products since it is hazardous to health.* If in doubt, assume that they *do* contain asbestos.

Fire

Remember at all times that petrol (gasoline) is highly flammable. Never smoke, or have any kind of naked flame around, when working on the vehicle. But the risk does not end there – a spark caused by an electrical short-circuit, by two metal surfaces contacting each other, by careless use of tools, or even by static electricity built up in your body under certain conditions, can ignite petrol vapour, which in a confined space is highly explosive.

Always disconnect the battery earth (ground) terminal before working on any part of the fuel or electrical system, and never risk spilling fuel on to a hot engine or exhaust.

It is recommended that a fire extinguisher of a type suitable for fuel and electrical fires is kept handy in the garage or workplace at all times. Never try to extinguish a fuel or electrical fire with water.

Note: *Any reference to a 'torch' appearing in this manual should always be taken to mean a hand-held battery-operated electric lamp or flashlight. It does* **not** *mean a welding/gas torch or blowlamp.*

Fumes

Certain fumes are highly toxic and can quickly cause unconsciousness and even death if inhaled to any extent. Petrol (gasoline) vapour comes into this category, as do the vapours from certain solvents such as trichloroethylene. Any draining or pouring of such volatile fluids should be done in a well ventilated area.

When using cleaning fluids and solvents, read the instructions carefully. Never use materials from unmarked containers – they may give off poisonous vapours.

Never run the engine of a motor vehicle in an enclosed space such as a garage. Exhaust fumes contain carbon monoxide which is extremely poisonous; if you need to run the engine, always do so in the open air or at least have the rear of the vehicle outside the workplace.

The battery

Never cause a spark, or allow a naked light, near the vehicle's battery. It will normally be giving off a certain amount of hydrogen gas, which is highly explosive.

Always disconnect the battery earth (ground) terminal before working on the fuel or electrical systems.

If possible, loosen the filler plugs or cover when charging the battery from an external source. Do not charge at an excessive rate or the battery may burst.

Take care when topping up and when carrying the battery. The acid electrolyte, even when diluted, is very corrosive and should not be allowed to contact the eyes or skin.

If you ever need to prepare electrolyte yourself, always add the acid slowly to the water, and never the other way round. Protect against splashes by wearing rubber gloves and goggles.

Mains electricity and electrical equipment

When using an electric power tool, inspection light etc, always ensure that the appliance is correctly connected to its plug and that, where necessary, it is properly earthed (grounded). Do not use such appliances in damp conditions and, again, beware of creating a spark or applying excessive heat in the vicinity of fuel or fuel vapour. Also ensure that the appliances meet the relevant national safety standards.

Ignition HT voltage

A severe electric shock can result from touching certain parts of the ignition system, such as the HT leads, when the engine is running or being cranked, particularly if components are damp or the insulation is defective. Where an electronic ignition system is fitted, the HT voltage is much higher and could prove fatal.

Tools and working facilities

The first priority when undertaking maintenance or repair work of any sort on a motorcycle is to have a clean, dry, well-lit working area. Work carried out in peace and quiet in the well-ordered atmosphere of a good workshop will give more satisfaction and much better results than can usually be achieved in poor working conditions. A good workshop must have a clean flat workbench or a solidly constructed table of convenient working height. The workbench or table should be equipped with a vice which has a jaw opening of at least 4 in (100 mm). A set of jaw covers should be made from soft metal such as aluminium alloy or copper, or from wood. These covers will minimise the marking or damaging of soft or delicate components which may be clamped in the vice. Some clean, dry, storage space will be required for tools, lubricants and dismantled components. It will be necessary during a major overhaul to lay out engine/gearbox components for examination and to keep them where they will remain undisturbed for as long as is necessary. To this end it is recommended that a supply of metal or plastic containers of suitable size is collected. A supply of clean, lint-free, rags for cleaning purposes and some newspapers, other rags, or paper towels for mopping up spillages should also be kept. If working on a hard concrete floor note that both the floor and one's knees can be protected from oil spillages and wear by cutting open a large cardboard box and spreading it flat on the floor under the machine or workbench. This also helps to provide some warmth in winter and to prevent the loss of nuts, washers, and other tiny components which have a tendency to disappear when dropped on anything other than a perfectly clean, flat, surface.

Unfortunately, such working conditions are not always available to the home mechanic. When working in poor conditions it is essential to take extra time and care to ensure that the components being worked on are kept scrupulously clean and to ensure that no components or tools are lost or damaged.

A selection of good tools is a fundamental requirement for anyone contemplating the maintenance and repair of a motor vehicle. For the owner who does not possess any, their purchase will prove a considerable expense, offsetting some of the savings made by doing-it-yourself. However, provided that the tools purchased meet the relevant national safety standards and are of good quality, they will last for many years and prove an extremely worthwhile investment.

To help the average owner to decide which tools are needed to carry out the various tasks detailed in this manual, we have compiled three lists of tools under the following headings: *Maintenance and minor repair, Repair and overhaul,* and *Specialized.* The newcomer to practical mechanics should start off with the simpler jobs around the vehicle. Then, as his confidence and experience grow, he can undertake more difficult tasks, buying extra tools as and when they are needed. In this way, a *Maintenance and minor repair* tool kit can be built-up into a *Repair and overhaul* tool kit over a considerable period of time without any major cash outlays. The experienced home mechanic will have a tool kit good enough for most repair and overhaul procedures and will add tools from the specialized category when he feels the expense is justified by the amount of use these tools will be put to.

It is obviously not possible to cover the subject of tools fully here. For those who wish to learn more about tools and their use there is a book entitled *Motorcycle Workshop Practice Manual* available from the publishers of this manual.

As a general rule, it is better to buy the more expensive, good quality tools. Given reasonable use, such tools will last for a very long time, whereas the cheaper, poor quality, item will wear out faster and need to be renewed more often, thus nullifying the original saving. There is also the risk of a poor quality tool breaking while in use, causing personal injury or expensive damage to the component being worked on.

For practically all tools, a tool factor is the best source since he will have a very comprehensive range compared with the average garage or accessory shop. Having said that, accessory shops often offer excellent quality tools at discount prices, so it pays to shop around. There are plenty of tools around at reasonable prices, but always aim to purchase items which meet the relevant national safety standards. If in doubt, seek the advice of the shop proprietor or manager before making a purchase.

The basis of any toolkit is a set of spanners. While open-ended spanners with their slim jaws, are useful for working on awkwardly-positioned nuts, ring spanners have advantages in that they grip the nut far more positively. There is less risk of the spanner slipping off the nut and damaging it, for this reason alone ring spanners are to be preferred. Ideally, the home mechanic should acquire a set of each, but if expense rules this out a set of combination spanners (open-ended at one end and with a ring of the same size at the other) will provide a good compromise. Another item which is so useful it should be considered an essential requirement for any home mechanic is a set of socket spanners. These are available in a variety of drive sizes. It is recommended that the $\frac{1}{2}$-inch drive type is purchased to begin with as although bulkier and more expensive than the $\frac{3}{8}$-inch type, the larger size is far more common and will accept a greater variety of torque wrenches, extension pieces and socket sizes. The socket set should comprise sockets of sizes between 8 and 24 mm, a reversible ratchet drive, an extension bar of about 10 inches in length, a spark plug socket with a rubber insert, and a universal joint. Other attachments can be added to the set at a later date.

Maintenance and minor repair tool kit

Set of spanners 8 – 24 mm
Set of sockets and attachments
Spark plug spanner with rubber insert – 10, 12, or 14 mm as appropriate
Adjustable spanner
C-spanner/pin spanner
Torque wrench (same size drive as sockets)
Set of screwdrivers (flat blade)
Set of screwdrivers (cross-head)
Set of Allen keys 4 – 10 mm
Impact screwdriver and bits
Ball pein hammer – 2 lb
Hacksaw (junior)
Self-locking pliers – Mole grips or vice grips
Pliers – combination
Pliers – needle nose
Wire brush (small)
Soft-bristled brush
Tyre pump
Tyre pressure gauge
Tyre tread depth gauge
Oil can
Fine emery cloth
Funnel (medium size)
Drip tray
Grease gun
Set of feeler gauges
Brake bleeding kit
Strobe timing light
Continuity tester (dry battery and bulb)
Soldering iron and solder
Wire stripper or craft knife
PVC insulating tape
Assortment of split pins, nuts, bolts, and washers

Repair and overhaul toolkit

The tools in this list are virtually essential for anyone undertaking major repairs to a motorcycle and are additional to the tools listed above. Concerning Torx driver bits, Torx screws are encountered on some of the more modern machines where their use is restricted to fastening certain components inside the engine/gearbox unit. It is therefore recommended that if Torx bits cannot be borrowed from a local dealer, they are purchased individually as the need arises. They are not in regular use in the motor trade and will therefore only be available in specialist tool shops.

Plastic or rubber soft-faced mallet
Torx driver bits
Pliers – electrician's side cutters
Circlip pliers – internal (straight or right-angled tips are available)
Circlip pliers – external
Cold chisel
Centre punch
Pin punch
Scriber
Scraper (made from soft metal such as aluminium or copper)
Soft metal drift
Steel rule/straight edge
Assortment of files
Electric drill and bits
Wire brush (large)
Soft wire brush (similar to those used for cleaning suede shoes)
Sheet of plate glass
Hacksaw (large)
Valve grinding tool
Valve grinding compound (coarse and fine)
Stud extractor set (E-Z out)

Specialized tools

This is not a list of the tools made by the machine's manufacturer to carry out a specific task on a limited range of models. Occasional references are made to such tools in the text of this manual and, in general, an alternative method of carrying out the task without the manufacturer's tool is given where possible. The tools mentioned in this list are those which are not used regularly and are expensive to buy in view of their infrequent use. Where this is the case it may be possible to hire or borrow the tools against a deposit from a local dealer or tool hire shop. An alternative is for a group of friends or a motorcycle club to join in the purchase.

Valve spring compressor
Piston ring compressor
Universal bearing puller
Cylinder bore honing attachment (for electric drill)
Micrometer set
Vernier calipers
Dial gauge set
Cylinder compression gauge
Vacuum gauge set
Multimeter
Dwell meter/tachometer

Care and maintenance of tools

Whatever the quality of the tools purchased, they will last much longer if cared for. This means in practice ensuring that a tool is used for its intended purpose; for example screwdrivers should not be used as a substitute for a centre punch, or as chisels. Always remove dirt or grease and any metal particles but remember that a light film of oil will prevent rusting if the tools are infrequently used. The common tools can be kept together in a large box or tray but the more delicate, and more expensive, items should be stored separately where they cannot be damaged. When a tool is damaged or worn out, be sure to renew it immediately. It is false economy to continue to use a worn spanner or screwdriver which may slip and cause expensive damage to the component being worked on.

Fastening systems

Fasteners, basically, are nuts, bolts and screws used to hold two or more parts together. There are a few things to keep in mind when working with fasteners. Almost all of them use a locking device of some type; either a lock washer, lock nut, locking tab or thread adhesive. All threaded fasteners should be clean, straight, have undamaged threads and undamaged corners on the hexagon head where the spanner fits. Develop the habit of replacing all damaged nuts and bolts with new ones.

Rusted nuts and bolts should be treated with a rust penetrating fluid to ease removal and prevent breakage. After applying the rust penetrant, let it 'work' for a few minutes before trying to loosen the nut or bolt. Badly rusted fasteners may have to be chiseled off or removed with a special nut breaker, available at tool shops.

Flat washers and lock washers, when removed from an assembly should always be replaced exactly as removed. Replace any damaged washers with new ones. Always use a flat washer between a lock washer and any soft metal surface (such as aluminium), thin sheet metal or plastic. Special lock nuts can only be used once or twice before they lose their locking ability and must be renewed.

If a bolt or stud breaks off in an assembly, it can be drilled out and removed with a special tool called an E-Z out. Most dealer service departments and motorcycle repair shops can perform this task, as well as others (such as the repair of threaded holes that have been stripped out).

Spanner size comparison

Jaw gap (in)	Spanner size
0.250	$\frac{1}{4}$ in AF
0.276	7 mm
0.313	$\frac{5}{16}$ in AF
0.315	8 mm
0.344	$\frac{11}{32}$ in AF; $\frac{1}{8}$ in Whitworth
0.354	9 mm
0.375	$\frac{3}{8}$ in AF
0.394	10 mm
0.433	11 mm
0.438	$\frac{7}{16}$ in AF
0.445	$\frac{3}{16}$ in Whitworth; $\frac{1}{4}$ in BSF
0.472	12 mm
0.500	$\frac{1}{2}$ in AF
0.512	13 mm
0.525	$\frac{1}{4}$ in Whitworth; $\frac{5}{16}$ in BSF
0.551	14 mm
0.563	$\frac{9}{16}$ in AF
0.591	15 mm
0.600	$\frac{5}{16}$ in Whitworth; $\frac{3}{8}$ in BSF
0.625	$\frac{5}{8}$ in AF
0.630	16 mm
0.669	17 mm
0.686	$\frac{11}{16}$ in AF
0.709	18 mm
0.710	$\frac{3}{8}$ in Whitworth; $\frac{7}{16}$ in BSF
0.748	19 mm
0.750	$\frac{3}{4}$ in AF
0.813	$\frac{13}{16}$ in AF
0.820	$\frac{7}{16}$ in Whitworth; $\frac{1}{2}$ in BSF
0.866	22 mm
0.875	$\frac{7}{8}$ in AF
0.920	$\frac{1}{2}$ in Whitworth; $\frac{9}{16}$ in BSF
0.938	$\frac{15}{16}$ in AF

Jaw gap (in)	Spanner size
0.945	24 mm
1.000	1 in AF
1.010	$\frac{9}{16}$ in Whitworth; $\frac{5}{8}$ in BSF
1.024	26 mm
1.063	$1\frac{1}{16}$ in AF; 27 mm
1.100	$\frac{5}{8}$ in Whitworth; $\frac{11}{16}$ in BSF
1.125	$1\frac{1}{8}$ in AF
1.181	30 mm
1.200	$\frac{11}{16}$ in Whitworth; $\frac{3}{4}$ in BSF
1.250	$1\frac{1}{4}$ in AF
1.260	32 mm
1.300	$\frac{3}{4}$ in Whitworth; $\frac{7}{8}$ in BSF
1.313	$1\frac{5}{16}$ in AF
1.390	$\frac{13}{16}$ in Whitworth; $\frac{15}{16}$ in BSF
1.417	36 mm
1.438	$1\frac{7}{16}$ in AF
1.480	$\frac{7}{8}$ in Whitworth; 1 in BSF
1.500	$1\frac{1}{2}$ in AF
1.575	40 mm; $\frac{15}{16}$ in Whitworth
1.614	41 mm
1.625	$1\frac{5}{8}$ in AF
1.670	1 in Whitworth; $1\frac{1}{8}$ in BSF
1.688	$1\frac{11}{16}$ in AF
1.811	46 mm
1.813	$1\frac{13}{16}$ in AF
1.860	$1\frac{1}{8}$ in Whitworth; $1\frac{1}{4}$ in BSF
1.875	$1\frac{7}{8}$ in AF
1.969	50 mm
2.000	2 in AF
2.050	$1\frac{1}{4}$ in Whitworth; $1\frac{3}{8}$ in BSF
2.165	55 mm
2.362	60 mm

Standard torque settings

Specific torque settings will be found at the end of the specifications section of each chapter. Where no figure is given, bolts should be secured according to the table below.

Fastener type (thread diameter)	kgf m	lbf ft
5mm bolt or nut	0.45 – 0.6	3.5 – 4.5
6 mm bolt or nut	0.8 – 1.2	6 – 9
8 mm bolt or nut	1.8 – 2.5	13 – 18
10 mm bolt or nut	3.0 – 4.0	22 – 29
12 mm bolt or nut	5.0 – 6.0	36 – 43
5 mm screw	0.35 – 0.5	2.5 – 3.6
6 mm screw	0.7 – 1.1	5 – 8
6 mm flange bolt	1.0 – 1.4	7 – 10
8 mm flange bolt	2.4 – 3.0	17 – 22
10 mm flange bolt	3.0 – 4.0	22 – 29

Choosing and fitting accessories

The range of accessories available to the modern motorcyclist is almost as varied and bewildering as the range of motorcycles. This Section is intended to help the owner in choosing the correct equipment for his needs and to avoid some of the mistakes made by many riders when adding accessories to their machines. It will be evident that the Section can only cover the subject in the most general terms and so it is recommended that the owner, having decided that he wants to fit, for example, a luggage rack or carrier, seeks the advice of several local dealers and the owners of similar machines. This will give a good idea of what makes of carrier are easily available, and at what price. Talking to other owners will give some insight into the drawbacks or good points of any one make. A walk round the motorcycles in car parks or outside a dealer will often reveal the same sort of information.

The first priority when choosing accessories is to assess exactly what one needs. It is, for example, pointless to buy a large heavy-duty carrier which is designed to take the weight of fully laden panniers and topbox when all you need is a place to strap on a set of waterproofs and a lunchbox when going to work. Many accessory manufacturers have ranges of equipment to cater for the individual needs of different riders and this point should be borne in mind when looking through a dealer's catalogues. Having decided exactly what is required and the use to which the accessories are going to be put, the owner will need a few hints on what to look for when making the final choice. To this end the Section is now sub-divided to cover the more popular accessories fitted. Note that it is in no way a customizing guide, but merely seeks to outline the practical considerations to be taken into account when adding aftermarket equipment to a motorcycle.

Fairings and windscreens

A fairing is possibly the single, most expensive, aftermarket item to be fitted to any motorcycle and, therefore, requires the most thought before purchase. Fairings can be divided into two main groups: front fork mounted handlebar fairings and windscreens, and frame mounted fairings.

The first group, the front fork mounted fairings, are becoming far more popular than was once the case, as they offer several advantages over the second group. Front fork mounted fairings generally are much easier and quicker to fit, involve less modification to the motorcycle, do not as a rule restrict the steering lock, permit a wider selection of handlebar styles to be used, and offer adequate protection for much less money than the frame mounted type. They are also lighter, can be swapped easily between different motorcycles, and are available in a much greater variety of styles. Their main disadvantages are that they do not offer as much weather protection as the frame mounted types, rarely offer any storage space, and, if poorly fitted or naturally incompatible, can have an adverse effect on the stability of the motorcycle.

The second group, the frame mounted fairings, are secured so rigidly to the main frame of the motorcycle that they can offer a substantial amount of protection to motorcycle and rider in the event of a crash. They offer almost complete protection from the weather and, if double-skinned in construction, can provide a great deal of useful storage space. The feeling of peace, quiet and complete relaxation encountered when riding behind a good full fairing has to be experienced to be believed. For this reason full fairings are considered essential by most touring motorcyclists and by many people who ride all year round. The main disadvantages of this type are that fitting can take a long time, often involving removal or modification of standard motorcycle components, they restrict the steering lock and they can add up to about 40 lb to the weight of the machine. They do not usually affect the stability of the machine to any great extent once the front tyre pressure and suspension have been adjusted to compensate for the extra weight, but can be affected by sidewinds.

The first thing to look for when purchasing a fairing is the quality of the fittings. A good fairing will have strong, substantial brackets constructed from heavy-gauge tubing; the brackets must be shaped to fit the frame or forks evenly so that the minimum of stress is imposed on the assembly when it is bolted down. The brackets should be properly painted or finished – a nylon coating being the favourite of the better manufacturers – the nuts and bolts provided should be of the same thread and size standard as is used on the motorcycle and be properly plated. Look also for shakeproof locking nuts or locking washers to ensure that everything remains securely tightened down. The fairing shell is generally made from one of two materials: fibreglass or ABS plastic. Both have their advantages and disadvantages, but the main consideration for the owner is that fibreglass is much easier to repair in the event of damage occurring to the fairing. Whichever material is used, check that it is properly finished inside as well as out, that the edges are protected by beading and that the fairing shell is insulated from vibration by the use of rubber grommets at all mounting points. Also be careful to check that the windscreen is retained by plastic bolts which will snap on impact so that the windscreen will break away and not cause personal injury in the event of an accident.

Having purchased your fairing or windscreen, read the manufacturer's fitting instructions very carefully and check that you have all the necessary brackets and fittings. Ensure that the mounting brackets are located correctly and bolted down securely. Note that some manufacturers use hose clamps to retain the mounting brackets; these should be discarded as they are convenient to use but not strong enough for the task. Stronger clamps should be substituted; car exhaust pipe clamps of suitable size would be a good alternative. Ensure that the front forks can turn through the full steering lock available without fouling the fairing. With many types of frame-mounted fairing the handlebars will have to be altered or a different type fitted and the steering lock will be restricted by stops provided with the fittings. Also check that the fairing does not foul the front wheel or mudguard, in any steering position, under full fork compression. Re-route any cables, brake pipes or electrical wiring which may snag on the fairing and take great care to protect all electrical connections, using insulating tape. If the manufacturer's instructions are followed carefully at every stage no serious problems should be encountered. Remember that hydraulic pipes that have been disconnected must be carefully re-tightened and the hydraulic system purged of air bubbles by bleeding.

Two things will become immediately apparent when taking a motorcycle on the road for the first time with a fairing – the first is the tendency to underestimate the road speed because of the lack of wind pressure on the body. This must be very carefully watched until one has grown accustomed to riding behind the fairing. The second thing is the alarming increase in engine noise which is an unfortunate but inevitable by-product of fitting any type of fairing or windscreen, and is caused by normal engine noise being reflected, and in some cases amplified, by the flat surface of the fairing.

Luggage racks or carriers

Carriers are possibly the commonest item to be fitted to modern motorcycles. They vary enormously in size, carrying capacity, and durability. When selecting a carrier, always look for one which is made specifically for your machine and which is bolted on with as few separate brackets as possible. The universal-type carrier, with its mass of brackets and adaptor pieces, will generally prove too weak to be of any real use. A good carrier should bolt to the main frame, generally using the two suspension unit top mountings and a mudguard mounting bolt as attachment points, and have its luggage platform as low and as far forward as possible to minimise the effect of any load on the machine's stability. Look for good quality, heavy gauge tubing, good welding and good finish. Also ensure that the carrier does not prevent opening of the seat, sidepanels or tail compartment, as appropriate. When using a carrier, be very careful not to overload it. Excessive weight placed so high and so far to the rear of any motorcycle will have an adverse effect on the machine's steering and stability.

Luggage

Motorcycle luggage can be grouped under two headings: soft and hard. Both types are available in many sizes and styles and have advantages and disadvantages in use.

Soft luggage is now becoming very popular because of its lower cost and its versatility. Whether in the form of tankbags, panniers, or strap-on bags, soft luggage requires in general no brackets and no modification to the motorcycle. Equipment can be swapped easily from one motorcycle to another and can be fitted and removed in seconds. Awkwardly shaped loads can easily be carried. The disadvantages of soft luggage are that the contents cannot be secure against the casual thief, very little protection is afforded in the event of a crash, and waterproofing is generally poor. Also, in the case of panniers, carrying capacity is restricted to approximately 10 lb, although this amount will vary considerably depending on the manufacturer's recommendation. When purchasing soft luggage, look for good quality material, generally vinyl or nylon, with strong, well-stitched attachment points. It is always useful to have separate pockets, especially on tank bags, for items which will be needed on the journey. When purchasing a tank bag, look for one which has a separate, well-padded, base. This will protect the tank's paintwork and permit easy access to the filler cap at petrol stations.

Hard luggage is confined to two types: panniers, and top boxes or tail trunks. Most hard luggage manufacturers produce matching sets of these items, the basis of which is generally that manufacturer's own heavy-duty luggage rack. Variations on this theme occur in the form of separate frames for the better quality panniers, fixed or quickly-detachable luggage, and in size and carrying capacity. Hard luggage offers a reasonable degree of security against theft and good protection against weather and accident damage. Carrying capacity is greater than that of soft luggage, around 15 – 20 lb in the case of panniers, although top boxes should never be loaded as much as their apparent capacity might imply. A top box should only be used for lightweight items, because one that is heavily laden can have a serious effect on the stability of the machine. When purchasing hard luggage look for the same good points as mentioned under fairings and windscreens, ie good quality mounting brackets and fittings, and well-finished fibreglass or ABS plastic cases. Again as with fairings, always purchase luggage made specifically for your motorcycle, using as few separate brackets as possible, to ensure that everything remains securely bolted in place. When fitting hard luggage, be careful to check that the rear suspension and brake operation will not be impaired in any way and remember that many pannier kits require re-siting of the indicators. Remember also that a non-standard exhaust system may make fitting extremely difficult.

Handlebars

The occupation of fitting alternative types of handlebar is extremely popular with modern motorcyclists, whose motives may vary from the purely practical, wishing to improve the comfort of their machines, to the purely aesthetic, where form is more important than function. Whatever the reason, there are several considerations to be borne in mind when changing the handlebars of your machine. If fitting lower bars, check carefully that the switches and cables do not foul the petrol tank on full lock and that the surplus length of cable, brake pipe, and electrical wiring are smoothly and tidily disposed of. Avoid tight kinks in cable or brake pipes which will produce stiff controls or the premature and disastrous failure of an overstressed component. If necessary, remove the petrol tank and re-route the cable from the engine/gearbox unit upwards, ensuring smooth gentle curves are produced. In extreme cases, it will be necessary to purchase a shorter brake pipe to overcome this problem. In the case of higher handlebars than standard it will almost certainly be necessary to purchase extended cables and brake pipes. Fortunately, many standard motorcycles have a custom version which will be equipped with higher handlebars and, therefore, factory-built extended components will be available from your local dealer. It is not usually necessary to extend electrical wiring, as switch clusters may be used on several different motorcycles, some being custom versions. This point should be borne in mind however when fitting extremely high or wide handlebars.

When fitting different types of handlebar, ensure that the mounting clamps are correctly tightened to the manufacturer's specifications and that cables and wiring, as previously mentioned, have smooth easy runs and do not snag on any part of the motorcycle throughout the full steering lock. Ensure that the fluid level in the front brake master cylinder remains level to avoid any chance of air entering the hydraulic system. Also check that the cables are adjusted correctly and that all handlebar controls operate correctly and can be easily reached when riding.

Crashbars

Crashbars, also known as engine protector bars, engine guards, or case savers, are extremely useful items of equipment which can contribute protection to the machine's structure if a crash occurs. They do not, as has been inferred in the US, prevent the rider from crashing, or necessarily prevent rider injury should a crash occur.

It is recommended that only the smaller, neater, engine protector type of crashbar is considered. This type will offer protection while restricting, as little as is possible, access to the engine and the machine's ground clearance. The crashbars should be designed for use specifically on your machine, and should be constructed of heavy-gauge tubing with strong, integral mounting brackets. Where possible, they should bolt to a strong lug on the frame, usually at the engine mounting bolts.

The alternative type of crashbar is the larger cage type. This type is not recommended in spite of their appearance which promises some protection to the rider as well as to the machine. The larger amount of leverage imposed by the size of this type of crashbar increases the risk of severe frame damage in the event of an accident. This type also decreases the machine's ground clearance and restricts access to the engine. The amount of protection afforded the rider is open to some doubt as the design is based on the premise that the rider will stay in the normally seated position during an accident, and the crash bar structure will not itself fail. Neither result can in any way be guaranteed.

As a general rule, always purchase the best, ie usually the most expensive, set of crashbars you an afford. The investment will be repaid by minimising the amount of damage incurred, should the machine be involved in an accident. Finally, avoid the universal type of crashbar. This should be regarded only as a last resort to be used if no alternative exists. With its usual multitude of separate brackets and spacers, the universal crashbar is far too weak in design and construction to be of any practical value.

Exhaust systems

The fitting of aftermarket exhaust systems is another extremely popular pastime amongst motorcyclists. The usual motive is to gain more performance from the engine but other considerations are to gain more ground clearance, to lose weight from the motorcycle, to obtain a more distinctive exhaust note or to find a cheaper alternative to the manufacturer's original equipment exhaust system. Original equipment exhaust systems often cost more and may well have a relatively short life. It should be noted that it is rare for an aftermarket exhaust system alone to give a noticeable increase in the engine's power output. Modern motorcycles are designed to give the highest power output possible allowing for factors such as quietness, fuel economy, spread of power, and long-term reliability. If there were a magic formula which allowed the exhaust system to produce more power without affecting these other considerations you can be sure that the manufacturers, with their large research and development facilities, would have found it and made use of it. Performance increases of a worthwhile and noticeable nature only come from well-tried and properly matched modifications to the entire engine, from the air filter, through the carburettors, port timing or camshaft and valve design, combustion chamber shape, compression ratio, and the exhaust system. Such modifications are well outside the scope of this manual but interested owners might refer to specialist books produced by the publisher of this manual which go into the whole subject in great detail.

Whatever your motive for wishing to fit an alternative exhaust system, be sure to seek expert advice before doing so. Changes to the carburettor jetting will almost certainly be required for which you must consult the exhaust system manufacturer. If he cannot supply adequately specific information it is reasonable to assume that insufficient development work has been carried out, and that particular make should be avoided. Other factors to be borne in mind are whether the exhaust system allows the use of both centre and side stands, whether it allows sufficient access to permit oil and filter changing and whether modifications are necessary to the standard exhaust system. Many two-stroke expansion chamber systems require the use of the standard exhaust pipe; this is all very well if the standard exhaust pipe and silencer are separate units but can cause problems if the two, as with so many modern two-strokes, are a one-piece unit. While the exhaust pipe can be removed easily by means of a hacksaw it is not so easy to refit the original silencer should you at any time wish to return the machine to standard trim. The same applies to several four-stroke systems.

On the subject of the finish of aftermarket exhausts, avoid black-painted systems unless you enjoy painting. As any trail-bike owner will tell you, rust has a great affinity for black exhausts and re-painting or rust removal becomes a task which must be carried out with monotonous regularity. A bright chrome finish is, as a general rule, a far better proposition as it is much easier to keep clean and to prevent rusting. Although the general finish of aftermarket exhaust systems is not always up to the standard of the original equipment the lower cost of such systems does at least reflect this fact.

When fitting an alternative system always purchase a full set of new exhaust gaskets, to prevent leaks. Fit the exhaust first to the cylinder head or barrel, as appropriate, tightening the retaining nuts or bolts by hand only and then line up the exhaust rear mountings. If the new system is a one-piece unit and the rear mountings do not line up exactly, spacers must be fabricated to take up the difference. Do not force the system into place as the stress thus imposed will rapidly cause cracks and splits to appear. Once all the mountings are loosely fixed, tighten the retaining nuts or bolts securely, being careful not to overtighten them. Where the motorcycle manufacturer's torque settings are available, these should be used. Do not forget to carry out any carburation changes recommended by the exhaust system's manufacturer.

Electrical equipment

The vast range of electrical equipment available to motorcyclists is so large and so diverse that only the most general outline can be given here. Electrical accessories vary from electronic ignition kits fitted to replace contact breaker points, to additional lighting at the front and rear, more powerful horns, various instruments and gauges, clocks, anti-theft systems, heated clothing, CB radios, radio-cassette players, and intercom systems, to name but a few of the more popular items of equipment.

As will be evident, it would require a separate manual to cover this subject alone and this section is therefore restricted to outlining a few basic rules which must be borne in mind when fitting electrical equipment. The first consideration is whether your machine's electrical system has enough reserve capacity to cope with the added demand of the accessories you wish to fit. The motorcycle's manufacturer or importer should be able to furnish this sort of information and may also be able to offer advice on uprating the electrical system. Failing this, a good dealer or the accessory manufacturer may be able to help. In some cases, more powerful generator components may be available, perhaps from another motorcycle in the manufacturer's range. The second consideration is the legal requirements in force in your area. The local police may be prepared to help with this point. In the UK for example, there are strict regulations governing the position and use of auxiliary riding lamps and fog lamps.

When fitting electrical equipment always disconnect the battery first to prevent the risk of a short-circuit, and be careful to ensure that all connections are properly made and that they are waterproof. Remember that many electrical accessories are designed primarily for use in cars and that they cannot easily withstand the exposure to vibration and to the weather. Delicate components must be rubber-mounted to insulate them from vibration, and sealed carefully to prevent the entry of rainwater and dirt. Be careful to follow exactly the accessory manufacturer's instructions in conjunction with the wiring diagram at the back of this manual.

Accessories – general

Accessories fitted to your motorcycle will rapidly deteriorate if not cared for. Regular washing and polishing will maintain the finish and will provide an opportunity to check that all mounting bolts and nuts are securely fastened. Any signs of chafing or wear should be watched for, and the cause cured as soon as possible before serious damage occurs.

As a general rule, do not expect the re-sale value of your motorcycle to increase by an amount proportional to the amount of money and effort put into fitting accessories. It is usually the case that an absolutely standard motorcycle will sell more easily at a better price than one that has been modified. If you are in the habit of exchanging your machine for another at frequent intervals, this factor should be borne in mind to avoid loss of money.

Fault diagnosis

Contents

1 Introduction

● This Section provides an easy reference-guide to the more common problems that are likely to afflict your machine. Obviously, the opportunities are almost limitless for faults to occur as a result of obscure failures, and to try and cover all eventualities would require a book. Indeed, a number have been written on the subject.
● Successful fault diagnosis is not a mysterious 'black art' but the application of a bit of knowledge combined with a systematic and logical approach to the problem. Approach any fault diagnosis by first accurately identifying the symptom and then checking through the list of possible causes, starting with the simplest or most obvious and progressing in stages to the most complex. Take nothing for granted, but above all apply liberal quantities of common sense.
● The main symptom of a fault is given in the text as a major heading below which are listed, as Section headings, the various systems or areas which may contain the fault. Details of each possible cause for a fault and the remedial action to be taken are given, in brief, in the paragraphs below each Section heading. Further information should be sought in the relevant Chapter.

Starter motor problems

2 Starter motor not rotating

● Engine stop switch off.
● Fuse blown. Check the main fuse located behind the battery side cover.
● Battery voltage low. Switching on the headlamp and operating the horn will give a good indication of the charge level. If necessary recharge the battery from an external source.
● Neutral gear not selected.
● Faulty neutral indicator switch or clutch interlock switch (where fitted). Check the switch wiring and switches for correct operation.
● Ignition switch defective. Check switch for continuity and connections for security.
● Engine stop switch defective. Check switch for continuity in 'Run' position. Fault will be caused by broken, wet or corroded switch contacts. Clean or renew as necessary.
● Starter button switch faulty. Check continuity of switch. Faults as for engine stop switch.
● Starter relay (solenoid) faulty. If the switch is functioning correctly a pronounced click should be heard when the starter button is depressed. This presupposes that current is flowing to the solenoid when the button is depressed.
● Wiring open or shorted. Check first that the battery terminal connections are tight and corrosion free. Follow this by checking that all wiring connections are dry, tight and corrosion free. Check also for frayed or broken wiring. Occasionally a wire may become trapped between two moving components, particularly in the vicinity of the steering head, leading to breakage of the internal core but leaving the softer but more resilient outer cover intact. This can cause mysterious intermittent or total power loss.
● Starter motor defective. A badly worn starter motor may cause high current drain from a battery without the motor rotating. If current is found to be reaching the motor, after checking the starter button and starter relay, suspect a damaged motor. The motor should be removed for inspection.

3 Starter motor rotates but engine does not turn over

● Starter motor clutch defective. Suspect jammed or worn engagement rollers, plungers and springs.
● Damaged starter motor drive train. Inspect and renew component where necessary. Failure in this area is unlikely.

4 Starter motor and clutch function but engine will not turn over

● Engine seized. Seizure of the engine is always a result of damage to internal components due to lubrication failure, or component breakage resulting from abuse, neglect or old age. A seizing or partially seized component may go un-noticed until the engine has cooled down and an attempt is made to restart the engine. Suspect first seizure of the valves, valve gear and the pistons. Instantaneous seizure whilst the engine is running indicates component breakage. In either case major dismantling and inspection will be required.

Engine does not start when turned over

5 No fuel flow to carburettor

● No fuel or insufficient fuel in tank.
● Fuel tap lever position incorrectly selected.
● Float chambers require priming after running dry.
● Tank filler cap air vent obstructed. Usually caused by dirt or water. Clean the vent orifice.
● EECS fuel tank vent hose blocked or kinked (1984 California models only). Clear/straighten.
● Fuel tap or filter blocked. Blockage may be due to accumulation of rust or paint flakes from the tank's inner surface or of foreign matter from contaminated fuel. Remove the tap and clean it and the filter. Look also for water droplets in the fuel.
● Fuel line blocked. Blockage of the fuel line is more likely to result from a kink in the line rather than the accumulation of debris.

6 Fuel not reaching cylinder

● Float chambers not filling. Caused by float needle or floats sticking in up position. This may occur after the machine has been left standing for an extended length of time allowing the fuel to evaporate. When this occurs a gummy residue is often left which hardens to a varnish-like substance. This condition may be worsened by corrosion and crystaline deposits produced prior to the total evaporation of contaminated fuel. Sticking of the float needle may also be caused by wear. In any case removal of the float chambers will be necessary for inspection and cleaning.
● Blockage in starting circuit, slow running circuit or jets. Blockage of these items may be attributable to debris from the fuel tank by-passing the filter system or to gumming up as described in paragraph 1. Water droplets in the fuel will also block jets and passages. The carburettor should be dismantled for cleaning.
● Fuel level too low. The fuel level in each float chamber is controlled by float height. The float height may increase with wear or damage but will never reduce, thus a low float height is an inherent rather than developing condition. Check the float height and make any necessary adjustment.

7 Engine flooding

● Float valve needle worn or stuck open. A piece of rust or other debris can prevent correct seating of the needle against the valve seat thereby permitting an uncontrolled flow of fuel. Similarly, a worn needle or needle seat will prevent valve closure. Dismantle the carburettor float bowl for cleaning and, if necessary, renewal of the worn components.
● Fuel level too high. The fuel level is controlled by the float height which may increase due to wear of the float needle, pivot pin or operating tang. Check the float height, and make any necessary adjustment. A leaking float will cause an increase in fuel level, and thus should be renewed.
● Cold starting mechanism. Check the choke (starter mechanism) for correct operation. If the mechanism jams in the 'On' position subsequent starting of a hot engine will be difficult.
● Blocked air filter. A badly restricted air filter will cause flooding. Check the filter and clean or renew as required. A collapsed inlet hose will have a similar effect.

8 No spark at plug

● Ignition switch not on.
● Engine stop switch off.
● Fuse blown. Check fuse for ignition circuit. See wiring diagram.

● Battery voltage low. The current required by a starter motor is sufficiently high that an under-charged battery may not have enough spare capacity to provide power for the ignition circuit during starting. Bump starting is recommended until the battery has been recharged either by the machine's generator or from an external charger.
● Starter motor inefficient. A starter motor with worn brushes and a worn or dirty commutator will draw excessive amounts of current causing power starvation in the ignition system. See the preceding paragraph. Starter motor overhaul will be required.
● Spark plug failure. Clean the spark plugs thoroughly and reset the electrode gap. Refer to the spark plug section and the condition guide in Routine Maintenance. If a spark plug shorts internally or has sustained visible damage to the electrodes, core or ceramic insulator it should be renewed. On rare occasions, a plug that appears to spark vigorously will fail to do so when refitted to the engine and subjected to the compression pressure in the cylinder.
● Spark plug caps or high tension (HT) leads faulty. Check condition and security. Replace if deterioration is evident.
● Spark plug cap loose. Check that the spark plug caps fit securely over each plug and, where fitted, the screwed terminal on each plug end is secure.
● Shorting due to moisture. Certain parts of the ignition system are susceptible to shorting when the machine is ridden or parked in wet weather. Check particularly the area from each spark plug cap back to the ignition coil. A water dispersant spray may be used to dry out waterlogged components. Recurrence of the problem can be prevented by using an ignition sealant spray after drying out and cleaning.
● Ignition or stop switch shorted. May be caused by water, corrosion or wear. Water dispersant and contact cleaning sprays may be used. If this fails to overcome the problem dismantling and visual inspection of the switches will be required.
● Shorting or open circuit in wiring. Failure in any wire connecting any of the ignition components will cause ignition malfunction. Check also that all connections are clean, dry and tight.
● Ignition coil failure. Check the coils, referring to Chapter 3.
● Capacitor (condenser) failure. The capacitors may be checked most easily by direct substitution with replacement items. Blackened contact breaker points indicate capacitor malfunction but this may not always occur.
● Contact breaker points pitted, burned or closed up. Check the contact breaker points, referring to Routine Maintenance. Check also that the low tension leads at the contact breaker are secure and not shorting out.
● Electronic ignition component failure. See Chapter 3.

9 Weak spark at plug

● Feeble sparking at the plugs may be caused by any of the faults mentioned in the preceding Section other than those items in paragraphs 1 and 2. Check first the contact breaker assembly and the spark plugs, these being the most likely culprits.

10 Compression low

● Spark plug loose. This will be self-evident on inspection, and may be accompanied by a hissing noise when the engine is turned over. Remove each plug and check that the threads in the cylinder head are not damaged. Check also that the plug sealing washers are in good condition.
● Cylinder head gasket leaking. This condition is often accompanied by a high pitched squeak from around the cylinder head and oil loss, and may be caused by insufficiently tightened cylinder head fasteners, a warped cylinder head or mechanical failure of the gasket material. Re-torqueing the fasteners to the correct specification may seal the leak in some instances but if damage has occurred this course of action will provide, at best, only a temporary cure.
● Valve not seating correctly. The failure of a valve to seat may be caused by insufficient valve clearance, pitting of the valve seat or face, carbon deposits on the valve seat or seizure of the valve stem or valve gear components. Valve spring breakage will also prevent correct valve closure. The valve clearances should be checked first and then, if these are found to be in order, further dismantling will be required to inspect the relevant components for failure.
● Cylinder, piston and ring wear. Compression pressure will be lost if any of these components are badly worn. Wear in one component is invariably accompanied by wear in another. A top end overhaul will be required.
● Piston rings sticking or broken. Sticking of the piston rings may be caused by seizure due to lack of lubrication or heating as a result of poor carburation or incorrect fuel type. Gumming of the rings may result from lack of use, or carbon deposits in the ring grooves. Broken rings result from over-revving, overheating or general wear. In either case a top-end overhaul will be required.

Engine stalls after starting

11 General causes

● Improper cold start mechanism operation. Check that the operating controls function smoothly and, where applicable, are correctly adjusted. A cold engine may not require application of an enriched mixture to start initially but may baulk without choke once firing. Likewise a hot engine may start with an enriched mixture but will stop almost immediately if the choke is inadvertently in operation.
● Ignition malfunction. See Section 9, 'Weak spark at plug'.
● Carburettors incorrectly adjusted. Maladjustment of the mixture strength or idle speed may cause the engine to stop immediately after starting. See Chapter 2.
● Fuel contamination. Check for filter blockage by debris or water which reduces, but does not completely stop, fuel flow or blockage of the slow speed circuit in the carburettor by the same agents. If water is present it can often be seen as droplets in the bottom of the float bowl. Clean the filter and, where water is in evidence, drain and flush the fuel tank and float bowl.
● Intake air leak. Check for security of the carburettor mounting and hose connections, and for cracks or splits in the hoses. Check also that each carburettor top is secure and that each vacuum gauge adaptor plug (where fitted) is tight.
● Air filter blocked or omitted. A blocked filter will cause an over-rich mixture; the omission of a filter will cause an excessively weak mixture. Both conditions will have a detrimental affect on carburation. Clean or renew the filter as necessary.
● Fuel filler cap air vent blocked. Usually caused by dirt or water. Clean the vent orifice.
● Faulty EECS system (1984 California models only). See Chapter 2.

Poor running at idle and low speed

12 Weak spark at plugs or erratic firing

● Battery voltage low. In certain conditions low battery charge, especially when coupled with a badly sulphated battery, may result in misfiring. If the battery is in good general condition it should be recharged; an old battery suffering from sulphated plates should be renewed.
● Spark plug fouled, faulty or incorrectly adjusted. See Section 8 or refer to Chapter 3.
● Spark plug caps or high tension leads shorting. Check the condition of both these items ensuring that they are in good condition and dry and that the caps are fitted correctly.
● Spark plug types incorrect. Fit plugs of correct type and heat range as given in Specifications. In certain conditions plugs of hotter or colder type may be required for normal running.
● Contact breaker points pitted, burned or closed up. Check the contact breaker assembly, referring to Routine Maintenance.
● Igniting timing incorrect. Check the ignition timing statically and dynamically, ensuring that the advance is functioning correctly.
● Faulty ignition coil. Partial failure of the coil internal insulation will diminish the performance of the coil. No repair is possible, a new component must be fitted.
● Faulty capacitor (condenser), where fitted. A failure of capacitor will cause blackening of the contact breaker point faces and will allow excessive sparking at the points. A faulty capacitor may best be checked by substitution of a serviceable replacement item.
● Electronic ignition system component failure.

13 Fuel/air mixture incorrect

- Intake air leak. See Section 11.
- Mixture strength incorrect. Adjust slow running mixture strength using pilot adjustment screw if allowed. See Chapter 2.
- Carburettor synchronisation.
- Pilot jet or slow running circuit blocked. The carburettors should be removed and dismantled for thorough cleaning. Blow through all jets and air passages with compressed air to clear obstructions.
- Air cleaner clogged or omitted. Clean or fit air cleaner element as necessary. Check also that the element and air filter cover are correctly seated.
- Cold start mechanism in operation. Check that the choke has not been left on inadvertently and the operation is correct.
- Fuel level too high or too low. Check the fuel levels and adjust as necessary. See Section 7.
- Fuel tank air vent obstructed. Obstruction usually caused by dirt or water. Clean vent orifice.
- EECS fuel tank vent hose blocked or kinked (1984 California models only). Clear/straighten.
- Valve clearances incorrect. Check, and if necessary, adjust, the clearances.

14 Compression low

- See Section 10.

Acceleration poor

15 General causes

- All items as for previous Section.
- Timing not advancing. This is caused by a sticking or damaged automatic timing unit (ATU), where fitted. Cleaning and lubrication of the ATU will usually overcome sticking, failing this, and in any event if damage is evident, renewal of the ATU will be required.
- Sticking throttle vacuum piston. CD carburettors only.
- Brakes binding. Usually caused by maladajustment or partial seizure of the operating mechanism due to poor maintenance. Check brake adjustment (where applicable). A bent wheel spindle or warped brake disc can produce similar symptoms.

Poor running or lack of power at high speeds

16 Weak spark at plug or erratic firing

- All items as for Section 12.
- HT leads insulation failure. Insulation failure of the HT leads and spark plug caps due to old age or damage can cause shorting when the engine is driven hard. This condition may be less noticeable, or not noticeable at all at lower engine speeds.

17 Fuel/air mixture incorrect

- All items as for Section 13, with the exception of items 2 and 4.
- Main jet blocked. Debris from contaminated fuel, or from the fuel tank, and water in the fuel can block the main jet. Clean the fuel filter, the float bowl area, and if water is present, flush and refill the fuel tank.
- Main jet is the wrong size.
- Jet needle and needle jet worn. These can be renewed individually but should be renewed as a pair. Renewal of both items requires partial dismantling of the carburettors.
- Air bleed holes blocked. Dismantle carburettors and use compressed air to blow out all air passages.
- Reduced fuel flow. A reduction in the maximum fuel flow from the fuel tank to the carburettors will cause fuel starvation, proportionate to the engine speed. Check for blockages through debris or a kinked fuel line.
- Vacuum diaphragm split. Renew.

18 Compression low

- See Section 10.

Knocking or pinking

19 General causes

- Carbon build-up in combustion chamber. After high mileages have been covered large accumulation of carbon may occur. This may glow red hot and cause premature ignition of the fuel/air mixture, in advance of normal firing by the spark plugs. Cylinder head removal will be required to allow inspection and cleaning.
- Fuel incorrect. A low grade fuel, or one of poor quality may result in compression induced detonation of the fuel resulting in knocking and pinking noises. Old fuel can cause similar problems. A too highly leaded fuel will reduce detonation but will accelerate deposit formation in the combustion chamber and may lead to early pre-ignition as described in item 1.
- Spark plug heat range incorrect. Uncontrolled pre-ignition can result from the use of spark plugs the heat range of which is too hot.
- Weak mixture. Overheating of the engine due to a weak mixture can result in pre-ignition occurring where it would not occur when engine temperature was within normal limits. Maladjustment, blocked jets or passages and air leaks can cause this condition.

Overheating

20 Firing incorrect

- Spark plugs fouled, defective or maladjusted. See Section 6.
- Spark plugs type incorrect. Refer to the Specifications and ensure that the correct type plugs are fitted.
- Incorrect ignition timing. Timing that is far too much advanced or far too much retarded will cause overheating. Check the ignition timing is correct and that the advance mechanism/circuit is functioning.

21 Fuel/air mixture incorrect

- Slow speed mixture strength incorrect. Adjust pilot screw.
- Main jet wrong size.
- Air filter badly fitted or omitted. Check that the filter element is in place and that it and the air filter box cover are sealing correctly. Any leaks will cause a weak mixture.
- Induction air leaks. Check the security of the carburettor mountings and hose connections, and for cracks and splits in the hoses. Check also that the carburettor tops are secure and that each vacuum gauge adaptor plug (where fitted) is tight.
- Fuel level too low. See Section 6.
- Fuel tank filler cap air vent obstructed or EECS hose blocked or kinked. Clear blockage or straighten hose.

22 Lubrication inadequate

- Engine oil too low. Not only does the oil serve as a lubricant by preventing friction between moving components, but it also acts as a coolant. Check the oil level and replenish.
- Engine oil overworked. The lubricating properties of oil are lost slowly during use as a result of changes resulting from heat and also contamination. Always change the oil at the recommended interval.
- Engine oil of incorrect viscosity or poor quality. Always use the recommended viscosity and type of oil.
- Oil filter and filter by-pass valve blocked. Renew filter and clean the by-pass valve.

23 Miscellaneous causes

- Engine fins clogged. A build-up of mud in the cylinder head and cylinder barrel cooling fins will decrease the cooling capabilities of the fins. Clean the fins as required.

Clutch operating problems

24 Clutch slip

- No clutch lever play. Adjust clutch lever end play according to the procedure in Routine Maintenance.
- Friction plates worn or warped. Overhaul clutch assembly, replacing plates out of specification.
- Steel plates worn or warped. Overhaul clutch assembly, replacing plates out of specification.
- Clutch springs broken or worn. Old or heat-damaged (from slipping clutch) springs should be replaced with new ones.
- Clutch release not adjusted properly, where applicable. See the adjustments section of Routine Maintenance.
- Clutch inner cable snagging. Caused by a frayed cable or kinked outer cable. Replace the cable with a new one. Repair of a frayed cable is not advised.
- Clutch release mechanism defective. Replace parts as necessary.
- Clutch centre and outer drum worn. Severe indentation by the clutch plate tangs of the channels in the centre and drum will cause snagging of the plates preventing correct engagement. If this damage occurs, renewal of the worn components is required.
- Lubricant incorrect. Use of a transmission lubricant other than that specified may allow the plates to slip.

25 Clutch drag

- Clutch lever play excessive. Adjust lever at bars or at cable end if necessary.
- Clutch plates warped or damaged. This will cause a drag on the clutch, causing the machine to creep. Overhaul clutch assembly.
- Clutch spring tension uneven. Usually caused by a sagged or broken spring. Check and replace springs.
- Engine oil deteriorated. Badly contaminated engine oil and a heavy deposit of oil sludge and carbon on the plates will cause plate sticking. The oil recommended for this machine is of the detergent type, therefore it is unlikely that this problem will arise unless regular oil changes are neglected.
- Engine oil viscosity too high. Drag in the plates will result from the use of an oil with too high a viscosity. In very cold weather clutch drag may occur until the engine has reached operating temperature.
- Clutch centre and outer drum worn. Indentation by the clutch plate tangs of the channels in the centre and drum will prevent easy plate disengagement. If the damage is light the affected areas may be dressed with a fine file. More pronounced damage will necessitate renewal of the components.
- Clutch release mechanism defective. Worn or damaged release mechanism parts can stick and fail to provide leverage. Overhaul release mechanism components.
- Loose clutch centre nut. Causes drum and centre misalignment, putting a drag on the engine. Engagement adjustment continually varies. Overhaul clutch assembly.

Gear selection problems

26 Gear lever does not return

- Weak or broken return spring. Renew the spring.
- Gearchange shaft bent or seized. Distortion of the gearchange shaft often occurs if the machine is dropped heavily on the gear lever. Provided that damage is not severe straightening of the shaft is permissible.

27 Gear selection difficult or impossible

- Clutch not disengaging fully. See Section 25.
- Gearchange shaft bent. This often occurs if the machine is dropped heavily on the gear lever. Straightening of the shaft is permissible if the damage is not too great.
- Gearchange claw arms or pins worn or damaged. Wear or breakage of any of these items may cause difficulty in selecting one or more gears. Overhaul the selector mechanism.
- Gearchange arm spring broken. Renew spring.
- Selector drum, detent cam or detent plunger damage. Failure, rather than wear, of these items may jam the drum thereby preventing gearchanging. The damaged items must be renewed.
- Selector forks bent or seized. This can be caused by dropping the machine heavily on the gearchange lever or as a result of lack of lubrication. Though rare, bending of a shaft can result from a missed gearchange or false selection at high speed.
- Selector fork claw end and guide pin wear. Pronounced wear of these items and the grooves in the selector drum can lead to imprecise selection and, eventually, no selection. Renewal of the worn components will be required.
- Structural failure. Failure of any one component of the selector mechanism will result in improper or fouled gear selection.

28 Jumping out of gear

- Detent plunger assembly worn or damaged. Wear of the plunger and the cam with which it locates and breakage of the detent spring can cause imprecise gear selection resulting in jumping out of gear. Renew the damaged components.
- Gear pinion dogs worn or damaged. Rounding off the dog edges and the mating recesses in adjacent pinion can lead to jumping out of gear when under load. The gears should be inspected and renewed. Attempting to reprofile the dogs is not recommended.
- Selector forks, selector drum and pinion grooves worn. Extreme wear of these interconnected items can occur after high mileages especially when lubrication has been neglected. The worn components must be renewed.
- Gear pinions, bushes and shafts worn. Renew the worn components.
- Bent gearchange shaft. Often caused by dropping the machine on the gear lever.
- Gear pinion tooth broken. Chipped teeth are unlikely to cause jumping out of gear once the gear has been selected fully; a tooth which is completely broken off, however, may cause problems in this respect and in any event will cause transmission noise.

29 Overselection

- Claw arm spring weak or broken. Renew the spring.
- Detent plunger worn or broken. Renew the damaged items.
- Overshift limiter claw arm worn or damaged. Renew the damaged items.

Abnormal engine noise

30 Knocking or pinking

- See Section 19.

31 Piston slap or rattling from cylinders

- Cylinder bore/piston clearances excessive. Resulting from wear, partial seizure or improper boring during overhaul. This condition can often be heard as a high, rapid tapping noise when the engine is under little or no load, particularly when power is just beginning to be applied. Reboring to the next correct oversize should be carried out and new oversize pistons fitted.
- Connecting rod bent. This can be caused by over-revving, trying to start a very badly flooded engine (resulting in a hydraulic lock in the cylinder) or by earlier mechanical failure such as a dropped valve. Attempts at straightening a bent connecting rod from a high performance engine are not recommended. Careful inspection of the crankshaft should be made before renewing the damaged connecting rod.
- Gudgeon pin, piston boss bore or small-end bearing wear or seizure. Excess clearance or partial seizure between normal moving

parts of these items can cause continuous or intermittent tapping noises. Rapid wear or seizure is caused by lubrication starvation resulting from an insufficient engine oil level or oilway blockage.
- Piston rings worn, broken or sticking. Renew the rings after careful inspection of the pistons and bores.

32 Valve noise or tapping from the cylinder head

- Valve clearances incorrect. Adjust the clearances with the engine cold.
- Valve spring broken or weak. Renew the spring set.
- Camshaft or cylinder head worn or damaged. The camshaft lobes are the most highly stressed of all components in the engine and are subject to high wear if lubrication becomes inadequate. The bearing surfaces on the camshafts and cylinder head are also sensitive to a lack of lubrication. Lubrication failure due to blocked oilways can occur, but over-enthusiastic revving before engine warm-up is complete is the usual cause.
- Excessive camshaft end float. See Chapter 1.
- Worn camshaft drive components. A rustling noise or light tapping can be emitted by a defective tensioner, worn cam chain, or worn sprockets and chain. If uncorrected, subsequent cam chain breakage may cause extensive damage. The worn components must be renewed before wear becomes too far advanced.

33 Other noises

- Big-end bearing wear. A pronounced knock from within the crankcase which worstens rapidly is indicative of big-end bearing failure as a result of extreme normal wear or lubrication failure. Remedial action in the form of a bottom end overhaul should be taken; continuing to run the engine will lead to further damage including the possibility of connecting rod breakage.
- Main bearing failure. Extreme normal wear or failure of the main bearings is characteristically accompanied by a rumble from the crankcase and vibration felt through the frame and footrests. Renew the worn bearings and carry out a very careful examination of the crankshaft.
- Crankshaft excessively out of true. A bent crank may result from over-revving or damage from an upper cylinder component or gearbox failure. Damage can also result from dropping the machine on either crankshaft end. Straightening of the crankshaft is not possible in normal circumstances; a replacement item should be fitted.
- Engine mounting loose. Tighten all the engine mounting nuts and bolts.
- Cylinder head gasket leaking. The noise most often associated with a leaking head gasket is a high pitched squeaking, although any other noise consistent with gas being forced out under pressure from a small orifice can also be emitted. Gasket leakage is often accompanied by oil seepage from around the mating joint or from the cylinder head holding down bolts and nuts. Leakage into the cam chain tunnel or oil return passages will increase crankcase pressure and may cause oil leakage at joints and oil seals. Also, oil contamination will be accelerated. Leakage results from insufficient or uneven tightening of the cylinder head fasteners, or from random mechanical failure. Retightening to the correct torque figure will, at best, only provide a temporary cure. The gasket should be renewed at the earliest opportunity.
- Exhaust system leakage. Popping or crackling in the exhaust system, particularly when it occurs with the engine on the overrun, indicates a poor joint either at the cylinder port or at the exhaust pipe/silencer connection. Failure of the gasket or looseness of the clamp should be looked for, also failure of, or damage to, the Clean Air system (US models only).

Abnormal transmission noise

34 Clutch noise

- Clutch outer drum/friction plate tang clearance excessive.
- Clutch outer drum/spacer clearance excessive.
- Clutch outer drum/thrust washer clearance excessive.
- Secondary drive gear teeth worn or damaged.
- Clutch shock absorber assembly worn or damaged.

35 Transmission noise

- Bearing or bushes worn or damaged. Renew the affected components.
- Gear pinions worn or chipped. Renew the gear pinions.
- Metal chips jams in gear teeth.This can occur when pieces of metal from any failed component are picked up by a meshing pinion. The condition will lead to rapid bearing wear or early gear failure.
- Engine/transmission oil level too low. Top up immediately to prevent damage to gearbox and engine.
- Gearchange mechanism worn or damaged. Wear or failure of certain items in the selection and change components can induce misselection of gears (see Section 27) where incipient engagement of more than one gear set is promoted. Remedial action, by the overhaul of the gearbox, should be taken without delay.

Chain drive models
- Loose gearbox chain sprocket. Remove the sprocket and check for impact damage to the splines of the sprocket and shaft. Excessive slack between the splines will promote loosening of the securing nut; renewal of the worn components is required. When retightening the nut ensure that it is tightened fully and that, where fitted, the lock washer is bent up against one flat of the nut.
- Chain snagging on cases or cycle parts. A badly worn chain or one that is excessively loose may snag or smack against adjacent components.

Shaft drive models
- Worn or damaged bevel gear sets. A whine emitted from either bevel gear set is indicative of improper meshing. This may increase progressively as wear develops or suddenly due to mechanical failure. Drain the lubricant and inspect for metal chips prior to dismantling.
- Output shaft joint failure. This can cause vibration and noise. Renew the affected component.

Exhaust smokes excessively

36 White/blue smoke (caused by oil burning)

- Piston rings worn or broken. Breakage or wear of any ring, but particularly the oil scraper rings, will allow engine oil past the pistons into the combustion chamber. Overhaul the cylinder block and pistons.
- Cylinder block cracked, worn or scored. These conditions may be caused by overheating, lack of lubrication, component failure or advanced normal wear. The cylinder block should be renewed or rebored and the next oversize pistons fitted.
- Valve oil seals damaged or worn. This can occur as a result of valve guide failure or old age. The emission of smoke is likely to occur when the throttle is closed rapidly after acceleration, for instance, when changing gear. Renew the valve oil seals and, if necessary, the valve guides.
- Valve guides worn. See the preceding paragraph.
- Engine oil level too high. This increases the crankcase pressure and allows oil to be forced past the piston rings. Often accompanied by seepage of oil at joints and oil seals.
- Cylinder head gasket blown between cam chain tunnel or oil return passage. Renew the cylinder head gasket.
- Abnormal crankcase pressure. This may be caused by blocked breather passages or hoses causing back-pressure at high engine revolutions.

37 Black smoke (caused by over-rich mixture)

- Air filter element clogged. Clean or renew the element.
- Main jet loose. Remove the float chamber to check for tightness of the jet.
- Cold start mechanism jammed on. Check that the mechanism works smoothly and correctly.

● Fuel level too high. The fuel level is controlled by the float height which can increase as a result of wear or damage. Remove the float bowls and check the float heights. Check also that floats have not punctured; a punctured float will loose buoyancy and allow an increased fuel level.
● Float valve needle stuck open. Caused by dirt or a worn valve. Clean the float chambers or renew the needles and, the valve seats.

Oil pressure/level indicator lamp goes on

38 Engine lubrication system failure

● Engine oil defective. Oil pump shaft or locating pin sheared off from ingesting debris or seizing from lack of lubrication (low oil level).
● Engine oil screen clogged. Change oil and filter and service pickup screen.
● Engine oil level too low. Inspect for leak or other problem causing low oil level and add recommended lubricant.
● Engine oil viscosity too low. Very old, thin oil, or an improper weight of oil used in engine. Change to correct lubricant. This would also cause oil level warning to operate.
● Camshafts or journals worn. High wear causing drop in oil pressure. Replace cams and/or head. Abnormal wear could be caused by oil starvation at high rpm from low oil level, improper oil weight or type.
● Crankshaft and/or bearings worn. Same problems as paragraph 5. Overhaul lower end.
● Relief valve stuck open. This causes the oil to be dumped back into the sump. Clean or replace.

39 Electrical system failure

● Oil pressure/level switch defective. Check switch according to the procedures in Chapter 6. Replace if defective.
● Oil pressure/level indicator lamp wiring system defective. Check for pinched, shorted, disconnected or damaged wiring.

Poor handling or roadholding

40 Directional instability

● Steering head bearing adjustment too tight. This will cause rolling or weaving at low speeds. Re-adjust the bearings.
● Steering head bearing worn or damaged. Correct adjustment of the bearing will prove impossible to achieve if wear or damage has occurred. Inconsistent handling will occur including rolling or weaving at low speed and poor directional control at indeterminate higher speeds. The steering head bearing should be dismantled for inspection and renewed if required. Lubrication should also be carried out.
● Bearing races pitted or dented. Impact damage caused, perhaps, by an accident or riding over a pot-hole can cause indentation of the bearing, usually in one position. This should be noted as notchiness when the handlebars are turned. Renew and lubricate the bearings.
● Steering stem bent. This will occur only if the machine is subjected to a high impact such as hitting a curb or a pot-hole. The bottom yoke should be renewed; do not attempt to straighten the stem.
● Front or rear tyre pressures too low.
● Front or rear tyre worn. General instability, high speed wobbles and skipping over white lines indicates that tyre renewal may be required. Tyre induced problems, in some machine/tyre combinations, can occur even when the tyre in question is by no means fully worn.
● Swinging arm or suspension linkage bearings worn. Difficulties in holding line, particularly when cornering or when changing power settings indicates wear in the swinging arm or suspension linkage bearings. The rear suspension should be removed from the machine and the bearings renewed.
● Swinging arm flexing. The symptoms given in the preceding paragraph will also occur if the swinging arm fork flexes badly. This can be caused by structural weakness as a result of corrosion, fatigue or impact damage, or because the rear wheel spindle is slack.
● Wheel bearings worn. Renew the worn bearings.
● Swinging arm bearings badly adjusted (shaft drive models). Adjust. See Chapter 4

● Tyres unsuitable for machine. Not all available tyres will suit the characteristics of the frame and suspension, indeed, some tyres or tyre combinations may cause a transformation in the handling characteristics. If handling problems occur immediately after changing to a new tyre type or make, revert to the original tyres to see whether an improvement can be noted. In some instances a change to what are, in fact, suitable tyres may give rise to handling deficiences. In this case a thorough check should be made of all frame and suspension items which affect stability.

41 Steering bias to left or right

● Rear wheel out of alignment. On chain drive machines, caused by uneven adjustment of chain tensioner adjusters allowing the wheel to be askew in the fork ends. A bent rear wheel spindle will also misalign the wheel in the swinging arm.
● Wheels out of alignment. This can be caused by impact damage to the frame, swinging arm, wheel spindles or front forks. Although occasionally a result of material failure or corrosion it is usually as a result of a crash.
● Front forks twisted in the steering yokes. A light impact, for instance with a pot-hole or low curb, can twist the fork legs in the steering yokes without causing structural damage to the fork legs or the yokes themselves. Re-alignment can be made by loosening the yoke pinch bolts, wheel spindle and mudguard bolts. Re-align the wheel with the handlebars and tighten the bolts working upwards from the wheel spindle. This action should be carried out only when there is no chance that structural damage has occurred.

42 Handlebar vibrates or oscillates

● Tyres worn or out of balance. Either condition, particularly in the front tyre, will promote shaking of the fork assembly and thus the handlebars. A sudden onset of shaking can result if a balance weight is displaced during use.
● Tyres badly positioned on the wheel rims. A moulded line on each wall of a tyre is provided to allow visual verification that the tyre is correctly positioned on the rim. A check can be made by rotating the tyre; any misalignment will be immediately obvious.
● Wheels rims warped or damaged. Inspect the wheels for runout as described in Chapter 5.
● Swinging arm bearings worn or badly adjusted. Renew or adjust the bearings.
● Wheel bearings worn. Renew the bearings.
● Steering head bearings incorrectly adjusted. Vibration is more likely to result from bearings which are too loose rather than too tight. Re-adjust the bearings.
● Loosen fork component fasteners. Loose nuts and bolts holding the fork legs, wheel spindle, mudguards or steering stem can promote shaking at the handlebars. Fasteners on running gear such as the forks and suspension should be check tightened occasionally to prevent dangerous looseness of components occurring.
● Engine mounting bolts loose. Tighten all fasteners.

43 Poor front fork performance

● Damping fluid level incorrect. If the fluid level is too low poor suspension control will occur resulting in a general impairment of roadholding and early loss of tyre adhesion when cornering and braking. Too much oil is unlikely to change the fork characteristics unless severe overfilling occurs when the fork action will become stiffer and oil seal failure may occur.
● Damping oil viscosity incorrect. The damping action of the fork is directly related to the viscosity of the damping oil. The lighter the oil used, the less will be the damping action imparted. For general use, use the recommended viscosity of oil, changing to a slightly higher or heavier oil only when a change in damping characteristic is required. Overworked oil, or oil contaminated with water which has found its way past the seals, should be renewed to restore the correct damping performance and to prevent bottoming of the forks.
● Air pressure incorrect. An imbalance in the air pressure between

the fork legs can give rise to poor fork performance (models with non-linked forks). Similarly if the air pressure is outside the recommended range, problems can occur.

● Damping components worn or corroded. Advanced normal wear of the fork internals is unlikely to ocur until a very high mileage has been covered. Continual use of the machine with damaged oil seals which allows the ingress of water, or neglect, will lead to rapid corrosion and wear. Dismantle the forks for inspection and overhaul. See Chapter 4.

● Anti-dive mechanism inoperative or unbalanced. Check settings, then refer to Chapter 4.

● Weak fork springs. Progressive fatigue of the fork springs, resulting in a reduced spring free length, will occur after extensive use. This condition will promote excessive fork dive under braking, and in its advanced form will reduce the at-rest extended length of the forks and thus the fork geometry. Renewal of the springs as a pair is the only satisfactory course of action.

● Bent stanchions or corroded stanchions. Both conditions will prevent correct telescoping of the fork legs, and in an advanced state can cause sticking of the fork in one position. In a mild form corrosion will cause stiction of the fork thereby increasing the time the suspension takes to react to an uneven road surface. Bent fork stanchions should be attended to immediately because they indicate that impact damage has occurred, and there is a danger that the forks will fail with disastrous consequences.

44 Front fork judder when braking (see also Section 56)

● Wear between the fork stanchions and the fork legs. Renewal of the affected components is required.

● Slack steering head bearings. Re-adjust the bearings.

● Warped brake disc. If irregular braking action occurs fork judder can be induced in what are normally serviceable forks. Renew the damaged brake components.

45 Poor rear suspension performance

● Suspension settings incorrectly matched, either left to right, front to rear or spring preload to damping (as applicable), or unsuitable for the load and/or road surface.

● Rear suspension unit damper worn out or leaking. The damping performance of most rear suspension units falls off with age. This is a gradual process, and thus may not be immediately obvious. Indications of poor damping include hopping of the rear end when cornering or braking, and a general loss of positive stability. See Chapter 4.

● Weak rear springs. If the suspension unit springs fatigue they will promote excessive pitching of the machine and reduce the ground clearance when cornering. Renew suspension units as a pair (where applicable).

● Loss of air pressure (where applicable). The loss of air from a rear suspension unit, unless from a faulty valve or connecting hose, will mean that that unit is defective. Renew the units as a matched pair.

● Leaking or damaged remote preload adjuster (ZX550 A1 only). Damage resulting in loss of oil and reduction in hydraulic pressure will soften the rear suspension spring preload setting. Renew as a complete assembly.

● Worn suspension linkage bearings. See Chapter 4.

● Swinging arm flexing or bearings worn. See Sections 40 and 41.

● Bent suspension unit damper rod. This is likely to occur only if the machine is dropped or if seizure of the piston occurs. If either happens the suspension units should be renewed as a pair.

Abnormal frame and suspension noise

46 Front end noise

● Oil level low or too thin. This can cause a 'spurting' sound and is usually accompanied by irregular fork action.

● Spring weak or broken. Makes a clicking or scraping sound. Fork oil will have a lot of metal particles in it.

● Steering head bearings loose or damaged. Clicks when braking. Check, adjust or replace.

● Fork clamps loose. Make sure all fork clamp pinch bolts are tight.

● Fork stanchion bent. Good possibility if machine has been dropped. Repair or replace tube.

47 Rear suspension noise

● Fluid level too low. Leakage of a suspension unit, usually evident by oil on the outer surfaces, can cause a spurting noise. The suspension units should be renewed, as a pair where applicable.

● Defective rear suspension unit with internal damage. Renew the suspension units, as a pair where applicable.

● Worn or corroded rear suspension linkage bearings. Renew defective components.

Brake problems

48 Brakes are spongy or ineffective – disc brakes

● Air in brake circuit. This is only likely to happen in service due to neglect in checking the fluid level or because a leak has developed. The problem should be identified and the brake system bled of air.

● Pad worn. Check the pad wear against the wear lines provided and renew the pads if necessary.

● Contaminated pads. Cleaning pads which have been contaminated with oil, grease or brake fluid is unlikely to prove successful; the pads should be renewed.

● Pads glazed. This is usually caused by overheating. The surface of the pads may be roughened using glass-paper or a fine file.

● Brake fluid deterioration. A brake which on initial operation is firm but rapidly becomes spongy in use may be failing due to water contamination of the fluid. The fluid should be drained and then the system refilled and bled.

● Master cylinder seal failure. Wear or damage of master cylinder internal parts will prevent pressurisation of the brake fluid. Overhaul the master cylinder unit.

● Caliper seal failure. This will almost certainly be obvious by loss of fluid, a lowering of fluid in the master cylinder reservoir and contamination of the brake pads and caliper. Overhaul the caliper assembly.

● Brake lever or pedal improperly adjusted. Adjust the clearance between the lever end and master cylinder plunger to take up lost motion, as recommended in Routine maintenance.

49 Brakes drag – disc brakes

● Disc warped. The disc must be renewed.

● Caliper piston, caliper or pads corroded. The brake caliper assembly is vulnerable to corrosion due to water and dirt, and unless cleaned at regular intervals and lubricated in the recommended manner, will become sticky in operation.

● Piston seal deteriorated. The seal is designed to return the piston in the caliper to the retracted position when the brake is released. Wear or old age can affect this function. The caliper should be overhauled if this occurs.

● Brake pad damaged. Pad material separating from the backing plate due to wear or faulty manufacture. Renew the pads. Faulty installation of a pad also will cause dragging.

● Wheel spindle bent. The spindle may be straightened if no structural damage has occurred.

● Brake lever or pedal not returning. Check that the lever or pedal works smoothly throughout its operating range and does not snag on any adjacent cycle parts. Lubricate the pivot if necessary.

● Twisted caliper support bracket. This is likely to occur only after impact in an accident. No attempt should be made to re-align the caliper; the bracket should be renewed.

50 Brake lever or pedal pulsates in operation – disc brakes

● Disc warped or irregularly worn. The disc must be renewed.

● Wheel spindle bent. The spindle may be straightened provided no structural damage has occurred.

51 Disc brake noise

- Brake squeal. This can be caused by the omission or incorrect installation of the anti-squeal shim. Squealing can also be caused by dust on the pads, usually in combination with glazed pads, or other contamination from oil, grease, brake fluid or corrosion. Persistent squealing which cannot be traced to any of the normal causes can often be cured by applying a thin layer of high temperature silicone grease to the rear of the pads. Make absolutely certain that no grease is allowed to contaminate the braking surface of the pads.
- Glazed pads. This is usually caused by high temperatures or contamination. The pad surfaces may be roughened using glass-paper or a fine file. If this approach does not effect a cure the pads should be renewed.
- Disc warped. This can cause a chattering, clicking or intermittent squeal and is usually accompanied by a pulsating brake lever or pedal or uneven braking. The disc must be renewed.
- Brake pads fitted incorrectly or undersize. Longitudinal play in the pads due to omission of the locating springs (where fitted) or because pads of the wrong size have been fitted will cause a single tapping noise every time the brake is operated. Inspect the pads for correct installation and security.

52 Brakes are spongy or ineffective – drum brakes

- Worn brake linings. Determine lining wear using the external brake wear indicator on the brake backplate, or by removing the wheel and withdrawing the brake backplate. Renew the shoe/lining units as a pair if the linings are worn below the recommended limit.
- Worn brake camshaft. Wear between the camshaft and the bearing surface will reduce brake feel and reduce operating efficiency. Renewal of one or both items will be required to rectify the fault.
- Worn brake cam and shoe ends. Renew the worn components.
- Linings contaminated with dust or grease. Any accumulations of dust should be cleaned from the brake assembly and drum using a petrol dampened cloth. Do not blow or brush off the dust because it is asbestos based and thus harmful if inhaled. Light contamination from grease can be removed from the surface of the brake linings using a solvent; attempts at removing heavier contamination are less likely to be successful because some of the lubricant will have been absorbed by the lining material which will severely reduce the braking performance.

53 Brake drag – drum brakes

- Incorrect adjustment. Re-adjust the brake operating mechanism.
- Drum warped or oval. This can result from overheating, impact or wear. The condition is difficult to correct, although if slight ovality only occurs, skimming the surface of the brake drum can provide a cure. This is work for a specialist engineer. Renewal of the complete wheel is normally the only satisfactory solution.
- Weak brake shoe return springs. This will prevent the brake lining/shoe units from pulling away from the drum surface once the brake is released. The springs should be renewed.
- Brake camshaft or lever pivot poorly lubricated. Failure to attend to regular lubrication of these areas will increase operating resistance which, when compounded, may cause tardy operation and poor release movement.

54 Brake lever or pedal pulsates in operation – drum brakes

- Drums warped or oval. This can result from overheating, impact or uneven wear. This condition is difficult to correct, although if slight ovality only occurs skimming the surface of the drum can provide a cure. This is work for a specialist engineer. Renewal of the wheel is normally the only satisfactory solution.

55 Drum brake noise

- Drum warped or oval. This can cause intermittent rubbing of the brake linings against the drum. See the preceding Section.
- Brake linings glazed. This condition, usually accompanied by heavy lining dust contamination, often induces brake squeal. The surface of the linings may be roughened using glass-paper or a fine file.

56 Brake induced fork judder

- Worn front fork stanchions and legs, or worn or badly adjusted steering head bearings. These conditions, combined with uneven or pulsating braking as described in Section 50 will induce more or less judder when the brakes are applied, dependent on the degree of wear and poor brake operation. Attention should be given to both areas of malfunction. See the relevant Sections.

Electrical problems

57 Battery dead or weak

- Battery faulty. Battery life should not be expected to exceed 3 to 4 years, particularly where a starter motor is used regularly. Gradual sulphation of the plates and sediment deposits will reduce the battery performance. Plate and insulator damage can often occur as a result of vibration. Complete power failure, or intermittent failure, may be due to a broken battery terminal. Lack of electrolyte will prevent the battery maintaining charge.
- Battery leads making poor contact. Remove the battery leads and clean them and the terminals, removing all traces of corrosion and tarnish. Reconnect the leads and apply a coating of petroleum jelly to the terminals.
- Load excessive. If additional items such as spot lamps, are fitted, which increase the total electrical load above the maximum alternator output, the battery will fail to maintain full charge. Reduce the electrical load to suit the electrical capacity.
- Regulator/rectifier failure.
- Alternator generating coils open-circuit or shorted.
- Charging circuit shorting or open circuit. This may be caused by frayed or broken wiring, dirty connectors or a faulty ignition switch. The system should be tested in a logical manner. See Section 60.

58 Battery overcharged

- Rectifier/regulator faulty. Overcharging is indicated if the battery becomes hot or it is noticed that the electrolyte level falls repeatedly between checks. In extreme cases the battery will boil causing corrosive gases and electrolyte to be emitted through the vent pipes.
- Battery wrongly matched to the electrical circuit. Ensure that the specified battery is fitted to the machine.

59 Total electrical failure

- Fuse blown. Check the main fuse. If a fault has occurred, it must be rectified before a new fuse is fitted.
- Battery faulty. See Section 57.
- Earth failure. Check that the main earth strap from the battery is securely affixed to the crankcase and is making a good contact.
- Ignition switch or power circuit failure. Check for current flow through the battery positive lead to the ignition switch. Check the ignition switch for continuity.

60 Circuit failure

- Cable failure. Refer to the machine's wiring diagram and check the circuit for continuity. Open circuits are a result of loose or corroded connections, either at terminals or in-line connectors, or because of broken wires. Occasionally, the core of a wire will break without there being any apparent damage to the outer plastic cover.
- Switch failure. All switches may be checked for continuity in each switch position, after referring to the switch position boxes incorporated in the wiring diagram for the machine. Switch failure may be a result of mechanical breakage, corrosion or water.

● Fuse blown. Refer to the wiring diagram to check whether or not a circuit fuse is fitted. Replace the fuse, if blown, only after the fault has been identified and rectified.

61 Bulbs blowing repeatedly

● Vibration failure. This is often an inherent fault related to the natural vibration characteristics of the engine and frame and is, thus, difficult to resolve. Modification of the lamp mounting to change the damping characteristics may help.

● Intermittent earth. Repeated failure of one bulb indicates that a poor earth exists somewhere in the circuit. Check that a good contact is available at each earthing point in the circuit.

● Reduced voltage. Where a quartz-halogen bulb is fitted the voltage to the bulb should be maintained or early failure of the bulb will occur. Do not overload the system with additional electrical equipment in excess of the system's power capacity and ensure that all circuit connections are maintained clean and tight.

KAWASAKI 400, 500 & 550 FOURS

Refer to Chapter 7 for 1984 on model data

Check list

Daily

1 Check the engine oil level via the sight glass
2 Ensure you have enough fuel to complete your journey
3 Check the tyre pressures and inspect the tyres for damage
4 Check the final drive chain adjustment (where applicable)
5 Inspect the drive units for oil leakage (where applicable)
6 Check the operation of the brakes
7 Check that all controls operate smoothly and are securely fastened
8 Check the electrical system for correct operation
9 Check the suspension settings

Monthly, or every 500 miles (800 km)

1 Check the battery electrolyte level
2 Check the brake fluid level
3 Lubricate and adjust the final drive chain (where applicable)

Six monthly, or every 3000 miles (5000 km)

1 Clean and gap the spark plugs
2 Check the emission control system components – US models
3 Check and adjust the valve clearances
4 Check and adjust the contact breaker points (where applicable)
5 Check the ignition timing
6 Clean the air filter
7 Adjust the carburettors
8 Change the engine oil
9 Adjust the clutch
10 Check the brake adjustment and wear
11 Check and adjust the steering head bearings
12 Check the suspension components for wear and inspect the tyres for signs of wear and damage
13 Lubricate all control cables and pivot points
14 Check around the machine, looking for loose nuts, bolts or screws, retightening them as necessary

Annually, or every 6000 miles (10 000 km)

1 Grease the automatic timing unit (where applicable)
2 Renew the air filter element
3 Renew the oil filter element
4 Change the brake fluid
5 Change the front fork oil
6 Grease the rear suspension
7 Check the final drive gear case oil level and grease the prop shaft joints (where applicable)

Every two years or 12 000 miles (20 000 km)

1 Grease the wheel bearings, speedometer drive gearbox and rear brake camshaft
2 Grease the steering head bearings

Every three years or 18 000 miles (30 000 km)

1 Change the final gear case oil (where applicable)

Additional routine maintenance items – see manual for intervals

1 Renew all brake seals and anti-dive plunger units
2 Renew all brake hoses and pipes
3 Renew the fuel pipe
4 Cleaning the machine
5 Clean the oil pump pick-up filter gauze

Adjustment data

Valve clearances (cold)

Inlet	0.10 – 0.20 mm (0.004 – 0.008 in)
Exhaust	0.15 – 0.25 mm (0.006 – 0.010 in)

Contact breaker gap 0.3 – 0.4 mm (0.012 – 0.016 in)

Spark plug gap

ZR400 B1, ZX550 A1, A1L	0.8 – 0.9 mm (0.032 – 0.035 in)
All other models	0.6 – 0.7 mm (0.024 – 0.028 in)

Spark plug type

ZX550 A1, A1L (US)	NGK DP9EA-9 or ND X27EP-U9
ZR400 B1, ZX550 A1 (UK)	NGK DPR9EA-9 or ND X27EPR-U9
Z400 J3, ZR400 A1, Z550 A3, A4, H1, H2, G1, G2, ZR550 A1, A2	NGK DR8ES or ND X24ESR-U
All other models	NGK D8EA or ND X24ES-U

Idle speed

All 400 models, KZ550 F2L and ZX550 A1, A1L (US)	1150 – 1250 rpm
All other models	1000 – 1100 rpm

Tyre pressures – pressures vary according to individual models and weights – see manual

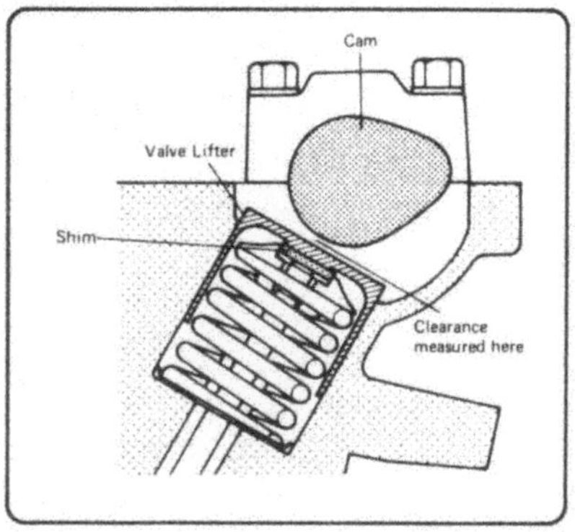

Valve clearance measurement

Recommended lubricants

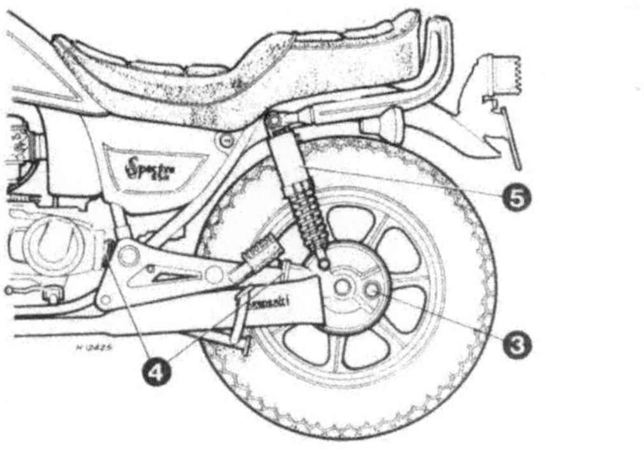

Component	Quantity	Type/viscosity
1 Engine: Oil change △	2.6 lit (4.6/2.8 Imp pt/US qt)	SAE 10W/40, 10W/50, 20W/40 or 20W/50 SE or SF class or equivalent
Oil and filter change △	3.0 lit (5.3/3.2 Imp pt/US qt)	
2 Final drive chain	As required	SAE 90 gear oil or aerosol lubricant suitable for O-ring chains
3 *Final drive case	190 cc (6.9/6.4 Imp/US fl oz)	API GL-5 (or GL-6) hypoid gear oil, SAE 90 above 5°C (41°F) and SAE 80 below 5°C (41°F)
4 *Final drive prop shaft joints	As required	High melting point grease
5 Suspension – see manual for oil capacity, oil level and oil viscosity		
6 Steering head bearings	As required	General purpose grease
7 Wheel bearings	As required	High melting point grease
8 Rear suspension pivots	As required	Molybdenum disulphide grease
9 Hydraulic brake	As required	SAE J1703 or DOT 3 hydraulic fluid
10 Automatic timing unit	As required	High melting point grease
11 Control cables	As required	Light machine oil
12 Speedometer drive gearbox	As required	High melting point grease
13 Rear brake camshaft	As required	High melting point grease

* shaft drive models only

△ models fitted with an oil cooler will require approximately an additional 0.2 lit (0.4 Imp pt/0.2 US qt) of oil

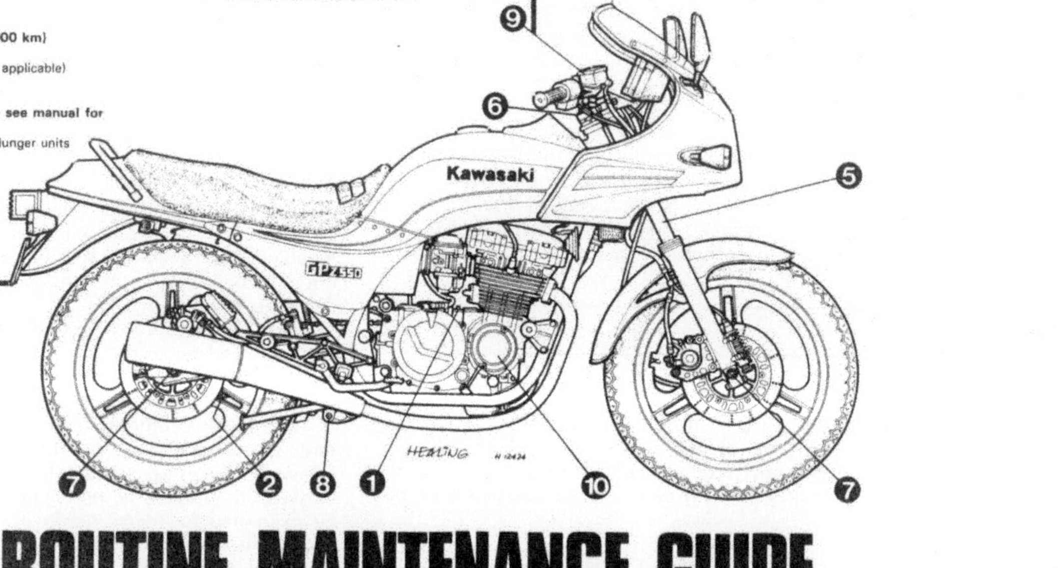

ROUTINE MAINTENANCE GUIDE

Routine maintenance

For information relating to the 1984 on models, see Chapter 7

Periodic routine maintenance is a continuous process which should commence immediately the machine is used. The object is to maintain all adjustments and to diagnose and rectify minor defects before they develop into more extensive, and often more expensive, problems.

It follows that if the machine is maintained properly, it will both run and perform with optimum efficiency, and be less prone to unexpected breakdowns. Regular inspection of the machine will show up any parts which are wearing, and with a little experience, it is possible to obtain the maximum life from any one component, renewing it when it becomes so worn that it is liable to fail.

Regular cleaning can be considered as important as mechanical maintenance. This will ensure that all the cycle parts are inspected regularly and are kept free from accumulations of road dirt and grime.

Cleaning is especially important during the winter months, despite its appearance of being a thankless task which very soon seems pointless. On the contrary, it is during these months that the paintwork, chromium plating, and the alloy casings suffer the ravages of abrasive grit, rain and road salt. A couple of hours spent weekly on cleaning the machine will maintain its appearance and value, and highlight small points, like chipped paint, before they become a serious problem.

The various maintenance tasks are described under their respective mileage and calendar headings, and are accompanied by diagrams and photographs where pertinent.

It should be noted that the intervals between each maintenance task serve only as a guide. As the machine gets older, or if it is used under particularly arduous conditions, it is advisable to reduce the period between each check.

For ease of reference, most service operations are described in detail under the relevant heading. However, if further general information is required, this can be found under the pertinent Section heading and Chapter in the main text.

Although no special tools are required for routine maintenance, a good selection of general workshop tools is essential. Included in the tools must be a range of metric ring or combination spanners, a selection of crosshead screwdrivers, and two pairs of circlip pliers, one external opening and the other internal opening. Additionally, owing to the extreme tightness of most casing screws on Japanese machines, an impact screwdriver, together with a choice of large or small crosshead screw bits, is absolutely indispensable. This is particularly so if the engine has not been dismantled since leaving the factory.

It will be noted that Allen screws are used extensively on the engine outer covers of some models, and it follows that a set of metric Allen keys (wrenches) will be required. These are not expensive and can be obtained from most auto accessory or tool suppliers.

Daily

A daily check of the motorcycle is essential both from mechanical and safety aspects. It is a good idea to develop this checking procedure in a specific sequence so that it will ultimately become as instinctive as actually riding the machine. Done properly, this simple checking sequence will give advanced warning of impending mechanical failures and any condition which may jeopardise the safety of the rider.

1 *Oil level*

The level of the engine oil is quickly checked by way of the oil sight glass set in the right-hand outer casing. With the machine standing on level ground, the oil should be visible half way up the plastic window. Marks are provided on the rim of the window, indicating the maximum and minimum oil levels. If necessary, top up the oil by way of the filler cap at the rear of the casing. Should too much oil have been added, it should be removed, using a syringe or an empty plastic squeeze pack such as that used for gear oils. Note that it is best to check the level with the machine at normal operating temperature so that the oil level is accurate; wait a few minutes for the oil to settle before checking the level if the machine has just been run.

Oil level is checked via sight glass – level must be between cast marks

2 *Fuel level*

Check that you have enough to complete your journey or at least enough to get to the nearest filling station. Do not overfill the tank; top up to the base of the filler neck only.

Refer to Chapter 7, Section 5 for information on fuel recommendations for all models.

3 *Tyre pressures*

Check the tyre pressures with a pressure gauge that is known to be accurate. Always check the pressure when the tyres are cold. If the machine has travelled a number of miles, the tyres will have become hot and consequently the pressure will have increased. A false reading will therefore result.

It is well worth purchasing a small pocket pressure gauge which can be relied on to give consistent readings, and which will remove any reliance on garage forecourt gauges which tend to be less dependable.

At the same time as the tyre pressures are checked, examine the tyres themselves. Check them for damage, especially splitting of the sidewalls. Remove any small stones or other road debris caught between the treads. When checking the tyres for damage, they should be examined for tread depth in view of both the legal and safety aspects. It is vital to keep the tread depth within the UK legal limits of 1 mm of depth over three-quarters of the tread breadth around the entire circumference with no bald patches in evidence. Many riders, however, consider nearer 2 mm to be the limit for secure roadholding, traction, and braking, especially in adverse weather conditions, and it should be noted that Kawasaki recommend a minimum tread depth of 1 mm for the front tyre, 2 mm for the rear tyre when used at low speed (under 130 km/h, 80 mph), or 3 mm for the rear tyre when used at speeds above 130 km/h (80 mph).

4 *Final drive*

On chain drive models check that the chain is correctly adjusted, as described under the monthly/500 mile heading, and that it is properly lubricated. If the chain rollers look dry it is in need of lubrication, which may be needed daily or even more often if the weather or road surfaces are poor; there is no fixed interval for chain lubrication.

On shaft drive models, it will suffice to check that there are no oil leaks from the drive units, that the gaiter at the shaft front end is undamaged, and that there are no strange noises coming from the mechanism.

5 *Brakes*

Check that the brakes are functioning correctly, that control levers and pedals are working smoothly and well lubricated and that the brake pad friction material or brake shoe wear indicators are within limits. Check that the hydraulic fluid in the master cylinder(s) is above the 'Minimum' mark.

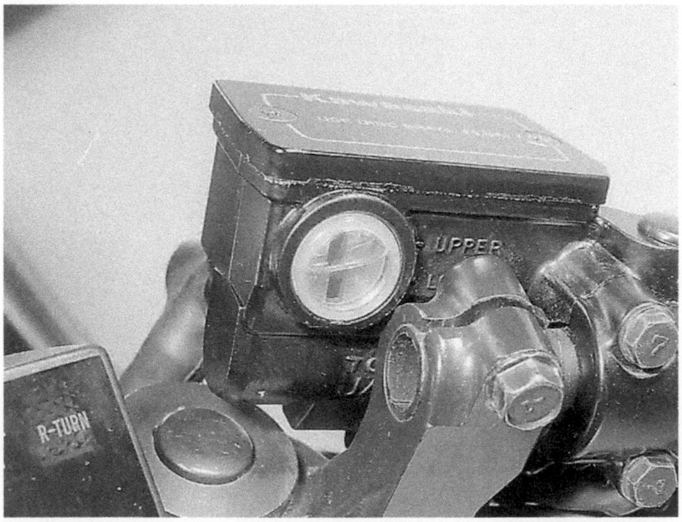

Brake fluid level is checked via sight glass in reservoir body ...

... or via translucent body, as applicable – level must be above lower mark

6 *General check*

Check that all controls are working smoothly and that they are correctly adjusted, also that the steering has no free play and moves from lock to lock without fouling any component, that the stands pivot smoothly and are held securely in the raised position by their return springs, and that all nuts, bolts and fasteners are securely tightened.

7 *Electrical check*

Ensure that all electrical components such as all lamps, horns, turn signals and the engine kill switch are working properly. If any bulb has failed or if it is flickering, it should be renewed before the journey commences. Note that it is a legal requirement that the above components are in working condition at all times.

8 *Suspension*

The suspension of modern motorcycles has become more and more sophisticated to cater for the needs of individual riders. While this may be a good thing for the enthusiast it does mean that far greater attention must be paid to hitherto almost forgotten components. All models except the Z400 J1, Z500 B1, B2 and KZ/Z550 A1 are fitted with air assisted front forks; of these all KZ/Z550 A, C, and D models and the Z400 J2, J3 models are fitted with an air valve at the top of each stanchion, while all remaining models are fitted with a linking tube with a valve fitted beneath the top yoke on the left-hand side (chain drive models) or above the bottom yoke on the right-hand side (shaft drive models).

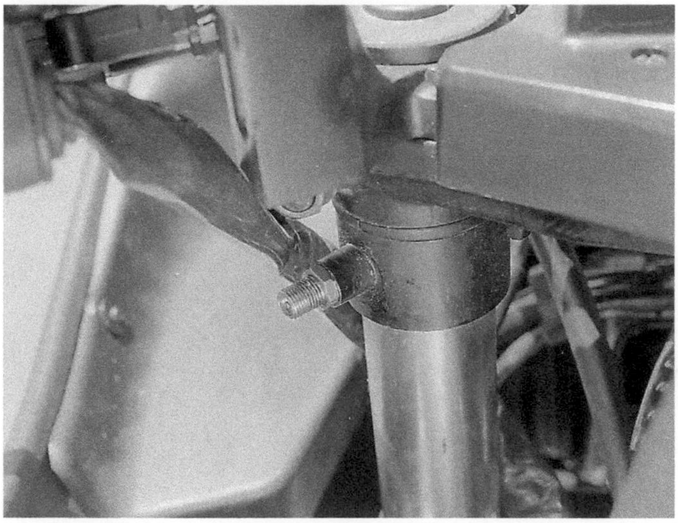

Front fork air valve – ZX550 A1 – use only accurate gauge when checking pressure

The air pressure needs to be checked before any journey, when the fork is cold, and with the machine on its centre stand with the front wheel clear of the ground so that the pressure is not artificially increased. Support the machine with blocks of wood under the crankcase, remove the valve cap(s) and check the pressure with an accurate gauge. Note that a tyre pressure gauge must not be used; they are not accurate enough and lose too much air when removed. Kawasaki produce a suitable gauge under Part Number 52005-1003.

The pressure can be adjusted to suit the owner's convenience within a range of 0.5–0.7 kg/cm^2 (7–10 psi) on all KZ/Z550 C, F and M models (ie the 'Ltd' range), 0.6–0.8 kg/cm^2 (8.5–11 psi) on all other models. On those models with non-linked stanchions the difference in pressure between stanchions must not exceed 0.1 kg/cm^2 (1.5 psi); in practice this is difficult to achieve and it is worth purchasing one of the many aftermarket kits now available to link the fork legs. Add air using only a bicycle tyre pump or a special mini-pump; air lines work at far too high a pressure and may damage the fork seals. The pressure should never exceed 2.5 kg/cm^2 (35.5 psi). Be very careful when adding air or checking the pressure; the air volume is so small that the least amount added or removed makes a large difference in pressure.

ZX550 A1 models are fitted with anti-dive units adjustable through three positions. The standard setting is the softest, Number 1, with the adjusting knobs both rotated fully anti-clockwise. At the rear all machines, except those with 'Unitrak' rear suspension and the Z550 G1, G2 and KZ550 F1, F2 and F2L models, are fitted with coil-spring suspension units on which the spring preload is adjustable through five positions using a C-spanner or a screwdriver. In general the softest (lowest) position is for a solo rider using the machine at low speed on good roads, while the hardest (highest) is for two-up riding or when the machine is carrying a heavy load of luggage, or being used at high speed. The positions in between are to suit needs between the two extremes. ZR400/550 models have no provision for preload adjustment but all other 'Unitrak'-equipped models have springs adjustable through a 10 mm range. A threaded collar is fitted at the top of the unit and is rotated anti-clockwise using a C-spanner to stiffen the spring setting, clockwise to soften it. So much dismantling was required to gain access to the unit (removal of the left-hand side panel, left-hand air filter housing and chainguard) on KZ/Z550 H1, H2 models that the ZX550 A1 models are fitted with an hydraulically-operated remote adjuster behind the right-hand side panel. This has an adjusting knob which is rotated to vary the spring setting through the range L-M-H marked on the adjuster side. In all cases, the standard spring setting for a single rider is the softest, ie at the top.

Remote rear suspension spring preload adjuster – ZX550 A1

All models except the Z400 J1, Z500 B1, B2 and KZ/Z550 A1 are fitted with damping adjustment. The commonest type is adjusted by the rotation of the suspension unit top collar, aligning the number of the desired setting with the arrow cast on the unit top mounting. Either four or five damping settings are available. On all other models so equipped, the damping is adjusted through four positions by the rotation of a small knurled wheel located in the unit top or bottom mounting; the number of the desired position being aligned with the arrow stamped in the unit mounting. The standard damping setting is number 2, but the settings should, in general, be increased or decreased to match the spring preload as it is altered. Number 1 setting has the least damping effect, number 4 (or 5) the most.

Z550 G1, G2, and KZ550 F1, F2L models are fitted with air/oil suspension units. If the oil viscosity and quantity are varied as well as the air pressure, units of this type have almost infinitely variable characteristics. With the standard grade and quantity of oil the air pressure can be adjusted to suit the rider's needs within a range of 0.8-1.5 kg/cm^2 (11-21 psi). Never exceed 5.0 kg/cm^2 (71 psi). When checking the air pressure note that the units are linked by a connecting hose with the single valve being on the left-hand unit. As with the front forks, the pressure should only be checked before a journey when the units are cold and only when the machine is supported on its centre stand, and the procedure is carried out using the equipment and observing the same precautions as given above for front forks.

It is **essential** that any suspension settings are exactly the same on the left- and right-hand sides of the machine (where applicable); any discrepancy in air pressure, spring preload or damping settings may upset the performance of the suspension to the point that the machine becomes unstable. This worsens markedly if the machine is carrying a heavy load or is being used on a rough road or at high speed. Similarly settings must be matched front to rear, if the front is stiffened, the rear must be stiffened by a proportionate amount and vice-versa. If any handling or stability problems are encountered at any time, check the suspension settings first. Similarly if any suspension faults are discovered while adjusting the settings, they must be put right immediately. Refer to the relevant Sections of Chapter 4.

Damping adjustment is provided by rotating knurled wheel to desired position

Monthly, or every 500 miles (800 km)

The real task of routine maintenance begins at this interval, as the daily checks merely serve to keep the machine in a safe and legal condition. However they are useful in giving advance warning of impending trouble; commence work, therefore, by repeating all the daily checks, then carry out the following.

1 *Check the battery electrolyte level*

The battery is located under the seat on all models except the ZR400/550 models and KZ/Z550 H1, H2 models, on which it is located behind the right-hand sidepanel. It is necessary to check the electrolyte level at regular intervals, even on those models with a low electrolyte level warning circuit; the sensor operates on only one cell. The electrolyte level can be seen in most cases without moving the battery but if this is not the case it can be lifted out after the retaining strap (secured either by two screws, two nuts, one bolt or two bolts as appropriate) has been removed. On all Z400, Z500 and Z550 A and D models the tool tray must be removed to expose the retaining strap securing bolt and on ZX550 A1, A1L and Z550 G1, G2 models the fuel tank rear mounting must be dismantled so that the tank can be raised high enough at the rear to allow the battery to be removed. If the battery is disconnected, remember to disconnect the negative (–) terminal first, to prevent the possibility of short circuits damaging the IC ignitor unit (electronic ignition models only). The electrolyte level, visible through the translucent casing, must be between the two level marks. If necessary remove the cell caps and top up to the upper level using only distilled water. Check that the terminals are clean and apply a thin smear of petroleum jelly (not grease) to each to prevent corrosion. On refitting, check that the vent hose is not blocked and that it is correctly routed with no kinks, also that it hangs well below any other component, particularly the chain or exhaust system. Remember always to connect the negative (–) terminal last when refitting the battery.

2 *Check the brake fluid level*

Ensuring that the master cylinder assembly is horizontal to the ground, check that the fluid level has not dropped below the 'Lower' level mark adjacent to the reservoir sight glass. The procedure is exactly the same for the rear master cylinder reservoir (mounted behind the right-hand sidepanel). If topping up is necessary, use only fresh hydraulic fluid to DOT 3 or SAE J1703 specification. The fluid must be of the same type as that already used and must come from a sealed container. Top up only to the lower level mark; it is quite natural for the level to fall as more fluid is drawn into the system to compensate for pad wear. Check that no fluid leaks can be seen in any part of the system. If a sudden drop in the level occurs, investigate the cause immediately.

3 *Lubricate and adjust the final drive chain*

Although the chain fitted as standard equipment is of the O-ring type, grease being sealed into the internal bearing surfaces by O-rings at each end of the rollers, lubrication is still required to prevent the rollers from wearing on the sprocket teeth and to prevent the O-rings from drying up. A heavy (SAE 90) gear oil is best; it will stay on the rollers longer than a lighter engine oil. Spinning the back wheel, allow oil to dribble on to the rollers until all are oily, then apply a small amount to the O-rings on each side. The only alternative to this is to use one of the proprietary aerosol-applied chain lubricants. **Warning**: some propellants used in aerosols cause the O-rings to deteriorate very rapidly, so make certain that the product is marked as suitable for use with O-ring chains; there are an increasing number now available.

To check the chain adjustment, place the machine on its centre stand, check the transmission is in neutral and measure the free play in terms of total up-and down movement midway between the two sprockets on the chain lower run. As chains do not wear evenly, the free play must be measured at points along the chain's whole length until the tightest spot is found. Free play at the tightest spot must be between 35 – 40 mm (1.4 – 1.6 in) on ZX550 A1, A1L models, and 20 – 35 mm (0.8 – 1.4 in) on all other models.

If adjustment is required, slacken the drawbolt adjuster locknuts and slacken the rear brake torque arm/stay retaining nut or bolt (noting that if an open-ended spanner is used it will not be necessary to displace the retaining split pin, where fitted), then remove its split pin and slacken the rear wheel spindle nut by just enough to permit the wheel to move in the swinging arm fork ends. Adjust the drawbolts by an equal amount to preserve wheel alignment. The fork ends are marked with a series of vertical lines, the same one of which on both sides must be aligned with the notch in the adjuster.

When adjustment is correct, apply hard the rear brake (drum brakes only) to centralise the shoes on the drum while the spindle nut is tightened securely. Refit the spindle nut split pin, spreading its end securely and tighten the adjuster locknuts and the torque arm/stay retaining nut or bolt. On ZX550 A1, A1L models, the torque stay clamp bolt must be removed and refitted in the second, rearmost, hole when the chain has worn sufficiently to warrant this. On all models with disc rear brakes, apply the brake pedal several times until the pads come back into full contact with the disc and full pedal pressure is restored. On models with drum rear brakes check the rear brake adjustment and stop lamp rear switch setting, adjusting both if necessary. **Note**: torque settings are given in the Specifications Section of Chapter 5 for spindle nuts, torque stay fasteners etc; these should be adhered to, where possible, to ensure the security of all components.

Drive chain wear can be assessed in one of two ways; the chain must be correctly lubricated and adjusted beforehand. If the chain can be pulled backwards off the rear of the rear wheel sprocket far enough to expose the whole of any sprocket tooth, it is worn out. Alternatively, hang a weight of approximately 10 kg (22 lb) on the chain lower run and measure the length of any 20 links, ie from any one pin to the 21st pin along; with the chain thus stretched taut, if the distance measured exceeds 323 mm (12.72 in) the chain is worn out and must be renewed.

Chain renewal will usually require the removal of the swinging arm, unless a chain with a connecting link has been fitted subsequently. Refer to the relevant Sections of Chapter 4. If the chain is to be renewed, this must always be done in conjunction with both sprockets to prevent the increased wear that would result from the running together of new and part-worn components. If a connecting link is fitted, the clip must be refitted with its closed end facing the direction of chain travel.

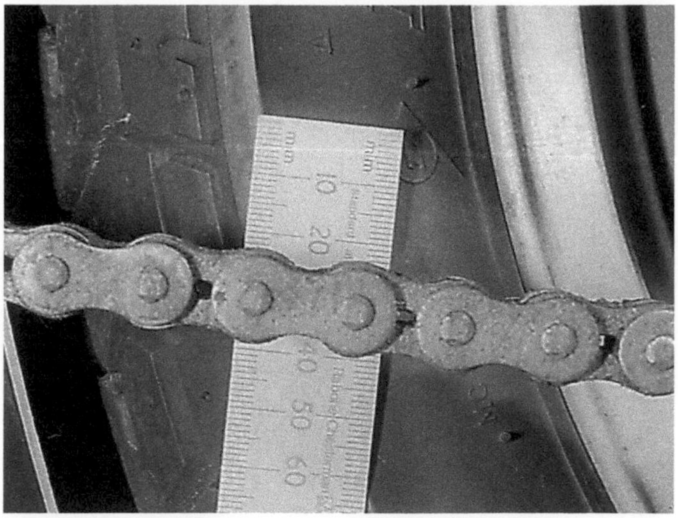

Measure chain free play at tightest point in position shown

When adjusting chain rotate drawbolts by equal amount ...

... and use reference marks to preserve wheel alignment

Tighten spindle nut securely and refit split pin as shown to prevent slackening

ZX550 A1 – slacken torque stay clamp bolt as shown when adjusting chain – move to rearmost hole when necessary

Six monthly, or every 3000 miles (5000 km)

Repeat the daily and monthly tasks, then carry out the following:

1 Clean and gap the spark plugs

Remove the spark plugs and clean them, using a wire brush. Clean the electrode points using emery paper or a fine file, and then reset the gaps. To reset the gap, bend the outer electrode to bring it closer to or further from the central electrode, until a feeler gauge of the correct size can just be slid between the electrodes. Never bend the central electrode or the insulator will crack, causing engine damage if the particles fall in whilst the engine is running. Note that although Kawasaki do not make specific recommendations, it is usual to renew the plugs every 6000 miles. If signs of excessive electrode wear or other faults are noted (see colour section on electrode conditions) the plugs should be renewed to preserve engine performance, economy and exhaust cleanliness.

On refitting, smear the threads with graphite grease to aid subsequent removal and refit the plugs using only the correct size and type of spanner. Tighten the plug by hand only to just nip tight the gasket, then a further quarter of a turn using a tommy bar or handle. If a torque wrench is available, the recommended setting is 1.4 kgf m (10 lbf ft).

2 Check the emission control system components – US models only

Two types of emission control system are fitted, these being described in full in Sections 13 and 14 of Chapter 2. Refer to those Sections for details of the components to be checked.

3 Check and adjust the valve clearances

This operation is described in two sub-sections since it is felt that while checking the clearances is within the scope of any owner, adjusting them is a different matter. The work necessary in each case is outlined below; refer to Sections 5, 7 and 44 of Chapter 1 for a more detailed description, if required. It is recommended that the owner reads the complete Section first to get some idea of what is required, then decides whether to undertake one or both parts of the task or to take the machine to a dealer.

Valve clearance adjustment shims are underneath cam follower buckets – checking clearances is easy, resetting them is not – see text

Checking the valve clearances (engine cold)

Remove the sidepanels, the seat and the fuel tank, the Clean Air System and Evaporative Emission Control System components (where applicable) and the ignition HT coils. Remove the contact breaker/ignition pickup cover and the spark plugs, then the cylinder head cover, which is secured by a total of 24 bolts. Applying a spanner to the larger, engine turning hexagon (never use the smaller retaining bolt head) rotate the engine clockwise until the '1.4' T mark stamped on the ATU/ignition rotor is aligned with the fixed index mark cast on the crankcase wall, all timing marks being visible via the aperture in the ignition backplate. Check which two inlet cams, numbers 1 and 3 or 2 and 4, have their lobes pointing away from their respective valves, as shown in the accompanying illustrations, this being the position of maximum clearance. Using feeler gauges, measure the clearance between each cam and its cam follower (also known as valve lifters or buckets) and record them carefully. Rotate the crankshaft clockwise through 180° until the '2.3' T mark (the plain T mark on electronic ignition models) aligns with the index mark, then measure and record the clearances of numbers 1 and 3 or 2 and 4 exhaust valves. Turn the crankshaft clockwise through 180° until the '1.4' T mark aligns again, then measure and record the clearances of the remaining pair of inlet valves, and rotate it again until the '2.3' T mark aligns so that the clearances of the remaining pair of exhaust valves can be measured and recorded. The specified clearances are 0.10-0.20 mm (0.004-0.008 in) for the inlet valves and 0.15-0.25 mm (0.006-0.010 in) for the exhaust valves; all valve clearances must be within this range. If any are less than that specified, action must be taken immediately to prevent damage to the valve and its seat. If any are larger than the specified limit the error must still be corrected but this problem is not quite so serious.

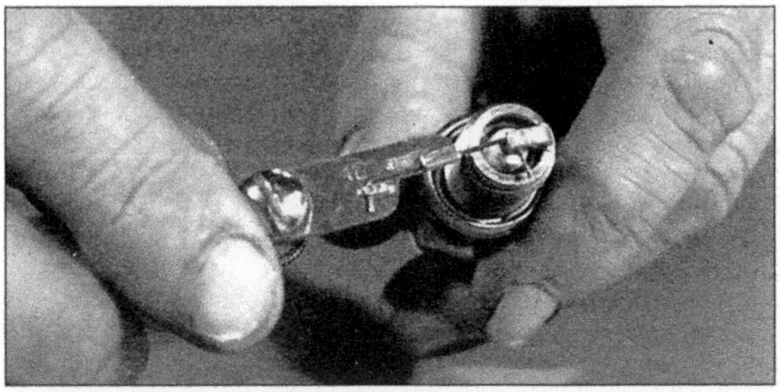

Electrode gap check - use a wire type gauge for best results

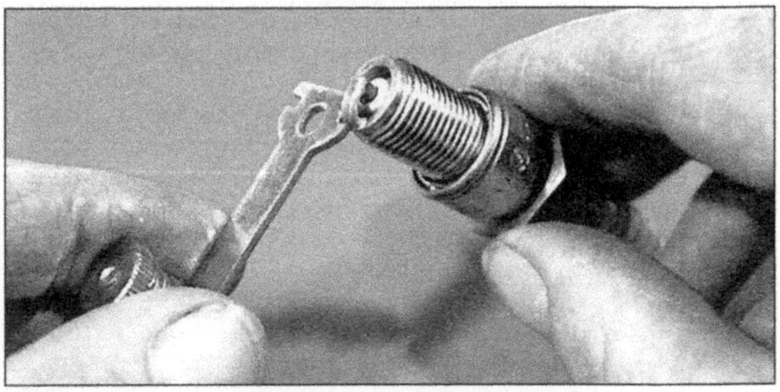

Electrode gap adjustment - bend the side electrode using the correct tool

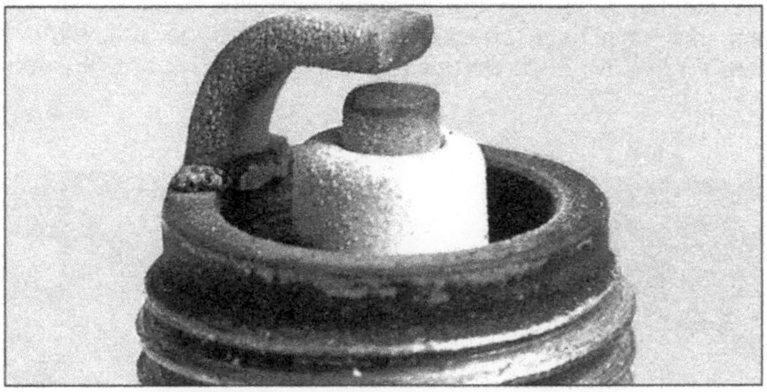

Normal condition - A brown, tan or grey firing end indicates that the engine is in good condition and that the plug type is correct

Ash deposits - Light brown deposits encrusted on the electrodes and insulator, leading to misfire and hesitation. Caused by excessive amounts of oil in the combustion chamber or poor quality fuel/oil

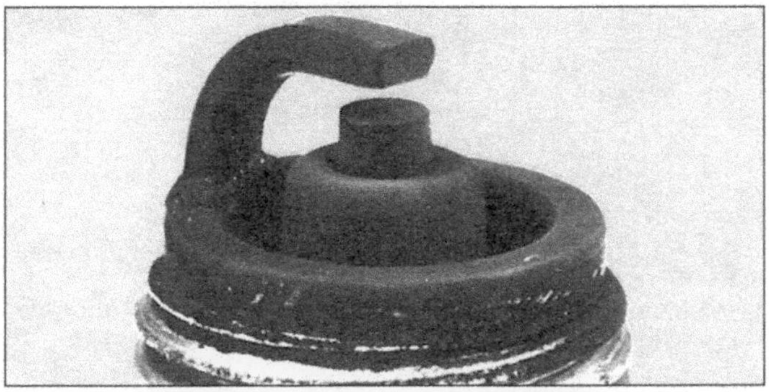

Carbon fouling - Dry, black sooty deposits leading to misfire and weak spark. Caused by an over-rich fuel/air mixture, faulty choke operation or blocked air filter

Oil fouling - Wet oily deposits leading to misfire and weak spark. Caused by oil leakage past piston rings or valve guides (4-stroke engine), or excess lubricant (2-stroke engine)

Overheating - A blistered white insulator and glazed electrodes. Caused by ignition system fault, incorrect fuel, or cooling system fault

Worn plug - Worn electrodes will cause poor starting in damp or cold weather and will also waste fuel

	PRESENT SHIM																								
PART NUMBER (92025-)	1090	1091	1092	1093	1094	1095	1096	1097	1098	1099	1100	1101	1102	1103	1104	1105	1106	1107	1108	1109	1110	1111	1112	1113	1114
THICKNESS (mm)	2.00	2.05	2.10	2.15	2.20	2.25	2.30	2.35	2.40	2.45	2.50	2.55	2.60	2.65	2.70	2.75	2.80	2.85	2.90	2.95	3.00	3.05	3.10	3.15	3.20
VALVE CLEARANCE (mm)																									
0.00 ~ 0.02				2.00	2.05	2.10	2.15	2.20	2.25	2.30	2.35	2.40	2.45	2.50	2.55	2.60	2.65	2.70	2.75	2.80	2.85	2.90	2.95	3.00	3.05
0.03 ~ 0.07			2.00	2.05	2.10	2.15	2.20	2.25	2.30	2.35	2.40	2.45	2.50	2.55	2.60	2.65	2.70	2.75	2.80	2.85	2.90	2.95	3.00	3.05	3.10
0.08 ~ 0.09		2.00	2.05	2.10	2.15	2.20	2.25	2.30	2.35	2.40	2.45	2.50	2.55	2.60	2.65	2.70	2.75	2.80	2.85	2.90	2.95	3.00	3.05	3.10	3.15
0.10 ~ 0.20	SPECIFIED CLEARANCE / NO CHANGE REQUIRED																								
0.21 ~ 0.22	2.05	2.10	2.15	2.20	2.25	2.30	2.35	2.40	2.45	2.50	2.55	2.60	2.65	2.70	2.75	2.80	2.85	2.90	2.95	3.00	3.05	3.10	3.15	3.20	
0.23 ~ 0.27	2.10	2.15	2.20	2.25	2.30	2.35	2.40	2.45	2.50	2.55	2.60	2.65	2.70	2.75	2.80	2.85	2.90	2.95	3.00	3.05	3.10	3.15	3.20		
0.28 ~ 0.32	2.15	2.20	2.25	2.30	2.35	2.40	2.45	2.50	2.55	2.60	2.65	2.70	2.75	2.80	2.85	2.90	2.95	3.00	3.05	3.10	3.15	3.20			
0.33 ~ 0.37	2.20	2.25	2.30	2.35	2.40	2.45	2.50	2.55	2.60	2.65	2.70	2.75	2.80	2.85	2.90	2.95	3.00	3.05	3.10	3.15	3.20				
0.38 ~ 0.42	2.25	2.30	2.35	2.40	2.45	2.50	2.55	2.60	2.65	2.70	2.75	2.80	2.85	2.90	2.95	3.00	3.05	3.10	3.15	3.20					
0.43 ~ 0.47	2.30	2.35	2.40	2.45	2.50	2.55	2.60	2.65	2.70	2.75	2.80	2.85	2.90	2.95	3.00	3.05	3.10	3.15	3.20						
0.48 ~ 0.52	2.35	2.40	2.45	2.50	2.55	2.60	2.65	2.70	2.75	2.80	2.85	2.90	2.95	3.00	3.05	3.10	3.15	3.20							
0.53 ~ 0.57	2.40	2.45	2.50	2.55	2.60	2.65	2.70	2.75	2.80	2.85	2.90	2.95	3.00	3.05	3.10	3.15	3.20								
0.58 ~ 0.62	2.45	2.50	2.55	2.60	2.65	2.70	2.75	2.80	2.85	2.90	2.95	3.00	3.05	3.10	3.15	3.20									
0.63 ~ 0.67	2.50	2.55	2.60	2.65	2.70	2.75	2.80	2.85	2.90	2.95	3.00	3.05	3.10	3.15	3.20										
0.68 ~ 0.72	2.55	2.60	2.65	2.70	2.75	2.80	2.85	2.90	2.95	3.00	3.05	3.10	3.15	3.20											
0.73 ~ 0.77	2.60	2.65	2.70	2.75	2.80	2.85	2.90	2.95	3.00	3.05	3.10	3.15	3.20												
0.78 ~ 0.82	2.65	2.70	2.75	2.80	2.85	2.90	2.95	3.00	3.05	3.10	3.15	3.20													
0.83 ~ 0.87	2.70	2.75	2.80	2.85	2.90	2.95	3.00	3.05	3.10	3.15	3.20														
0.88 ~ 0.92	2.75	2.80	2.85	2.90	2.95	3.00	3.05	3.10	3.15	3.20															
0.93 ~ 0.97	2.80	2.85	2.90	2.95	3.00	3.05	3.10	3.15	3.20																
0.98 ~ 1.02	2.85	2.90	2.95	3.00	3.05	3.10	3.15	3.20																	
1.03 ~ 1.07	2.90	2.95	3.00	3.05	3.10	3.15	3.20																		
1.08 ~ 1.12	2.95	3.00	3.05	3.10	3.15	3.20																			
1.13 ~ 1.17	3.00	3.05	3.10	3.15	3.20																				
1.18 ~ 1.22	3.05	3.10	3.15	3.20																					
1.23 ~ 1.27	3.10	3.15	3.20																						
1.28 ~ 1.32	3.15	3.20																							
1.33 ~ 1.37	3.20																								

INSTALL THE SHIM OF THIS THICKNESS (mm)

Inlet valve clearances shim selection chart

	PRESENT SHIM																								
PART NUMBER (92025-	1090	1091	1092	1093	1094	1095	1096	1097	1098	1099	1100	1101	1102	1103	1104	1105	1106	1107	1108	1109	1110	1111	1112	1113	1114
THICKNESS (mm)	2.00	2.05	2.10	2.15	2.20	2.25	2.30	2.35	2.40	2.45	2.50	2.55	2.60	2.65	2.70	2.75	2.80	2.85	2.90	2.95	3.00	3.05	3.10	3.15	3.20
VALVE CLEARANCE (mm)																									
0.00 ~ 0.02					2.00	2.05	2.10	2.15	2.20	2.25	2.30	2.35	2.40	2.45	2.50	2.55	2.60	2.65	2.70	2.75	2.80	2.85	2.90	2.95	3.00
0.03 ~ 0.07				2.00	2.05	2.10	2.15	2.20	2.25	2.30	2.35	2.40	2.45	2.50	2.55	2.60	2.65	2.70	2.75	2.80	2.85	2.90	2.95	3.00	3.05
0.08 ~ 0.12			2.00	2.05	2.10	2.15	2.20	2.25	2.30	2.35	2.40	2.45	2.50	2.55	2.60	2.65	2.70	2.75	2.80	2.85	2.90	2.95	3.00	3.05	3.10
0.13 ~ 0.14		2.00	2.05	2.10	2.15	2.20	2.25	2.30	2.35	2.40	2.45	2.50	2.55	2.60	2.65	2.70	2.75	2.80	2.85	2.90	2.95	3.00	3.05	3.10	3.15
0.15 ~ 0.25	SPECIFIED CLEARANCE / NO CHANGE REQUIRED																								
0.26 ~ 0.27	2.05	2.10	2.15	2.20	2.25	2.30	2.35	2.40	2.45	2.50	2.55	2.60	2.65	2.70	2.75	2.80	2.85	2.90	2.95	3.00	3.05	3.10	3.15	3.20	
0.28 ~ 0.32	2.10	2.15	2.20	2.25	2.30	2.35	2.40	2.45	2.50	2.55	2.60	2.65	2.70	2.75	2.80	2.85	2.90	2.95	3.00	3.05	3.10	3.15	3.20		
0.33 ~ 0.37	2.15	2.20	2.25	2.30	2.35	2.40	2.45	2.50	2.55	2.60	2.65	2.70	2.75	2.80	2.85	2.90	2.95	3.00	3.05	3.10	3.15	3.20			
0.38 ~ 0.42	2.20	2.25	2.30	2.35	2.40	2.45	2.50	2.55	2.60	2.65	2.70	2.75	2.80	2.85	2.90	2.95	3.00	3.05	3.10	3.15	3.20				
0.43 ~ 0.47	2.25	2.30	2.35	2.40	2.45	2.50	2.551	2.60	2.65	2.70	2.75	2.80	2.85	2.90	2.95	3.00	3.05	3.10	3.15	3.20					
0.48 ~ 0.52	2.30	2.35	2.40	2.45	2.50	2.55	2.60	2.65	2.70	2.75	2.80	2.85	2.90	2.95	3.00	3.05	3.10	3.15	3.20						
0.53 ~ 0.57	2.35	2.40	2.45	2.50	2.55	2.60	2.65	2.70	2.75	2.80	2.85	2.90	2.95	3.00	3.05	3.10	3.15	3.20							
0.58 ~ 0.62	2.40	2.45	2.50	2.55	2.60	2.65	2.70	2.75	2.80	2.85	2.90	2.95	3.00	3.05	3.10	3.15	3.20								
0.63 ~ 0.67	2.45	2.50	2.55	2.60	2.65	2.70	2.75	2.80	2.85	2.90	2.95	3.00	3.05	3.10	3.15	3.20									
0.68 ~ 0.72	2.50	2.55	2.60	2.65	2.70	2.75	2.80	2.85	2.90	2.95	3.00	3.05	3.10	3.15	3.20										
0.73 ~ 0.77	2.55	2.60	2.65	2.70	2.75	2.80	2.85	2.90	2.95	3.00	3.05	3.10	3.15	3.20											
0.78 ~ 0.82	2.60	2.65	2.70	2.75	2.80	2.85	2.90	2.95	3.00	3.05	3.10	3.15	3.20												
0.83 ~ 0.87	2.65	2.70	2.75	2.80	2.85	2.90	2.95	3.00	3.05	3.10	3.15	3.20													
0.88 ~ 0.92	2.70	2.75	2.80	2.85	2.90	2.95	3.00	3.05	3.10	3.15	3.20														
0.93 ~ 0.97	2.75	2.80	2.85	2.90	2.95	3.00	3.05	3.10	3.15	3.20															
0.98 ~ 1.02	2.80	2.85	2.90	2.95	3.00	3.05	3.10	3.15	3.20																
1.03 ~ 1.07	2.85	2.90	2.95	3.00	3.05	3.10	3.15	3.20																	
1.08 ~ 1.12	2.90	2.95	3.00	3.05	3.10	3.15	3.20																		
1.13 ~ 1.17	2.95	3.00	3.05	3.10	3.15	3.20																			
1.18 ~ 1.22	3.00	3.05	3.10	3.15	3.20																				
1.23 ~ 1.27	3.05	3.10	3.15	3.20																					
1.28 ~ 1.32	3.10	3.15	3.20																						
1.33 ~ 1.37	3.15	3.20																							
1.38 ~ 1.45	3.20																								

INSTALL THE SHIM OF THIS THICKNESS (mm)

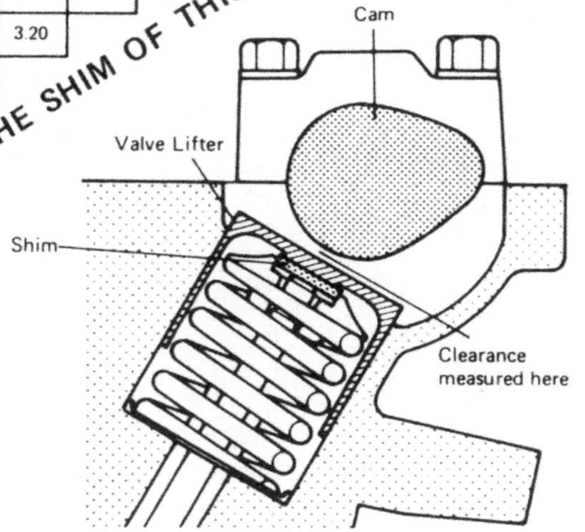

Exhaust valve clearances shim selection chart

Adjusting the valve clearances

Since the clearances are set by a shim placed under each cam follower, as shown in the accompanying illustrations, the clearances are adjusted by removing the camshafts and by replacing the shim with one of different thickness.

Remove the tachometer driven gear and its housing, where a mechanically-driven tachometer is fitted, then remove the carburettors, rotate the crankshaft clockwise until the '1.4' mark is aligned and remove the cam chain tensioner. Withdraw the camshaft(s) and secure the cam chain to prevent it dropping into the crankcase. Very carefully remove the cam followers from the valves to be adjusted. This requires care and the use of a rubber sucker (a valve grinding tool) or of a strong magnet; the shim can then be displaced from its recess in the valve spring top collar. Place the followers (and shims) in marked containers so that they can be refitted to their original valves.

The thickness of each shim should be etched on its underside, but if the numbers are illegible, the thickness can be measured using a micrometer or vernier caliper and recorded. Using the appropriate table (inlet or exhaust) follow the vertical column down until the measured clearance is found (expressed in millimetres), then follow the top horizontal column across until the existing shim thickness is found; where the two columns intersect the size of the required shim is given. Match this with the part numbers given in the top horizontal column when ordering new shims.

Having recorded all the above information it will be necessary to purchase new shims from a Kawasaki dealer, but note that it may be possible to swap shims between the valves to be adjusted to reduce the cost. Do not forget to purchase a new cylinder head cover gasket, cam chain tensioner O-ring and any other parts required.

On reassembly, insert each shim into its recess with the marked surface downwards to preserve the markings as long as possible. Smear oil over each follower and refit it to its original valve, taking great care. If the followers are not absolutely square in their housings they will jam; removal is then very difficult.

Clean both gasket surfaces very carefully, removing all traces of old gasket material without marking the surface. Unless one is extremely lucky it will be necessary to renew the gasket whenever the cylinder head cover is disturbed; do not apply any jointing compound unless the gasket surface is marked, in which case a very thin film of RTV or liquid gasket will be sufficient. In normal use a new gasket will be sufficient, but a smear of grease should be applied to stick it in place.

Refitting the camshafts involves resetting the valve timing; as the procedure is rather involved, although basically simple, it is given in full in Section 44 of Chapter 1. Recheck all the valve clearances to ensure that the shims have been correctly selected and refitted to the correct valve.

Fortunately the valve clearances, once set up properly, will not go out of adjustment for thousands of miles and the adjusting procedure will not be required very often. They must still be checked at every 3000 mile interval to preserve engine performance and to prevent the risk of engine damage. Always record all information (original clearance and shim thickness, new clearance and shim thickness) so that an accurate picture can be built up of the valve gear and its rate of wear. This will greatly assist subsequent adjustment.

4 Check and adjust contact breaker points – Z400 J1, Z500 B1, B2, KZ/Z550 A1

To gain access to the contact breaker it is necessary to remove the cover plate screws and the cover on the right-hand front of the crankcase. Two methods of adjustment are possible.

Adjustment using feeler gauges

Rotate the engine by applying a spanner to the engine-turning hexagon until one set of points is fully open. Examine the faces of the contacts for pitting and burning. If badly pitted or burnt they should be renewed. Undo the two screws that hold the fixed contact of each set of points, slacken the retaining nut, and remove the two low-tension wires, which will allow the points to be lifted off. Removal of the circlip on the end of the pivot pin will permit the moving contact point to be detached. Note the arrangement of the insulating washers.

The points should be dressed with an oil stone or fine emery cloth to remove deposits due to arcing. Keep them absolutely square throughout the dressing operation, otherwise they will make angular contact on reassembly, and rapidly burn away. If emery cloth is used, it should be backed by a flat strip of steel. If it is necessary to remove a substantial amount of material before the faces can be restored, the points should be renewed.

Refit the contacts with the peg on each fixed contact rear face inserted into the hole in the ignition backplate, then reverse the dismantling procedure, making quite certain that the insulating washers are fitted in the correct way. In order for the ignition system to function at all, the moving contact and the low-tension leads must be perfectly insulated from the fixed contact. Apply a very light smear of grease to the pivot pin, before refitting the moving contact.

Adjustment is carried out by slackening the screws on the base of the fixed contact, and adjusting the gap within the range 0.3 - 0.4 mm (0.012 - 0.016 in) when the points are fully open, by moving the fixed contact with a screwdriver between the screw head and the two raised ears. Retighten the two screws after adjustment with the feeler gauge and re-check the gap, then repeat the same operation for the other set of points. Do not forget to double check after you have tightened the setting screws, in case the setting has altered.

Before replacing the cover and gasket, place a slight smear of grease on the cam and a few drops of oil on the felt pad. Do not over lubricate for fear of oil getting on the points, and causing poor electrical contact.

Adjustment using a dwell meter

Dwell meters, available from most auto accessory shops, provide a much more accurate setting by measuring, in terms of degrees (or percentage) of crankshaft rotation, the time that the points remain closed; this being done while the engine is running. Since most dwell meters are designed for use with car distributors with more than one points cam lobes, the true reading must be calculated from that measured as follows, the specified dwell angle being 192.5 ± 7.5° (53.5 ± 2.5%):

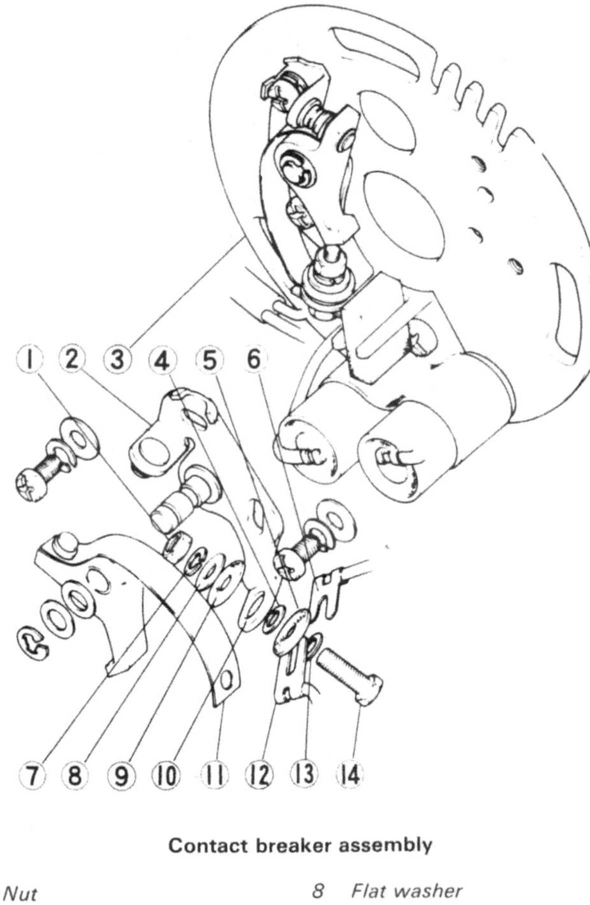

Contact breaker assembly

1	*Nut*	*8*	*Flat washer*
2	*Fixed contact*	*9*	*Large insulator*
3	*Mounting plate*	*10*	*Fixed contact baseplate*
4	*Small insulator*	*11*	*Moving contact*
5	*Large insulator*	*12*	*Contact breaker lead*
6	*Capacitor lead*	*13*	*Flat washer*
7	*Lock washer*	*14*	*Bolt*

Meter setting	Reading
1 cylinder	185 - 200° (51 - 56%)
2 cylinders	92.5 - 100° (25.5 - 28%)
3 cylinders	61.5 - 67° (17 - 19%)
4 cylinders	46 - 50° (12.5 - 14%)

Set the dwell tester in the appropriate position, then connect the positive (+) probe to the moving contact terminal, and the negative (–) probe to the crankcase. With the engine running at idle speed, check that the reading is within the limits given. If this is not the case, slacken the fixed contact securing screw just enough to permit movement, then adjust the gap until the reading is within the specified tolerances. Tighten the securing screw, then check that the setting has not altered. The test sequence should be repeated on the remaining set of contact breaker points. Note that the points must be examined, cleaned and renewed, if necessary, as described above.

5 Checking the ignition timing – all models

Contact breaker ignition system

Note that while the timing can be checked statically using a bulb and two lengths of wire to make a test circuit, it is recommended that the timing is checked with the engine running, using a strobe timing light. Use only the more expensive xenon-tube type of lamp which uses an external power source such as the household mains supply or a separate battery (do not connect a strobe to the machine's own battery or a spurious reading may result).

First check that the points are clean and correctly adjusted as described above, then connect the strobe, following its manufacturer's instructions, to number 4 spark plug (or its HT lead). Start the engine and allow it to idle, then aim the light at the aperture in the ignition backplate. The '1.4' F mark (see accompanying illustration) should be aligned with the fixed index mark on the crankcase wall. If not, slacken the three mounting screws by just enough to permit backplate movement and rotate the complete backplate using a screwdriver in the notches provided until the marks align. Tighten the screws and recheck the timing. If the setting is correct switch off the engine. Moving the backplate will almost certainly alter the points gap; check the gap again as described in the previous Section. If adjustment is required, it will be necessary to check the timing again, but if the gap is within tolerances, numbers 1 and 4 cylinders are correctly set and the timing lamp should be connected to number 3 cylinder spark plug and the engine restarted. Check the timing as described above but using '2.3'F mark (exactly the same as those shown but stamped on the opposite side of the ATU). If adjustment is required it is made by opening or closing the gap of the right-hand contact breaker; with care this can be done while the engine is running and is easiest if a dwell meter is used, as described in the previous Section; do not set the gap outside the specified range.

If the ignition timing on 2 and 3 cylinders cannot be set with the contact breaker gap within tolerances, set the gap to exactly 0.35 mm (0.014 in) and adjust the timing setting by rotating the complete backplate, as described above. With 2 and 3 cylinders thus set, attempt to correct the ignition timing on 1 and 4 cylinders by opening or closing the gap; ensuring that it remains within the specified range. Thus the final setting may well be a compromise with both points gaps at the extremes of their tolerance and the ignition backplate in the midway position.

If the correct settings cannot be achieved, it is likely that the contact breakers are worn out. Renew both, regardless of their apparent condition, set the backplate to the middle of its adjustment slots and set both points gaps at exactly 0.35 mm (0.014 in). Repeat the procedure to check and reset the ignition timing.

When the timing is correct at idle speed, use the timing light to check the operation of the ATU. Connect the lamp to number 4 spark plug and slowly increase speed. At approximately 1500 rpm the T mark should appear to start moving to the left to be replaced by the two parallel lines of the full advance mark which will align with the crankcase mark at 3-3400 rpm. The movement should be smooth and even, also in the reverse direction when the twistgrip is closed. Stop the engine and repeat the test on numbers 2 and 3 cylinders.

If movement appears jerky and uncontrolled, or if its range is too much or too little, so that the advance marks do not align, the ATU must be removed and dismantled for lubrication as described under the annual/6000 mile heading. If excessive wear is found or if lubrication fails to cure the fault, the ATU must be renewed. When the ignition timing is correct, stop the engine, disconnect all equipment and refit all disturbed components.

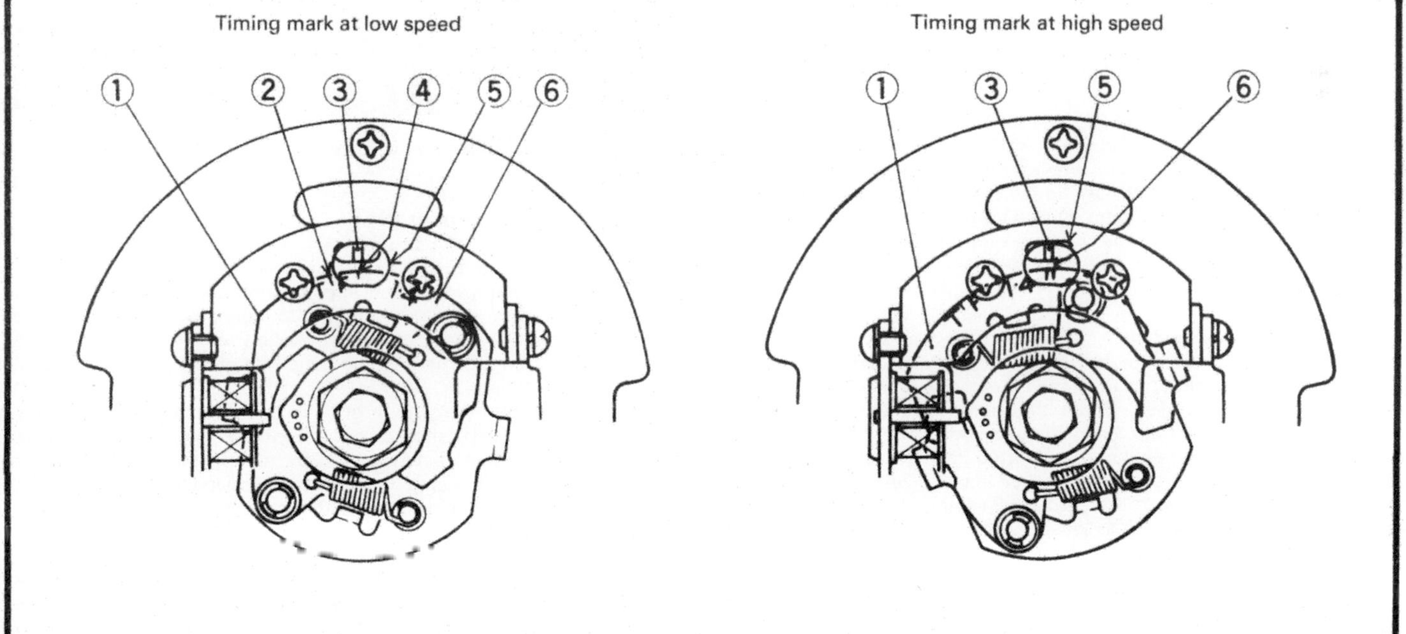

Ignition timing marks – typical

1 Timing advancer
2 'T' mark (not used for timing test)
3 Timing mark on crankcase
4 'F' mark
5 Inspection window
6 Advanced timing marks

Electronic ignition system

While this is not strictly necessary as an item of routine maintenance, it should be carried out at regular intervals to check that all is well and that the advance mechanism or circuits are functioning correctly. The ignition timing can only be checked with the engine running, using a strobe, and again the use of the xenon-tube type is recommended (see above).

Remove the ignition pickup cover from the crankcase front right-hand end and connect the strobe to number 4 cylinder's spark plug (or its HT lead). Start the engine, allow it to idle and aim the strobe at the aperture in the ignition backplate; the '1.4' F mark should align with the crankcase index mark, as shown in the accompanying illustration. If not, some ignition backplates have slots at their mounting points for adjustment; others have not. Slacken the three mounting screws and move the backplate in the direction required until the marks align. Note that this problem should not arise unless the backplate has been refitted wrongly for any reason; in normal use the ignition timing will not vary.

Increase engine speed to 3500 - 4000 rpm; on all models with ATU-advance the timing marks should behave exactly as described above for contact breaker ignition systems. If not, examine and lubricate or renew the ATU as necessary. On all models fitted with electronic advance, ie an ignition rotor is fitted, not an ATU, the crankcase mark should be halfway between the F mark and the single line of the full advance mark and the advance should still be operating. In fact full advance is at 10 000 rpm although this is impossible to check as an engine should never be run at that sort of speed when not under load; serious engine damage may occur if an attempt is made to check the timing at full advance. If the advance does not appear to be functioning correctly, the IC ignitor may be at fault and should be tested as described in Chapter 3. Since it can only be renewed, have your findings confirmed by a competent Kawasaki dealer before such an expensive solution is adopted.

On all electronic ignition-equipped models, there is no need to check the remaining pair of cylinders as all relevant components are the same for both pairs.

6 *Clean the air filter*

The air filter fitted to KZ/Z550 H1, H2, and all ZR400/550 models is in two parts. Behind the left-hand sidepanel will be found a circular cover retained by a wing nut; remove the cover and withdraw the dry pleated-paper filter element. The other filter is behind the right-hand sidepanel; remove the panel, the battery (taking care to disconnect first the negative (–) terminal) and the battery case, then remove the element housing from the main filter casing, the housing being secured by three bolts or screws. Detach the foam filter element from the supporting frame.

On all other models the air filter is found under the seat at the rear of the fuel tank. Except for ZX550 A1, A1L and all shaft drive models, the filter cap is retained by two screws; remove these and withdraw the cap and filter element. It may be necessary to remove the baffle plate that is secured to the frame by two bolts immediately above the filter housing. The procedure is the same for ZX550 A1, A1L and all shaft drive models except that the tank rear mounting must be dismantled and the tank raised sufficiently to permit the removal of the element.

Kawasaki recommend that all elements be cleaned in a bath of non-oily high flash-point solvent such as Stoddard solvent (white spirit). This is somewhat unusual in the case of dry paper elements and great care should be taken. When the element is clean, dry it by shaking it or by blowing from the inside outwards with compressed air. When dry, all paper elements can be refitted. In the case of the foam filter mentioned above, it should be soaked in clean SAE 30 SE engine oil and squeezed to remove the surplus, finishing off by squeezing it when wrapped in a clean rag to leave the element as dry as possible and only slightly oily to the touch. Never wring out a foam element as this will damage it. Refit the element to its supporting frame then to its housing and refit the housing to the machine.

If any element is severely clogged, worn out, or has any holes or splits in it, it should be renewed. Check also that the foam sealing band is intact on all paper elements and that the element is seated correctly in its housing so that unfiltered air cannot leak past.

Warning: Solvents of the type recommended are not only inflammable, they are toxic if inhaled for prolonged periods. Always wear hand and eye protection, work in a well-ventilated area and take all precautions to prevent the risk of fire when using such solvents.

Air filter must be removed at regular intervals for cleaning or renewal

Ensure air filter housing cover is correctly refitted and securely fastened

7 *Adjust the carburettors*

Before attempting to perform any form of carburettor adjustment it should be realised that uneven running may be caused by a number of other factors, many of which appear unrelated to carburation. In particular, attention should be given to spark plug and air filter condition, contact breaker points and ignition timing, valve clearance adjustment, and on US models, the condition of the emission control system components.

Start by ensuring that the throttle cable operates smoothly and is adjusted properly. There should be 2 - 3 mm of free play in the cable, and this can be adjusted as required by turning the in-line adjuster below the twistgrip housing (constant depression carburettors). On models fitted with slide carburettors there should be 2 - 3 mm of free play measured in terms of twistgrip rotation. If not, push the twistgrip fully closed and use the closing cable adjusters (the rear cable at the twistgrip) to ensure that there is 2 - 3 mm (0.08 - 0.12 in) clearance between the right-hand cable pulley and the cable catch. It may be necessary to slacken their locknuts and to screw in fully the opening cable (front cable at the twistgrip) adjusters to gain sufficient free play for the closing cable to be set. When the closing cable is set, release the twistgrip and use the opening cable adjusters to set the twistgrip free play specified above. Tighten all adjuster locknuts. Note that if the cables are not set correctly and free play is excessive throttle control will be sloppy; if there is insufficient free play the throttle action will be heavy and imprecise and the idling speed will be unreliable. Once adjusted, check that the prescribed amount of free play is evident with the handlebar turned to full lock in both directions. If the amount of free play varies with the position of the handlebar, check the routing of the throttle cable(s) and improve it as required.

Idle speed should be checked with the engine at its normal operating temperature, and thus is best dealt with after a run. Allow the engine to idle and check that it runs at the specified speed. If adjustment is necessary, use the idle adjusting screw.

If, after idle speed adjustment, the engine is not running smoothly then it will be necessary to check carburettor synchronisation. This requires the use of a set of vacuum gauges; refer to the Section on synchronisation in Chapter 2. Failing this, have the synchronisation checked by a Kawasaki dealer. Do not touch the pilot screws. (See Chapter 2.)

Finish off carburettor maintenance by placing a suitable container under the ends of the carburettor drain hoses (where these are not fitted hoses should be connected to the spigot on the bottom of each float chamber). Unscrew by one or two full turns each carburettor drain plug and turn the fuel tap to the 'Pri' position to flush out any dirt or water. Allow fuel to flow for a few seconds, then turn off. If a drain plug is fitted to the fuel tap (early models only) this should be removed to clear water or dirt from the tap and tank. Ensure that all drain plugs are securely fastened, that the hoses are routed correctly (or removed) and that all surplus fuel is mopped up. Be very careful to prevent the risk of fire while this is being done. If a large amount of dirt or water is evident in the drained fuel, the system must be cleaned out thoroughly. See Chapter 2.

8 Change the engine oil

It is important that the engine/transmission oil is changed at the recommended intervals to ensure adequate lubrication of the engine components. If regular oil changes are overlooked, the prolonged use of degraded and contaminated oil will lead to premature engine wear. Note that where mileages are unusually low, the engine oil should be changed annually irrespective of mileage readings.

The oil should be changed with the engine at its normal operating temperature, preferably after a run. This ensures that the oil is relatively thin and will drain more quickly and completely. Obtain a container of at least 3.5 litres (6.2 Imp pint/3.7 US qt) capacity, and arrange it beneath the crankcase drain plug. Slacken and remove the plug, noting that the oil filter cover should be left undisturbed, and allow the oil to drain.

When the crankcase is completely emptied, clean the drain plug orifice and refit the plug, tightening it to 27.5 lbf ft (3.8 kgf m). Remove the filler plug, and add sufficient oil of the recommended grade to bring the level half way up the window in the outer cover. This will take approximately 2.6 litres (4.6 Imp pint/2.8 US qt).

Start the engine and allow it to idle for a few minutes to distribute the new oil through the system. Switch off and allow the machine to stand for a while, then check that the oil level comes between the upper and lower level marks on the sight glass.

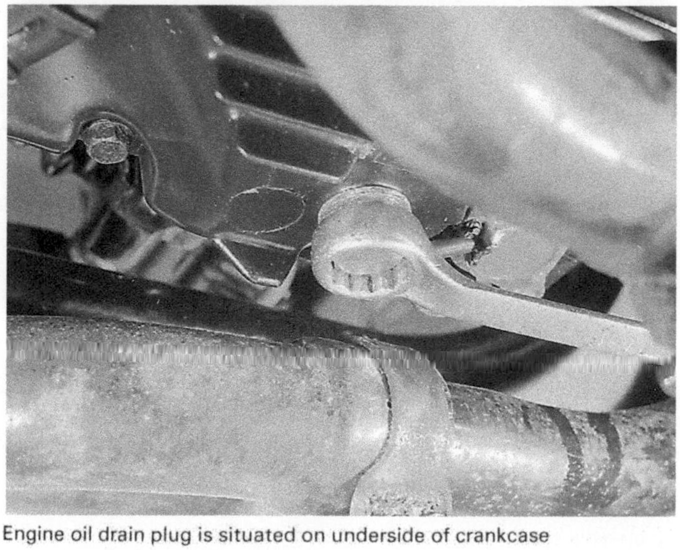

Engine oil drain plug is situated on underside of crankcase

Use only good quality engine oil of specified grade when topping up

9 Adjust the clutch

There should be 2 - 3 mm (0.08 - 0.12 in) of free play in the clutch cable, measured between the handlebar lever butt end and the handlebar clamp. If the free play is not correct the clutch must be adjusted. The procedure varies and instructions are given below for the different models, but owners of Z400 J1, J2, and KZ/Z550 A1, A2, C1, C2 models should note that their machines may be fitted with one of two different types of release mechanism and that the procedure required may be either the second or the third of those given. Refer to Section 29 of Chapter 1 for the details available.

ZX550 A1, A1L and all shaft drive models

Slacken both upper and lower adjuster locknuts and screw in the handlebar adjuster as far as possible before tightening its locknut. Using the adjuster on the crankcase top, rotate the adjusting nut until the specified free play is set, then tighten the locknut. Apply a few drops of oil to all cable nipples, lever pivots and exposed lengths of inner cable. There is no means of adjusting the clutch itself on these models, and the handlebar adjuster is used for quick settings.

Z400 J1, Z500 B1, B2, KZ/Z550 A1, C1, C2 and KZ550 A2 models

Slacken both cable adjuster locknuts and screw in fully the cable mid-way adjuster sleeve nut to gain the maximum free play. Turn the handlebar adjuster until there is a gap of 5 - 6 mm (0.20 - 0.24 in) between the locknut and the inner edge of the adjuster cable abutment, ie the exposed length of adjuster thread.

Remove from the gearbox sprocket cover the clutch adjusting cover and its gasket, these being retained by two screws. Slacken the locknut and rotate the adjusting screw anti-clockwise by one or two turns to check that there is no pressure on it, then rotate it clockwise until it seats firmly. Do not apply excessive pressure; this is the point at which the release mechanism is starting to lift the pressure plate. Rotate the screw anti-clockwise by $\frac{1}{2}$ turn ($\frac{1}{4}$ turn on 550 models) to set the necessary free play, and hold it in that position while the locknut is tightened.

Check that the lower end of the clutch outer cable is securely fitted in its recess in the sprocket cover, then unscrew the sleeve nut on the cable mid-way adjuster until all cable free play is eliminated. Tighten the mid-way adjuster locknut and use the handlebar adjuster to set the specified free play at the lever. Tighten the locknut, apply a few drops of oil to all cable nipples, lever pivots, adjusters and exposed lengths of inner cable. Refit the adjuster cover and its gasket.

All other models

Remove from the gearbox sprocket cover the clutch adjusting cover and its gasket, these being retained by two screws, then slacken both cable adjuster locknuts and screw in both adjusters to gain the maximum cable free play. Slacken the locknut and rotate the adjusting screw clockwise by one turn to check that there is no pressure on it, then rotate it anti-clockwise until it seats firmly. Do not apply excessive pressure; this is the point at which the release mechanism is starting to lift the pressure plate. Rotate the screw clockwise by $\frac{1}{4}$ turn to set the necessary free play and hold it in that position while the locknut is tightened.

Tighten the handlebar adjuster locknut and use the cable mid-way adjuster to remove all but the specified free play from the cable, then tighten its locknut. Apply a few drops of oil to all cable nipples, lever pivots, adjusters and exposed lengths of inner cable. Refit the adjuster cover and its gasket.

Side-stand retracting mechanism

A few UK models are fitted with an automatic side-stand retracting mechanism which is operated by a short cable branching off the main clutch cable. Whenever the clutch is adjusted, check the adjustment of this mechanism, which should be set so that all cable free play is just eliminated when the stand is in the down position. Secure the cable adjuster locknut and apply a few drops of oil to the mechanism pivots, cable nipples and bearing surfaces.

10 Check the brakes

Hydraulic disc brakes require no adjustment; all that is necessary is to maintain a check on the fluid level and pad wear, the fluid level check being described under the monthly/500 mile heading. On those models with disc rear brakes the pedal setting should be checked. Z500 B1, B2 and KZ/Z550 D1 models are fitted with adjustment for pedal height, the recommended setting being anything up to 30 mm (1.2 in) below the level of the footrest top surface. The actual setting can be altered to suit the rider's convenience by slackening its locknut and rotating the pedal/master cylinder pushrod to shorten it and provide the necessary free play, then by screwing up or down the stop screw provided. When the setting is correct, tighten the stop screw locknut. Alter the pushrod length so that there is 8 - 10 mm (0.32 - 0.40 in) of pedal movement before the pushrod end contacts the master cylinder piston, as shown in the accompanying illustration, then tighten its locknut. Free play is discernible by feeling for side-to-side play at the point where the pushrod is linked to the pedal rear end. On KZ/Z550 H1, H2 and ZX550 A1 and A1L models coarse pedal height adjustment can be made by refitting the pedal in a different place on its shaft splines (the standard position being indicated by punch marks to be aligned on both components) but the pedal/master cylinder pushrod is adjustable for length to provide adjustment for all normal use; the standard setting is between 19 - 23 mm (0.75 - 0.90 in) below the footrest level on KZ/Z550 H1, H2 models, 25 - 30 mm (1.0 - 1.2 in) on ZX550 A1, A1L models. Note that although no specific provision is made for free play adjustment, some free play must always be present or the pushrod will be applying pressure to the piston all the time, producing severe brake drag.

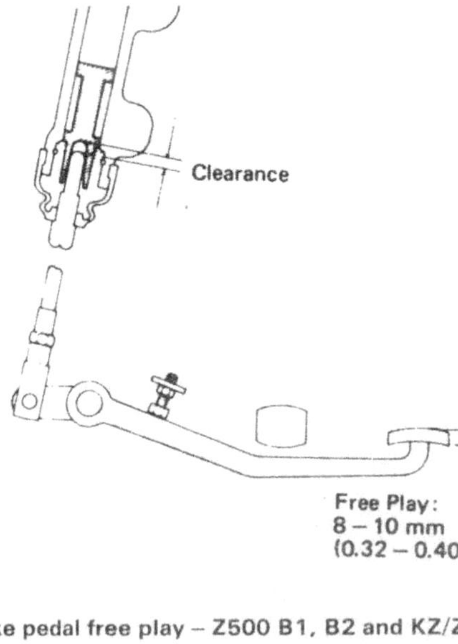

Brake pedal free play – Z500 B1, B2 and KZ/Z 550 D1

On machines fitted with drum rear brakes, the pedal height can be set to suit the rider's preference at anywhere from level with the footrest top surface to 30 mm (1.2 in) below it, the setting being made by slackening its locknut and rotating the pedal stop screw located near the pivot; it may be necessary to slacken off the brake adjustment to set the required pedal height. Tighten the locknut when the pedal height is correct. Before adjusting the rear brake, check that the angle between the brake rod and the operating arm is less than 90° when the brake is fully applied so that the linkage is working at maximum efficiency. If the angle exceeds 90°, disconnect the brake rod from the operating arm, remove the pinch bolt securing the arm to the brake camshaft and pull the arm off the camshaft, taking care not to disturb the wear indicator pointer. Reposition the arm on the shaft so that with the brake correctly adjusted and fully applied, the angle between arm and rod is 80 - 90°; this may require some experimentation, especially if the brake shoes are nearing the end of their useful life. Press the arm fully on to the shaft and tighten securely the pinch bolt. Drum rear brakes are adjusted using the nut at the rear end of the brake rod so that the tip of the brake pedal has between 20 - 30 mm (0.8 - 1.2 in) of travel between the at rest and fully applied positions.

On all models, adjust the stop lamp rear switch by means of its two metal nuts or single plastic sleeve nut (whichever is fitted) so that the lamp lights as soon as the pedal has taken up its free play and the brake can be felt to be engaging the drum or disc; this is usually after about 15 mm (0.6 in) of pedal movement. On KZ/Z550 H1 and H2 models the switch is concealed beneath the battery case; remove the battery (taking care to disconnect first the negative (–) terminal) and its

Do not forget to check stop lamp switch setting after brake adjustment

case to expose the switch. **Note**: On all models, do not rotate the switch body during adjustment; this will damage the lead wire terminals necessitating switch renewal.

The amount of friction material remaining on brake shoes or pads can be checked without dismantling any component. On all drum rear brakes a wear indicator pointer is set on the camshaft to align with an arc marked 'Usable range' stamped on the backplate; if, when the brake is fully applied, the pointer indicates outside the rear end of the arc, the shoes are worn out and must be renewed. This will involve the removal of the rear wheel and the dismantling of the brake; all work necessary is described in Chapter 5. On all disc brakes, the friction material of all brake pads must be at least 1 mm (0.04 in) thick; on some types of pad this may be indicated by a stepped portion machined in the friction material or by a red-painted groove, but on most pad types the remaining thickness of friction material must be measured. If any pad is worn to less than the specified thickness of friction material at any point, then both pads of that caliper must be renewed as a set. Note: on models with twin front disc brakes it is preferable to renew both pairs of pads together, even if only one pad is excessively worn. On some models, an inspection window is provided in the top surface of each caliper, so that the pads can be

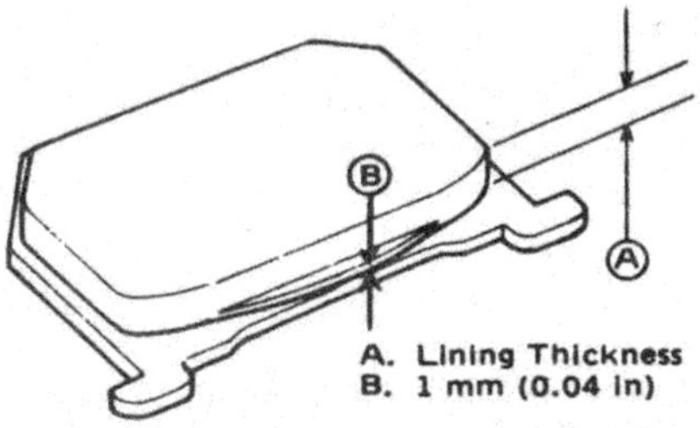

Brake pad friction material wear limits – typical

checked with a torch, and the rear caliper of Z500 B1 and B2 models is fitted with a plastic top cover which must be unclipped to expose the pads, but in most cases it will be necessary to look at the pads from above or below the caliper. If the pads are found to be so fouled with road dirt that the friction material cannot be seen, they must be removed for cleaning and inspection.

Pad removal and refitting varies according to model. Proceed as follows:

Z400 J1, Z500 B1, B2 (front) and KZ/Z550 A1

Remove the two bolts securing the caliper to the fork leg and withdraw the caliper from the disc, taking care not to twist the brake hose. Remove the screw and lockwasher securing the fixed pad, then displace the metal plate and tap out the pad. Press the caliper mounting bracket as far as possible towards the piston to displace the moving pad. On reassembly, first see the general notes below. Check that the caliper slides across smoothly and easily on its axle bolts. Refit the moving pad with its stepped portion pointing upwards, and the fixed pad with its tongue located in the groove in the caliper body. Refit the metal plate and the screw and lockwasher, applying thread locking compound to its threads and tightening the screw to a torque setting of 0.3 kgf m (2 lbf ft). Refit the caliper, tightening its mounting bolts to a torque setting of 3.0 kgf m (22 lbf ft). Repeat for the other caliper, if fitted.

Z500 B1, B2-rear

Unclip the plastic cover and hook out the R-clip securing each pad pin, then pull out each pin in turn, taking care that the anti-rattle springs do not fly off. Lift out the pads. On reassembly, first see the general notes below. Thoroughly clean the interior of the caliper, removing all road dirt etc, and note that this is a twin-piston caliper; therefore it will be necessary to use two long, slim levers, so that both pistons can be pressed back at the same time, instead of using hand pressure. Lever against the disc centre/wheel hub, or the disc outer edge may be warped. The general notes still apply if undue stiffness is encountered. Clean carefully both pins and smear them with the recommended grease. Insert the outside (right-hand) pad first and partially refit both pins to retain it, noting that each pin's drilled end must be outwards, then refit the second pad and push one pin fully into place. Locate one end of both anti-rattle springs under this pin, check that the spring ears are correctly engaged on the pad metal backing, then refit the second pin, pushing down on each spring in turn so that both are held fast by the pins. Refit the two R-clips to the pins and ensure that the R-clips are inclined towards each other so they are well clear of, and cannot damage, the dust seal. Refit the plastic cover.

ZR400 A1, B1, ZR550 A1, A2, Z550 G1, G2 and ZX550 A1, A1L (front and rear)

Remove the two bolts securing the caliper mounting bracket to the fork lower leg or torque stay (as appropriate) and withdraw the caliper from the disc, taking care not to twist the brake hose. Press the mounting bracket as far as possible towards the piston and disengage the fixed pad, then displace the moving pad. Check that the caliper slides across smoothly and easily on its axle bolts. On reassembly, first see the general notes below. Check that the anti-rattle spring is correctly located then refit the moving pad. Press the mounting bracket towards the piston and refit the fixed pad, noting that its locating pins should also be greased to prevent corrosion. Refit the caliper, tightening its mounting bolts to a torque setting of 3.0 kgf m (22 lbf ft) on all models except for ZX550 A1 and A1L (front only) calipers, where the setting is 2.5 kgf m (18 lbf ft). Repeat for the other caliper, where applicable.

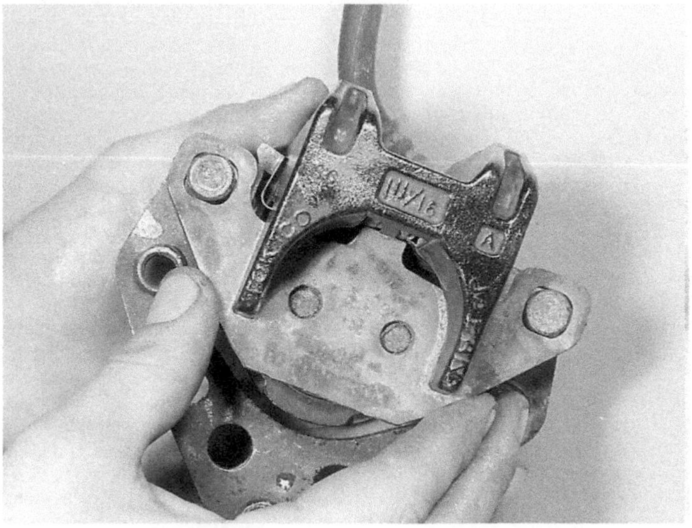

Pad removal, later type of caliper – press mounting bracket towards piston to release fixed pad ...

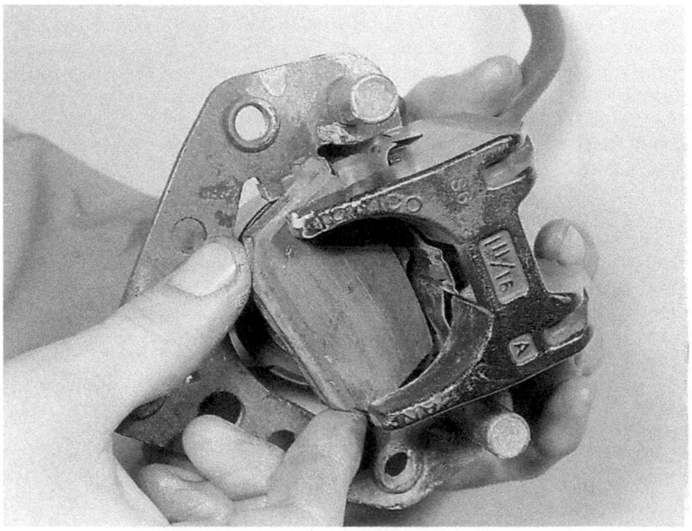

... then lift out moving pad

All other models – front (and rear)

Remove the nut and bolt securing the torque arm to the caliper (KZ/Z550 D1, H1, H2 rear brakes only). Remove the two axle bolts securing the caliper to its mounting bracket and lift away the caliper taking care not to twist its brake hose; this will leave the pads and mounting bracket in place on the disc. Temporarily refit the caliper with the pads removed to check that it slides smoothly and easily on its axle bolts. Before reassembly, see the general notes below. Reassembly is the reverse of the dismantling procedure, tighten the axle bolts to a torque setting of 1.8 kgf m (13 lbf ft) and the torque arm retaining nut (where applicable) to 3.0 kgf m (22 lbf ft).

Brake pads and calipers – general

If the friction material of any brake pad is worn at any point to a thickness of 1.0 mm (0.04 in) or less, or if the friction material is fouled with oil or grease, both pads (of that caliper) must be renewed as a set. There is no satisfactory way of degreasing friction material. If the pads are not excessively worn, clean them thoroughly with a wire brush, paying careful attention to the sides and rear of the pad metal backing. Dig out any embedded particles from the friction material and clean out any grooves or wear indicator shoulders. Areas of glazing can be removed by rubbing the pad with emery paper. Where brake pads are the same on each side of the caliper (eg Z500 rear brake) and one has worn more than the other, it is possible to prolong their effective life by swapping them over on refitting, to even out the wear. Moving pads (those next to the caliper piston) must slide in the caliper body; thoroughly clean the edges of the pad backing and the matching surface in the caliper body and apply a thin smear of grease to both. Also apply a thin smear of grease to the backs of each pad to prevent brake squeal. Kawasaki recommend the use of PBC (Poly Butyl Cuprysil) based grease which is easily obtainable from auto accessory shops; **do not** use ordinary high-melting point grease. This will melt and foul the friction material, rendering the brake ineffective. Except for Z500 rear calipers, check that the caliper slides smoothly and easily on its axle bolts; if not, the caliper must be checked as described in the relevant Section of Chapter 5.

If new pads are to be fitted, the piston must be pushed back to make room. Remove the master cylinder reservoir cover and diaphragm and pack clean rag around the reservoir to catch any spilt fluid. If working on one caliper of a twin disc system, check that the other caliper is in place on the disc or wedge a piece of wood between its pads so that the piston cannot move. Using hand pressure only, push the piston fully back into the caliper body, watching carefully the fluid level in the reservoir; if it rises above the 'Upper' level mark soak up the surplus fluid using a clean rag. If the piston cannot be pushed back using hand pressure, it must be dismantled as described in Chapter 5 for cleaning and examination. On reassembly, apply a thin smear of PBC grease to all moving parts, but take great care to keep it away from the friction material. All single-piston calipers are dependent for their efficiency on being able to move smoothly so that equal pressure is applied to both pads. When the caliper is refitted, (both calipers, where applicable), apply the brake lever or pedal several times to bring the pads back into contact with the disc. Check the hydraulic fluid level; if the original pads have been refitted, it will suffice that the level is above the 'Lower' level mark on the reservoir body, next to the sight glass. If new pads have been fitted, the level must be topped up to the 'Upper' level mark; if this is not marked on the reservoir as described above, it is in the form of a line cast on the inside of the reservoir. Dry the diaphragm with a clean cloth, fold it into its compressed state, and refit it. Refit the cover, tightening it or its retaining screws securely. Before using the machine on the road, check that the brakes and stop lamp work properly and that full lever (or pedal) pressure is restored. New pads should be used lightly but firmly, avoiding excessively light or heavy pressures, to enable them to bed in properly for the first 50 miles.

Note: *brake fluid is an extremely effective paint stripper; do not spill any onto any plastic or painted component, and use copious quantities of fresh water to flush away any fluid that is spilt.*

11 Check and adjust the steering head bearings

Place the machine on its centre stand on level ground and raise the front wheel clear of the ground by placing a suitable support under the crankcase.

Check the bearing adjustment by grasping the bottom of both fork lower legs, then pulling and pushing in a fore and aft direction; any free play should be felt between the fork bottom yoke and the frame head lug. Check for overtightened bearings by placing the forks in the straight ahead position and tapping lightly on one handlebar end; the forks should fall away smoothly and easily to the opposite lock, taking into account the effect of cables and wiring, with no trace of notchiness. If adjustment is required, proceed as follows:

All Z400 J, Z500 and KZ/Z550 A, C, and D models

Either remove the fuel tank or protect its paintwork with padding such as an old blanket, then remove the handlebars and lay them on the frame or tank clear of the steering head, taking care not to stretch or kink any control cables, brake hoses or wiring. Slacken fully the bottom yoke pinch bolts and apply a few drops of penetrating fluid to the stanchions so that they are free to slide in the bottom yoke. Slacken fully the steering stem head clamp bolt, then the large steering stem bolt at the centre of the top fork yoke, and use a C-spanner to rotate the slotted adjuster nut immediately below the top yoke. As a guide to adjustment, tighten the slotted nut until a light resistance is felt, then back it off by $\frac{1}{8}$ turn. The object is to remove all discernible play without applying any appreciable preload. It should be noted that it is possible to apply a loading of several tons on the small steering head bearings without this being obvious when turning the handlebars. This will cause an accelerated rate of wear, and causes the machine to weave from side to side at low speeds. When adjustment is correct, tighten the steering stem bolt to a torque setting of 4.5 kgf m (32.5 lbf ft) and recheck the adjustment. If all is well, tighten the clamp bolt to a torque setting of 1.8 kgf m (13 lbf ft) and the bottom yoke pinch bolts to 1.8 kgf m (13 lbf ft). The handlebar clamp bolts should be tightened evenly to a torque setting of 1.8 kgf m (13 lbf ft) with the handlebars at the same angle as the stanchions and the gap between the clamps and the top yoke equal at front and rear.

KZ/Z550 H1, H2, ZR400/550, ZX550 A1, A1L and all shaft drive models

These machines are fitted with taper roller steering head bearings. The procedure for adjustment is exactly as that given above, except that it is necessary to remove the handlebars only on KZ550 F1, F2, F2L and M1 models; on all other models, detach the ignition switch/fork top yoke cover which is retained by two screws. The machine must be positioned as described in the first paragraph of this Section and the stanchions must be free to move in the fork bottom yoke. With the steering stem head bolt slackened (there is no clamp bolt fitted) use a slim C-spanner to slacken the (upper) slotted locknut before making the adjustment using the (lower) slotted adjuster nut. When the adjustment is correct, hold the adjusting nut and use a second C-spanner to tighten lightly the locknut, then tighten the head bolt to a torque setting of 4.3 kgf m (31 lbf ft) and recheck the adjustment. If all is well, tighten the bottom yoke pinch bolts to 1.8 kgf m (13 lbf ft) on chain drive models, 2.8 kgf m (20 lbf ft) on shaft drive models. Refit the ignition switch/top yoke cover, where applicable. On KZ550 F1, F2, F2L and M1 models tighten the handlebar clamp bolts to 1.8 kgf m (13 lbf ft), noting that the handlebars must be at the same angle as the stanchions, that the arrow marks in each clamp must face forwards and that the front clamp bolts are tightened first, to leave a gap at the rear between clamp and top yoke.

Note: Do not confuse this procedure with that given in Chapter 4, which is for refitting only; this is the correct procedure for normal adjustments. Note also that the steering head lower bearing is **not sealed** and is completely open to the road dirt, water and debris thrown up by the front wheel. Take every opportunity to pack fresh grease into the bearing to prolong as much as possible its life. The more enterprising owner may feel able to fabricate a suitable seal, even if this is only a thick bead of RTV jointing compound.

12 Check the suspension and tyres

Ensure that the front forks work smoothly and progressively by pumping them up and down whilst the front brake is held on; set the anti-dive unit to the minimum setting on ZX550 A1 and A1L models. Any faults revealed by this check should be investigated immediately; first check the suspension settings, where applicable, then refer to Chapter 4, particularly if the oil seals are leaking. On all models except the Z550 G1 and G2 and KZ550 F1 and M1, inspect the stanchions, looking for signs of chips or other damage, then lift the dust excluder at the top of each fork lower leg and wipe away any dirt from its sealing lips or above the fork oil seal. Pack grease above the seal and refit the dust excluder. Note that none of this would be necessary, and fork stiction would be reduced, if gaiters are fitted; they are available from any good motorcycle dealer.

The rear suspension can be checked with the machine on the centre stand. Check that all the suspension components are securely attached to the frame. Check for free play in the swinging arm by pushing and pulling it horizontally. If free play is found, refer to the relevant Sections of Chapter 4, but note that on shaft drive machines the pivot bearings are adjustable.

On all models with Unitrak rear suspension stand the machine on its wheels and ask an assistant to bounce several times the rear of the machine up and down. Examine closely the suspension linkage, looking for free play or stiffness; any signs of corrosion or wear necessitate the removal of the swinging arm/suspension linkage assembly for cleaning and examination. On all models, if grease nipples are fitted to the rear suspension, use a suitable grease gun to pump in grease until the old grease is expelled and new grease can be seen issuing from both sides of the bearing.

Owners of machines fitted with Unitrak rear suspension should note that the suspension linkage consists of a number of highly-stressed bearings, none of which are fitted with grease nipples, but all of which are exposed to road dirt, salt and water thrown up by the rear wheel. Regular cleaning and greasing is essential. UK owners should note that wear in the suspension linkage will almost certainly cause the machine to fail its DOT test certificate. Refer to the annual/6000 mile heading, or to Chapter 4, if any doubts arise about the condition of any part of the linkage.

Referring to the instructions given under the daily (pre-ride) heading, check that the tyres are undamaged and still have adequate tread depth remaining, renewing them if they are worn out.

13 General checks and lubrication

Check around the machine, looking for loose nuts, bolts or screws, retightening them as necessary. Check the stand and lever pivots for security and lubricate them with light machine oil or engine oil. Make sure that each stand spring is in good condition.

It is advisable to lubricate the handlebar switches, sidestand switch (where fitted), and stoplamp switches with WD40 or similar water dispersant lubricant at regular intervals, and this is a convenient time to do it. This will keep the switches working properly and prolong their life, especially if the machine is used in adverse weather conditions.

Examine closely all control cables, checking the outer cables for signs of damage, then examine the exposed portions of the inner cables. Any signs of kinking or fraying will indicate that renewal is required. To obtain maximum life and reliability from the cables they should be thoroughly lubricated. To do the job properly and quickly use one of the hydraulic cable oilers available from most motorcycle shops. Free one end of the cable and assemble the cable oiler as described by the manufacturer's instructions. Operate the oiler until oil emerges from the lower end, indicating that the cable is lubricated throughout its length. This process will expel any dirt or moisture and will prevent its subsequent ingress.

If a cable oiler is not available, an alternative is to remove the cable from the machine. Hang the cable upright and make up a small funnel arrangement using plasticene or by taping a plastic bag around the upper end. Fill the funnel with oil and leave it overnight to drain through. Note that where nylon-lined cables are fitted, they should be used dry or lubricated with a silicone-based lubricant suitable for this application. On no account use ordinary engine oil because this will cause the liner to swell, pinching the cable.

When removing cables, slacken all adjuster locknuts and screw in the adjuster nuts to gain the maximum cable free play. Remove the clutch cable by aligning the slots in the handlebar clamp, adjuster and locknut so that the inner can be pulled clear. The cable lower end is disconnected as described in Section 5 of Chapter 1. Disconnect the throttle cable(s) first at the lower end, as described in Section 5 of Chapter 2, then separate the two halves of the twistgrip by removing the clamping screws; the cable(s) can then be released from the twistgrip drum and the adjuster unscrewed. Grease the twistgrip on reassembly.

Check all pivots and control levers, cleaning and lubricating them to prevent wear or corrosion. Where necessary, dismantle and clean any moving part which may have become stiff in operation.

Remove also the speedometer and tachometer drive cables, slide out their inners from the lower end, check them for signs of wear or damage which would require their renewal, then smear grease over all but the upper six inches or so of the inner cable (to prevent grease from working its way into the instrument head and ruining it) and refit the cable to the machine.

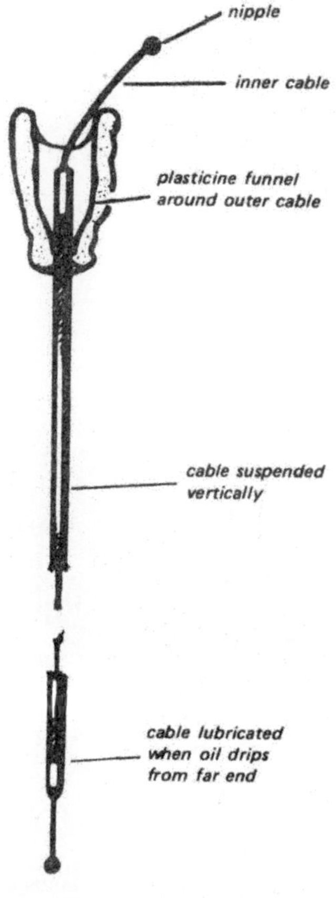

Oiling a control cable

Annually, or every 6000 miles (10 000 km)

Repeat all the tasks described under the previous mileage/time headings, then carry out the following:

1 Grease the automatic timing unit (ATU)

The ATU, where fitted, should be removed at this interval for dismantling and lubrication, also whenever a check of its operation reveals a fault. Removal and refitting are described in Sections 9 and 42 respectively of Chapter 1. On models fitted with contact breaker ignition, note that the removal and refitting of the ignition backplate, however careful, will alter the points gaps and ignition timing; these should be checked and reset on reassembly.

Holding the centre, twist the cam clockwise and pull it off. Displace the circlips, remove the first thrust washers and lift off the bobweights, unhooking each spring to release the weights. Remove the second thrust washers.

Wash each part in petrol or a similar solvent, taking care that all residual grease is removed from the groove inside the cam. Check the pivot pins and the corresponding holes in the bobweights. If these are badly worn, inaccurate timing will be unavoidable and a new ATU must be fitted. Similarly, the complete unit must be renewed if the return springs are weak or broken, or if there is excessive wear on the inside or outside of the cam.

Fill the cam internal groove with high melting point grease and apply a light smear to the pivot posts and cam bearing surface. Refit the cam. On contact breaker ignition systems, align the mark on the

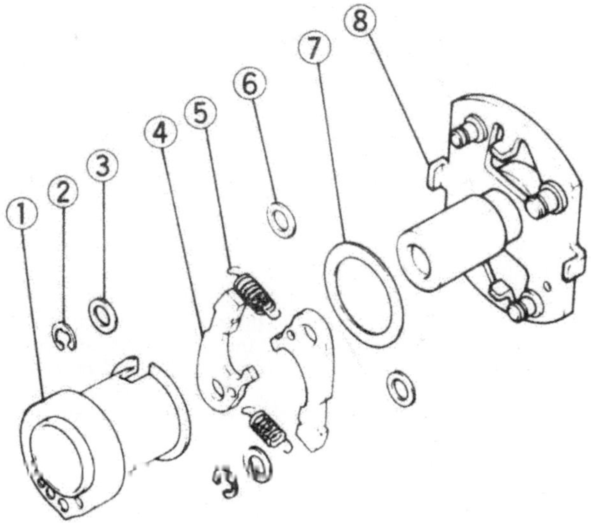

Automatic timing unit (ATU) – typical

1 *Cam*
2 *Circlip*
3 *Washer*
4 *Bobweight*
5 *Spring*
6 *Washer*
7 *Washer*
8 *ATU centre*

cam outer end with the small hole drilled in the ATU centre's backplate (adjacent to one of the spring posts or 'TEC' mark); on electronic ignition systems align the projection on the cam with the 'TEC' mark stamped on the backplate near one of the raised bobweight stops. Fit a thrust washer over each pivot post, followed by the bobweights, ensuring that their springs are correctly engaged. Fit the second thrust washer to each post and secure it by refitting a circlip to each. Lubricate the bobweight bearing surfaces with a few drops of light machine oil.

Check the unit's operation. The cam should move smoothly and easily clockwise on the ATU centre until the bobweights contact their raised stops, and should return quickly and smoothly under spring pressure to the full retard position when released.

2 *Renew the air filter element*

Irrespective of its apparent condition, the element must be renewed after it has been cleaned five times or at 6000 mile intervals. To keep check, mark the element with a permanent marker at each cleaning interval. When the machine is used in dusty conditions, the cleaning and renewal intervals should be reduced to compensate. Note that this does not apply to the foam element fitted to some models; this should last much longer and should only be renewed if found to be damaged or split.

3 *Renew the oil filter element*

The oil filter element should be renewed at every second oil change. Drain the engine oil as detailed in the 6 monthly/3000 mile service interval, then remove the single central bolt which retains the filter cover, noting that about 0.4 litre of oil will be released as the cover is lifted away. Clean the cover, filter components and filter recess and renew the O-rings if marked or broken. Using the accompanying photographic sequence to ensure that all components are correctly refitted, screw a new element on to the centre bolt, preventing any damage to its mounting grommets, and refit the filter assembly tightening the centre bolt to 2.0 kgf m (14.5 lbf ft). Complete the oil change, noting that at least 3.0 litres (5.3 Imp pint, 3.2 US qt) of oil will be required for topping up. Run the engine and check for leaks, then check, and if necessary top-up to the correct oil level.

4 *Change the brake fluid*

Brake fluid must be renewed at regular intervals as it absorbs moisture from the air. Although the system is sealed, the fluid will gradually deteriorate and must be renewed before contamination lowers its boiling point to an unsafe level.

Before starting work, obtain a full can of new DOT 3 or SAE J1703 hydraulic fluid and read Chapter 5, Section 14. Prepare the clear plastic tube and glass jar in the same way as for bleeding the hydraulic system, open the bleed nipple by unscrewing it $\frac{1}{4}$ - $\frac{1}{2}$ a turn with a spanner and apply the brake lever or pedal gently and repeatedly. This will pump out the old fluid. Keep the master cylinder reservoir topped up at all times, otherwise air may enter the system and greatly lengthen the operation. The old brake fluid is invariably much darker in colour than the new, making it easier to see when the old fluid is pumped out and the new fluid has completely replaced it.

When the new fluid appears in the clear plastic tubing completely uncontaminated by traces of old fluid, close the bleed nipple, remove the plastic tubing and replace the rubber cap on the nipple. Top up the master cylinder reservoir to above the 'Lower' level mark, unless the brake pads have been renewed, in which case the reservoir should be topped up to its upper level. Clean and dry the rubber diaphragm, fold it into its compressed state and refit the diaphragm and reservoir cover, tightening securely the two retaining screws or cover (as applicable).

The above procedure applies to all single disc brake systems; on twin disc systems work first on one caliper, then repeat the operation on the other. On ZX550 A1 and A1L models, work on one side at a time, using first the caliper nipple, then the anti-dive unit nipple, then the junction block nipple. Again repeat the operation on the other side.

Be careful at all times not to let air into the system or it will be necessary to carry out the bleeding procedure; once the fluid has been completely renewed, check that full pressure is available at the lever or pedal. If any trace of sponginess is felt, it must be assumed that air is in the system.

Wash off any surplus fluid and check that the brake is operating correctly before taking the machine out on the road.

Engine oil filter element must be renewed at interval specified

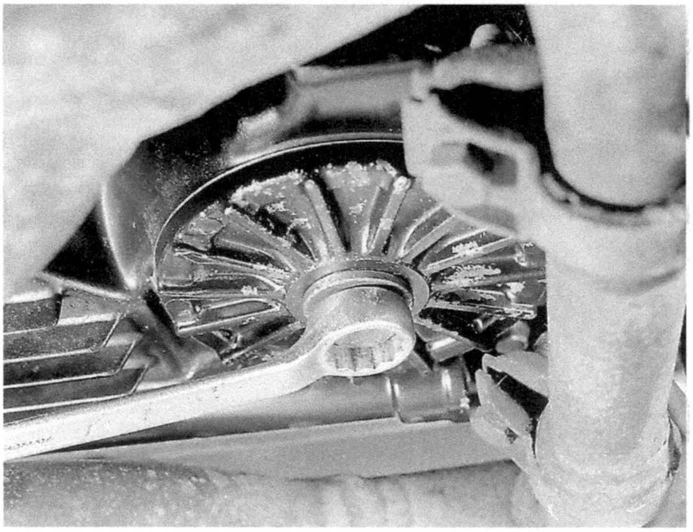

Some oil will be released as filter cover is unscrewed

Renew O-ring around centre bolt if worn or damaged

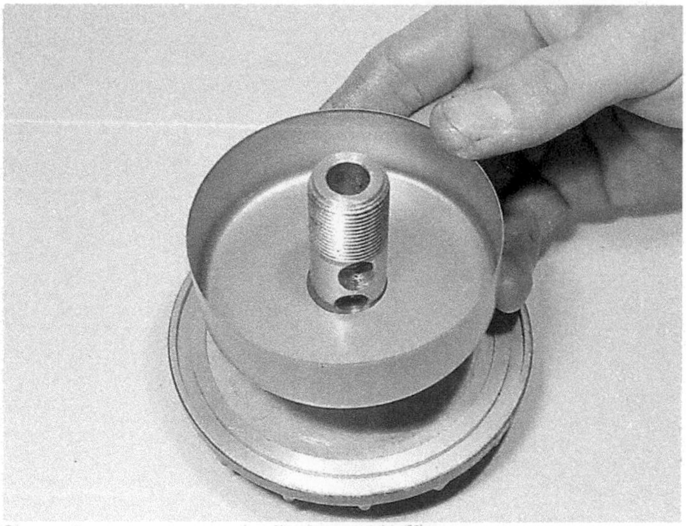

Clean all components and refit element baffle ...

... coil spring is fitted next to ensure correct operation of bypass valve ...

... followed by plain washer to prevent damage to filter sealing grommets

Do not omit sealing O-rings – renew if worn or damaged

5 Change the fork oil

The oil provided for damping and lubrication in the front forks deteriorates in use like any other oil and can become contaminated with water. If this is allowed to go unchecked the front suspension performance will suffer and the machine's handling and stability will deteriorate to the point where it can become unsafe to ride. This can be prevented by regular changes of the fork oil at this interval.

Remove the drain plug from one fork lower leg and place a container underneath it with a sheet of cardboard to keep oil off the wheel and tyre. Remove the valve cap(s) and depress the valve(s) to expel the air from the forks, on models with air-assisted forks. Set the anti-dive units to their softest (number 1) setting on ZX550 A1 and A1L models. Apply the front brake and pump the forks vigorously up and down until all oil is expelled, then repeat the operation on the remaining fork leg. Allow the machine to stand for a while to allow as much residual oil as possible to drain to the bottom, then repeat the pumping action. When as much oil as possible is expelled, wipe the surplus of the fork legs and clean the drain plugs and their threads. Renew the sealing washers if they are too flattened to be of further use, apply a smear of jointing compound to their threads and refit the drain plugs. Do not overtighten them; use the recommended torque settings if possible.

Remove the fuel tank or cover it with a thick layer of padding to protect its paintwork and place the machine on its centre stand with the front wheel raised clear of the ground by blocks of wood wedged under the front of the crankcase. On all models fitted with four-piece handlebars, remove the ignition switch/fork top yoke cover, prise off the rubber or plastic cap or plug from the top of each leg and remove the bolt securing each handlebar locating plate; remove the plates, noting the direction of their arrow marks. Remove the large threaded bolt or plug securing each handlebar half to the top yoke; these are fastened to a high torque setting and care must be taken to hold steady the forks while they are slackened. Ensuring that all control cables, brake hoses and wiring are not stretched, kinked or trapped, allow the handlebar halves to hang down on each side of the forks. On all models with conventional chromed-tubing handlebars, check whether the tops of the fork stanchions are obscured; if so remove the clamp bolts and clamps and draw the handlebars backwards to rest on the tank padding or frame top tube, taking care that control cables, brake hose and wiring are not kinked or stretched. Remove the rubber or plastic cap from the top of each leg.

Remove the fork top plug. On all shaft drive models use an Allen key to unscrew the threaded top plug, noting its sealing O-ring, then remove the spacer, thick washer and the fork spring. Compress the forks sufficiently to reach the spring and note which way round it is fitted. On all chain drive models an assistant will be required. Placing a suitable implement in its central recess or around the air valve, press down on the top plug while the wire circlip is removed from its groove in the stanchion, then slowly allow fork spring pressure to drive out the plug, noting its sealing O-ring. On KZ/Z550 D1 models only remove the thick washer above the fork spring. On all models, withdraw the fork spring, noting which way round it was fitted.

The grade of oil recommended for each machine, and the approximate amount to be poured into each leg at an oil change, is given in the Specifications Section of Chapter 4. Add all but the last 10 - 20cc, using a finely graduated vessel, then slowly pump the forks up and down several times to distribute the oil around the components and to expel any air pockets, the pumping being done by lifting the front wheel. With the forks fully extended and the springs removed, measure the oil level by inserting a dipstick (a length of welding rod would be ideal) into the leg. The correct level is measured from the top of the stanchion to the top of the oil, and is adjusted by adding or removing oil. Note that the amount of oil does not matter; the level is the more important factor, especially on models with air-assisted forks. The level must be exactly the same in both legs.

On reassembly, the fork springs should be refitted with their tapered ends downwards, so that the close-pitched coils are uppermost, not forgetting any washers or spacers fitted above them. Renew the O-ring around both top plugs if necessary. On shaft drive models, refit the plug tightening securely by hand using only the correct size Allen key. On chain drive models, press the plug down into its stanchion until the circlip can be refitted in its stanchion groove. Release the plug and check that it is securely fastened. Where applicable, fill the forks with air to the recommended pressure then refit the valve cap(s) and each fork leg top cap or plug. If removed, refit the handlebars. On models with conventional handlebars, if the clamps are marked with arrows, these marks must face forwards and the front clamp bolts must be tightened first, then the rear, to leave a gap between the clamp rear and the top yoke; where no arrow marks are found the clamp bolts are tightened evenly to leave equal gaps at front and rear. The recommended torque setting is 1.8 kgf m (13 lbf ft). Check that all cables, hoses and wiring are correctly routed and that the handlebars are at the same angle as the fork legs. On models with four-piece handlebars, position each half on the top yoke and secure it with the large threaded bolt or plug, then refit the locating plates with their arrow marks pointing to the rear. Tighten the large bolt to a torque setting of 10.0 kgf m (72 lbf ft) on all models except the ZX550 A1, A1L, where the plug is tightened to 7.5 kgf m (54 lbf ft). Tighten the locating plate Allen screws securely and refit the rubber or plastic cap or plug to the top of each leg, followed by the ignition switch/top yoke cover. Check that all disturbed components are securely refitted, that all controls are correctly adjusted and that the brakes and front suspension are operating correctly.

Z550 G1, G2, KZ550 F1, F2, F2L rear suspension: owners of these machines should note that the oil in the rear suspension units will also require renewal at this interval, as it is subject to the same conditions as that in the front forks. Refer to the relevant Section of Chapter 4.

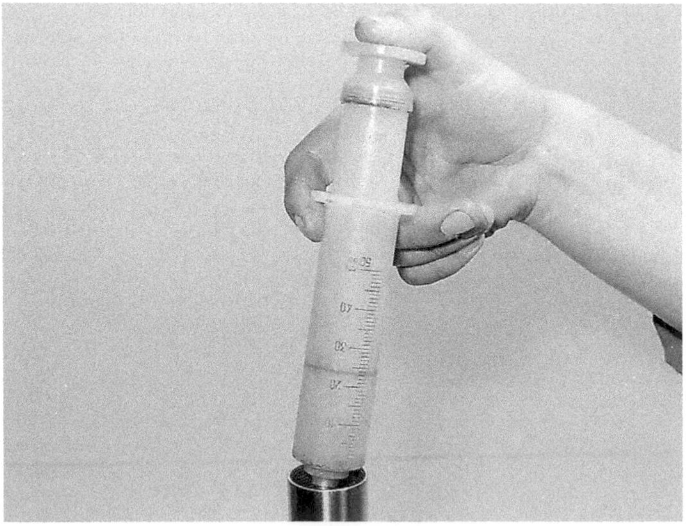

Add exactly the same amount of oil to each fork leg ...

... and use a dipstick to ensure that oil level is correct

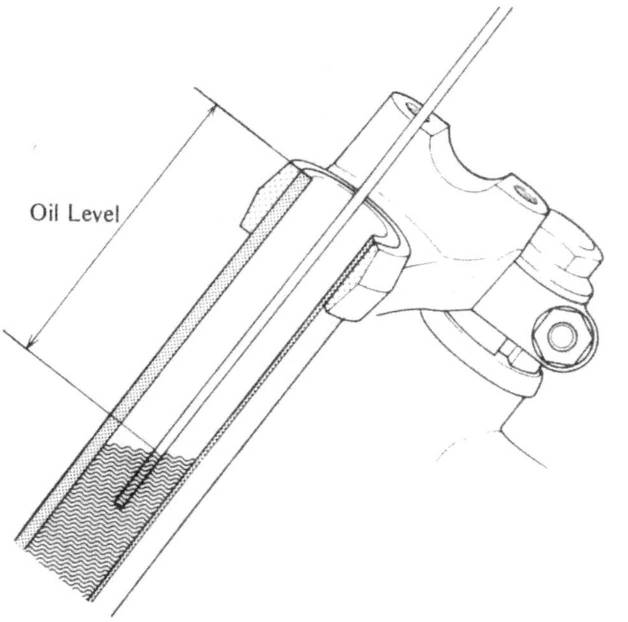

Measuring the fork leg oil level

6 *Grease the rear suspension*

On all machines with conventional swinging arm rear suspension, those fitted with grease nipples can be lubricated as described under the six monthly/3000 mile heading. Where no grease nipples are fitted, the swinging arm must be removed as described in Chapter 4 and its bearings cleaned and repacked with grease.

On all models with Unitrak rear suspension, no grease nipples are fitted; therefore the suspension must be removed as an assembly to ensure that it can be fully dismantled and all components cleaned, checked for wear and reassembled. All bearings and bearing surfaces should be packed with grease. If the machine is used in bad weather or on dirty or salty roads, this operation should be performed with much greater frequency. Refer to Chapter 4.

7 *Check final gear case oil level and grease the prop shaft joints*

The oil level must be checked at this interval to ensure that the correct amount of oil is always present, daily checks should reveal immediately any leaks.

Place the machine on its centre stand on level ground and remove the larger level/filler plug from the rear of the casing; oil should be seen level with the bottom threads of the plug orifice. If necessary, top up with the recommended type of oil and use only the same brand as that already in the case. Note that oil is not consumed, it can only leak out; if the level drops noticeably at any time, the gear case should be checked thoroughly for signs of oil leakage and the fault rectified by a competent Kawasaki dealer. Tighten the plug securely.

To allow for the changes in its effective length as the swinging arm moves up and down, the driveshaft is fitted with a sliding joint at its rear end. Both this joint and the splines at the shaft forward end must be greased regularly as follows:

Remove the rear wheel as described in Section 4 of Chapter 5, then remove the left-hand suspension unit bottom mounting nut and pull the unit off its mounting stud. Remove its four retaining nuts and withdraw the final drive casing from the swinging arm, taking care not to lose the coil spring. Slacken its clamp screw and pull the gaiter back off the front gear case unit. Rotate the shaft until a small hole is located in the shaft end, then insert a metal rod into the hole to depress the locking pin inside. The shaft can then be pulled rearwards off its splines.

Wipe all old grease off the shaft male and female splines and apply a thin coat of new high melting point grease to all of them. The universal joint is sealed for life; check that there are no signs of free play, indicating wear, or of a lack of lubrication. If any of this is found, the propeller shaft unit must be renewed. If all is well, check that the locking pin and its hole are aligned and press the shaft on to its splines; the pin should be heard to click into place. Refit the rubber gaiter and tighten its clamp. At the rear, wipe all old grease from the gear case splines and from the inside of the sliding joint, then pack it with high melting-point grease. A specific amount, 17 ml, is recommended, to be packed around the outside of the joint, as shown in the accompanying illustration.

Refit the final drive casing, placing the coil spring over the pinion nut as shown in the accompanying illustration. If smears of grease have appeared at the swinging arm/drive casing joint face, apply a thin smear of jointing compound to the mating surfaces. Apply thread locking compound to the stud threads and tighten the nuts to a torque setting of 2.3 kgf m (16.5 lbf ft). Refit the suspension unit, tightening its nut to 2.5 kgf m (18 lbf ft), then refit the rear wheel as described in Chapter 5.

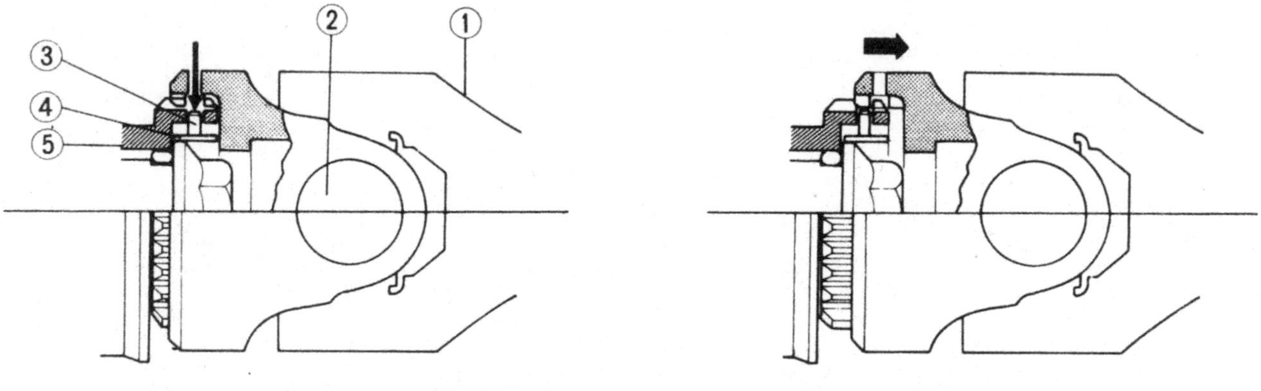

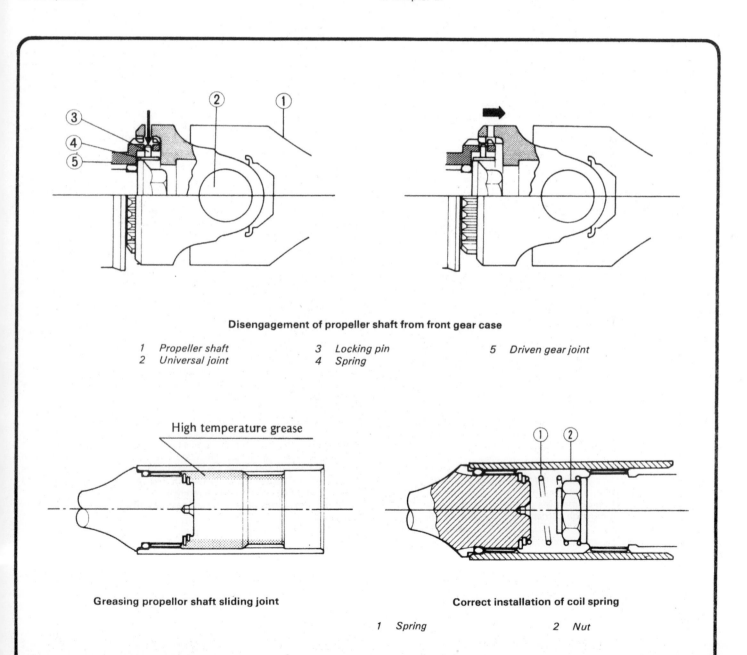

Disengagement of propeller shaft from front gear case

1 Propeller shaft
2 Universal joint
3 Locking pin
4 Spring
5 Driven gear joint

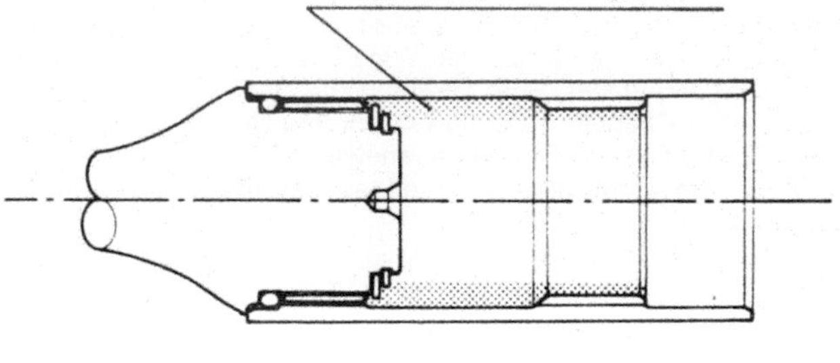

Greasing propellor shaft sliding joint

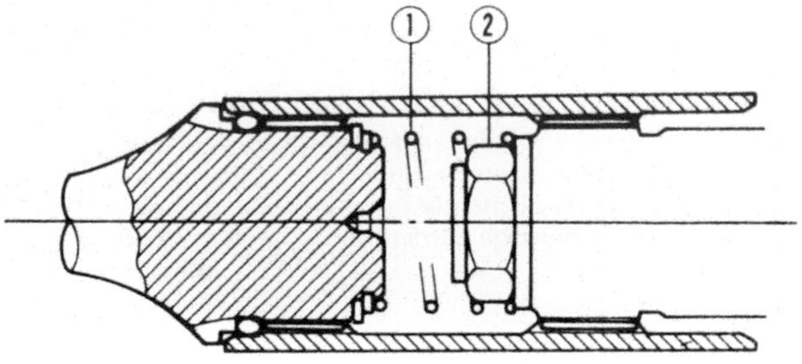

Correct installation of coil spring

1 Spring
2 Nut

Every two years or 12 000 miles (20 000 km)

Repeat all previous tasks, then carry out the following:

1 Grease wheel bearings, speedometer drive gearbox and rear brake camshaft

To prolong their life as much as possible, the front and rear wheels should be removed from the machine so that their bearings can be driven out, cleaned, checked for wear and renewed if necessary, then repacked with fresh high-melting point grease and refitted. At the same time the speedometer drive gearbox should be cleaned and packed with grease, and on machines with drum rear brakes only, the brake camshaft should be removed, cleaned and greased. Refer to the relevant Sections of Chapter 5.

2 Grease the steering head bearings

As already mentioned, the steering head bearings should be checked and adjusted at the six-monthly/3000 mile inspection. If, however, they have not been dismantled during routine maintenance or for accident repair they should be dismantled at this interval for examination and greasing. The work necessary is described in the relevant Sections of Chapter 4.

Every three years or 18 000 miles (30 000 km)

Repeat the tasks listed under the monthly/500 mile, six-monthly/3000 mile and annual/6000 mile headings, then carry out the following:

Change the final gear case oil

When ready, take the machine for a journey of sufficient length to warm up fully the oil in the final drive unit. The oil is thick and will not drain quickly or remove any impurities until it is fully warmed up.

Place the machine on its centre stand and place a container under the gear case with a sheet of paper or cardboard to keep oil off the wheel and tyre. Remove the oil level/filler and drain plugs and allow the oil to drain fully. Renew the plug sealing washer (or O-ring, as appropriate) if it is damaged or worn. When draining is complete, refit the drain plug, tightening it securely. Add sufficient oil of the recommended grade and viscosity to bring the level up to the bottom of the filler/level plug orifice; the amount required should be 190 cc (6.9 Imp fl oz, 6.4 US fl oz). Refit the filler/level plug, wash off any surplus oil and take the machine for a short journey to warm up the oil and distribute it, then stop the engine and allow a few minutes for the level to settle before rechecking it, top up as necessary. Tighten the filler/level plug securely and wash off all traces of oil from the outside of the swinging arm and casing.

Additional routine maintenance

Some items which do not wear in the usual sense must be renewed for safety reasons as they deteriorate with age, whether the machine is ridden a great deal, or hardly at all.

1 Renew all brake seals and anti-dive plunger units

Every two years the master cylinder and caliper(s) of both front and rear brakes must be dismantled and all seals renewed. This is a safety measure which must be carried out regardless of the apparent condition of any of the components, unless a particular set of seals has recently been renewed. For the same reasons the anti-dive plunger units (ZX550 A1, A1L only) must be renewed also. Refer to the relevant Sections of Chapter 5.

2 Renew all brake hoses and pipes

Every four years all brake hoses must be renewed, as described in Chapter 5. Also on ZX550 A1, A1L models only, the metal brake pipes from the junction blocks to the plunger units must be renewed. Again this is a safety measure which must be carried out regardless of the apparent condition of the components.

3 Renew the fuel pipe

Every four years the fuel pipe must be renewed to prevent the risk of fuel spillage due to an age-hardened hose splitting or cracking. Refer to Chapter 2. Note that other hoses such as vacuum pipes or emission control system tubing should be inspected closely and renewed. If necessary, at the same time.

4 Cleaning the machine

Keeping the motorcycle clean should be considered as an important part of the routine maintenance, to be carried out whenever the need arises. A machine cleaned regularly will not only succumb less speedily to the inevitable corrosion of external surfaces, and hence maintain its market value, but will be far more approachable when the time comes for maintenance or service work. Furthermore, loose or failing components are more readily spotted when not partially obscured by a mantle of road grime and oil.

Surface dirt should be removed using a sponge and warm, soapy water, the latter being applied copiously to remove the particles of grit which might otherwise cause damage to the paintwork and polished surfaces.

Oil and grease are removed most easily by the application of a cleaning solvent such as 'Gunk' or 'Jizer'. The solvent should be applied when the parts are still dry and worked in with a stiff brush. Large quantities of water should be used when rinsing off, taking care that water does not enter the carburettors, air cleaners or electrics.

If desired, a polish such as Solvol Autosol can be applied to the aluminium alloy parts to restore the original lustre. This does not apply in instances, much favoured by Japanese manufacturers, where the components are lacquered. Application of a wax polish to the cycle parts and a good chrome cleaner to the chrome parts will also give a good finish. Always wipe the machine down if used in the wet, and make sure the chain is well oiled, when fitted. There is less chance of water getting into control cables if they are regularly lubricated, which will prevent stiffness of action.

5 Clean the oil pump pick-up filter gauze

Although not specified as a regular maintenance item, the gauze must be cleaned with reasonable frequency.

Drain the transmission oil and remove the sump as described in Chapter 1, Section 11. There is no need to remove the oil pump itself, the filter gauze can be carefully picked out of the pump inlet. Clean the filter in solvent, removing any deposits with a soft-bristled brush, then refit the filter to the pump inlet. Wipe out the inside of the sump and clean the gasket faces. Refit the sump and all disturbed components as described in Chapter 1, Section 40, noting that the sump gasket and three sealing O-rings should be renewed.

Refill the crankcases with oil as described earlier in Routine maintenance.

Chapter 1 Engine, clutch and gearbox

For information relating to the 1984 on models, see Chapter 7

Contents

Specifications

Engine

Type	DOHC, 4-cylinder, air-cooled
Capacity:	
400 models	399 cc (24.3 cu in)
500 models	497 cc (30.3 cu in)
550 models	553 cc (33.7 cu in)
Bore:	
400 models except ZR400 B1	52.0 mm (2.05 in)
500 models and ZR400 B1	55.0 mm (2.17 in)
550 models	58.0 mm (2.28 in)
Stroke:	
400 models except ZR400 B1	47.0 mm (1.85 in)
ZR400 B1	42.0 mm (1.65 in)
500 and 550 models	52.4 mm (2.06 in)

Compression ratio:		
KZ/Z550 D1, H1, H2, ZX550 A1, A1L	10.0 : 1	
ZR400 A1, B1	9.7 : 1	
All other models	9.5 : 1	
Engine output:	**Maximum power – bhp @ rpm**	**Maximum torque – kgf m (lbf ft) @ rpm**
Z400 J1, J2, J3	43 @ 9500	3.5 (25.3) @ 7500
ZR400 A1	48 @ 10500	3.5 (25.3) @ 8500
ZR400 B1	49 @ 10500	3.5 (25.3) @ 8500
Z500 B1, B2	52 @ 9000	4.5 (32.6) @ 7500
KZ550 A1, A2, A3, A4, C1, C2, C3, C4	53 @ 8500	4.8 (34.7) @ 7000
Z550 A1, A2, A3, C1, C2	53 @ 8500	4.9 (35.4) @ 7000
KZ550 M1	55 @ 9000	4.9 (35.4) @ 7500
ZR550 A1, A2, Z550 G1, G2, KZ550 F1, F2, F2L	56 @ 9000	4.9 (35.4) @ 7500
KZ550 D1	57 @ 9000	4.9 (35.4) @ 8000
Z550 D1	58 @ 9000	4.9 (35.4) @ 8000
KZ550 H1, H2	60 @ 9500	4.9 (35.4) @ 8500
Z550 H1, H2	61 @ 9500	4.9 (35.4) @ 8500
ZX550 A1, A1L	65 @ 10500	4.9 (35.4) @ 8500
Cylinder identification	Left to right, 1-2-3-4	
Firing order	1-2-4-3	
Compression pressure – at cranking speed, engine fully warmed up:		
ZR400 A1, B1	7.7 - 12.0 kg/cm² (109 - 171 psi)	
ZX550 A1, A1L	8.4 - 13.0 kg/cm² (119 - 185 psi)	
KZ550 A4	8.8 - 13.5 kg/cm² (125 - 192 psi)	
Z400 J1, J2, J3, ZR550 A1, A2, Z550 G1, G2, KZ550 F1, F2, F2L, M1	9.0 - 13.9 kg/cm² (128 - 198 psi)	
All other models	10.2 - 15.5 kg/cm² (145 - 220 psi)	

Note: compression pressure must not vary by more than 1.0 kg/cm² (14.2 psi) between any two cylinders

Valve clearances

Inlet	0.10 - 0.20 mm (0.004 - 0.008 in)
Exhaust	0.15 - 0.25 mm (0.006 - 0.010 in)

Valve timing

	Inlet opens	Inlet closes	Exhaust opens	Exhaust closes
Z400 J1, J2, J3	33° BTDC	41° ABDC	51° BBDC	23° ATDC
ZR400 A1, B1	36° BTDC	44° ABDC	54° BBDC	26° ATDC
KZ/Z550 D1, H1, H2	31° BTDC	59° ABDC	59° BBDC	31° ATDC
ZX550 A1, A1L	46° BTDC	54° ABDC	64° BBDC	36° ATDC
All other models	20° BTDC	48° ABDC	48° BBDC	20° ATDC
Duration – inlet and exhaust:				
Z400 J1, J2, J3	254°			
ZR400 A1, B1	260°			
KZ/Z550 D1, H1, H2	270°			
ZX550 A1, A1L	280°			
All other models	248°			

Camshafts, drive chain and tensioner

Cam lobe height:	
Z400 J1, J2, J3	35.746 – 35.854 mm (1.4073 – 1.4116 in)
Service limit	35.65 mm (1.4035 in)
ZR400 A1, B1, KZ/Z550 D1, H1, H2, ZX550 A1, A1L	36.246 – 36.354 mm (1.4270 – 1.4313 in)
Service limit	36.15 mm (1.4232 in)
All other models	35.546 - 35.654 mm (1.3995 - 1.4036 in)
Service limit	35.45 mm (1.3957 in)
Camshaft journal OD	21.890 - 21.912 mm (0.8618 - 0.8627 in)
Service limit	21.860 mm (0.8606 in)
Camshaft journal/cylinder head bearing clearance	0.088 - 0.131 mm (0.0034 - 0.0052 in)
Service limit	0.220 mm (0.0087 in)
Distance between camshaft thrust flanges	180.50 - 180.60 mm (7.1063 - 7.1102 in)
Camshaft endfloat	0.28 – 0.82 mm (0.0110 – 0.0323 in)
Camshaft standard runout	0 - 0.02 mm (0 - 0.0008 in)
Service limit	0.10 mm (0.0039 in)
Cam chain - standard length of 20 links	127.00 - 127.36 mm (4.9999 - 5.0142 in)
Service limit	128.90 mm (5.0748 in)
Cam chain tensioner/guide blade maximum groove depth:	
Upper guide blade	4.0 mm (0.1575 in)
Front guide blade	5.0 mm (0.1969 in)
Tensioner blade	3.0 mm (0.1181 in)

Cylinder head

Gasket surface maximum warpage	0.05 mm (0.002 in)
Combustion chamber volume:	
ZR400 B1, ZX550 A1, A1L	14.6 – 15.4 cc (0.49 – 0.52/0.51 – 0.54 US/Imp fl oz)
All other models	15.2 - 16.0 cc (0.51 - 0.54/0.53 - 0.56 US/Imp fl oz)

Camshaft bearing ID	22.000 - 22.021 mm (0.8661 - 0.8670 in)
Service limit	22.080 mm (0.8693 in)
Distance between outside faces of camshaft bearing caps	179.78 - 180.22 mm (7.0779 - 7.0953 in)
Valve seat OD:	
Inlet	27.0 mm (1.0630 in)
Exhaust	23.0 mm (0.9055 in)
Valve seat width	0.5 - 1.0 mm (0.0196 - 0.0394 in)
Valve installed height:	
Inlet	36.08 - 37.02 mm (1.4205 - 1.4575 in)
Exhaust	36.03 - 36.97 mm (1.4185 - 1.4555 in)

Valves, guides and springs

Valve head thickness	0.85 - 1.15 mm (0.0335 - 0.0453 in)
Service limit:	
Inlet	0.50 mm (0.0197 in)
Exhaust	0.70 mm (0.0276 in)
Valve stem maximum runout	0.05 mm (0.0020 in)
Valve stem OD:	
Inlet	5.475 - 5.490 mm (0.2156 - 0.2161 in)
Service limit	5.460 mm (0.2150 in)
Exhaust	5.455 - 5.470 mm (0.2148 - 0.2154 in)
Service limit	5.440 mm (0.2142 in)
Valve guide ID	5.500 – 5.512 mm (0.2165 – 0.2170 in)
Service limit	5.580 mm (0.2197 in)
Valve/guide clearance - by direct measurement:	
Inlet	0.010 - 0.037 mm (0.0004 - 0.0015 in)
Service limit	0.120 mm (0.0047 in)
Exhaust	0.030 - 0.057 mm (0.0012 - 0.0022 in)
Service limit	0.140 mm (0.0055 in)
Valve/guide clearance – valve installed, using dial gauge:	
Inlet	0.020 - 0.090 mm (0.0008 - 0.0035 in)
Service limit	0.220 mm (0.0087 in)
Exhaust	0.070 - 0.140 mm (0.0028 - 0.0055 in)
Service limit	0.270 mm (0.0106 in)
Valve spring free length:	
Inner	36.7 mm (1.4449 in)
Service limit	35.0 mm (1.3780 in)
Outer	38.7 mm (1.5236 in)
Service limit	37.1 mm (1.4606 in)
Valve spring minimum pressure:	
Inner	16.2 kg at 23.1 mm (35.7 lb at 0.9095 in)
Outer	31.3 kg at 25.1 mm (69.0 lb at 0.9882 in)
Maximum tilt from vertical	1.5 mm (0.0591 in) at top of spring

Cylinder block

Cylinder bore ID:	
Z400 J1, J2, J3, ZR400 A1	52.000 – 52.012 mm (2.0472 – 2.0477 in)
Service limit	52.10 mm (2.0512 in)
Z500 B1, B2, ZR400 B1	55.000 - 55.012 mm (2.1654 - 2.1658 in)
Service limit	55.10 mm (2.1693 in)
550 models	58.000 - 58.012 mm (2.2835 - 2.2839 in)
Service limit	58.10 mm (2.2874 in)
Taper and ovality	0 - 0.010 mm (0 - 0.0004 in)
Service limit	0.050 mm (0.0020 in)
Piston/cylinder clearance	0.020 – 0.047 mm (0.0008 – 0.0019 in)

Pistons

Piston OD:	
Z400 J1, J2, J3, ZR400 A1	51.965 - 51.980 mm (2.0459 - 2.0465 in)
Service limit	51.83 mm (2.0406 in)
Z500 B1, B2, ZR400 B1	54.965 - 54.980 mm (2.1640 - 2.1646 in)
Service limit	54.83 mm (2.1587 in)
550 models	57.965 - 57.980 mm (2.2821 - 2.2827 in)
Service limit	57.83 mm (2.2768 in)
Gudgeon pin bore maximum ID	14.07 mm (0.5539 in)
Piston/gudgeon pin clearance	0.005 - 0.016 mm (0.0002 - 0.0006 in)
Top compression ring groove width:	
ZR400 B1	1.02 - 1.04 mm (0.0402 - 0.0410 in)
Service limit	1.12 mm (0.0441 in)
ZR400 A1, ZR550 A1, A2, ZX550 A1, A1L, Z550 G1, G2, KZ550 F1, M1, F2, F2L	1.21 - 1.23 mm (0.0476 - 0.0484 in)
Service limit	1.31 mm (0.0516 in)
All other models	N/Av
Service limit	1.33 mm (0.0524 in)

Second compression ring groove width:	
ZR400 B1	1.22 - 1.24 mm (0.0480 - 0.0488 in)
Service limit	1.32 mm (0.0520 in)
ZR400 A1, ZX550 A1, A1L	1.21 - 1.23 mm (0.0476 - 0.0484 in)
Service limit	1.31 mm (0.0516 in)
ZR550 A1, A2, Z550 G1, KZ550 F1, M1, F2, F2L	1.23 – 1.25 mm (0.0484 – 0.0492 in)
Service limit	1.33 mm (0.0524 in)
All other models	N/Av
Service limit	1.33 mm (0.0524 in)
Oil scraper ring groove width:	
ZR400 A1, B1, ZR550 A1, A2, ZX550 A1, A1L, Z550 G1, G2, KZ550 F1, M1, F2, F2L	2.51 - 2.53 mm (0.0988 - 0.0996 in)
Service limit	2.61 mm (0.1028 in)
All other models	N/Av
Service limit	2.60 mm (0.1024 in)
Top compression ring/groove clearance:	
Z550 G1, G2, KZ550 F1, M1, F2, F2L	0.010 - 0.045 mm (0.0004 - 0.0018 in)
Service limit	0.150 mm (0.0059 in)
ZR400 A1, ZR550 A1	0.020 - 0.060 mm (0.0008 - 0.0024 in)
Service limit	0.160 mm (0.0063 in)
ZR400 B1	0.030 – 0.070 mm (0.0012 – 0.0028 in)
Service limit	0.170 mm (0.0067 in)
ZR550 A2, ZX550 A1, A1L	0.020 - 0.055 mm (0.0008 - 0.0022 in)
Service limit	0.160 mm (0.0063 in)
All other models	N/Av
Service limit	0.150 mm (0.0059 in)
Second compression ring/groove clearance:	
ZR550 A1, A2, Z550 G1, G2, KZ550 F1, M1, F2,F2L	0.040 - 0.075 mm (0.0016 - 0.0030 in)
Service limit	0.180 mm (0.0071 in)
ZR400 A1	0.020 - 0.060 mm (0.0008 - 0.0024 in)
Service limit	0.160 mm (0.0063 in)
ZR400 B1	0.030 – 0.070 mm (0.0012 – 0.0028 in)
Service limit	0.170 mm (0.0067 in)
ZX550 A1, A1L	0.020 - 0.055 mm (0.0008 - 0.0022 in)
Service limit	0.160 mm (0.0063 in)
All other models	N/Av
Service limit	0.150 mm (0.0059 in)
Oil scraper ring/groove clearance:	
Z400 J1, J2, J3 - service limit	0.150 mm (0.0059 in)
All other models	N/App

Piston rings

Top compression ring thickness:	
Z550 G1, G2, ZR550 A1, A2, ZX550 A1, A1L, KZ550 F1, M1, F2, F2L	1.175 – 1.190 mm (0.0463 – 0.0469 in)
Service limit	1.100 mm (0.0433 in)
ZR400 A1	1.170 - 1.190 mm (0.0461 - 0.0469 in)
Service limit	1.100 mm (0.0433 in)
ZR400 B1	0.970 - 0.990 mm (0.0382 - 0.0390 in)
Service limit	0.900 mm (0.0354 in)
All other models	N/Av
Service limit	1.100 mm (0.0433 in)
Second compression ring thickness:	
Z550 G1, G2, ZR550 A1, A2, ZX550 A1, A1L, KZ550 F1, M1, F2, F2L	1.175 - 1.190 mm (0.0463 - 0.0469 in)
Service limit	1.100 mm (0.0433 in)
ZR400 A1, B1	1.170 - 1.190 mm (0.0461 - 0.0469 in)
Service limit	1.100 mm (0.0433 in)
All other models	N/Av
Service limit	1.100 mm (0.0433 in)
Oil scraper ring thickness:	
Z400 J1, J2, J3 – service limit	2.40 mm (0.0945 in)
All other models	N/App
Compression rings end gap - installed:	
Z550 G1, G2, ZR550 A1, A2, ZX550 A1, A1L, KZ550 F1, M1, F2, F2L	0.15 - 0.30 mm (0.0059 - 0.0118 in)
Service limit	0.60 mm (0.0236 in)
ZR400 A1, B1	0.15 - 0.35 mm (0.0059 - 0.0138 in)
Service limit	0.70 mm (0.0276 in)
All other models	N/Av
Service limit	0.70 mm (0.0276 in)
Oil scraper ring end gap - installed:	
Z400 J1, J2, J3 - service limit	0.70 mm (0.0276 in)
All other models	N/App

Connecting rod and bearings

Gudgeon pin OD service limit	13.96 mm (0.5496 in)		
Small-end bearing maximum ID	14.05 mm (0.5532 in)		
Gudgeon pin/small-end bearing clearance	0.003 - 0.019 mm (0.0001 - 0.0008 in)		
Connecting rod maximum distortion	0.2/100 mm (0.008/3.94 in)		
Big-end bearing standard ID	36.000 - 36.016 mm (1.4173 - 1.4180 in)		
Size groups:			
Connecting rod unmarked	36.000 - 36.008 mm (1.4173 - 1.4176 in)		
Connecting rod marked 'O'	36.009 - 36.016 mm (1.4177 - 1.4180 in)		
Crankpin standard OD	32.984 - 33.000 mm (1.2986 - 1.2992 in)		
Service limit	32.970 mm (1.2980 in)		
Size groups:			
Crankshaft unmarked	32.984 - 32.994 mm (1.2986 - 1.2989 in)		
Crankshaft marked 'O'	32.995 - 33.000 mm (1.2990 - 1.2992 in)		
Big-end bearing insert size:	**Thickness**	**Part number**	**Colour code**
Thin	1.480 - 1.485 mm (0.0583 - 0.0585 in)	13034 - 1006	Brown
Medium	1.485 - 1.490 mm (0.0585 - 0.0587 in)	13034 - 1005	Black
Thick	1.489 - 1.494 mm (0.0586 - 0.0588 in)	13034 - 1004	Blue*
Bearing insert/crankpin clearance	0.031 - 0.059 mm (0.0012 - 0.0023 in)		
Service limit	0.100 mm (0.0039 in)		
Big-end bearing side clearance	0.13 - 0.33 mm.(0.0051 - 0.0130 in)		
Service limit	0.50 mm (0.0197 in)		

* *P/No. 13034 – 1004 may also be coded green*

Crankshaft

Runout	0 - 0.02 mm (0 - 0.0008 in)
Service limit	0.05 mm (0.0020 in)
Endfloat	0.05 - 0.20 mm (0.0020 - 0.0079 in)
Service limit	0.40 mm (0.0158 in)

Main bearings

Crankcase main bearing ID	36.000 - 36.016 mm (1.4173 - 1.4180 in)		
Size groups:			
Crankcase marked 'O'	36.000 – 36.008 mm (1.4173 – 1.4176 in)		
Crankcase unmarked	36.009 - 36.016 mm (1.4177 - 1.4180 in)		
Crankshaft journal OD	31.984 - 32.000 mm (1.2592 - 1.2598 in)		
Service limit	31.960 mm (1.2583 in)		
Size groups:			
Crankshaft unmarked	31.984 - 31.992 mm (1.2592 - 1.2595 in)		
Crankshaft marked 'I'	31.993 - 32.000 mm (1.2596 - 1.2598 in)		
Main bearing insert size:	**Thickness**	**Part number**	**Colour code**
Thin	1.991 - 1.995 mm (0.0784 - 0.0785 in)	13034 - 1016	Brown
Medium	1.995 - 1.999 mm (0.0785 - 0.0787 in)	13034 - 1017	Black
Thick	1.999 - 2.003 mm (0.0787 - 0.0789 in)	13034 - 1018	Blue*
Bearing insert/journal clearance	0.014 - 0.038 mm (0.0006 - 0.0015 in)		
Service limit	0.08 mm (0.0032 in)		

* *P/No. 13034 – 1018 may also be coded green*

Primary drive

Type	Hy-Vo chain and sprockets, gear
Reduction ratio:	
400 models	3.277 : 1 (27/23 x 67/24T)
500 and 550 models	2.935 : 1 (27/23 x 65/26T)
Hy-Vo chain:	
Size	MA0364 ($\frac{3}{8}$ x $\frac{3}{4}$ in) x 50 links
Make	Tsubaki 63 - 139
Maximum wear	1.4%
Maximum free play	25 mm (0.98 in)

Clutch

Type	Wet, multi-plate
Friction plates:	
Number	7
Thickness	2.9 - 3.1 mm (0.1142 - 0.1221 in)
Service limit	2.7 mm (0.1063 in)
Maximum warpage	0.3 mm (0.0118 in)
Friction plate tang/outer drum maximum clearance	0.7 mm (0.0276 in)
Plain plates:	
Number	6
Maximum warpage	0.3 mm (0.0118 in)

Springs:	
Number	5
Free length	32.6 mm (1.2835 in)
Service limit	31.7 mm (1.2480 in)
Minimum pressure	18.5 kg at 23.5 mm (40.8 lb at 0.9252 in)
Outer drum/secondary shaft gear maximum backlash	0.13 mm (0.0051 in)
Outer drum centre maximum ID	37.03 mm (1.4579 in)
Clutch bearing sleeve minimum OD	31.96 mm (1.2583 in)

Gearbox

Type	6-speed, constant mesh
Reduction ratio:	
1st	2.571 : 1 (36/14T)
2nd	1.778 : 1 (32/18T)
3rd	1.381 : 1 (29/21T)
4th	1.125 : 1 (27/24T)
5th	0.962 : 1 (25/26T)
6th	0.852 : 1 (23/27T)
Gear backlash - all pinions	0 - 0.17 mm (0 - 0.0067 in)
Service limit	0.25 mm (0.0098 in)
Gear pinion selector fork groove width	5.05 - 5.15 mm (0.1988 - 0.2028 in)
Service limit	5.30 mm (0.2087 in)
Selector fork claw end thickness	4.90 - 5.00 mm (0.1929 - 0.1969 in)
Service limit	4.80 mm (0.1890 in)
Selector fork guide pin diameter:	
Fork on selector drum	7.985 – 8.000 mm (0.3144 – 0.3150 in)
Service limit	7.900 mm (0.3110 in)
Forks on selector fork shaft	7.900 - 8.000 mm (0.3110 - 0.3150 in)
Service limit	7.800 mm (0.3071 in)
Selector drum groove width	8.05 - 8.20 mm (0.3169 - 0.3228 in)
Service limit	8.30 mm (0.3268 in)
Gear pinion/shaft maximum clearance	0.16 mm (0.0063 in)
Minimum diameter of input shaft left-hand end, output shaft right-hand end	19.96 mm (0.7858 in)
Bearing outer race maximum ID	26.04 mm (1.0252 in)
Neutral detent spring minimum free length	30.70 mm (1.2087 in)

Final drive

Type:	
Z550 G1, G2, KZ550 F1, M1, F2, F2L	Shaft
All other models	Chain and sprockets
Reduction ratio:	
Shaft drive models	2.523 : 1 (15/22 x 37/10T)
400 and 500 models, KZ/Z550 A1, A2, A3, A4	2.500 : 1 (40/16T)
All other models	2.375 : 1 (38/16T)
Chain size:	
ZX550 A1, A1L	520 ($\frac{5}{8}$ x $\frac{1}{4}$) x 104 links
ZR400 A1, B1, ZR550 A1, A2, KZ/Z550 H1, H2	530 ($\frac{5}{8}$ x $\frac{3}{8}$) x 104 links
All other chain drive models	530 ($\frac{5}{8}$ x $\frac{3}{8}$) x 100 links
Standard length of 20 links	317.50 - 318.40 mm (12.50 - 12.54 in)
Service limit	323.00 mm (12.72 in)

Torque wrench settings

Component	kgf m	lbf ft	Remarks*
Air suction valve cover bolts – US models only	1.0	7.0	
Cylinder head cover bolts	1.0	7.0	
Intake stub:			
Allen screws	1.5	11.0	A
Hexagon-headed screws	1.2	9.0	A
Cross-head screws	N/App	N/App	A
Spark plugs	1.4	10.0	
Camshaft cap bolts	1.2	9.0	C
Camshaft sprocket bolts	1.5	11.0	A
Cylinder head:			
Nuts	2.3	16.5	C
Bolts	1.2	9.0	C
Cylinder block nuts	1.2	9.0	C
Cam chain tensioner cap	2.5	18.0	
Breather cover bolts	0.6	4.0	
Clutch centre nut	13.5	97.5	
Clutch spring bolts	0.9	6.5	
Advance/retard unit or ignition rotor mounting bolt	2.5	18.0	
Alternator rotor bolt	7.0	50.5	
Alternator stator Allen screws	1.0	7.0	A
Secondary shaft nut	6.0	43.0	
Neutral switch	1.5	11.0	

Component	kgf m	lbf ft	Remarks*
Oil pressure switch - where fitted	1.5	11.0	
Gearchange shaft return spring post:			
ZR400 A1, B1, ZR550 A1, A2, ZX550 A1, A1L	2.0	14.5	A
All other models	2.5	18.0	A
Gearbox sprocket retaining bolts - chain drive models only	1.0	7.0	A
Damper cam retaining nut - shaft drive models only	12.0	87.0	D
Front gear case mounting bolts - shaft drive models only	0.9	6.5	
Connecting rod big-end cap nuts	2.4	17.0	
Starter clutch Allen screws	3.5	25.0	A
Crankcase fastening bolts:			
6 mm	1.0	7.0	C
8 mm	2.5	18.0	C
Oil pressure relief valve	1.5	11.0	A
Sump (oil pan) bolts	1.0	7.0	
Engine oil drain plug	3.8	27.5	
Oil filter mounting bolt	2.0	14.5	
Engine mounting bolts - 10 mm - engine/frame or mounting bracket:			
KZ/Z550 H1, H2, ZR400 A1, B1, ZR550 A1, A2, Z550 G1, G2, KZ550 F1, M1, F2, F2L	3.5	25.0	
All other models	4.0	29.0	
Engine mounting bolts - 8 mm - frame/mounting bracket	2.4	17.0	
Gearchange pedal pivot post - models fitted with gearchange linkage only	2.5	18.0	A
Starter motor terminal nut	1.1	8.0	
Oil cooler - where fitted:			
Mounting bolts	1.0	7.0	
Pipe gland nuts	2.3	16.5	
Bottom union mounting bolts	1.0	7.0	

***Remarks**

A Use thread locking compound
B Apply sealant
C Follow specified slackening/tightening sequence
D Stake to prevent slackening

1 General description

The engine/gearbox unit fitted to the Kawasaki 400, 500 and 550 fours is of the 4 cylinder in-line type, fitted transversely across the frame. The valves are operated by double overhead camshafts driven off the crankshaft by a Hy-Vo chain. The engine/gearbox unit is of aluminium alloy construction, with the crankcase divided horizontally.

The crankcase incorporates a wet sump, pressure fed lubrication system, which incorporates a gear driven dual rotor oil pump, an oil filter, a safety by-pass valve, and an oil level or pressure switch.

The engine is built in unit with the gearbox. This means that when the engine is completely dismantled, the clutch and gearbox are dismantled too. This task is made easy by arranging the crankcase to separate horizontally.

Power from the crankshaft is transmitted via a Hy-Vo chain to the secondary shaft which runs to the rear of the crankshaft. The secondary shaft carries the starter clutch assembly and provides gear drive to the clutch. From the clutch, power is transmitted to the input shaft of the six-speed constant mesh gearbox.

On most of the models covered in this manual the output shaft carries an external sprocket which drives a final drive chain. In the case of the Z550 G1, G2 and KZ550 M1, F1, F2 and F2L models, the output shaft engages a bolted-on bevel gearbox unit, transferring drive to the rear wheel by shaft.

2 Operations with the engine/gearbox unit in the frame

The components and assemblies listed below can be removed without having to remove the engine unit first. If, however, a number of areas require attention, removal of the engine is recommended. Note that although the cylinder block and pistons can be attended to with the engine in the frame, this is inadvisable because accumulated road dirt around the cylinder barrel holding studs is likely to drop into the crankcase as the block is lifted. Refer to Section 7 for more details. details.

The following components/assemblies can be removed and refitted with the engine unit in place:

a) Sprocket/front gear case cover
b) Sprocket/front gear case
c) Clutch release mechanism
d) Neutral switch
e) Gear selector mechanism (external)
f) Alternator
g) Secondary shaft assembly and starter motor
h) Ignition system components
i) Clutch assembly
j) Oil filter
k) Bypass valve, sump and oil pump
l) Cylinder head cover, camshafts and tensioner
m) Cylinder head
n) Cylinder block and pistons*

**Not recommended – see above*

3 Operations with the engine/gearbox unit removed from the frame

Removal of any component or assembly contained within the crankcase halves, except the secondary shaft components and the oil pump, will require the removal of the engine unit and separation of the crankcase halves. These can be summarised as follows:

a) Crankshaft assembly
b) Main and big-end bearings
c) Connecting rods
d) Gearbox shafts and pinions
e) Gear selector drum and forks
f) All crankcase bearings
g) Primary drive chain
h) Camshaft chain

Note that if the removal of any of the gearbox components is required, the engine/gearbox unit can be removed from the frame and inverted so that the crankcase lower half can be withdrawn. This will permit the examination of the gearbox without disturbing the engine top half. See Section 15, paragraph 1.

4 Likely problem areas: general

1 In the course of the workshop project, a number of minor problems were encountered which would normally require the use of a factory service tool. It is, however, possible to avoid this provided that the correct approach is taken.

Clutch centre nut

2 This nut will be **very** tight having been tightened to 97.5 lbf ft (13.5 kgf m) and in addition has a flattened section which will exert considerable drag on the shaft threads. Kawasaki produce a universal holding tool Part Number 57001-305/1243 with which the clutch centre can be held during removal of the nut. A good home-made alternative is described in Section 10 of this Chapter. The only other possible course of action is to strip the clutch while the engine unit is in the frame. This will allow the clutch centre to be held by selecting top gear and applying the rear brake while it is slackened. The nut must be renewed whenever it is disturbed.

Secondary shaft nut

3 This may also prove to be rather tight, and will require the crankshaft to be held while it is removed. To this end, remove the nut **before** removing the alternator rotor, using a strap wrench around the outside of the rotor to hold the crankshaft.

Output shaft shock absorber cam retaining nut – shaft-drive models only

4 This is tightened to 87 lbf ft (12.0 kgf m) and is staked in place. Kawasaki produce a holding tool Part Number 57001-1025 which engages on the cam while the nut is removed. Should the removal of the cam be required, the only alternative to the use of this tool, unless a copy can be fabricated, is to engage top gear and to apply a holding tool either to the clutch centre or to the alternator rotor, as described above. As with the clutch centre nut, this nut must be renewed whenever it is disturbed.

General

5 Cylinders are numbered 1, 2, 3 and 4, starting from the left, for identification purposes. Remember that if the engine must be rotated during the course of work, it revolves clockwise when viewed from its right-hand side, ie when using the hexagon in the contact breaker/ignition pickup assembly, or anti-clockwise when viewed from the left, ie using the alternator rotor retaining bolt. Always turn the engine in its normal direction of rotation, especially when setting the valve timing, and always recheck the valve timing if the engine is turned over with the camchain tensioner removed.

6 Remember that the cam chain tensioner is automatic in operation. If its mounting bolts are slackened (or found to be slack) it must be completely removed and reset. If this is not done, the pushrod will move out slightly and will press too hard against the chain when the tensioner mounting bolts are retightened. If not corrected, this will cause severe damage to the chain and tensioner.

4.2 Drastic measures may prove necessary to remove clutch centre nut – always have a replacement available

5 Removing the engine/gearbox unit from the frame

1 If possible, raise the machine to a comfortable working height on a hydraulic ramp or a platform made from stout planks and strong blocks. Place the machine on its centre stand and secure it with ropes so that it is in no danger of toppling over. Place a drain tray or bowl of about one gallon (5 litre) capacity beneath the sump and remove the drain plug. Leave the machine until most of the oil has drained, then remove the centre bolt which retains the circular oil filter cover to the underside of the unit. Remove and discard the filter element but check that the cup, spring and plain washer are retained.

2 Unlock and raise the seat, or unlock it, lift it at the rear and remove it (as applicable). On ZX550A1, A1L and Z550G1, G2 models a latch must be pushed forward to release the seat after it has been unlocked. Where the sidepanels engage the fuel tank base, pull them away carefully. On all models the fuel tank is located at the front on two rubber buffers and, on all except the ZX550A1, A1L and shaft drive models, is located at the rear by a prong on the tank underside which engages in a grommet set in the frame. Since the tank is only retained by the closed seat, all that is necessary is to lift it at the rear and pull it backwards to remove it, once any wiring and the fuel pipes are disconnected. On ZX550A1 and A1L models, the battery retaining strap passes across a tongue projecting from the rear of the tank; remove the two nuts which secure the retaining strap to release the tank. Z550 G1 and G2 models have a similar arrangement except that a single bolt fastens the rear of the tank to the battery retaining strap; remove the three bolts to release the tank and strap. On KZ550 F1, F2, F2L and M1 models, remove the two bolts which secure the tank rear mounting.

3 Check that the fuel tap is in the 'On' or 'Res' position, then free the fuel pipe and the smaller diameter vacuum pipe by squeezing together the ears of the retaining clip and sliding it down the pipe. The pipes can now be worked off the tap stubs with the aid of a small screwdriver. Where a fuel gauge or low fuel level warning lamp is fitted, lift the tank at the rear and disconnect the sender unit wiring at its connectors. On ZX550A1 and A1L models disconnect the fuel tank console wiring, and on ZX550A1L and KZ550F2L models disconnect the Evaporative Emission Control System pipes from their tank unions. Work as described above for the fuel and vacuum pipes and note that the red (fuel return) hose goes to the left-hand union, which should be plugged to prevent the escape of fuel, while the blue (vent) hose goes to the right-hand union. Having made a final check that all pipes and wiring are disconnected, lift the tank at the rear and pull it backwards to remove it. Remember that the tank represents a significant fire hazard and store it accordingly, also take precautions to protect its paintwork.

4 Removing any retaining screws, where fitted (eg models with Unitrak rear suspension), carefully pull away the sidepanels, if not already removed. Disconnect the battery (remembering always to disconnect first the negative (-) terminal) and remove it as described in the relevant Section of Chapter 6. If the machine is to be out of service for some time, arrangements must be made to give the battery regular refresher charges as described in Chapter 6. Store the battery in a safe place.

5 On all US models only, remove the Clean Air System components. Working as described above for the fuel pipes, release the hoses from the cylinder head cover valve housings, from Number 1 and 4 carburettor intake stubs and from the air filter casing. Releasing any clamps or ties securing it to the frame, withdraw as one unit the vacuum switch valve and its hoses. On ZX550A1L and KZ550F2L models only, release the remaining Evaporative Emission Control System pipes from the carburettor bank and from the air filter front casing. The white (vacuum) hose goes to Number 3 carburettor intake stub, the yellow (carburettor vent) hose goes to the carburettor bank and the green (canister purge) hose goes to the union on the upper rear right-hand end of the air filter front casing.

6 Noting carefully their correct connections, disconnect the HT leads from the spark plugs and the ignition coil wires at their respective connectors. Release the coil mounting nuts and withdraw the HT coils.

7 On all Z400J, Z500 and KZ/Z550A, C and D models, remove the air filter casing top cover (secured by two screws) and lift out the filter element. Remove the small baffle plate secured to the frame by two bolts above the air filter casing, then remove the filter casing mounting brackets from the rear of the casing. Each bracket is secured to the frame by a single bolt on each side; it may be necessary to remove the electrical component inner cover and/or the fuse casing to gain access

to the left-hand side bracket. On ZX550A1, A1L and all shaft drive models, remove the two bolts and mounting brackets which secure the rear of the air filter element housing and the two bolts which secure the filter casing front half to the frame, thus separating both halves of the filter casing and allowing them the maximum possible movement. On ZR400, ZR550 and KZ/Z550 H1 and H2 models, remove both left- and right-hand air filter housings, noting that it will be necessary to remove the battery case to gain access to one of the right-hand side housing retaining screws.

8 Slacken the carburettor/intake stub clamp screws, roll fully rearwards the spring retainers clamping the air filter hoses to the carburettors, pull the carburettor drain hoses (where fitted) clear of the frame on the right-hand side and check that all vacuum hoses and other pipes have been disconnected. Slacken the adjuster locknuts and screw in fully the throttle cable twistgrip adjusters.

9 The throttle cable arrangement at the carburettor end varies slightly according to the model. Where a cable adjuster is fitted, slacken the adjuster locknut so that the adjuster can be disengaged from the support bracket. Where no lower adjuster is fitted, lift the cable outer clear of its locating recess and slide the inner cable clear of the bracket via the slot provided. In both cases the cable nipples can be released from the pulley by aligning the cable inner with the slot and sliding it clear. Since access is limited, this operation can be left until the carburettors are partly removed, if desired.

10 The carburettors are a tight fit between the two sets of mounting rubbers and a certain amount of careful manipulation will be required to extricate them. Pull the carburettor bank rearwards, twisting it clear of the intake stubs. Once these have been freed, manoeuvre the assembly clear of the air filter rubbers and withdraw from the right-hand side. As the carburettor bank clears the mountings, unhook each throttle cable inner.

11 Displace the clip which secures the breather hose to the breather cover by squeezing the clip ears together and sliding the clip up the hose. The hose can now be pulled off its stub. Pull the air filter casing or casing front half forward and remove it from the frame.

12 Check whether the footrests are likely to hinder the removal of any component such as the gearbox sprocket/front gear case cover or the lower rear engine mounting bolt, or if they will prevent the removal of the engine/gearbox unit itself. If this is the case, as for example on all 'Ltd' models (KZ/Z550 C1, C2, KZ550 C3, C4, F1, F2, F2L, M1) remove the footrest mounting bolts and withdraw each footrest as a unit with its mounting brackets. This step will not be necessary on Z550 A3, G1, G2, KZ/Z550 D1, H1, H2 and ZX550 A1 and A1L models.

13 Slacken the clamps which secure the joints between Numbers 2 and 3 exhaust pipes and their respective exhaust pipe/silencer assemblies, remove the eight nuts securing the four exhaust pipe retaining collars at the cylinder head, allow the collars to drop down the pipes and remove the eight split collet halves to release the exhaust pipes. Using only a soft-faced mallet, tap forwards Numbers 2 and 3 exhaust pipes until they can be removed. Remove the silencer rear mounting nuts and bolts (combined with the pillion footrest mountings on most models), then push the exhaust system carefully forwards until the remaining pipes are clear of the engine and can drop to the floor. The system can then be manoeuvred around the centre stand and withdrawn from the machine as a single unit. If the aid of a second person is not available to control the movement of the system, place a thick layer of rags under the front of the engine so that the pipes are not damaged as they drop clear. **Note**: the exhaust clamps will almost always be dirty and corroded and therefore difficult to remove. If this is the case, apply a good quantity of penetrating fluid to all clamps, nuts and bolts and allow plenty of time for the fluid to work before attempting the removal of the system.

14 Releasing its two mounting bolts, remove the starter motor cover. Use a felt marker or sharp tool to mark or scratch the position of the gearchange lever on its shaft, remove fully the pinch bolt and pull the lever off the shaft splines. On those models with a gearchange linkage, displace the plastic cap from the linkage pivot shaft end, remove the circlip and plain washer and pull the linkage off the pivot shaft. Do not forget the second plain washer on the pivot shaft and note carefully which way round the linkage is fitted, making notes or marks to assist on reassembly.

15 On all models except ZX550 A1, A1L and shaft drive machines, slacken all clutch cable adjuster locknuts and screw in fully the adjuster to gain maximum cable free play. Where applicable, disconnect the clutch cable from the side stand automatic retracting mechanism. Remove the sprocket cover mounting bolts and withdraw the cover, complete with the clutch release mechanism. Place the cover out of the way; if removed it can be detached from the cable, or the cable disconnected from the machine. Unless they came away with the cover, remove the two locating dowel pins from their crankcase recesses and store them with the cover; similarly remove the clutch pushrod from the input shaft end, in front of the gearbox sprocket and store this where it cannot be bent. Plug the shaft end (oil seal) with grease to prevent the loss of the steel ball. On ZX550 A1, A1L and shaft drive models, remove the cover retaining bolts, withdraw the cover (and its dowel pins, if necessary) and put it to one side.

Chain drive models

16 Apply the rear brake hard to prevent the sprocket from rotating, then unscrew both sprocket retaining bolts, withdraw the retaining plate, rotating it to align it with the shaft splines, and pull the sprocket off the output shaft end. Disengage the sprocket from the chain and allow the chain to hang over the swinging arm/subframe pivot. Note that it may be necessary to slacken the rear wheel spindle nut and rear brake torque nut so that the wheel can be moved forward to gain sufficient chain free play for the above to be achieved.

Shaft drive models

17 Remove the rear wheel as described in Section 4 of Chapter 5, then remove the left hand rear suspension unit bottom mounting nut and pull the unit off its mounting stud. Slacken and remove the four nuts which secure the final drive casing to the swinging arm, then lift the casing away, taking care not to lose the spring which will be released as the casing comes free. Slacken the screw clip which retains the rubber gaiter (boot) to the front gear case. Pull the gaiter back to expose the plain cylindrical end of the driveshaft. Rotate the shaft, by turning the crankshaft with a gear engaged, until a small hole is located in the shaft end. Insert a thin metal rod or a small screwdriver into the hole, depress the locking pin inside and pull the drive shaft rearwards until it disengages with the front gear case output splines.

18 Slacken evenly and remove the eight bolts which secure the front gear case to the crankcase, detach the neutral switch lead, and lift the unit away taking care to press on the end of the gearchange shaft so that it does not become dislodged. Use a soft-faced mallet to tap gently on the gear case/crankcase joint area to ensure that it separates easily, and be careful that the selector fork shaft does not come away with the gear case. If it is necessary to remove the damper cam from the output shaft, do so at this stage. Refer to Section 4.

All models

19 Disconnect, at the connectors joining them to the main loom, the following leads: alternator wires, neutral switch wire (at the switch), oil pressure/level switch wire and side stand switch wire (where fitted). Peel back the rubber grommet and disconnect the starter motor lead from the terminal on the underside of the starter motor body, the lead being retained by a single nut. Release all the leads from any clamps or guides and tie them to the frame so that they are well clear of the engine/gearbox unit and will not hinder its removal.

20 On all models except those with Unitrak rear suspension, disconnect the wires leading to the stop lamp rear switch and unhook the switch spring at its lower end. Where the rear brake pedal is retained by a pinch bolt on its shaft, mark the position of the lever on the shaft, remove fully the pinch bolt and pull the brake pedal off the shaft splines. For the remaining models, slacken fully the rear brake adjusting nut (on Z500 B1 and B2 models it will suffice to remove the master cylinder mounting bolts), unscrew fully its locknut and screw in as far as possible the brake pedal height adjusting screw to depress the pedal fully, out of the way. On shaft drive models only, if the stop lamp rear switch is likely to hinder engine/gearbox unit removal, it should be removed; the switch bracket is retained by two bolts.

21 Remove its fastening screw to disconnect the battery earth lead from the crankcase immediately below the engine right-hand upper rear mounting. On ZX550 A1, A1L and all shaft drive machines, slacken its adjuster locknuts, screw in fully the adjusters and release the clutch cable lower end nipple from the operating lever. Disengage the cable from the adjuster casting on the crankcase top surface and place the cable over the frame top tubes.

22 Trace the ignition contact breaker/pick-up coil wires from the housing on the crankcase right-hand side up to the connector joining

them to the main loom, disconnect the wires and release them from any clamps securing them to the frame. Secure the wires around the cam chain tensioner to keep them out of the way. On models equipped with a mechanically-driven tachometer, unscrew the knurled nuts at the cable lower end and pull it out of its housing.

23 On those models fitted with an oil cooler, it is sufficient to release the oil pipe/crankcase unions before the engine is removed, but it was found to be much easier in practice if the whole cooler is removed, working space being minimal. Place a bowl under the front of the engine and remove the two bolts which secure each pipe union, noting that an O-ring seals each joint. Remove the four oil cooler mounting bolts and withdraw the cooler with both covers and with its pipes as a single unit once all residual oil has finished draining into the bowl.

24 The engine/gearbox unit should now be ready for removal, being retained only by its mounting bolts. Check carefully that all components have been removed which might hinder this, and that all electrical leads, hoses and control cables have been secured well out of the way. Check particularly the footrests; as stated above their removal is very necessary on some models but not all. If in doubt, remove them. Note that the working space is minimal and that at least two people will be required to move the engine/gearbox unit in safety; it would be helpful if a third person was available to steady the frame while the unit is removed.

25 Taking this into consideration, it was felt advisable to remove the fairing of the ZX550 A1 model featured in the photographs, also its horns, thus permitting greater freedom of movement for the mechanics. Fairing removal is a simple matter and is described in the relevant Section of Chapter 4.

26 Remove the retaining nuts from the rear upper and lower mounting bolts then take the weight of the engine unit with a jack, being careful to use a large piece of wood to spread the load over the sump casting. Alternatively have an assistant ready with a pair of large tyre levers so that the unit can be levered up slightly to ease bolt removal and to prevent damage to the bolt threads or other components. Unscrew the two front mounting bolts, noting that these are retained by nuts captive in recesses in the crankcase casting. On all models except shaft-drive machines and those with Unitrak rear suspension, remove in a similar fashion the two lower front mounting

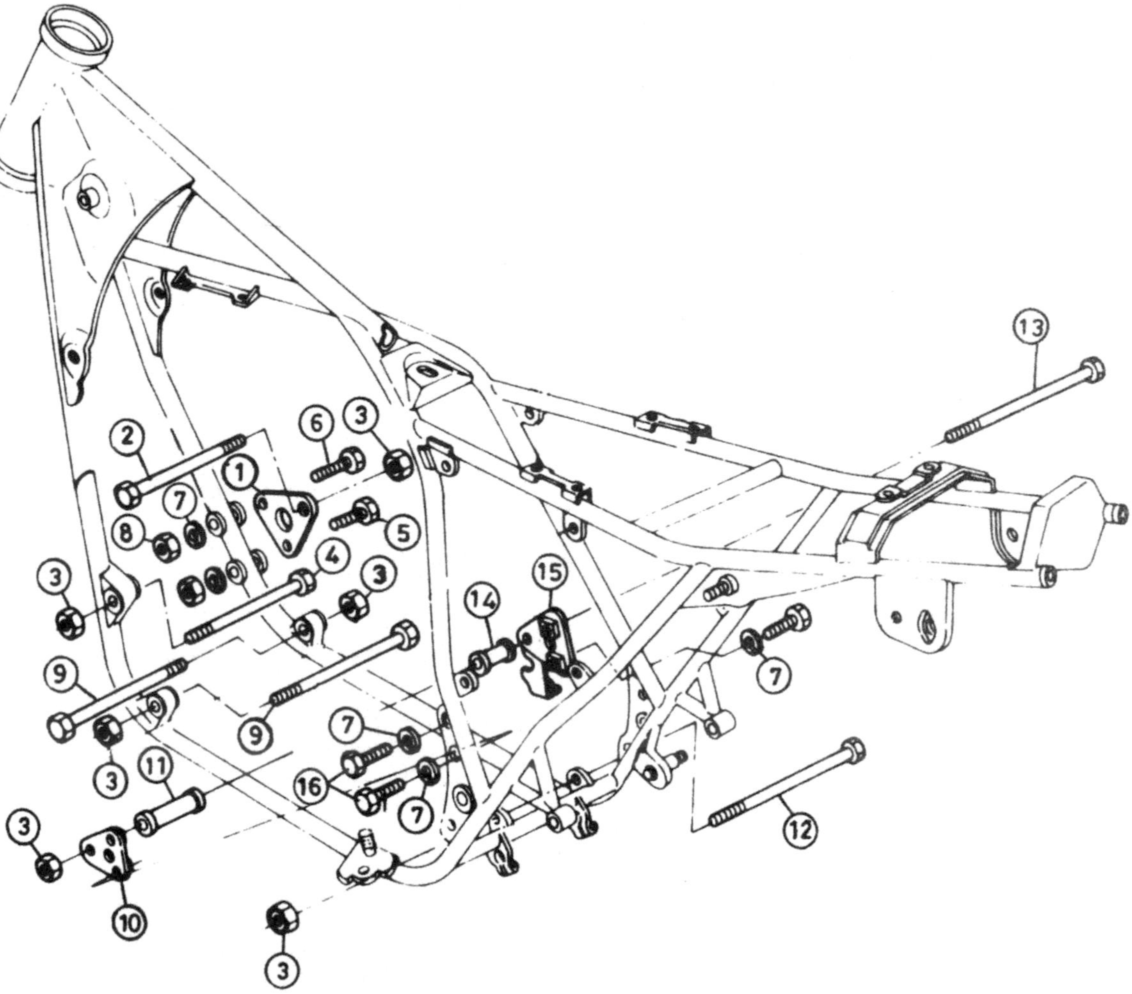

Fig. 1.1 Engine mountings – all 400, 500 and 550 A, C and D models

1 Front right-hand mounting bracket
2 Right-hand upper bolt
3 Nut
4 Left-hand upper bolt
5 Bolt
6 Bolt
7 Lock washer
8 Nut
9 Front lower mounting bolt
10 Rear upper left-hand mounting bracket
11 Spacer
12 Rear lower mounting bolt
13 Rear upper mounting bolt
14 Spacer
15 Rear upper right-hand mounting bracket
16 Bolt

bolts. On all models except ZX550 A1 and A1L machines and those with shaft drive, remove the front right-hand engine mounting bracket, which is retained by two bolts to the frame downtube, and on ZX550 A1, A1L and shaft drive models, remove the two similarly-fastened rubber mounting assemblies, taking care not to lose the small side damper from each assembly's inner end.

27 Tap out the rear upper mounting bolt, taking care not to lose the spacer at each end (one spacer only, on the right-hand side, on ZR400 and ZR550 models, none on ZX550 A1, A1L, and one only on the left-hand side on shaft drive models). Two bolts fasten each rear upper mounting bracket; depending on the model being worked on, there will be two plain metal brackets, the right-hand one of which may have the stop lamp rear switch attached, two rubber mounting assemblies with small side dampers, one rubber mounting assembly without a side damper or a single plain metal bracket. Where only one separate bracket is fitted, this will be found on the right-hand side. Whichever model is being worked on, be careful to identify and remove all mounting components to leave the engine resting on the single lower rear mounting bolt and the jack or levers.

28 Tap out the single remaining mounting bolt and establish a firm hold on the engine. Keeping it level, lift it straight up until the sump clears the mountings, work it to the right until the sump can be lifted over the frame right-hand tube, then carefully lift it out.

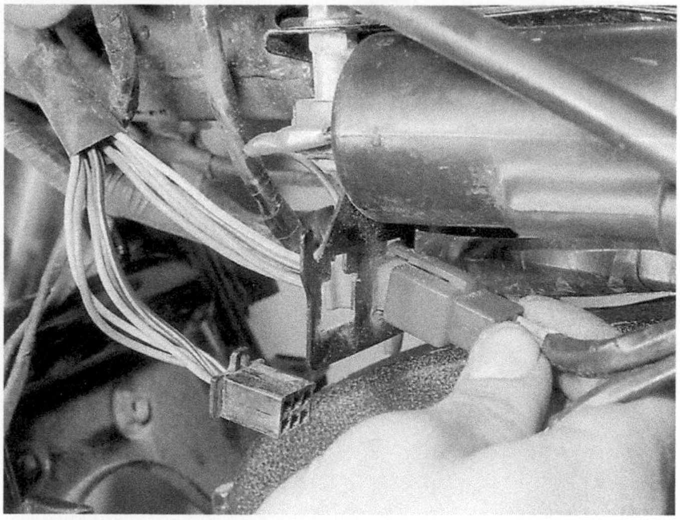

5.6 Disconnect HT coils at spark plugs and wiring connectors

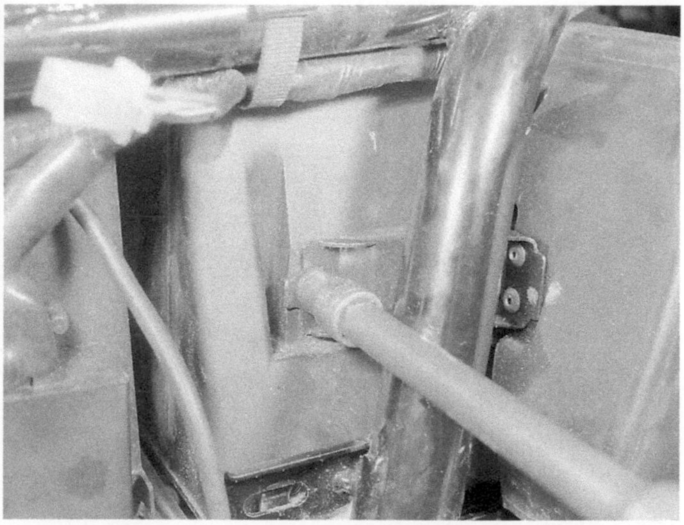

5.7 Remove all air filter casing mounting bolts before attempting carburettor removal

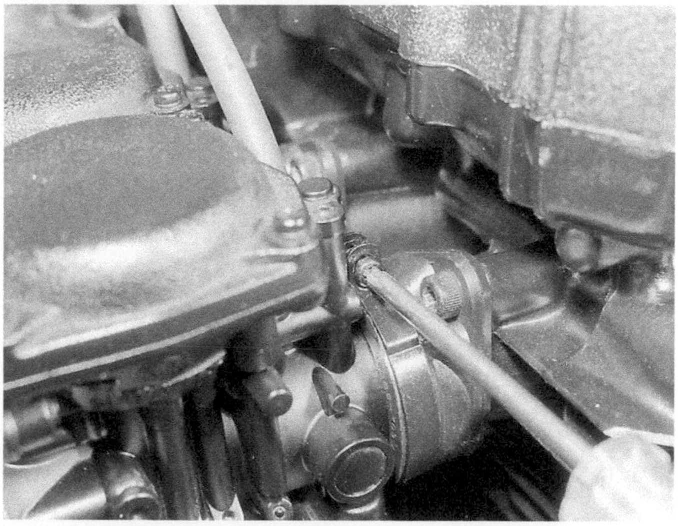

5.8a Slacken fully carburettor/intake stub clamp screws ...

5.8b ... and roll back air filter hose spring retainers

5.11 Do not forget to release breather hose before removing filter casing or casing front half

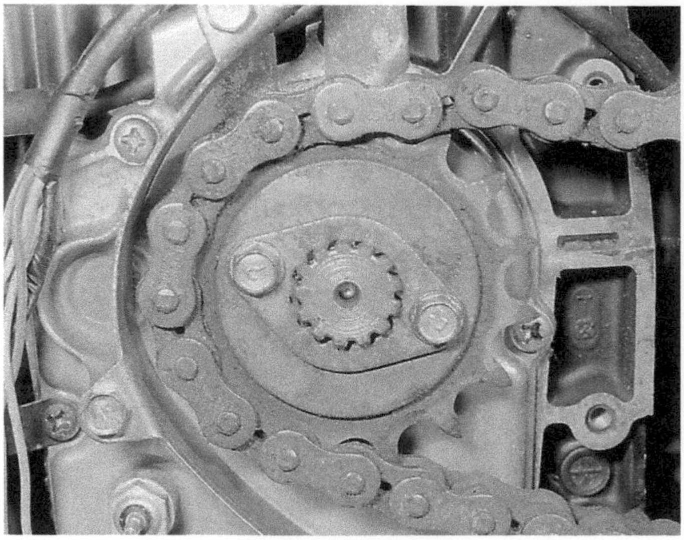

5.16 Gearbox sprocket is retained by two bolts and retainer plate – chain drive models

5.19 Starter motor lead is retained by single nut to motor terminal

5.21 Battery earth lead is attached to crankcase as shown – do not omit to disconnect

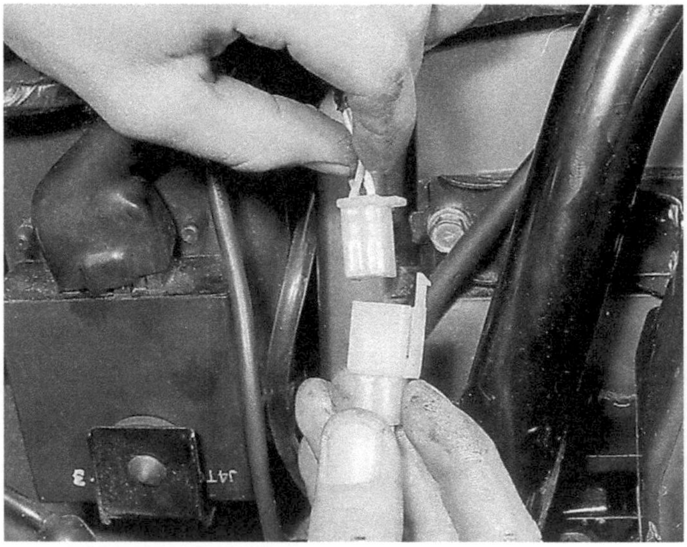

5.22 Trace and disconnect ignition system wires from engine unit

5.25 Removal of additional components may be necessary to provide clearance for engine removal – see text

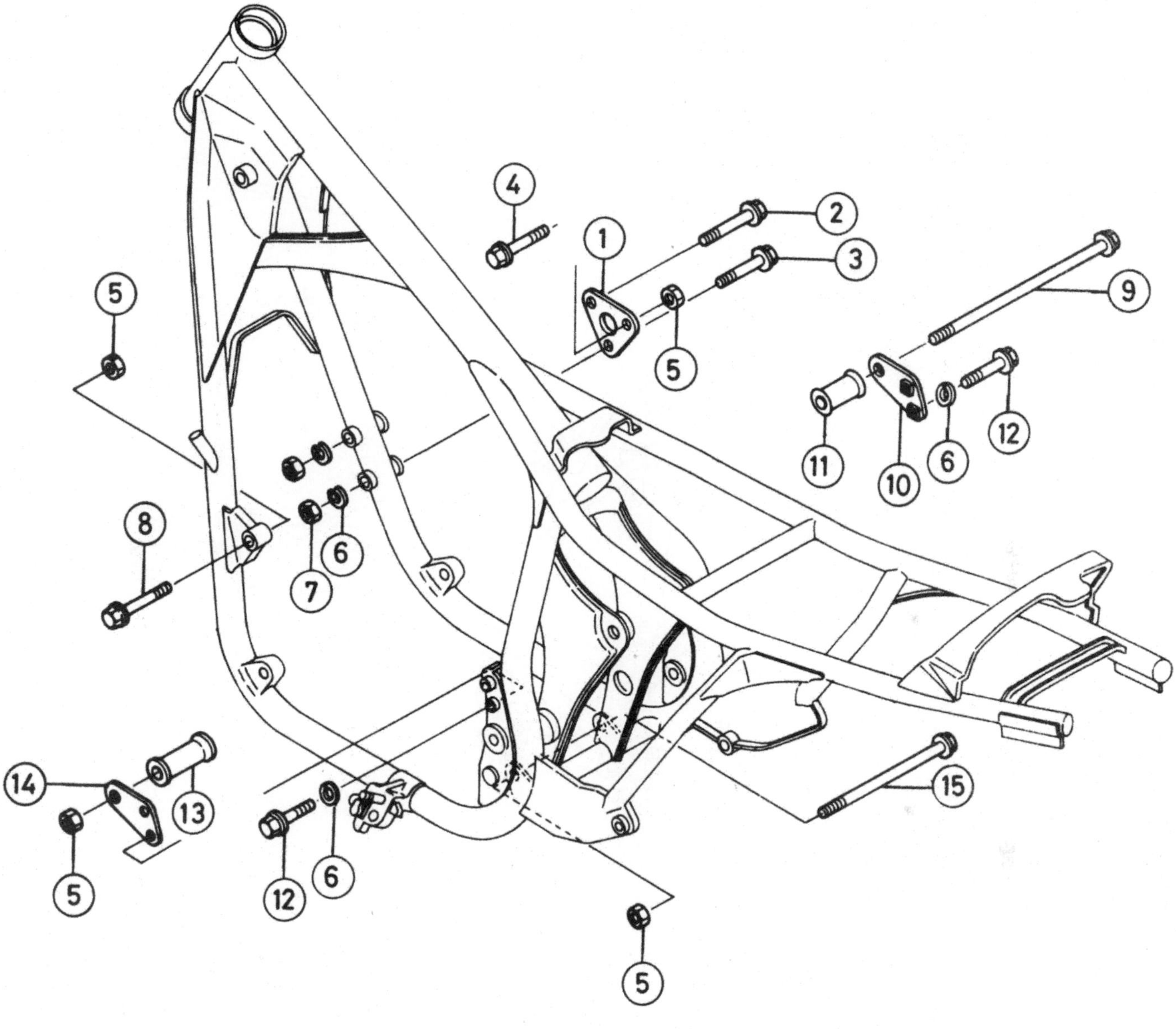

Fig. 1.2 Engine mountings – KZ/Z550 H1, H2 and all ZR400/550 models

1 *Front right-hand mounting bracket*
2 *Bolt*
3 *Bolt*
4 *Front right-hand bolt*
5 *Lock nut*
6 *Lock washer*
7 *Nut*
8 *Front left-hand bolt*
9 *Rear upper mounting bolt*
10 *Rear upper right-hand bracket*
11 *Spacer*
12 *Bolt*
13 *Spacer**
14 *Rear upper left-hand bracket**
15 *Rear lower mounting bolt*
* *KZ/Z550 H1 and H2 only*

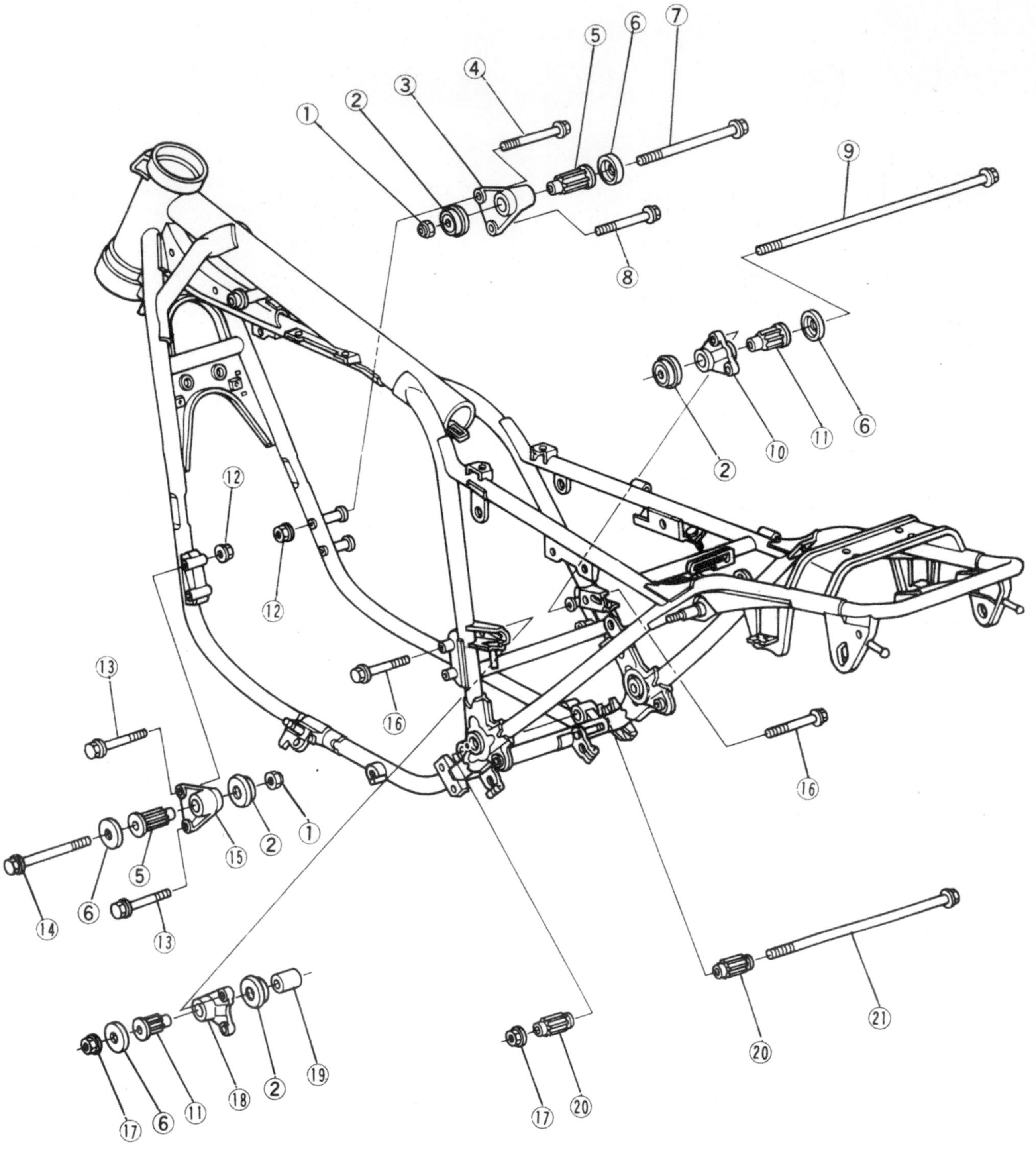

Fig. 1.3 Engine mountings – ZX550 A1, A1L and all shaft drive models

1 Nut
2 Side damper
3 Bracket marked 1193
4 Bolt
5 Damper
6 Cap
7 Bolt
8 Bolt
9 Bolt
*10 Bracket marked 1196**
11 Damper
12 Nut
13 Bolt
14 Bolt
15 Bracket marked 1192
16 Bolt
17 Nut
*18 Bracket marked 1195**
*19 Collar**
*20 Damper**
21 Bolt
**Shaft drive models only*

6 Dismantling the engine/gearbox unit: preliminaries

1 Before any dismantling work is undertaken, the external surfaces of the unit should be thoroughly cleaned and degreased. A high flash point solvent, such as paraffin (kerosene) can be used, or better still, a proprietary engine degreaser such as Gunk. Use old paintbrushes and toothbrushes to work the solvent into the various recesses of the engine castings. Take care to exclude solvent or water from the electrical components and inlet and exhaust ports. The use of petrol (gasoline) as a cleaning medium should be avoided, because the vapour is explosive and can be toxic if used in a confined space.
2 When clean and dry, arrange the unit on the workbench, leaving a suitable clear area for working. Gather a selection of small containers and plastic bags so that parts can be grouped together in an easily identifiable manner. Some paper and a pen should be on hand to permit notes to be made and labels attached where necessary. A supply of clean rag is also required.
3 Before commencing work, read through the appropriate section so that some idea of the necessary procedure can be gained. When removing the various engine components it should be noted that great force is seldom required, unless specified. In many cases, a component's reluctance to be removed is indicative of an incorrect approach or removal method. If in any doubt, re-check with the text.

7 Dismantling the engine/gearbox unit: removing the cylinder head cover and camshafts

1 These components can be removed with the engine/gearbox unit in or out of the frame, but in the former case the seat, the side panels, the fuel tank, the ignition HT coils, the carburettors and, on US models only, the Clean Air System and (where fitted) the Evaporative Emission Control System components must all be removed first as described in Section 5 of this Chapter.
2 Remove the spark plugs and the contact breaker/ignition pick-up cover and gasket. Applying a spanner to the larger engine-turning hexagon especially provided (**never** use the smaller retaining bolt hexagon), rotate the crankshaft clockwise until the pistons of 1 and 4 cylinders are at TDC (top dead centre). Turn the crankshaft just past the '1.4' and 'F' marks stamped on the ATU/ignition rotor until the 'T' mark aligns exactly with the raised index mark on the crankcase wall. The aperture in the ignition backplate allows the marks to be seen.
3 On US models only remove the eight bolts which retain the two Clean Air System suction valve covers, noting that two bolts per cover are very short while the other two are long. Be very careful when removing the covers; tap them gently with a soft-faced mallet to break the seal and do not attempt to lever them off. If they are stubborn, the reed valve assemblies can be lifted away by gripping the projection on the top surface of each with a pair of pliers.
4 Working evenly from the outside inwards, progressively slacken, then remove the cylinder head cover retaining bolts (24 on UK models, 20 remaining on US models). Note that the outside four are longer than the remainder. Tap gently all around the joint area with a soft-faced mallet to break the seal and lift away the cover. Again, never attempt to lever it away. Do not try to save the gasket; it will leak unless in perfect condition and should be renewed as a matter of course.
5 On models with a mechanically driven tachometer, remove the retaining screw and stop plate, then withdraw the housing and driven gear. The housing is usually tight but will submit to very careful and gentle leverage. Remove the camchain tensioner cap and its sealing washer, then pull out the small spring and the push rod stop and store them with the cap. Remove the two tensioner mounting bolts and withdraw the tensioner.
6 Slacken each of the camshaft cap bolts evenly by about one turn at a time. The camshafts are under pressure from the valve springs and will be pushed clear of the bearing surfaces in the cylinder head. Once valve spring pressure has been relieved, remove the bolts and place them with the bearing caps in a safe place, keeping separate the inlet and exhaust caps. Place a bar or a length of wire between the camshaft sprockets and under the chain to prevent the latter from dropping into the crankcase when the camshafts are removed. Lift each camshaft clear of the cylinder head and disengage it from the chain. The exhaust camshaft on machines with mechanical tachometers can be identified by its tachometer drive gear. Although on electronic tachometer models a raised blank will be found in place of the machined gear, identification is far from obvious. To avoid any possible mistake during assembly, tie a marked label to the exhaust camshaft to identify it. Also, do not remove the sprockets from the camshafts unless absolutely necessary, since these will also serve as a means of identification. The exhaust camshaft fitted to the project machine was found to be marked with a splash of pink paint; similar markings may be found on other machines.
7 If the camshafts are to be removed for some time, it is worthwhile to cut two lengths of wooden dowel of a diameter similar to the camshafts and to bolt these lightly in place using the camshaft caps. This would avoid the loss of any bolts or dowels and would rule out the possibility of the cam followers and shims falling out and getting mixed up.
8 If necessary, the cam chain upper guide blade can be prised out of its recess in the cylinder head cover.

8 Dismantling the engine/gearbox unit: removing the cylinder head, block, pistons and cam chain tensioner components

1 These components can be removed with the engine in or out of the frame, but in the former case the exhaust system must first be removed and the work described in the previous Section carried out.
2 Remove the five bolts (three at the front, two at the back) which pass upwards through the cylinder block into the head, and the three cap nuts (one at the front, two at the back) which retain the cylinder block to the crankcase, these nuts and bolts are to be found in the centre of the block, around the cam chain tunnel.
3 Working in the **reverse** order of the tightening sequence shown in Fig. 1.21 slacken by about one turn at a time the twelve cylinder head retaining nuts until all pressure is released, then remove the nuts. The cylinder head should now lift off the holding studs, but may prove to be stuck in place by the gasket. Avoid the temptation to lever between the cylinder head and block fins; they are brittle and are easily broken off. Tap around the joint faces using a wooden block and a hammer, or a soft-faced mallet, to jar the head free. Once the seal has been broken, lift the head clear, feeding the camshaft chain through the tunnel.
4 Secure the cam chain again, then remove the cylinder head gasket, followed by the two dowels from the front holding-down studs, and the O-rings and oil feed nozzles from their recesses at each end of the block gasket surface.
5 **Important note:** Before removing the cylinder block, it should be noted that there is likely to be an accumulation of road dirt around the base of the holding down studs. Unless great care is taken, this will drop down into the crankcase during removal, necessitating crankcase separation. Try to arrange the unit so that the block faces downwards, permitting the debris to drop clear of the crankcase mouths. Clean the studs carefully before turning the unit up the right way again. If the cylinder block is to be removed with the engine unit in the frame, the above approach will obviously be impracticable. The only alternative here is to clean the area around the front studs as carefully as possible, using a vacuum cleaner to remove as much of the loose dirt as can be reached. Remove the small rubber bungs in the drillings on each side of the stud holes, then spray the area with aerosol chain lubricant and leave it to solidify. The grease-like consistency of the chain lubricant should engulf and hold the dirt particles, but great care should be taken when lifting the block. If any of the debris enters the crankcase it is vital that it is removed, even if it proves necessary to strip the engine bottom end to be certain.
6 Lift the cylinder block by an inch or two and carefully remove any residual road dirt from the vicinity of the studs. Before the pistons are allowed to emerge from the bores, guard against pieces of broken ring or any further debris entering the crankcase by packing clean rag into the crankcase mouths. The block can now be lifted clear of the holding studs and pistons. Secure the camchain once it is clear.
7 Remove the circlips from the pistons by inserting a screwdriver (or a piece of welding rod chamfered one end) through the groove in each piston boss. Discard them. Never re-use old circlips during the rebuild.
8 Using a drift of suitable diameter, tap each gudgeon pin out of position, supporting each piston and connecting rod in turn. Using a spirit-based marker or a scriber, mark each piston inside the skirt so that it is replaced in the appropriate bore. If the gudgeon pins are a tight fit in the piston bosses, it is advisable to warm the pistons. One way is to soak a rag in very hot water, wring the water out and wrap

the rag round the piston very quickly. The resultant expansion should ease the grip of the piston bosses on the steel pins.

9 Remove the cylinder base gasket, followed by the two O-rings and oil feed nozzles from their recesses at each end of the gasket surface, and the two O-rings from around the holding-down studs next to the cam chain tunnel right-hand side. Both cylinder head and base gaskets and all O-rings should be renewed on reassembly.

10 The cam chain front guide blade can be removed by carefully twisting it sideways and working it out of the cylinder block cam chain tunnel, while the tensioner blade assembly is retained by two Allen screws to the crankcase top surface. Remove the screws and withdraw the assembly; the blade can be separated from its mounting by pressing out the pivot pin.

11 The piston rings can be removed by holding the piston in both hands and gently prising apart the ring ends with the thumbnails until the rings can be lifted out of their grooves and onto the piston lands, one side at a time. The rings can then be slipped off the piston and put to one side, noting carefully which way round and in what order they are fitted. If the rings are stuck in their grooves, use three strips of thin metal sheet to remove them, as shown in the accompanying illustration.

8.2 Do not forget to remove all bolts and cap nuts from front (and rear) of cylinder head and barrel when dismantling

8.7 Remove gudgeon pin circlips as shown – note rag packing crankcase mouth

9 Dismantling the engine/gearbox unit: removing the contact breaker/ignition pickup assembly

1 This can be done whether the engine is in the frame or not; the only preliminary dismantling required in the former case is to remove the right-hand side panel so that the ignition lead can be disconnected.

2 Remove the retaining screws and withdraw the circular cap and its gasket from the front right-hand side of the crankcase.

3 Lightly scratch or mark a line between the ignition backplate and the nearest part of the crankcase so that the backplate can be returned to its original position on reassembly, then remove the three retaining screws and withdraw the backplate, releasing the ignition lead from any clamps which secure it to the engine (or frame). Do not disturb any of the contact breaker/pick-up coil retaining screws.

4 Remove the ATU/ignition rotor retaining bolt; no special tools are necessary, a sharp tap on the end of an ordinary ring spanner being sufficient to slacken it. Withdraw the ATU or ignition rotor (as applicable) and check that the locating pin in the crankshaft end is firmly fixed. If not, remove it and store it with the ATU/rotor.

10 Dismantling the engine/gearbox unit: removing the clutch

1 The clutch can be removed for inspection or overhaul with the engine unit in or out of the frame. In the former case it will first be necessary to drain the engine oil, and on some machines such as all 'Ltd' models to remove the right-hand footrest assembly and/or the brake pedal to gain working space around the clutch cover. In addition, on ZX550A1, A1L, and all shaft drive machines the clutch cable must be disconnected. Refer to Section 5 of this Chapter.

2 **Note**: If it is necessary to remove the clutch centre nut, a holding tool will be required. The nut is designed to resist loosening in service by the simple expedient of crushing it slightly during manufacture. Added to the drag that this produces is its 98 lbf ft (13.5 kgf m) torque figure, the combination of these two factors calling for a lot of force to be applied during removal. It follows that a secure method of holding the clutch centre is essential if damage is to be avoided. Kawasaki produce an excellent holding tool, Part Number 57001-305/1243, which is rather like a self-locking wrench, but having extended blade-like jaws turned through 90° to engage in the clutch centre splines. In the absence of the correct tool, a simple alternative is shown in the accompanying photograph. It was fabricated from 1/8 steel strip and uses a nut and bolt as a pivot. The jaws should be ground or filed to suit the clutch centre splines, and the handles should be left at about 2 -3 feet in length to provide a secure grip.

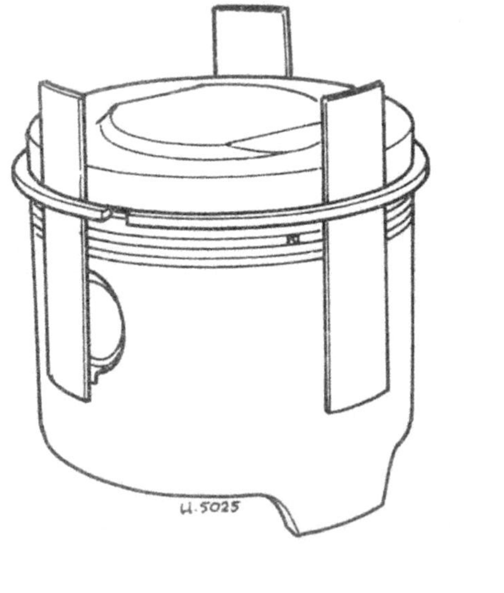

Fig. 1.4 Method of removing gummed piston rings

3 Remove the clutch cover screws and detach the cover, catching any residual oil which is released and noting the two locating dowel pins. Remove the five hexagon-headed bolts which retain the clutch springs, slackening them progressively until spring pressure has been released. Remove the washers and springs, then lift the pressure plate away. Pull out the mushroom-headed pushrod from the centre of the input shaft and displace and remove the single steel ball which is fitted behind it, except on ZX550A1, A1L and all shaft drive machines, on which the clutch release mechanism, a pull-rod and balljournal thrust bearing, are removed with the pressure plate. Remove as a single assembly all the clutch plates.

4 Using the Kawasaki holding tool or substitute, and a socket spanner with a long extension bar, hold the clutch centre stationary while the clutch centre nut is slackened. Once the nut has been slackened, remove it and the large plain washer or Belville washer beneath it.

5 Pull off the clutch centre followed by the plain thrust washer. Remove the clutch outer drum, followed by its needle roller bearing and inner sleeve. Finally, remove the thick thrust washer noting which way round it is fitted.

10.2 Holding tool must be fabricated, as shown, to permit clutch centre nut removal

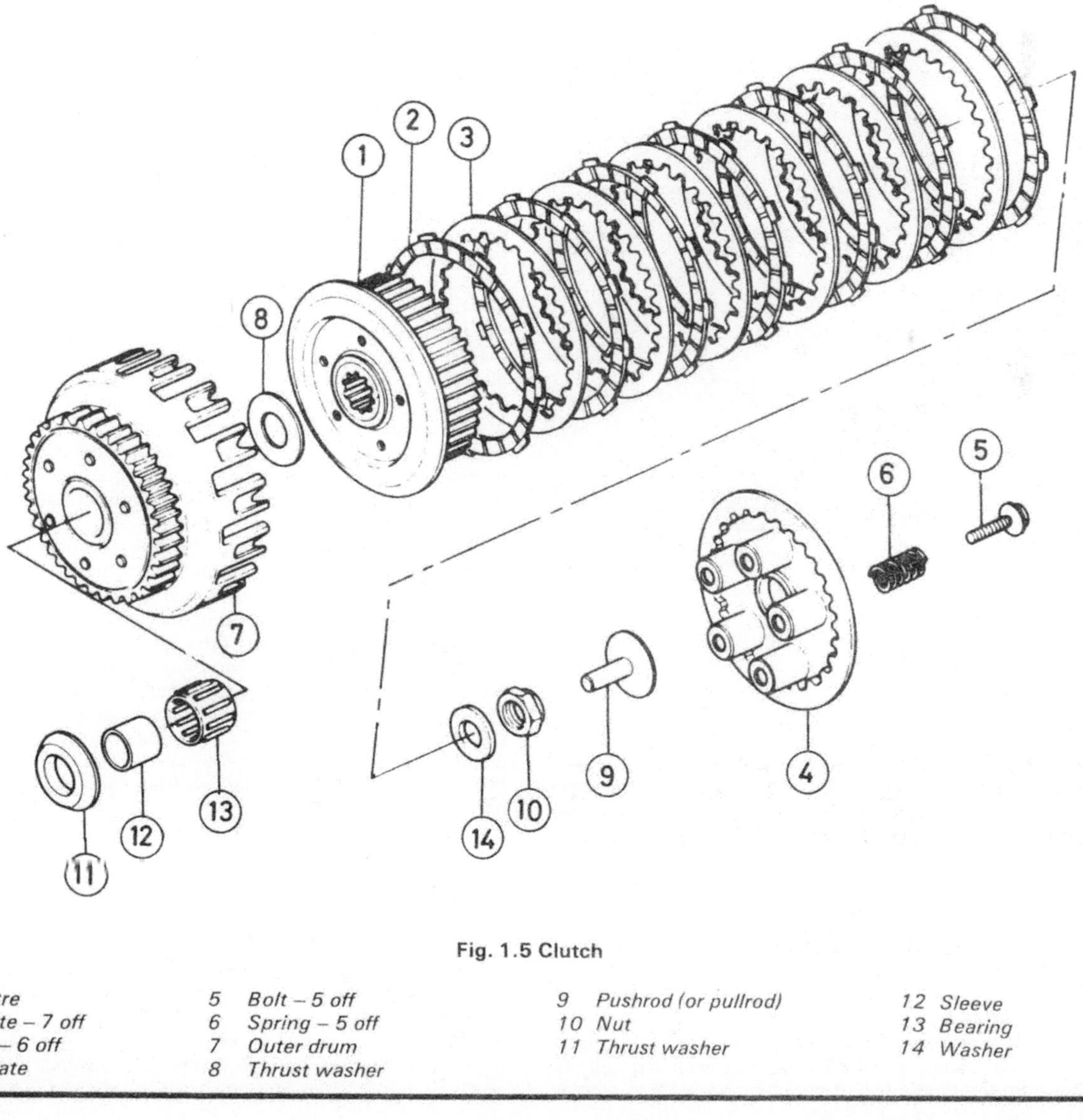

Fig. 1.5 Clutch

1 *Clutch centre*
2 *Friction plate – 7 off*
3 *Plain plate – 6 off*
4 *Pressure plate*
5 *Bolt – 5 off*
6 *Spring – 5 off*
7 *Outer drum*
8 *Thrust washer*
9 *Pushrod (or pullrod)*
10 *Nut*
11 *Thrust washer*
12 *Sleeve*
13 *Bearing*
14 *Washer*

11 Dismantling the engine/gearbox unit: removing the sump and oil pump

1 The oil pump can be removed with the engine unit installed in the frame or on the workbench. If the operation is to be carried out with the engine installed, it will be necessary to drain the engine oil and remove the oil filter and exhaust system first. The clutch cover and clutch assembly should be dismantled as described in Section 10 of this Chapter. Disconnect the oil pressure/level switch lead.
2 Slacken and remove the sump retaining screws and lift the sump away from the crankcase. Make provision to catch the residual engine oil. The pump is secured by two screws and a single bolt which pass through the crankcase wall and screw into the pump body. The screws also form the mounting points for the secondary shaft right-hand bearing retainer, whilst the bolt is located just to the rear of the retainer. Remove the screws and bolt, noting that the former are staked against the retainer for security. Use an impact driver to free these screws to avoid damage to the screw heads.
3 The pump can now be removed by disengaging it from the two dowel pins which locate it. These may come away with the pump, or can be pushed out of the crankcase but must be removed and stored with the pump.
4 Remove the sump gasket, and the three O-rings from around the oilways. These should never be re-used and must be renewed on reassembly.

11.2 Oil pump is retained by two screws and one bolt (arrowed)

12 Dismantling the engine/gearbox unit: removing the alternator

1 The alternator components can be removed with the engine in or out of the frame, but in the former case the left-hand footrest (where applicable), the gearchange pedal or linkage and the gearbox sprocket/front gear case cover must be removed first to gain access to the stator coil lead connectors. See Section 5 of this Chapter. If the secondary shaft is to be removed, refer to Section 13 before removing the rotor.
2 Remove the screws securing the alternator cover to the left-hand side of the engine, and lift away the cover and gasket complete with the alternator stator, noting the two locating dowels. The alternator rotor is secured to the crankshaft end by a single retaining bolt. To remove the rotor from the tapered shaft end it is necessary to employ an extractor. Use only the Kawasaki rotor removal tool, Part Number 57001-254, 57001-1099 or 57001-1216. An alternative is to obtain an 18 mm metric bolt having the necessary fine-pitched thread.
3 Hold the rotor using either the Kawasaki rotor holding tool, Part Number 57001-308/1248, or a heavy duty strap wrench. Slacken and remove the retaining bolt and screw the extractor into the large thread provided. Hold the rotor, and tighten the extractor firmly. If the rotor does not draw off the shaft, strike smartly the end of the extractor to jar it free. If necessary, tighten the extractor a little more and repeat the above, but beware of overtightening and never tap directly on the rotor itself.
4 To remove the stator from the alternator cover, remove its two retaining screws and withdraw the wiring guide plate, then remove the three Allen screws and displace the stator.

13 Dismantling the engine/gearbox unit: removing the secondary shaft and starter clutch

1 The secondary shaft, incorporating the starter clutch unit can be removed with the engine installed in the frame after the exhaust system, engine sprocket cover (or front case), clutch, oil pump and alternator cover have been removed. Refer to Sections 5, 10, 11 and 12 of this Chapter.
2 Before the secondary shaft is removed check the primary chain for wear by measuring the total amount of up and down movement midway between the sprockets. If free play exceeds 25 mm (0.98 in) at any point, the chain must be renewed. Rotate the crankshaft to repeat the test at points all along the chain's length. This is the only way that actual chain wear can be assessed and so the test must be carried out at this stage.
3 Remove the two screws which secure the bearing cap to the crankcase at the left-hand end of the secondary shaft, noting that the upper screw also retains a wiring clip. The secondary shaft nut is removed next, noting that its torque setting of 43 lbf ft (6.0 kgf m) means that the crankshaft will have to be held securely while this is done. Kawasaki produce an alternator rotor holding tool, Part Number 57001-308, for this purpose, but most owners will have to find an alternative. If the cylinder head, cylinder block and pistons have been removed, it is permissible to run a smooth round metal bar through one of the connecting rod small ends, provided that the ends of the bar are supported on wooden blocks to avoid damage to the crankcase mouth. Alternatively, use a heavy duty strap wrench around the alternator rotor, noting that the lightweight types sold for oil filter removal will probably be too weak for this job.
4 Check that the bearing retainer at the right-hand end of the shaft has been removed, then tap the shaft through from the left-hand side to displace the right-hand bearing noting the presence of a sleeve in the left-hand bearing. (Note: On the model used for the workshop project, the shaft had been coated with Loctite or some similar compound. This made removal of the shaft more difficult.) Support the starter clutch/sprocket assembly with one hand and withdraw the shaft and bearing noting the thrust washer on the clutch assembly right-hand end. The clutch unit can now be disengaged from the primary chain and lifted out of the crankcase. This is difficult, owing to the minimal clearance but can be done with care and patience.
5 The starter driven gear should be pulled away from the clutch/sprocket assembly, followed by a second thrust washer, so that the starter clutch rollers can be prised out of their locations and the plunger and spring behind each can then be withdrawn. If necessary, the clutch body can be detached from the shock absorber centre by removing the three Allen screws.
6 To dismantle the shock absorber, remove the circlip from the assembly's left-hand end and pull the sprocket off the shock absorber centre to release the rubbers. It may be necessary to hold the sprocket carefully in a vice and to tap out the centre using a hammer and drift.
7 The secondary shaft gear and right-hand bearing can only be withdrawn, after the retaining circlip has been removed, using a conventional knife-edged bearing puller or possibly a heavy-duty sprocket puller. There is a thick spacer between the two which must not be lost on reassembly.
8 Once the secondary shaft and starter clutch have been removed, the starter idler gear can be withdrawn. Displace the retaining circlip and push out its shaft, then manoeuvre the gear out of the crankcase.

13.6 Remove circlip to dismantle secondary shaft shock absorber assembly

13.8 Remove circlip as shown to release shaft and starter idler gear

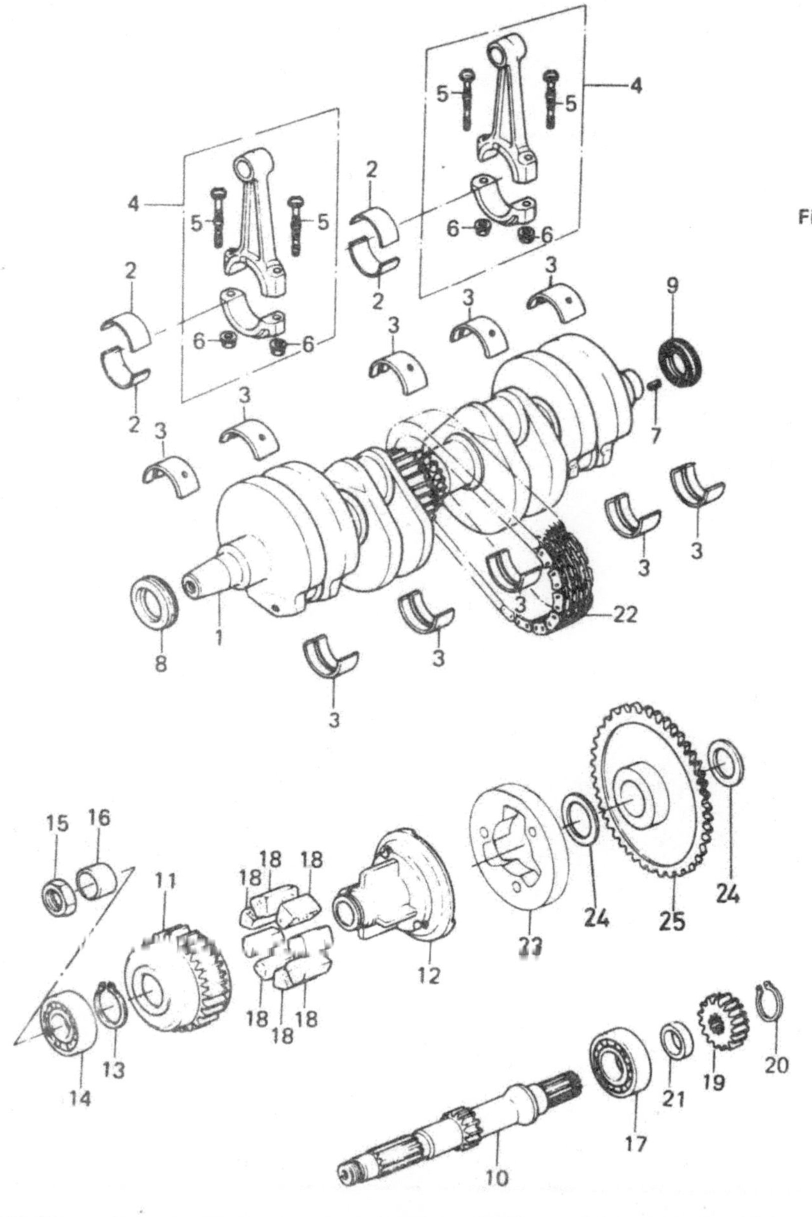

Fig. 1.6 Crankshaft and secondary shaft components

1 *Crankshaft*
2 *Big-end bearing shell*
3 *Main bearing shell*
4 *Connecting rod*
5 *Bolt*
6 *Nut*
7 *Pin*
8 *Oil seal*
9 *Oil seal*
10 *Secondary shaft*
11 *Primary drive sprocket*
12 *Shock absorber centre*
13 *Circlip*
14 *Bearing*
15 *Nut*
16 *Sleeve*
17 *Bearing*
18 *Damper rubber*
19 *Gear*
20 *Circlip*
21 *Collar*
22 *Primary drive chain*
23 *Starter clutch*
24 *Thrust washer*
25 *Starter driven gear*

14 Dismantling the engine/gearbox unit: removing the external gearchange components

1 Access to the external components of the gearchange mechanism is gained after removal of the gearbox sprocket (chain drive models) or front gear case (shaft drive models), and may be carried out with the engine unit in or out of the frame. Refer to Section 5 for details of the above tasks.
2 In the case of chain drive machines, remove the three bolts which secure the final drive chain guard to the crankcase, disconnect the neutral switch lead and remove the gearchange mechanism cover bolts. Detach the cover and remove the two dowels. On shaft drive models, the front gear case takes the place of the cover and will already have been removed.
3 Holding the selector and overshift limiter claws clear of the end of the selector drum, grasp the end of the gearchange shaft and pull the assembly clear of the crankcase. Do not disturb the selector fork shaft or the output shaft sleeve.

15 Dismantling the engine/gearbox unit: separating the crankcase halves

1 The crankcase halves cannot be separated until the engine/gearbox unit has been removed from the frame as described in Section 5 of this Chapter, and after the necessary preliminary dismantling has been carried out. If a full engine/gearbox strip is being undertaken, then all operations described in Sections 6-14 inclusive must be carried out first. If, however, it is wished merely to examine the gearbox components, then the instructions in Sections 7 and 8 can be ignored and it is only necessary to remove the alternator cover (Section 12) and the gearchange mechanism cover (chain drive models, Section 14) and to carry out the tasks described in Sections 6, 9, 10, 11 and 13. Remove the starter motor as described in Chapter 6.
2 The crankcase halves are secured by thirteen 6 mm bolts fitted from the upper crankcase, plus a further seven 6 mm bolts fitted from the underside. In addition, the area around the main bearings is closed by ten 8 mm bolts, fitted from the underside. With the unit the right way up on the work surface, progressively slacken, then remove the top thirteen bolts. Invert the unit, supporting it securely on wooden blocks to prevent distortion of the cylinder studs, then progressively slacken and remove the seven 6 mm bolts. Working in the reverse of the sequence indicated by the numbers cast into the crankcase, repeated in Fig. 1.19, slacken and remove the ten 8 mm bolts. As each bolt is removed, store it in its correct relative position in a cardboard template so that it can be refitted in its correct position on reassembly.
3 Leverage points are provided at the front and rear on both sides of the jointing face, and a screwdriver can be used here to break the joint. Alternatively, a hammer may be used with a block of wood interposed between it and the casing, to jar the casing halves apart. Usually the joint will break fairly easily, but it may be found that separation will be impaired during the first half inch or so of removal. Check carefully to determine which component or components is sticking, and take steps to release it before proceeding further. Separate the crankcase halves with the unit inverted on the bench, drawing the lower half off the upper half. The crankshaft assembly and gearshafts and clusters will remain in position in the upper half but take care not to dislodge or lose any main bearing inserts or similar small components.

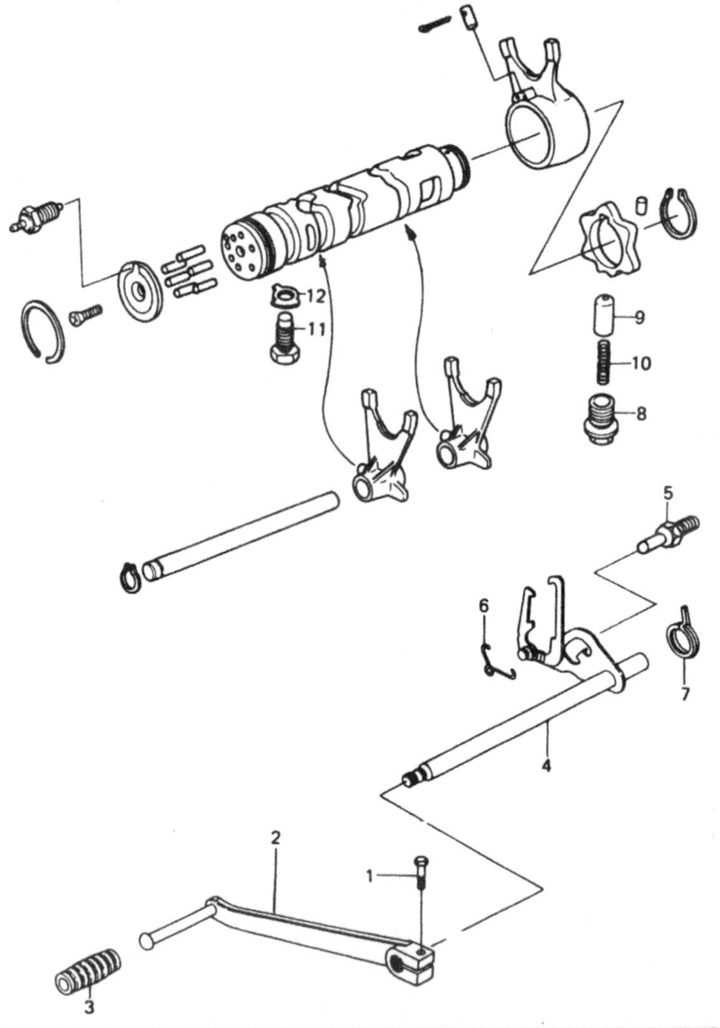

Fig. 1.7 Gear selector mechanism

1 Bolt
2 Pedal
3 Pedal rubber
4 Gearchange shaft
5 Return spring post
6 Return spring
7 Claw arm tensioning spring
8 Cap bolt
9 Neutral detent plunger
10 Spring
11 Guide bolt
12 Tab washer

15.3 Leverage points are provided to assist crankcase separation – be very careful when using them

16 Dismantling the engine/gearbox unit: removing the upper crankcase half components

1 The crankshaft assembly can be lifted out of the upper casing half, and the camshaft drive chain and Hy-Vo primary drive chain disengaged and removed. Take care not to dislodge any bearing inserts.

2 The gearbox input and output shafts can also be lifted out of the casing, noting that these have half ring retainers fitted to the ball journal bearing grooves, and dowel pins to locate the needle roller bearing outer races on the opposite end of each shaft. If the dowel pins are loose in their crankcase recesses they should be removed for safe keeping.

3 Note also the presence of two dowel pins in the upper crankcase gasket surface, and of a ball-shaped rubber plug blanking off an oilway in the input shaft right-hand bearing recess. Unless firmly fixed in position, these should be removed for safe keeping.

4 The breather cover on the crankcase top surface need not be removed except for cleaning purposes. Remove its single retaining bolt to release the cover noting the O-rings under the bolt head and around the cover joint.

5 If the cylinder holding-down studs are to be removed, apply a good quantity of penetrating fluid and allow time for it to work before attempting removal, which should be carried out by locking two nuts together on the exposed thread. A spanner can then be applied to the lower nut so that the stud can be unscrewed. Studs are usually difficult to remove and if this is found to be the case the task should be entrusted to an expert as there is a high risk of shearing off the stud.

6 Before removing the connecting rods use a spirit-based felt marker to draw a line across the front flat-machined surface of each rod and its cap so that both can be refitted the correct way round and in their original positions. Be careful not to score the crankpins as the rods are removed, and keep each rod and cap together at all times.

17 Dismantling the engine/gearbox unit: removing the lower crankcase half components

1 The selector mechanism has three selector forks, two of which are supported by a selector fork shaft. Support the two forks, withdraw the shaft from the left-hand side and lift the forks from position. The third gear selector fork fits round the selector drum itself, and has a guide pin which runs in the selector drum track. Remove and discard the split pin which retains the guide pin, then remove the latter using a pair of pointed-nose pliers.

2 The selector drum is located by a special bolt which screws into the casing and engages in a locating groove. This bolt should be removed after bending back the locking tab. The detent plunger is located at the opposite end of the selector drum, and should also be removed. Detach the large external circlip which retains the detent cam to the end of the selector drum. The cam and locating dowel can now be removed, and the selector drum displaced and withdrawn, leaving the selector fork to be lifted clear of the casing half.

3 There are three oil feed nozzles screwed into oilways in the lower crankcase half; one large silver nozzle set in the crankcase wall behind the clutch, immediately below and to the rear of the selector drum right-hand end. One small silver nozzle is screwed into the output shaft right-hand bearing recess, and one small black nozzle is set on the left-hand side of the centre main bearing pillar. All can be unscrewed using Allen keys, but note that the small black nozzle is staked; this must be released by the careful use of a drill before it can be unscrewed.

4 In addition to the nozzles, two hexagon-headed plugs are screwed into each end of the main oil gallery which runs parallel to, and underneath, the crankshaft. These plugs are therefore found on the 'outside' immediately under the alternator and the contact breaker/ignition pickup housings. Each plug is fitted with a sealing O-ring.

5 If it is necessary to remove a bearing from the crankcase, the casting must be heated in an oven to about 100°C; this will expand the alloy sufficiently to release its grip on the bearing outer race, and the bearing can then be driven out using a hammer and a drift, such as a socket spanner. If the bearing is to be re-used, ensure that the drift bears only on its outer race to avoid damaging the bearing balls or rollers. If a large enough oven cannot be found, place the casting in a container and carefully pour boiling water on it; it is essential that the casting is heated evenly to avoid the distortion that would result from the application of fierce localised heat, ie from a blowtorch or similar. Take great care to avoid the risk of personal injury when heating components or when handling components that have been heated.

18 Examination and renovation: general

1 Before examining the parts of the dismantled engine unit for wear they should be cleaned thoroughly. Use a petrol/paraffin mix or a high flash-point solvent to remove all traces of old oil and sludge which may have accumulated within the engine. Where petrol is included in the cleaning agent normal fire precautions should be taken and cleaning should be carried out in a well ventilated place.

2 Examine carefully each part to determine the extent of wear, checking with the tolerance figures listed in the Specifications section of this Chapter or in the main text. If there is any doubt about the condition of a particular component, play safe and renew.

3 Use a clean lint-free rag for cleaning and drying the various components. This will obviate the risk of small particles obstructing the internal oilways, and causing the lubrication system to fail.

4 Various instruments for measuring wear are required, including a vernier gauge or external micrometer and a set of standard feeler gauges. The machine's manufacturer recommends the use of Plastigage for measuring radial clearance between working surfaces such as shell bearings and their journals. Plastigage consists of a fine strand of plastic material manufactured to an accurate diameter. A short length of Plastigage is placed between the two surfaces, the clearance of which is to be measured. The surfaces are assembled in their normal working positions and the securing nuts or bolts fastened to the correct torque loading; the surfaces are then separated. Be careful not to rotate the shaft while Plastigage is used. The amount of compression to which the gauge material is subjected and the resultant spreading indicates the clearance. This is measured directly, across the width of the Plastigage, using a pre-marked indicator supplied with the Plastigage kit. If Plastigage is not available, both an internal and external micrometer will be required to check wear limits. Additionally, although not absolutely necessary, a dial gauge with mounting bracket is invaluable for accurate measurement of end float, and play between components of very low diameter [illegible] – where a micrometer cannot reach.

5 After some experience has been gained, the state of wear of many components can be determined visually or by feel and a decision on their suitability for continued service made without resorting to direct measurement.

6 Gear backlash is measured with all relevant components temporarily reassembled and refitted in their correct positions. A dial gauge should be mounted on the crankcase so that its tip bears against a tooth of one of the pinions, then that pinion rotated as far as

possible back and forth while the other is held absolutely steady. The difference between highest and lowest gauge readings is the amount of backlash present.

19 Examination and renovation: engine cases and covers

1 Small cracks or holes in aluminium castings may be repaired with an epoxy resin adhesive, such as Araldite, as a temporary expedient. Permanent repairs can be effected only by welding and a specialist will be able to advise on the availability of a proposed repair.
2 Damaged threads can be economically reclaimed by using a diamond section wire insert of the Helicoil type, which is easily fitted after drilling and re-tapping the affected thread. Most motorcycle dealers and small engineering firms offer a service of this kind.
3 Sheared studs or screws can usually be removed with screw extractors, which consist of tapered left-hand thread screws of very hard steel. These are inserted by screwing anti-clockwise into a pre-drilled hole in the stud. If any problem arises which seems to be beyond your scope it is worthwhile consulting a professional engineering firm before condemning an otherwise sound casing; many such firms advertise in the motorcycle papers.
4 Note that Kawasaki recommend that the crankcase 8 mm securing bolts must be renewed after they have been removed and refitted five times.

20 Examination and renovation: bearings and oil seals

1 Ball bearings should be washed thoroughly to remove all traces of oil then tested as follows. Hold the outer race firmly and attempt to move the inner race up and down, then from side to side. Examine the bearing balls, cages, and tracks, looking for signs of pitting or other damage. Finally spin hard the bearings; any roughness caused by wear or damage will be felt and heard immediately. If any free play, roughness or other damage is found the bearing must be renewed.
2 Roller bearings are checked in much the same way, except that free play can be checked only in the up and down direction with the components concerned temporarily reassembled. Remember that if a roller bearing fails it may well mean having to replace, as well as the bearing itself, one or two components which form its inner and outer races. If in doubt about a roller bearing's condition, renew it.
3 Do not waste time checking oil seals; discard all seals and O-rings disturbed during dismantling work and fit new ones on reassembly. Considering their habit of leaking once disturbed, and the amount of time and trouble necessary to replace them, they are relatively cheap if renewed as a matter of course whenever they are disturbed.
4 Those seals which cannot be pulled off their shafts can be levered or drifted out providing care is taken not to scratch or damage their housings. On refitting, use a hammer and a tubular drift such as a socket spanner which bears only on the seal's hard outer edge; take great care to tap the seal squarely into its housing until it is just flush with the surrounding metal. A good smear of grease on the sealing lips will help prevent damage as the shaft is refitted.

21 Examination and renovation: camshafts and camshaft drive mechanism

1 Examine the camshaft lobes for signs of wear or scoring. Wear is normally evident in the form of visual flats worn on the peak of the lobes, and this may be checked by measuring each lobe at its widest point. If any lobe is worn to less than the service limit specified the camshaft must be renewed. Scoring or similar damage can usually be attributed to a partial failure of the lubrication system, possibly due to the oil filter element not having been renewed at the specified mileage, causing unfiltered oil to be circulated by way of the bypass valve. Before fitting new camshafts, examine the bearing surfaces of the camshafts, and cylinder head, and rectify the cause of the failure.
2 If the camshaft bearing surfaces are scored or excessively worn, it is likely that renewal of both the cylinder head and the camshafts will be the only solution. This is because the camshaft runs directly in the cylinder head casting, using the alloy as a bearing surface. Assemble the bearing caps and measure the internal bore using a bore micrometer. If the diameter is 22.08 mm (0.8693 in) or more, it will be necessary to renew the cylinder head and bearing caps. Note that it is not possible to renew the caps alone, because they are machined together with the cylinder head and are thus matched to it.
3 Measure the camshaft bearing journals, using a micrometer. If the journals have worn to 21.86 mm (0.8606 in) or less, the camshaft(s) should be renewed. The clearance between the camshafts and their bearing surfaces can be checked using Plastigage or by direct measurement. The clearance must not exceed the service limit of 0.22 mm (0.0087 in).
4 Camshaft run-out can be checked by supporting each end of the shaft on V-blocks, and measuring any run-out using a dial test indicator running on the camshaft sprocket boss (having first removed the sprocket). This should not normally be more than 0.02 mm (0.008 in). The camshaft must be renewed if run-out exceeds 0.1 mm (0.0039 in).
5 Excessive camshaft endfloat can produce a loud, regular, ticking noise noticeable mainly at idle speed. Each camshaft is located by two thrust flanges which bear against a thrust face on the outside of both inner bearing caps. End float is measured by mounting a dial gauge on the cylinder head so that it is parallel to the camshaft, with its tip touching one end. Push the camshaft as far as possible away from the gauge, zero the gauge, then push the camshaft towards the gauge as far as possible and note the reading. Note that this will be very difficult if the valve gear is still in place; it should be removed if possible. If the reading taken exceeds 0.82 mm (0.0323 in) remove the camshaft and refit the inner bearing caps to the cylinder head. Measure the distance between the camshaft thrust flanges and across the outside faces of the bearing caps, comparing the readings obtained with those given in the Specifications Section of this Chapter. If found to be worn, it will be necessary to renew the camshaft or the cylinder head or both; repairs to reclaim such wear are extremely difficult and may be undertaken only by an engineering expert, if at all.
6 The camshaft drive chain should be checked for wear, particularly if tensioner adjustment has failed to prevent chain noise, this latter condition being indicative that the chain is probably due for renewal. Lay the chain on a flat surface, and get an assistant to stretch it taut. Using a vernier caliper, measure a 20 link length of the chain, ie from any one pin to the 21st along. Repeat this check in one or two other places. The maximum length (service limit) is 128.9 mm (5.075 in).
7 The tensioner and guide blades and the tensioner assembly should be examined for wear or damage, which will normally be fairly obvious. Renew any parts which appear worn or are damaged, especially if a new chain is to be fitted. The same can be applied to the two camshaft sprockets. The push rod and push rod stop of the tensioner assembly must be smooth, unworn, free from dirt or corrosion and able to slide smoothly in the tensioner body. The two springs can be tested only by comparison with new components; renew them if in any doubt about their condition.
8 Z500B1, B2 and KZ/Z550A1 and C1 models (engine numbers up to KZ550AE009031) suffered from a cam chain tensioner fault. The push rod stop spring originally fitted proved to be weak, allowing the stop, and therefore the push rod, to move back out. This produced a slack cam chain with its characteristic light rattle, particularly audible at 2-3000 rpm. A series of modifications was tried on the Z500 models, the last of which (the fitting of another spring) was also necessary on the 550 models mentioned above. **All later** 550 models and all 400 models have been fitted with the modified parts and the tensioner is now reported to be completely reliable. For those owners with suspect machines the new parts are the push rod stop spring (all affected models) and the tensioner cap (Z500B1 and B2 models only); the new spring has a free length when new of 46.2 mm (1.8189 in) and the new cap has an overall length of 40 mm (1.5748 in). When ordering replacement parts ensure that only modified parts (the same as those fitted to all other 400/550 models) are fitted. Note, however, that it is unlikely that any machine has survived unmodified, or that any dealer still has the original (weak) parts in stock.
9 In the case of machines fitted with a mechanically driven tachometer, the worm drive to the tachometer is an integral part of the camshaft which meshes with a pinion attached to the cylinder head. If the worm is damaged or badly worn, it will be necessary to renew the camshaft complete.
10 The tachometer driven worm gear shaft is fitted in a housing which is a press fit in the cylinder head cover. If the worm gear is chipped or broken, the gear and integral shaft should be renewed.

22 Examination and renovation: cylinder head

1 Remove all traces of carbon from the cylinder head using a blunt ended scraper (the round end of an old steel rule will do). Finish by polishing with metal polish to give a smooth, shiny surface.
2 Check the condition of the spark plug threads. If the threads are worn or crossed they can be reclaimed by a Helicoil insert. Most motorcycle dealers operate this service which is very simple, cheap and effective.
3 Clean the cylinder head fins with a wire brush to prevent overheating through dirt blocking the fins.
4 Lay the cylinder head on a sheet of $\frac{1}{4}$ inch plate glass to check for distortion. Aluminium alloy cylinder heads distort very easily, especially if the cylinder head bolts are tightened down unevenly. If the amount of distortion is only slight, it is permissible to rub the head down until it is flat once again by wrapping a sheet of very fine emery cloth around the plate glass base and rubbing with a rotary motion.
5 If the cylinder head is distorted badly (one way of determining this is if the cylinder head gasket has a tendency to keep blowing), the head will have to be machined by a competent engineer experienced in this type of work. This will, of course, raise the compression of the engine, and if too much is removed can adversely affect the performance of the engine. If there is risk of this happening, the only remedy is a new replacement cylinder head.
6 Refer to Sections 21 and 23 of this Chapter for details of work concerning the camshaft bearings and valve seats respectively.

23 Examination and renovation: valves, valve seats and valve guides

1 Remove the cam follower buckets using a small rubber sucker, then the shims, keeping them separate for installation in their original locations. Compress the valve springs with a valve spring compressor, and remove the split valve collets, also the oil seals from the valve guides, as it is best to renew these latter components.
2 Remove the valves and springs, making sure to keep to the locations during assembly. Inspect the valves for wear, overheating or burning, and replace them as necessary. Normally, the exhaust valves will need renewal far more often than the inlet valves, as the latter run at relatively low temperatures. If any of the valve seating faces are badly pitted, do not attempt to cure this by grinding them, as this will invariably cause the valve seats to become pocketed. It is permissible to have the valve(s) refaced by a motorcycle specialist or small engineering works. The valve seating angle is 45°. The valve must be renewed if the head thickness (the area between the edge of the seating surface and the top of the head) is reduced to the service limit specified.
3 Measure the bore of each valve guide in at least four places using a small bore gauge and micrometer. If the measurement exceeds 5.58 mm (0.2197 in) the guide should be replaced with a new one. If a small bore gauge and micrometer are not available, insert a new valve into the guide, and set a dial gauge against the valve stem. Gently move the valve in each direction. The guide will have to be renewed if the clearance between the valve and guide exceeds the specified figures.
4 Valve guide renewal is not easy, and will require that the valve seats be recut after the guide has been fitted and reamed. It is also remarkably easy to damage the cylinder head unless great care is taken during these operations. It may, therefore, be considered better to entrust these jobs to a competent engineering company or to a Kawasaki Service Agent. For the more skilled and better equipped owner, the procedure is as follows:
5 Heat the cylinder head slowly and evenly, in an oven to prevent warpage, to 120 - 150°C (248 - 302°F). Using a stepped drift, tap the guide(s) lightly out of the head, taking care not to burn yourself on the hot casting. New guides should be fitted in a similar manner, being tapped down until the locating circlip seats against the head and ensuring that they seat squarely in the head casting. If a valve guide is loose in the head, it may be possible to have an oversize guide machined and fitted by a competent engineering works, noting that the cylinder head must be bored to suit the new guide. The popular 'dodge' of knurling the outside of the guide is crude and is not recommended.
6 After the guide has been fitted it must be reamed to 5.5 mm (0.2165 in) using a Kawasaki reamer (Part Number 57001 - 1079). Make sure that the reamer passes squarely through the valve guide bore, taking care not to accidentally gouge out too much material. The valve seat must now be re-cut in the following manner:
7 If a valve guide has been renewed, or a valve seat face is worn or pitted, it must be re-cut to ensure efficient sealing. The process requires the use of five cutters, with a pilot bar and T-handle, all of which are available in the Kawasaki valve seat cutter set (Part Number 57001 - 1110). The cutters actually required are as follows:

Part number	Valve	Cutting angle	Cutter OD
57001 - 1114	Exhaust	45°	27.5 mm (1.08 in)
57001 - 1115	Inlet	45°	32.0 mm (1.26 in)
57001 - 1119	Exhaust	32°	28.0 mm (1.10 in)
57001 - 1120	Inlet	32°	30.0 mm (1.18 in)
57001 - 1123	Both	60°	30.0 mm (1.18 in)

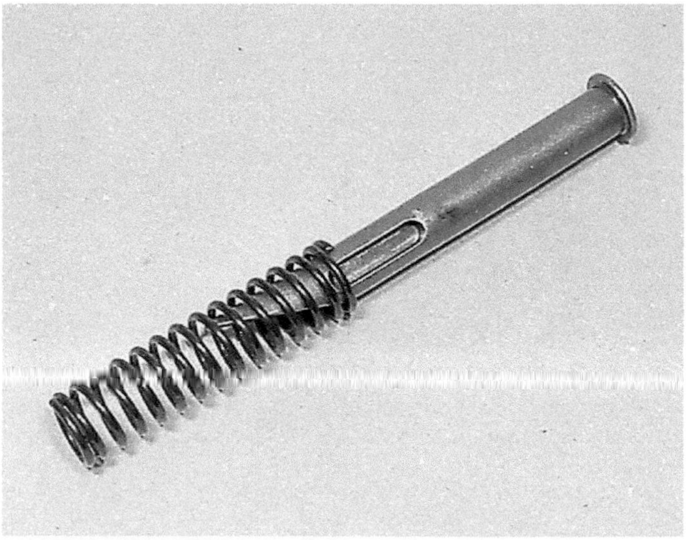
21.7 Cam chain tensioner pushrod must be smooth and unworn as shown – renew spring if in doubt about its condition

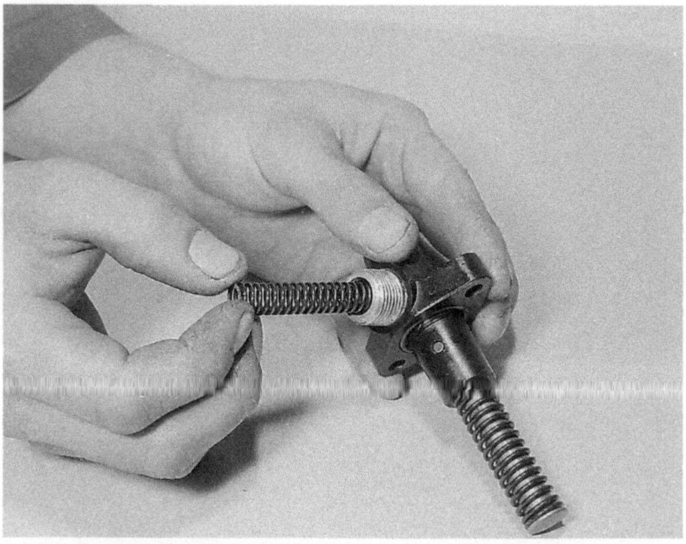
21.8 Tensioner push rod stop spring has proved to be weak in some cases – check carefully

8 Fit the appropriate 45° cutter to the pilot bar, fit the T-handle and insert the pilot bar into the guide until the cutter makes contact with the valve seat. Using firm hand pressure, rotate the cutter through one or two full turns to clean the seat then withdraw the cutter and examine the seat. If the seat is continuous and free from pitting, proceed to the next step, but if pitting is still evident, refit the cutter and repeat the procedure until all pitting has been removed. Be very careful to remove only the bare minimum of material necessary as Kawasaki do not supply valve seat inserts. If the seat becomes sunken through over-cutting, the complete cylinder head must be renewed.

9 With the seat face in good condition, it must now be trimmed to the correct size. Fit the appropriate 32° cutter and use it carefully as described above until the outside diameter of the seat upper edge (see accompanying illustration) is exactly 27.0 mm (1.0630 in) on all inlet valve seats or 23.0 mm (0.9055 in) on all exhaust seats. Finally, fit the 60° cutter and apply it until the seat width is 0.5 - 1.0 mm (0.02 - 0.04 in). It must be stressed that only the barest minimum of material should be removed during this operation.

10 The valves should be ground in, using ordinary oil-bound grinding paste, to remove any light pitting or to finish off a newly cut seat. Note that it is not normally essential to resort to using the coarse grade of paste which is supplied in dual-grade containers. Valve grinding is a simple task. Commence by smearing a trace of fine valve grinding compound (carborundum paste) on the valve seat and apply a suction tool to the head of the valve. Oil the valve stem and insert the valve in the guide so that the two surfaces to be ground in make contact with one another. With a semi-rotary motion, grind in the valve head to the seat, using a backward and forward motion. Lift the valve occasionally so that the grinding compound is distributed evenly. Repeat the application until an unbroken ring of light grey matt finish is obtained on both valve and seat. This denotes the grinding operation is now complete. Before passing to the next valve, make sure that all traces of the valve grinding compound have been removed from both the valve and its seat and that none has entered the valve guide. If this precaution is not observed, rapid wear will take place due to the highly abrasive nature of the carborundum paste.

11 To ensure that the valve seats have not been too deeply cut, insert each valve into its respective guide, retaining it against its seat by finger pressure and use the depth gauge of a vernier caliper to measure the distance between the bottom of the valve spring/cam follower recess and top of the stem. The distance should be 36.08 - 37.02 mm (1.4205 - 1.4575 in) for inlet valves and 36.03 - 36.97 mm (1.4185 - 1.4555 in) for exhaust valves. If the figure obtained is any less than that given, but only by a small amount, the valve seat can be recut to sink the valve into the head until it is at the correct height; this should be very carefully checked before any drastic action is taken, however. The manufacturer states that in such a case the height should be checked first by moving the valve to a deeper cut seat, then by substituting a new valve; if this fails to correct the height a new cylinder head is required. If the figure obtained is more than that given, meaning that the valve head is too thin or the seat too deeply cut, a new valve should be fitted and its height remeasured. If this fails to restore the correct height the only answer is to fit a new cylinder head. Note that the valve stem end should never be ground, either to repair damage or to reduce the valve installed height; the risk of the shim touching the retaining collar or split collets is too great.

12 Examine the spring retaining collars and split collets, renewing any that are marked, worn, or damaged in any way. Measure the free length of each valve spring. If any spring has settled to a length the same as, or shorter than the service limit given, then it should be renewed. Note that while it is possible to buy the springs individually it is considered good practice to renew them all as a set, and that many mechanics renew the springs as a matter of course to ensure good engine performance.

13 Owners of Z400 J1 machines (engine number up to KZ400EE024636) or of KZ/Z550 A1, A2, C1 and C2 machines (engine number up to KZ550AE018931) should note the following before renewing valves or split collets. The design of the valve stem upper end and of the split collets was slightly modified, the new parts being introduced at the engine numbers given. The modification was made in the interests of parts standardisation only, not because of any problem with the earlier type, and the two types are completely interchangeable provided that they are treated as a matched set (valve and collets). To ensure that the correct parts are purchased, take both valve and collets with the engine and frame numbers to a Kawasaki dealer so that the old items can be identified; this is to avoid the mixing of old and new style parts in the same engine. The above is also applicable to owners of Z500 machines when renewing valves or split collets, although the modified parts were not fitted as standard to the B1 or B2 models.

14 Place the spring seats over the guides and press new oil seals into place on each guide upper end. Liberally oil the guide bore and valve stem before refitting the valves. The springs are fitted with the closer-pitched coils downwards, next to the cylinder head. Refit the retaining collars, ensuring that the springs are correctly seated, compress the springs and refit the split collets. Give the end of each valve stem a light tap with a hammer to ensure that the collets have located correctly. Refit the shims, ensuring that the marked surface is downwards and using a smear of grease to retain each shim as the cam follower is oiled and refitted. Be careful to keep the followers absolutely square in their housings; the slightest tilt will jam them and make removal very difficult. Refit the dowels (if used) to retain the followers.

23.14a Always fit new valve guide oil seals

23.14b Fit spring seat as shown

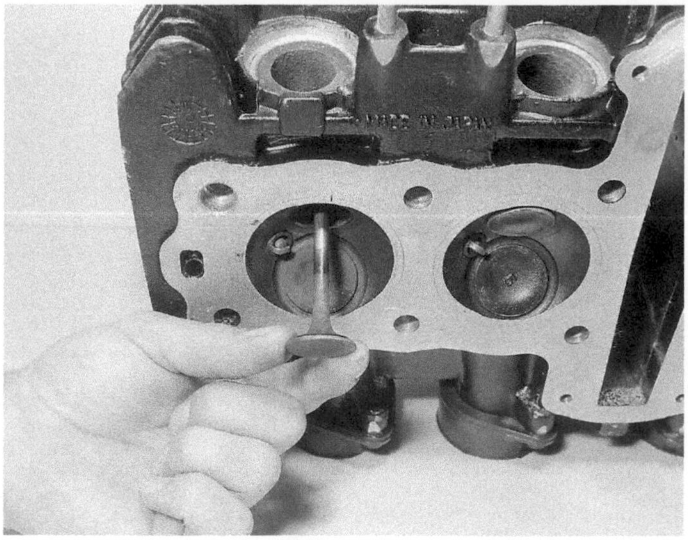

23.14c Oil valve stem and guide before refitting valve

23.14d Both springs are refitted with closer-pitched coils next to cylinder head – upper end may be identified by dab of paint, as shown

23.14e Spring retainer collar is refitted as shown

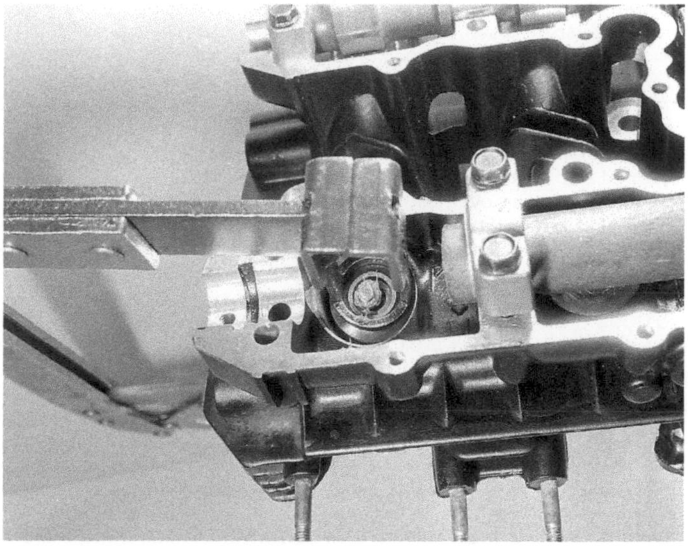

23.14f Use grease to retain split collets while applying valve spring compressor

23.14g Refit shim and oil cam follower before inserting – note use of wooden dowel to retain cam followers

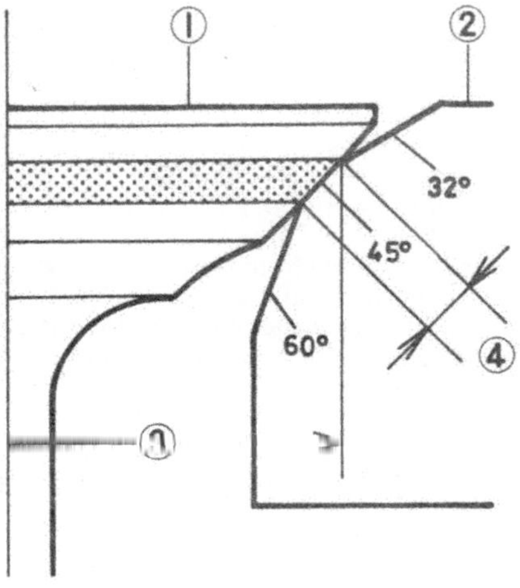

Fig. 1.8 Valve seat recutting

1 Valve
2 Cylinder head
3 Seating area outside diameter
4 Seating area width

24 Examination and renovation: cylinder block

1 The usual indication of badly worn cylinder bores and pistons is excessive smoking from the exhausts. This usually takes the form of blue haze tending to develop into a white haze as the wear becomes more pronounced.
2 The other indication is piston slap, a form of metallic rattle which occurs when there is little load on the engine. If the top of the bore is examined carefully, it will be found that there is a ridge on the thrust side, the depth of which will vary according to the rate of wear which has taken place. This marks the limit of travel of the top piston ring.
3 Measure the bore diameter just below the ridge both along the gudgeon pin axis and at right angles to it using an internal micrometer, or a dial gauge. Compare the readings you obtain with the readings at the bottom of the cylinder bore, which has not been subjected to any piston wear. If the difference in readings exceeds 0.05 mm (0.002 in) the cylinder block will have to be bored and honed, and fitted with the required oversize pistons.
4 If the measuring instrument is not available, the amount of cylinder bore wear can be measured by inserting a new piston (without rings) so that it is approximately $\frac{3}{4}$ inch from the top of the bore. If it is possible to insert a 0.05 mm (0.002 in) feeler gauge between the piston and cylinder wall on the thrust side of the piston, remedial action must be taken.
5 Kawasaki supply pistons in two oversizes: 0.5 mm (0.020 in) and 1.0 mm (0.040 in). If boring in excess of 1.0 mm oversize becomes necessary, the cylinder block must be renewed since new liners are not available from Kawasaki.
6 Make sure the external cooling fins of the cylinder block are free from oil and road dirt, as this can prevent the free flow of air over the engine and cause overheating problems.
7 On reboring, Kawasaki recommend that the block is bored in the order 2-4-1-3 (or 3-1-4-2) to avoid distortion. The size of the rebore is determined by measuring the OD of the new oversize piston and adding the piston/cylinder clearance figure of 0.020 - 0.047 mm (0.0008 - 0.0019 in). The finished bore, measured when cool, must not vary by more than 0.01 mm (0.0004 in) from this figure at any point.
8 When fitting new rings to be run in a used cylinder bore, the bore surface must first be prepared by honing, or glaze-busting. This process, which can also be used to remove marks caused by very light piston seizure, involves the use of a cylinder bore honing tool usually in conjunction with an electric drill to break down the glazed surface which forms on any bore in normal use. The prepared bore surface will then have a very lightly roughened finish which will assist the rings to bed in rapidly and fully. This is normally done as a matter of course after reboring. It also has the advantage of removing the lip from the top of the bore which would otherwise shatter the new top ring. Most motorcycle dealers have glaze-busting equipment and can carry out the work for a small charge.

25 Examination and renovation: pistons and piston rings

1 If a rebore becomes necessary, the existing pistons and piston rings can be disregarded because they will have to be replaced by their new oversizes. If, however, the bores have been checked as described in Section 24 and are to be reused, clean and check the pistons and rings as described below.
2 Remove all traces of carbon from the piston crowns, using a blunt ended scraper to avoid scratching the surface. Finish off by polishing the crowns of each piston with metal polish, so that carbon will not adhere so rapidly in the future. Never use emery cloth on the soft aluminium.
3 Piston wear usually occurs at the skirt or lower end of the piston and takes the form of vertical streaks or score marks on the thrust side of the piston. Damage of this nature will necessitate renewal and is checked by measuring the piston outside diameters at a point 5 mm ($\frac{3}{16}$ in approx) from the base of each piston skirt at right angles to its gudgeon pin axis. If any piston is worn to the service limit specified or less, it must be renewed.
4 After the engine has covered high mileages, it is possible that the ring grooves may have become enlarged. To check this, measure the clearance between the ring and groove with a feeler gauge. A clearance in excess of the service limits given will mean that the piston groove width and ring thickness must be measured and checked with the figures given in the Specifications Section of this Chapter; any component found to be excessively worn must be renewed. Note that this does not apply to three-piece oil scraper rings.
5 To measure the end gap, insert each piston ring into its cylinder bore, using the crown of the bare piston to locate it about 1 inch from the bottom of the bore. Make sure it is square in the bore and insert a feeler gauge in the end gap of the ring. If the end gap exceeds the service limits given, the ring must be renewed.
6 When refitting new piston rings, it is also necessary to check the end gap. If there is insufficient clearance, the rings will break up in the bore whilst the engine is running and cause extensive damage. The ring gap may be increased by filing the ends of the rings with a fine file.
7 The ring should be supported on the end as much as possible to avoid breakage when filing, and should be filed square with the end. Remove only a small amount of metal at a time and keep rechecking the clearance in the bore. Note that standard (new) ring end gaps are not given for the earlier models, only service limits; in such cases the gap should be measured carefully but not corrected unless obviously too tight or too large. It is reasonable to assume, however, that the correct gap will not be too different from that given for the later models and that this can be used as a guide.

26 Examination and renovation: connecting rods and big-end bearings

1 Examine the connecting rods for signs of cracking or distortion, renewing any rod which is not in perfect condition. Check the connecting rod side clearance, using feeler gauges. If the clearance exceeds the service limit of 0.50 mm (0.0197 in) it will be necessary to renew the crankshaft assembly. Connecting rod distortion, both bending and twisting, can only be measured using a great deal of special equipment and should therefore be checked only by an expert; otherwise the rods should be renewed if there is any doubt about their condition.
2 If the necessary equipment is available, the condition of the small-end assemblies can be checked by direct measurement, referring to the tolerances given in the Specifications Section of this Chapter. If the equipment is not available, it will suffice to ensure that the bearing surfaces in the connecting rod small-end bearing, in the piston bosses and over the entire gudgeon pin are smooth and unmarked by wear. The gudgeon pin should be a tight press fit in both connecting rod and piston, and there should be no free play discernible when the components are temporarily reassembled. If any wear is found, the component concerned should be renewed.
3 If a connecting rod is renewed at any time, it is essential that it is of the correct weight group to minimise vibration. Rods are supplied in two weight groups for the machines described in this Manual – group 'E' (286 – 289 grams) and group 'G' (294 – 297 grams). The weight group is indicated by the relevant letter marked on the machine surface across each rod and its cap; if a letter cannot be seen the rod should be weighed, complete with its cap, bolts and nuts but without the bearing inserts, and its group noted. Ideally, all rods should be of the same weight group but it is permissible for the rods of cylinders 1 and 2 to be of one weight group while those of cylinders 3 and 4 are of the second group. The maximum difference permissible between matched rods is 12 grams.
4 Examine closely the big-end bearing inserts (shells). The bearing surface should be smooth and of even texture, with no sign of scoring or streaking on its surface. If any insert is in less than perfect condition the complete set should be renewed. In practice, it is advisable to renew the bearing inserts during a major overhaul as a precautionary measure. The inserts are relatively cheap and it is false economy to reuse part worn components.
5 The crankshaft journals should be given a close visual examination, paying particular attention where damaged bearing inserts were discovered. If the journals are scored or pitted in any way, a new crankshaft will be required. Note that undersized inserts are not available, thus precluding re-grinding the crankshaft.
6 The standard connecting rod big-end eye inside diameter is 36.000 – 36.016 mm (1.4173 – 1.4180 in). To allow for manufacturing tolerances all connecting rods are divided into two size ranges; one indicated by an 'O' mark surrounding the weight group letter, the second indicated by an unmarked surface. Similarly the

standard crankpin diameter is 32.984 – 33.000 mm (1.2986 – 1.2992 in), subdivided into two groups; the group of each crankpin is indicated by its adjacent flywheel being unmarked or marked 'O'. (See accompanying illustrations.) The various sizes are given in the Specifications Section of this Chapter.

7 Measure the crankpins using a micrometer, making a written note of each. Note that each crankpin should be checked in several different places and the smallest diameter noted. If significant ovality is found, renew the crankshaft. Compare the readings obtained with those given. If any crankpin falls below the service limit of 32.97 mm (1.298 in), renew the crankshaft, otherwise mark the flywheel edges as shown in the accompanying figure to indicate its diameter range. Disregard the existing flywheel marks.

8 If necessary, the big-end eye diameter can be checked, with each connecting rod and cap assembled without inserts and with the cap nuts tightened to the correct torque setting, using a bore micrometer. It is unlikely that the diameter will have changed from that marked on the assemblies' machined faces because this area is not subjected to mechanical wear.

9 Using the sizings obtained, select the appropriate insert size from the table shown below.

Bearing insert selection

Connecting rod mark	Crankshaft mark	Insert colour
O	Unmarked	Blue
O	O	Black
Unmarked	Unmarked	Black
Unmarked	O	Brown

10 If the existing inserts are to be checked for wear with a view to re-using them, this is best carried out using Plastigage to check the bearing insert/crankpin clearance as described in Section 18 of this Chapter; the standard clearance is 0.031 – 0.059 mm (0.0012 – 0.0023 in). If the clearance is inside this range the inserts are in good enough condition to be re-used. If the clearance is between 0.059 mm (0.0023 in) and the service limit of 0.1 mm (0.0039 in) the inserts must be discarded and replaced by blue-coded inserts. If the clearance is beyond the service limit, the components must be measured, as described in paragraphs 7 and 8 above, to determine which is worn.

11 **Note**: the colour-coding of the thickest bearing insert, Part Number 13034-1004, is most often referred to as blue, but can be referred to as green. Either may be encountered when ordering replacement parts.

12 Finally, carefully examine the condition of the connecting rod bolts and nuts, renewing any that are damaged or worn in the slightest way; these are very highly-stressed components. Note that it is usually considered good practice to renew bolts and nuts of this type whenever they are disturbed although the manufacturer makes no specific recommendation to this effect.

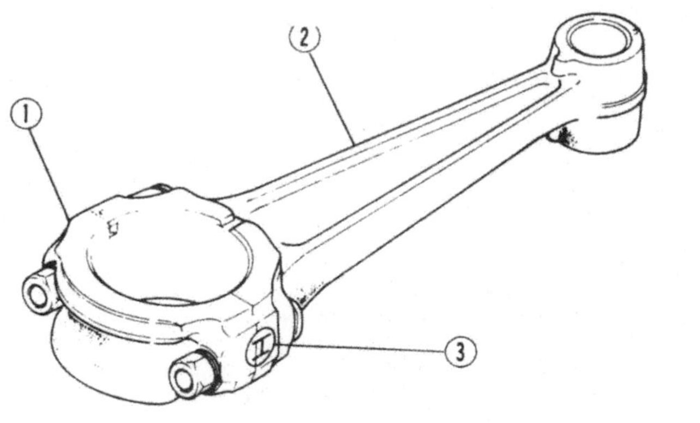

Fig. 1.9 Connecting rod size and weight marks

1 Big-end cap
2 Connecting rod
3 Weight group mark (letter) and big-end size range mark (O mark or unmarked)

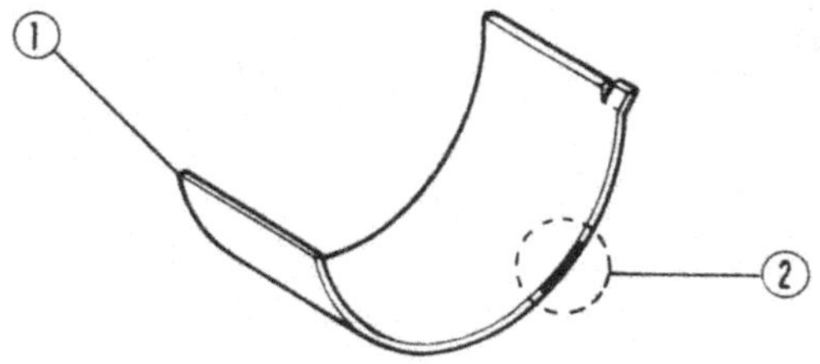

Fig. 1.10 Bearing insert colour code location

1 Bearing insert (shell)
2 Location of colour code

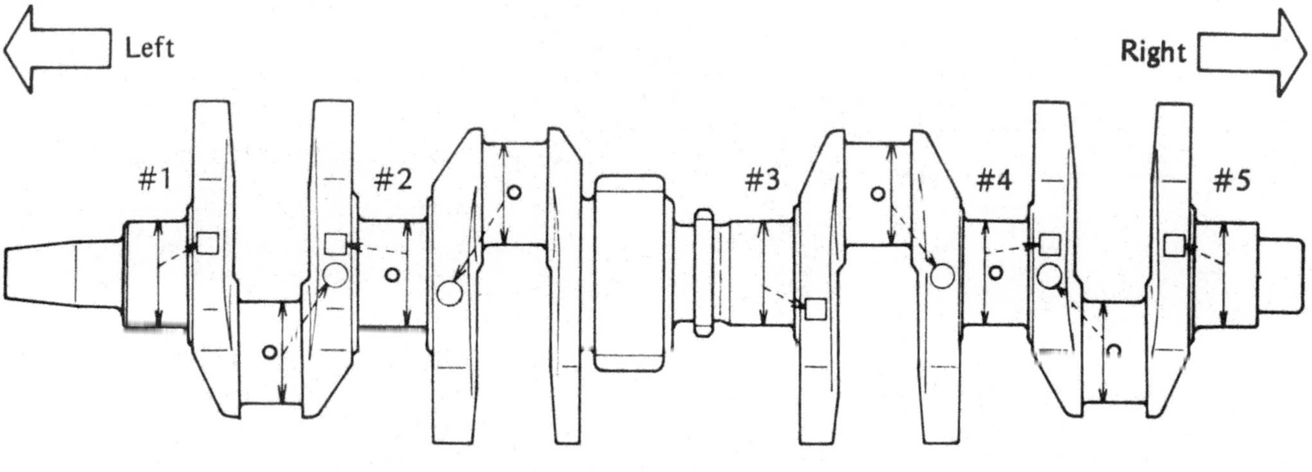

Fig. 1.11 Location of crankshaft size range marks

○ *Crankpin diameter mark – O mark or unmarked*

□ *Main bearing journal diameter mark – 1 mark or unmarked*

27 Examination and renovation: crankshaft and main bearings

1 The crankshaft should be cleaned, after the oil seals, drive chains and connecting rods have been removed, using copious quantities of solvent and compressed air. Be very careful to check that all oilways are completely free from dirt and other foreign matter.
2 Examine the crankshaft closely. Any obvious signs of damage such as marked bearing surfaces, damaged threads, worn tapers and damaged chain sprockets will mean that it must be renewed. There are, however, light engineering firms advertising in the motorcycle press who can undertake major crankshaft repairs; in view of the expense of a new component it is worth trying such firms provided they are competent.
3 Refit temporarily the crankshaft in either crankcase half and use feeler gauges to measure the clearance between the crankcase bearing pillars and their respective adjacent flywheels. Alternatively, a dial gauge can be mounted parallel to the crankshaft, and with its tip touching one end. Push the crankshaft fully away from the gauge, zero the gauge, push the crankshaft fully towards the gauge again and note the reading obtained. If the crankshaft endfloat (side clearance) exceeds 0.40 mm (0.0158 in) the crankcases must be renewed as a matched pair. Crankshaft runout is measured using a dial gauge with the crankshaft mounted on V-blocks at each outer main bearing journal; runout must not exceed 0.05 mm (0.002 in), measured at the centre main bearing journals.
4 The standard main bearing inside diameter is 36.000 – 36.016 mm (1.4173 – 1.4180 in) sub-divided into two groups; one is indicated by an 'O' mark on the crankcase upper or lower half, in the positions indicated in the illustrations accompanying the text. The second group is indicated by the crankcase being unmarked in those positions. Similarly, the crankshaft main bearing journal standard diameter is 31.984 – 32.000 mm (1.2592 – 1.2598 in) sub-divided into two size groups indicated by the crankshaft being unmarked or marked '1' in the positions indicated in the accompanying illustrations. The various dimensions are given in the Specifications Section of this Chapter.
5 The procedures for examining the bearing inserts and journals measuring the diameters of the crankcase main bearings and crankshaft journals, and measuring the amount of wear on existing bearing inserts are the same as for the big-end bearings. Refer to paragraphs 4 – 10 of Section 26 of this Chapter. It will of course be necessary to reassemble temporarily the crankcase halves, tightening all securing bolts to the correct torque setting, when using Plastigage or measuring the main bearing bosses' internal diameters. The crankshaft must be renewed if any journal is found to be oval or worn to the service limit of 31.96 mm (1.2583 in) or less.
6 Having checked the various dimensions, and altered the relevant markings where necessary, select the required main bearing inserts as follows:

Bearing insert selection

Crankcase mark	Crankshaft mark	Insert colour
0	1	Brown
No mark	No mark	Blue
0	No mark	Black
No mark	1	Black

7 When using Plastigage to check insert wear, note that inserts can be reused if the clearance is within the range of 0.014 – 0.038 mm (0.0006 – 0.0015 in) but if the clearance is between 0.038 mm (0.0015 mm) and the service limit of 0.08 mm (0.0032 in) they must be discarded and replaced by blue-coded inserts.
8 **Note**: the colour-coding of the thickest bearing insert, Part Number 13034-1018, is most often referred to as blue, but can be referred to as green. Either may be encountered when ordering replacement parts.

28 Examination and renovation: secondary shaft and starter drive components

1 The only practical check for wear in the primary drive chain is given in Section 13 of this Chapter. Check that the chain is not otherwise damaged, that all links are free to move but have no free play present and that its teeth are undamaged. Light wear on the teeth of the crankshaft and shock absorber sprockets may be corrected by the careful use of an oilstone, but excessive wear or damage will mean that the component concerned must be renewed and the chain checked even more carefully.
2 Wear of the secondary shaft itself should only occur after a very high mileage has been covered. Check carefully its various threads and splines.
3 The shock absorber consists of inner and outer parts with rubber segments to take up transmission shocks. Any damage will be self-evident once the unit is dismantled and should normally be confined to the rubber segments. These will become compressed and rounded off, or may even start to break up, but only after a high mileage. Renew the rubbers if in the slightest doubt about their condition.
4 Renew the starter idler gear and driven gear if their teeth are chipped or worn, or if their centre bearing surfaces are worn making them a sloppy fit on their respective shafts. Check also that the driven gear boss is smooth and unmarked by contact with the clutch rollers. The rollers themselves must be undamaged with no signs of wear such as pitting or flat spots. Check that the plungers and clutch body are in good condition so that the rollers can move smoothly. The springs can only be checked by comparison with a new component and should be renewed if compressed excessively or damaged.

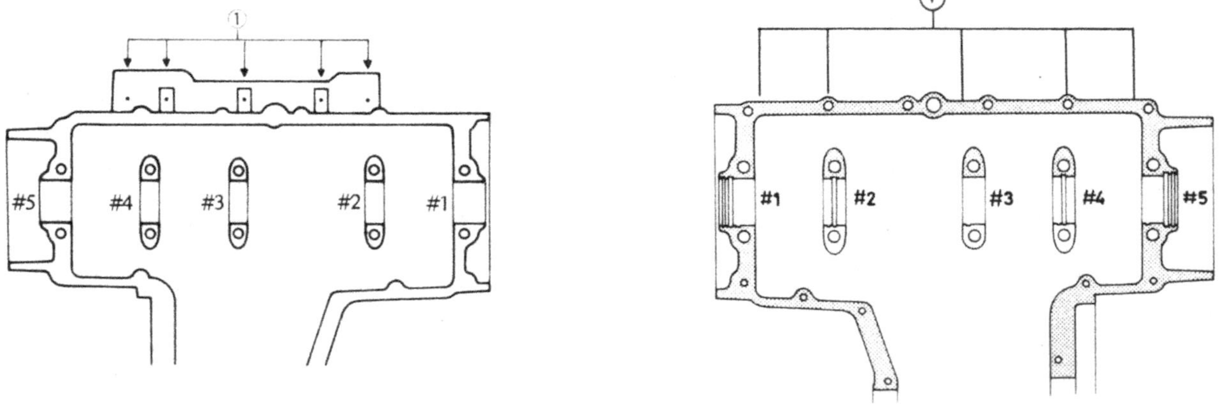

1 Diameter mark – O or unmarked
ZX550 A1, A1L and shaft drive models – upper crankcase

1 Diameter mark – O or unmarked
All other models – lower crankcase

Fig. 1.12 Location on crankcase of main bearing ID size range marks

29 Examination and renovation: clutch

1 After an extended period of service the clutch friction plates will wear and promote clutch slip. When the thickness of any plate reaches the limit given in Specifications, the friction plates must be renewed, preferably as a complete set.
2 The plain plates should not show any excess heating (blueing). Check the warpage of each plate using plate glass or surface plate and a feeler gauge. The maximum allowable warpage is 0.3 mm (0.0118 in).
3 Examine the clutch assembly for burrs or indentation on the edges of the protruding tangs of the friction plates and/or slots worn in the edges of the outer drum with which they engage. Similar wear can occur between the inner tongues of the plain clutch plates and the slots in the clutch inner drum. Wear of this nature will cause clutch drag and slow disengagement during gear changes, since the plates will become trapped and will not free fully when the clutch is withdrawn. A small amount of wear can be corrected by dressing with a fine file; more extensive wear will necessitate renewal of the worn parts. Note that the clearance between the clutch drum slots and the tangs of the friction plates must not exceed 0.7 mm (0.0276 in).
4 Check that the teeth of the secondary shaft gear and the clutch outer drum are unworn. If a dial gauge is available, the gear backlash can be measured; this should not exceed 0.13 mm (0.0051 in). If wear or damage of any sort is found, both components must be renewed as a matched pair.
5 Measure the outer drum centre inside diameter and the outside diameter of the clutch bearing sleeve; if either is worn to beyond the service limit given, it must be renewed, also if it is marked or damaged

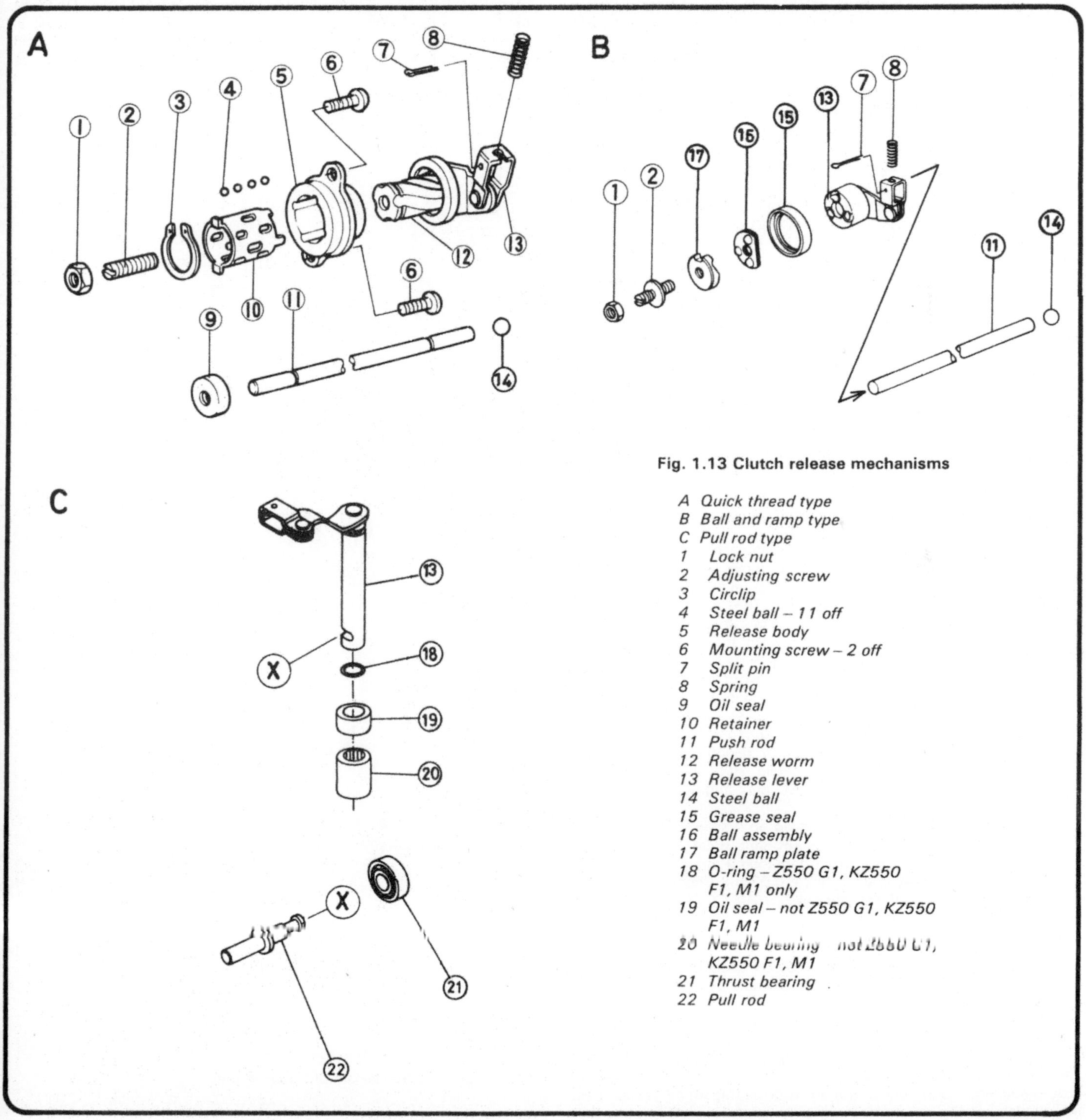

Fig. 1.13 Clutch release mechanisms

A Quick thread type
B Ball and ramp type
C Pull rod type
1 Lock nut
2 Adjusting screw
3 Circlip
4 Steel ball – 11 off
5 Release body
6 Mounting screw – 2 off
7 Split pin
8 Spring
9 Oil seal
10 Retainer
11 Push rod
12 Release worm
13 Release lever
14 Steel ball
15 Grease seal
16 Ball assembly
17 Ball ramp plate
18 O-ring – Z550 G1, KZ550 F1, M1 only
19 Oil seal – not Z550 G1, KZ550 F1, M1
20 Needle bearing – not Z550 G1, KZ550 F1, M1
21 Thrust bearing
22 Pull rod

in any way. Check for wear in the clutch roller bearing as described in Section 20.
6 Measure the free length of each clutch spring. If any one has settled to a length of 31.7 mm (1.248 in) or less, the springs must all be renewed as a set.
7 Three types of clutch release mechanism are fitted to the machines described in this Manual. ZX550 A1, A1L and all shaft drive machines are fitted with a simple pull-rod mechanism (referred to as type 'C' in the accompanying illustration) in which a release lever set in the clutch cover acts on a pull rod set in the input shaft right-hand end, the pull rod lifting the pressure plate via a thrust bearing. On Z550 G1 and KZ550 F1, M1 models the release lever pivots in the cover itself, an O-ring being fitted to seal the upper end, while on all later models a needle roller bearing is fitted in the clutch cover from above and an oil seal is provided. There is no provision for adjustment of the mechanism itself and no attention is required other than to grease the release lever whenever it is refitted. Wear is unlikely until a very high mileage has been covered and will be restricted to the bearing surfaces of the release lever and pull rod, which should be checked accordingly.
8 All other models use one of two types of mechanism which are both set in the gearbox sprocket cover and act on a push rod which passes through the input shaft centre to lift the pressure plate via a $\frac{3}{8}$ inch steel ball and a second, shaped push rod. Z500 models, early Z400 J1, J2 models and early KZ/Z550 A1, A2, C1, C2 models use the quick thread mechanism (referred to as type 'A' in the accompanying illustration), while later Z400 J1, J2 models, later KZ/Z550 A1, A2, C1, C2 models and all other 400 and 550 models use the ball and ramp mechanism (referred to as type 'B' in the accompanying illustration). Since it is not clear exactly when the change was made, owners of Z400 J1, J2 and KZ/Z550 A1, A2, C1, C2 machines will have to check which type is fitted by removing the sprocket cover and examining the mechanism. The maintenance requirements of each type are discussed separately.

Quick thread mechanism

9 With the sprocket cover removed, as described in Section 5, paragraphs 12 and 15, detach the clutch cable by removing the split-pin and releasing the cable end nipple from the release lever. Slacken the adjuster locknut then remove its two mounting screws and withdraw the mechanism. Hold it over a piece of clean rag on the work surface and remove the circlip. Slowly unscrew the release lever and worm from the body and catch the eleven $\frac{1}{8}$ inch steel balls which will drop clear, then remove the ball retainer. Wash all components, removing all traces of dirt and grease and reassemble them temporarily. Holding the body, attempt to move the release lever and worm in and out without rotating it; if excessive free play or sloppiness is felt, the assembly must be renewed. The component parts are available separately and can be renewed individually, if necessary.
10 On reassembly, insert the retainer into the body, aligning the retainer ears with the recesses in the body. Note that the body has one machined surface which faces to the left, to fit against the sprocket cover. Refit the 11 steel balls into the retainer, using grease to stick them in place. Before refitting the release lever and worm, check that the lever, when fully fitted, will be aligned correctly as shown in the accompanying illustration. Insert the worm, taking care not to disturb any of the steel balls, then refit the circlip. Pack the assembly with grease before refitting it to the sprocket cover and tightening the two screws.
11 The clutch push rod should be rolled on a sheet of plate glass or similar flat surface to check that it is not bent. It can be straightened, but if its hardened ends are worn or damaged it must be renewed. Its overall length when new is 270 mm (10.63 in). Check that the steel ball and shaped push rod are unworn.

Ball and ramp mechanism

12 With the sprocket cover removed, as described in Section 5, paragraphs 12 and 15, detach the clutch cable by removing the split-pin and releasing the cable end nipple from the release lever. Slacken the adjuster locknut and withdraw first the release lever, noting how it is aligned with the cover, then withdraw the ball assembly followed by the ramp plate and adjusting screw. Wash all components and the sprocket cover recess, removing all traces of dirt and grease. The grease seal should be renewed if worn or damaged; otherwise it can stay in place on the cover. Wear will be restricted normally to the balls and their ramp tracks; check carefully for these and any other signs of wear, renewing components where necessary.
13 On reassembly smear grease inside the sprocket cover recess and over each component as it is refitted. Insert the ramp plate with the adjusting screw and locknut into the recess, aligning the ramp plate cut-out with the lug on the inside of the recess. The ball assembly is fitted next, with its flat edge aligned with the lug so that the balls fit into the ramp plate, and is followed by the release lever which must also be aligned so that the balls engage correctly when it is at the correct angle.
14 The clutch pushrod should be rolled on a sheet of plate glass or similar flat surface to check that it is not bent. It can be straightened, but if its hardened ends are worn or damaged it must be renewed. Its overall length when new is 270 mm (10.63 in) on all models fitted with this type of mechanism except for some later KZ/Z550 A2, C2 models which had a slightly different gearbox sprocket cover. On these models the push rod length is 268 mm (10.55 in); if renewal is necessary the new push rod must be cut to this length and the tip rounded off. Refit the push rod with the cut end bearing on the release mechanism, not against the steel ball. Any Kawasaki dealer should have the information necessary to identify your machine and the type of mechanism fitted, should any problems arise when ordering replacement parts. Finally, check that the steel balls and shaped push rod are unworn.

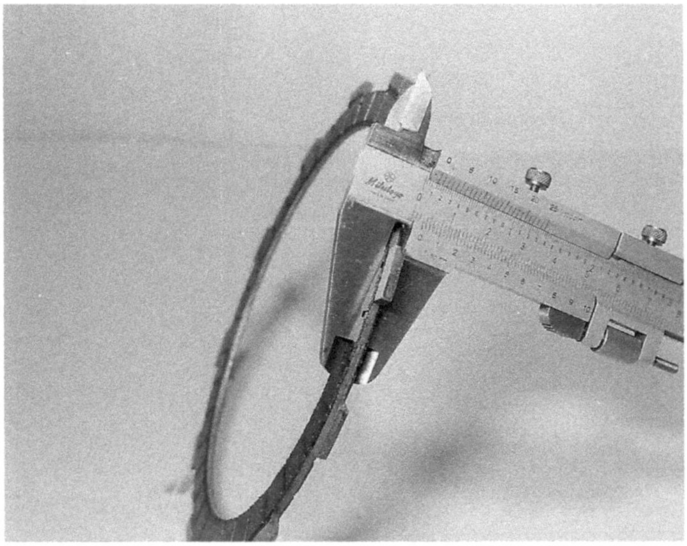

29.1 Measuring clutch friction plate thickness

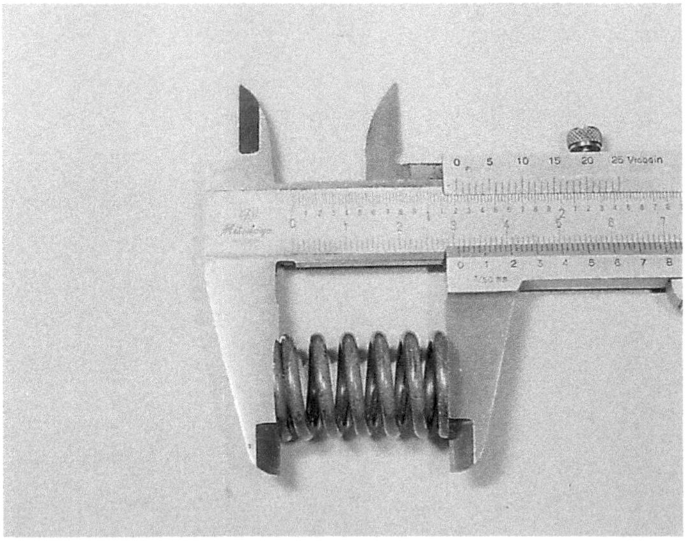

29.6 Measuring free length of clutch springs

29.7 Three different types of clutch release mechanism are fitted – see text for relevant details

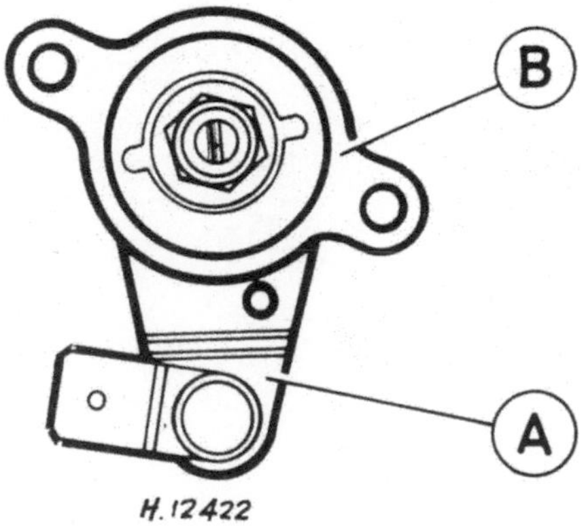

Fig. 1.14 Alignment of release lever and body – quick thread mechanism
A Release lever *B Release body*

30 Examination and renovation: gearbox components

1 Examine each of the gear pinions to ensure that there are no chipped or broken teeth and that the dogs on the end of the pinions are not rounded. Gear pinions with any of these defects must be renewed; there is no satisfactory method of reclaiming them. On those pinions which rotate around their shafts, ie not internally splined, check that the gear pinion/shaft clearance does not exceed 0.16 mm (0.0063 in). If a dial gauge is available the gear backlash can be checked to measure tooth wear; if any pair of gear pinions is found to have backlash of 0.25 mm (0.0098 in) or more, both must be renewed together. Measure the width at several points of each of the three pinion selector fork grooves: if any groove has worn to 5.3 mm (0.2087 in) or wider, the pinion must be renewed.

2 Check the shafts for damage or wear, looking carefully at all splines and bearing surfaces, threads and circlip grooves. Check also that they are straight; damage of any sort can only be cured by the renewal of the shaft concerned.

3 Check the condition of all bearings and renew all oil seals as described in Section 20 of this Chapter. Measure the outside diameter of the input shaft left-hand end and of the output shaft right-hand end; this must not be less than 19.96 mm (0.7858 in) and the bearing surface must be smooth and unmarked. Check also the inside diameter of the two bearing outer races; these must not exceed 26.04 mm (1.0252 in) and must also be unmarked.

4 Check the selector fork shaft for straightness by rolling it on a sheet of plate glass. A bent shaft will cause difficulty in selecting gears and will make the gearchange particularly heavy.

5 The selector forks should be examined closely, to ensure that they are not bent or badly worn. Measure the width of both claw ends of each fork, and the diameter of its guide pin; if found to be worn at any point to less than the service limits given in the Specifications Section of this Chapter, the forks must be renewed. Only the guide pin for the input shaft 3rd/4th gear selector fork can be renewed separately; the other two are part of their respective forks.

6 Measure the width at several points of each of the three selector drum grooves; if any is worn to 8.3 mm (0.3268 in) or more the drum must be renewed. Examine also the detent cam and plunger, the selector drum guide bolt, the selector pins, the drum end plate, the gearchange shaft and the three springs, renewing any component that shows signs of wear or damage. The gearchange shaft return spring and claw arm tensioning spring can be checked only by comparison with new components; renew them if there is any doubt about their condition. The neutral detent plunger spring should be renewed if it has settled to a length of 30.7 mm (1.2087 in) or less.

7 If it is found necessary to renew any of the gearbox components, the accompanying line drawing will give details of the assembly sequence. In addition, gear cluster reassembly is covered in the accompanying photographic sequence.

8 Removing the individual pinions presents no undue problems providing a good quality pair of circlip pliers is used. It is advisable to lay each pinion, thrust washer and circlip out in sequence and the correct way round to prevent mistakes during reassembly. The output shaft incorporates an ingenious neutral finder mechanism which consists of three steel balls running in radial drillings spaced 120° apart within the 5th gear pinion.

9 To remove the pinion the shaft must be held by the 2nd, or 3rd and 4th gear pinions in a vertical position so that it can be spun to throw the balls outwards under centrifugal force; the 5th gear pinion will then drop clear on to the work surface or can be lifted off the shaft from above, depending on which way up the shaft is held. Take great care not to lose any of the balls.

10 When reassembling the shafts always use new circlips, taking care not to open the clips any more than is essential to ease them into position. The clip ends should be supported by the raised areas of the splines for security. Where thrust washers with only two internal splines are used, arrange the circlip so that its ends do not coincide with either spline. The input shaft 6th gear pinion bush is drilled for lubrication purposes; when refitting, ensure that the holes in the bush align with those of the shaft or the oil supply will be blanked off.

11 On refitting the output shaft 5th gear pinion, note that it has six radial drillings, three of which are partially closed off on the outside; insert the three steel balls into these drillings only. Use only engine oil to retain the balls as grease would stick them in place and render the neutral finding device inoperative. Fit the pinion over the shaft end and slide it down into place with the shaft held absolutely vertical and with each ball aligned with a groove in the shaft; take great care that the balls remain fully in their drillings or it will not be possible to fit the pinion. When it is fitted check that the balls have dropped into their grooves by attempting to slide the pinion along the shaft.

12 By limiting the movement in this way of the 5th gear pinion, and therefore of the selector drum, it becomes impossible for any gear other than neutral to be selected from 1st gear. This is with the output shaft not revolving (ie the machine is stationary); as soon as the output shaft is revolving fast enough in 1st gear the balls are flung out under centrifugal force and full selector movement is restored. **Note**: this makes it impossible to check gear selection other than 1st and neutral with the engine/gearbox unit dismantled unless some means can be devised of rotating the output shaft; in practice this is very difficult. It follows that greater than usual care must be taken in rebuilding the gearbox.

13 Owners of machines fitted with a gearchange linkage should note that if this is only slightly worn or bent, the lost motion due to excess free play will seriously affect gear selection, also if the lever cannot move through its full travel because it is fouling some other component. Check the linkage carefully for wear and renew any worn components.

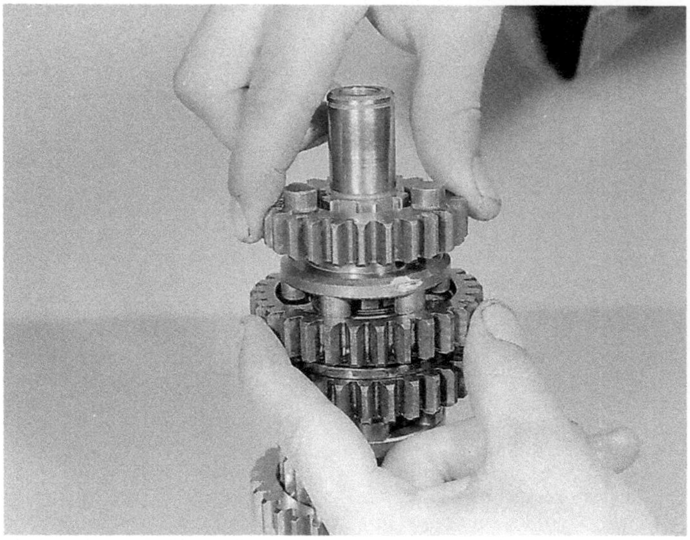

30.9 Hold output shaft assembly by any rotating pinions as shown and spin hard to release 5th gear pinion

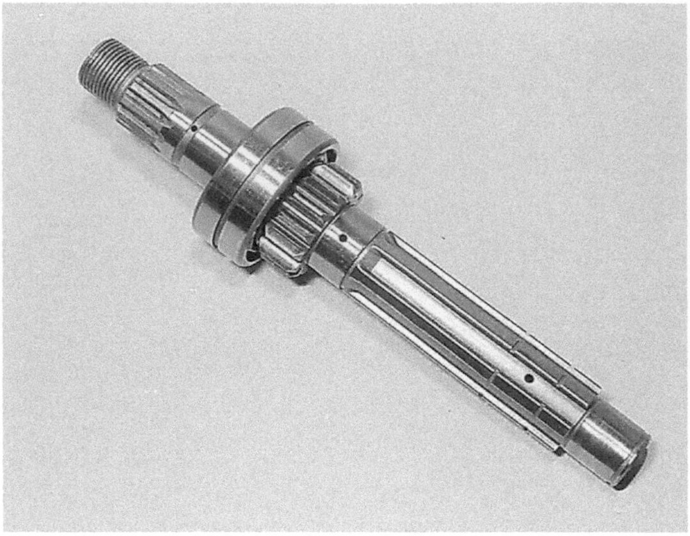

30.10a Input shaft is identified by integral 1st gear pinion

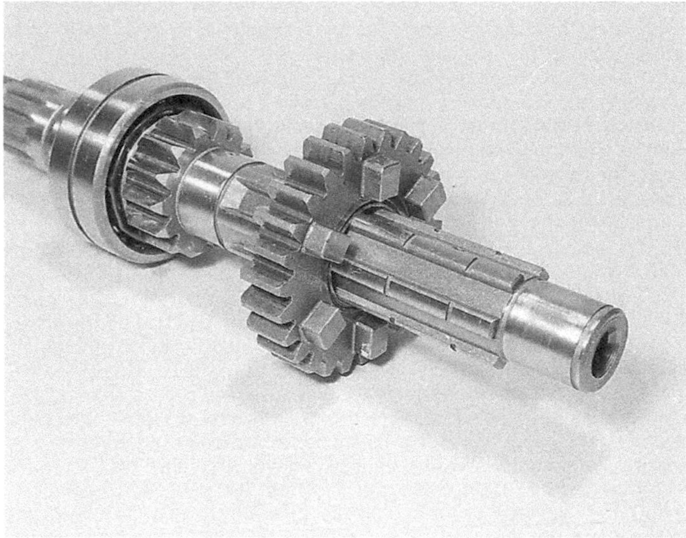

30.10b 5th gear pinion is fitted with selector dogs facing to left ...

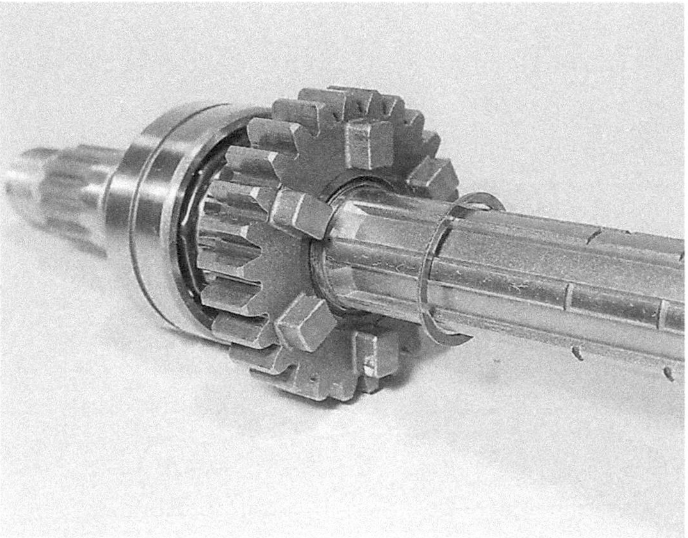

30.10c ... and is located by a plain thrust washer ...

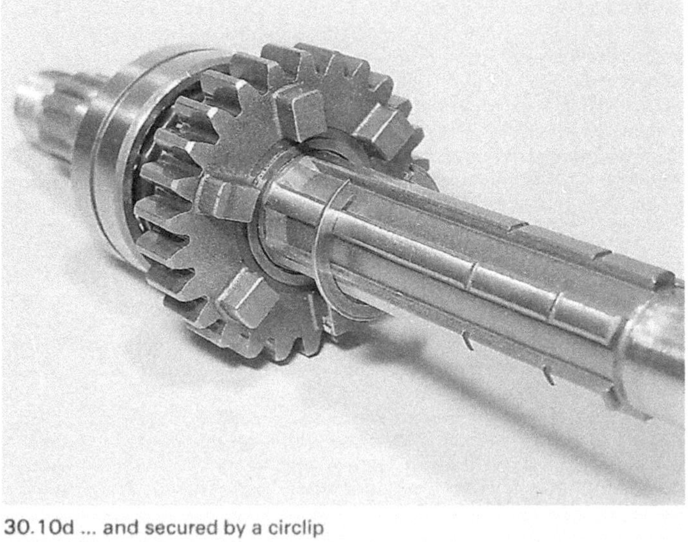

30.10d ... and secured by a circlip

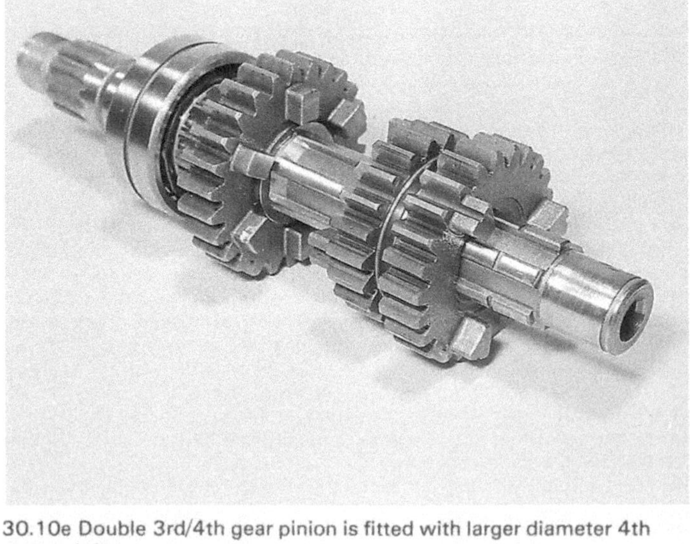

30.10e Double 3rd/4th gear pinion is fitted with larger diameter 4th gear to left

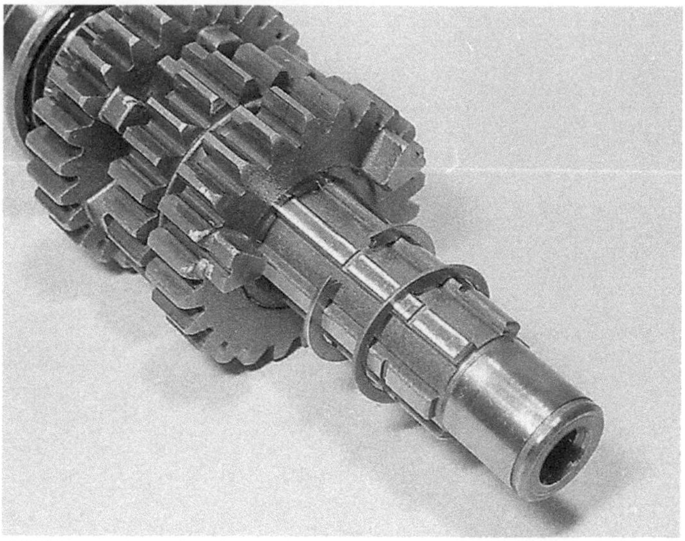

30.10f Fit circlip in groove shown, followed by splined thrust washer

30.10g Align oil holes in 6th gear bush with those in shaft ...

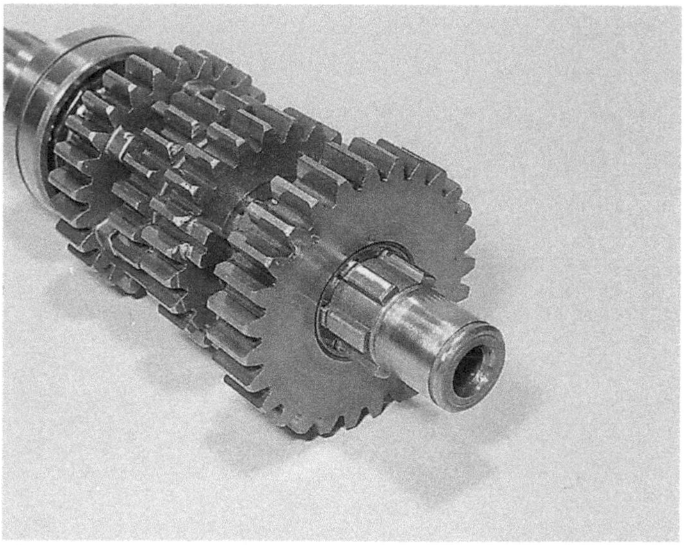

30.10h ... and refit 6th gear pinion as shown

30.10i Fit second splined thrust washer ...

30.10j ... followed by circlip – note correct circlip installation

30.10k Fit 2nd gear pinion ...

30.10l ... followed by thick thrust washer ...

30.10m ... needle roller bearing ...

30.10n ... which is retained by a circlip

30.10o Lubricate bearing before refitting outer race

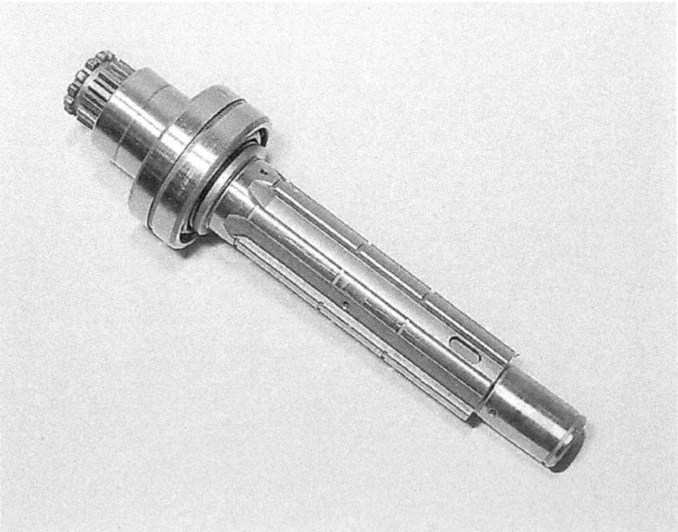

30.11a Take the bare output shaft ...

30.11b ... and refit 2nd gear pinion as shown

30.11c 2nd gear pinion is located by a splined thrust washer ...

30.11d ... and secured by a circlip

30.11e 6th gear pinion selector fork groove faces to right

30.11f Fit circlip in groove shown, followed by a splined thrust washer

30.11g 4th gear pinion is refitted as shown ...

30.11h ... and is followed by a splined thrust washer

30.11i Refit 3rd gear pinion as shown ...

30.11j ... followed by a splined thrust washer

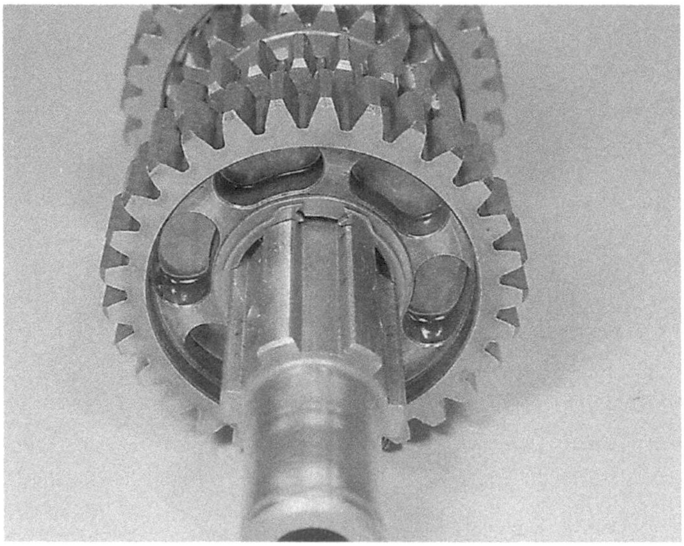
30.11k Fit a circlip as shown to retain gear pinions

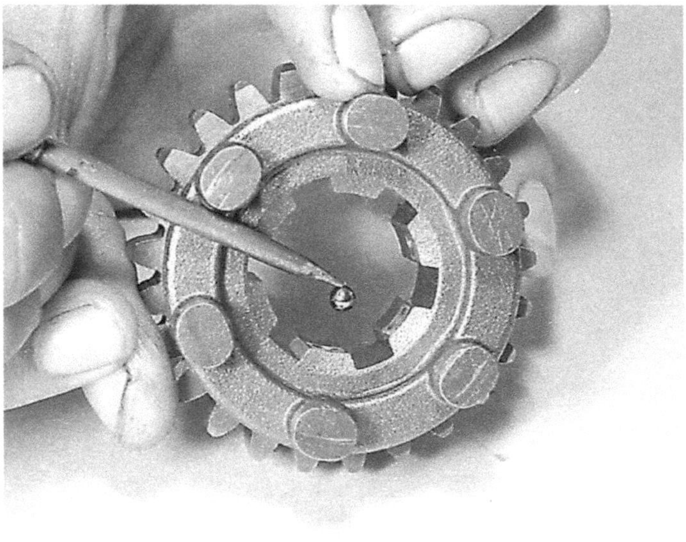
30.11l Insert neutral finder steel balls into drillings in 5th gear pinion ...

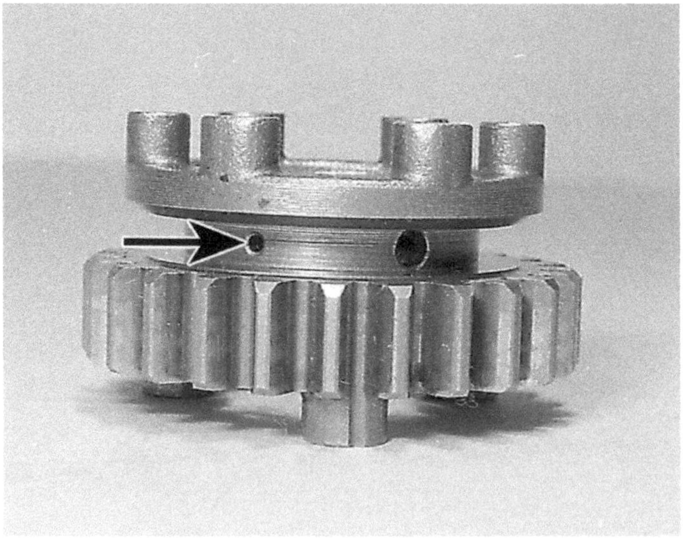
30.11m ... noting that only closed-off drillings (arrowed) should be used

30.11n Align balls with output shaft grooves (arrowed) on refitting 5th gear pinion

30.11o Refit 1st gear pinion as shown ...

30.11p ... followed by thick thrust washer

30.11q Needle roller bearing is retained ...

30.11r ... by a circlip

30.11s Lubricate bearing before refitting outer race

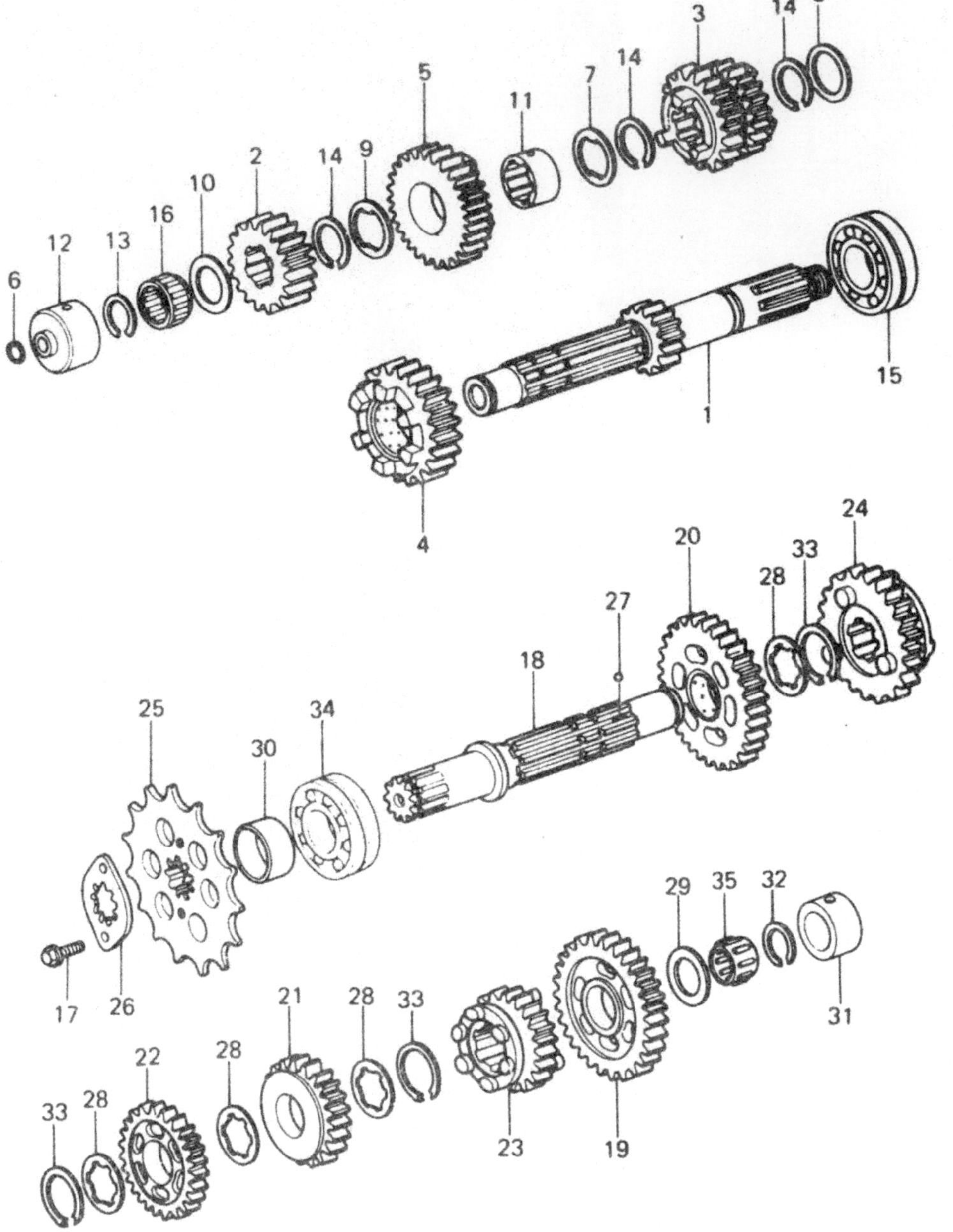

Fig. 1.15 Gearbox shafts

1 *Input shaft*
2 *Input shaft 2nd gear*
3 *Input shaft 3rd/4th gear*
4 *Input shaft 5th gear*
5 *Input shaft 6th gear*
6 *O-ring*
7 *Splined thrust washer*
8 *Plain thrust washer*
9 *Splined thrust washer*
10 *Plain washer*
11 *Bush*
12 *Needle bearing outer race*
13 *Circlip*
14 *Circlip – 3 off*
15 *Bearing*
16 *Needle bearing*
17 *Bolt – 2 off*
18 *Output shaft*
19 *Output shaft 1st gear*
20 *Output shaft 2nd gear*
21 *Output shaft 3rd gear*
22 *Output shaft 4th gear*
23 *Output shaft 5th gear*
24 *Output shaft 6th gear*
25 *Engine sprocket*
26 *Plate*
27 *Ball – 3 off*
28 *Splined thrust washer – 4 off*
29 *Plain washer*
30 *Sleeve*
31 *Needle bearing outer race*
32 *Circlip*
33 *Circlip – 3 off*
34 *Bearing*
35 *Needle bearing*

31 Examination and renovation: front gear case unit – shaft drive models only

1 The models featuring shaft final drive are equipped with a front gear case unit which replaces the final drive sprocket and gear selector mechanism cover of the chain drive machines. The front gear case forms a separate unit which can be detached from the crankcase proper.

2 The unit contains a pair of bevel gears through which the final drive is turned through 90° to meet the driveshaft. A cam-type shock absorber is also included, this being designed to absorb shock loadings which would otherwise be transmitted from the rear wheel.

3 The gear case unit is a robust assembly and should not normally require attention unless it has been set up incorrectly at some previous date, or one or more of the bearings or bevel gears has worn and requires renewal. Normal maintenance should be confined to careful cleaning of the unit and checking that the teeth are not wearing unevenly. If necessary, the manufacturer recommends that a checking compound is applied to the driven bevel gear teeth after cleaning and degreasing. If the gears are now rotated an impression of the tooth contact points will be visible. If all is well an elliptical contact patch will be evident at the centre of the teeth. If this is offset to either edge the need for adjustment is indicated. The checking compound is a specialist product and should be available from engineering companies which specialise in the reconditioning of car differential units.

4 The bearings are of the tapered roller type and are designed to operate under a carefully adjusted preload. If any discernible play is evident or if there appears to be excessive drag in the bearings, adjustment may be required.

5 Although dismantling and assembling the bevel gears and bearings is not beyond the capabilities of most owners, a number of specialist tools are required because it is vital that mesh depth, gear backlash and bearing preload are set with great precision. For this reason, it is recommended that in the event of a suspected fault the unit is taken to a Kawasaki dealer for checking and overhaul. The dealer will have the necessary tools and test equipment to carry out the work, plus the range of shims and preload collars which are essential during assembly.

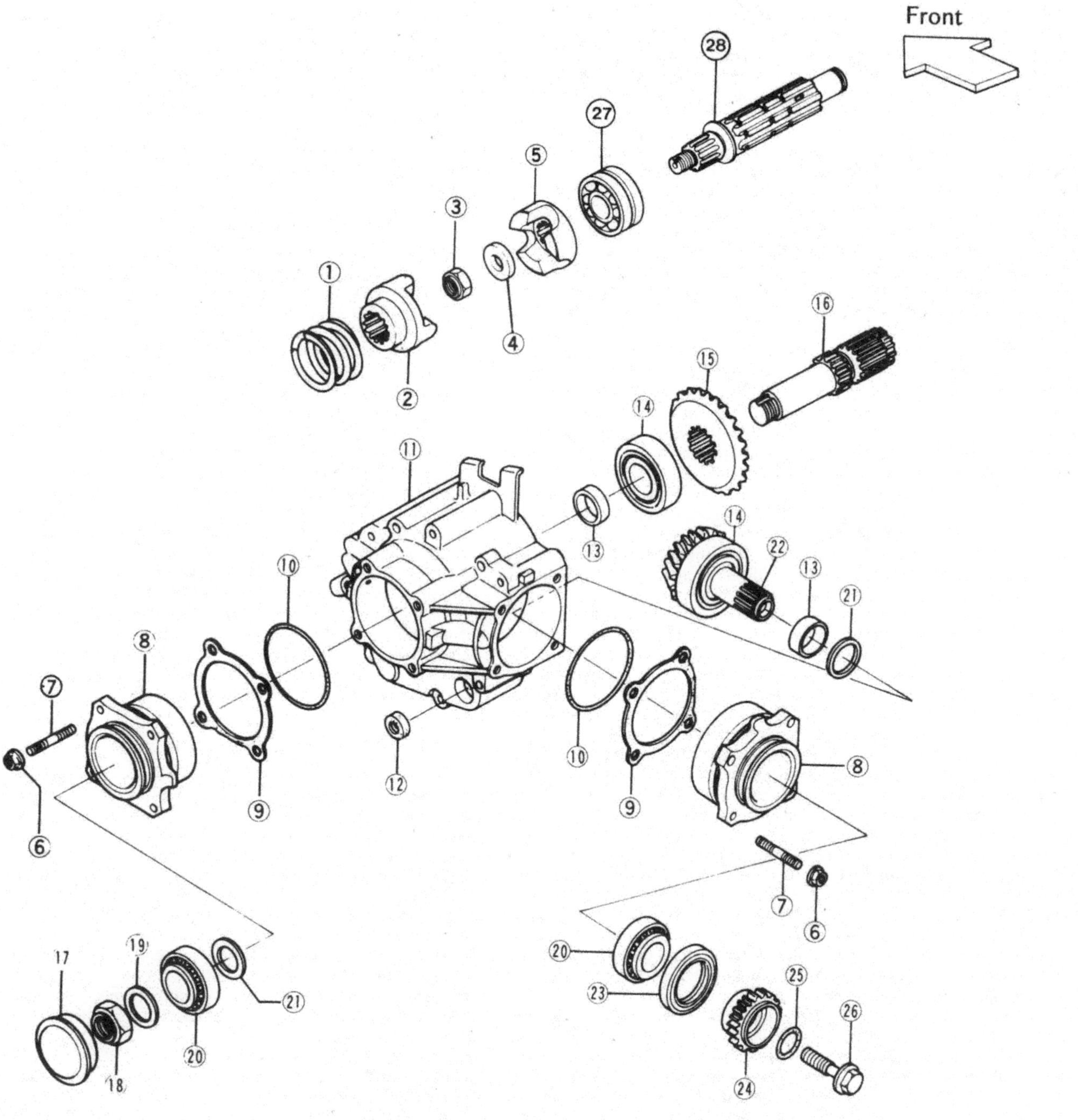

Fig. 1.16 Front gear case components – shaft drive models

1 *Damper spring*
2 *Cam follower*
3 *Nut*
4 *Washer*
5 *Damper cam*
6 *Nut – 4 off*
7 *Stud – 4 off*
8 *Bearing housing*
9 *Shim*
10 *O-ring*
11 *Case*
12 *Oil seal*
13 *Collar, preload adjustment*
14 *Taper roller bearing*
15 *Bevel gear drive*
16 *Bevel gear shaft*
17 *Cap*
18 *Nut*
19 *Washer*
20 *Taper roller bearing*
21 *Spacer, preload adjustment*
22 *Bevel gear, driven*
23 *Oil seal*
24 *Sliding joint*
25 *O-ring*
26 *Bolt*
27 *Bearing*
28 *Output shaft*

32 Curing oil leaks

1 Most oil leaks are caused by careless workmanship during dismantling and reassembly. The usual causes are the reuse of worn or damaged seals or gaskets, gasket surfaces damaged by levering apart or by being struck heavily, excessive use of jointing compound, retaining nuts, bolts and screws being over- or under-tightened and in the wrong sequence (where applicable) and insufficient care being taken to clean gasket surfaces on reassembly. All these can be avoided by following the instructions in this Manual, but some machines have

suffered from persistent leaks which appear to come from the cylinder head/cylinder block and/or the cylinder block/crankcase joints. These require special attention, if encountered, as described below.

2 The first task is to identify the source of the leak, which is normally caused by oil leaking past the threads of one of the cylinder holding-down studs. The oil works its way up into the stud hole under crankcase pressure and fills it up until it finds an escape route. Depending on whether the stud is in the centre or on the outside of the cylinder block, oil may escape at the base of the block, at the cylinder head/block joint area, or from around the cylinder head retaining nuts. Wash the whole engine/gearbox unit thoroughly, removing all traces of oil and dirt; a pressure washer or steam cleaner would be most suitable but should be used with care. Ride the machine for 100 – 200 miles examining frequently the suspect areas and noting immediately when and where oil appears. If a leak is particularly hard to spot, dust the area with French chalk or talcum powder to make any oil more obvious.

3 Although individual studs can be pinpointed in this way, and although the outer four studs can be removed if necessary without disturbing any other component, if oil is found to be seeping past any one of the studs it is recommended that all studs be treated as described to cure the problem completely. It is assumed that the leak is proved to be due to a leaking stud and not to be due to one of the other causes outlined in paragraph 1 of this Section.

4 Remove the cylinder head and block as described in Sections 5, 7 and 8 of this Chapter, then remove all twelve studs as described in Section 16. Taking care not to allow any solvent into the crankcase, thoroughly clean each stud hole with solvent or degreaser at least three times. Finally clean the holes with thinners, allow a few minutes for it to work, then dry the hole using compressed air.

5 Renew any studs that are damaged or worn, then clean all stud threads with a wire brush, followed by immersing them for a few minutes in thinners. Dry the threads using compressed air then refit each stud, applying Loctite 'Lock n' Seal' (blue) to its threads. This is a proprietary compound that is readily available but can be acquired under Kawasaki Part number K61079-001. Tighten securely all studs and reassemble the engine using new seals and gaskets.

6 If a persistent oil leak is positively identified as coming from the cylinder block/crankcase joint, the following action can be taken. Remove the cylinder head and block as described in Sections 5, 7 and 8 of this Chapter, then clean carefully the gasket surfaces of the block and crankcase, as described in Section 33. Be particularly careful to clean out the small groove at the base of each cylinder liner and finish off using acetone, thinners, or a similar non-oily solvent. Using applicator nozzle Part Number 92062-1052, apply a thin bead of liquid gasket '3-Bond 10' (Part Number 56019-120) to the groove around each liner, then leave it to cure for 24 hours at room temperature. Reassemble the engine using all new gaskets and seals. This treatment will prevent oil from working its way up between the cylinder liners and crankcase casting.

7 Depending on the model, there are one or two modified parts to assist sealing at the cylinder head/cylinder block joint. All Z400 and 500 models should be fitted with a modified cylinder head gasket whenever the head is refitted; the new type gasket is treated with a special liquid gasket. On all models described in this Manual, the oil feed nozzles that are inserted into recesses at each end of the cylinder block upper gasket surface have been modified; the new nozzles are longer to provide better support for the sealing O-rings and their tapered upper ends project noticeably above the surrounding gasket surface. Check with a good Kawasaki dealer if in doubt, taking the existing nozzles for comparison with new components to check whether they are of the modified type or not. These nozzles should be fitted to all engines at the earliest opportunity.

8 Finally, it must be stressed that all the above treatments apply only to the two areas mentioned, and only in the event of a persistent oil leak occurring at one or both of these areas. Oil leaks of this type occur only rarely. Remember also that these treatments are an addition to, and not a substitute for, the normal skill and care required in rebuilding any engine that is to remain oiltight.

Warning: *Solvents of the type described are harmful to health and extremely inflammable. Take great care to prevent any risk of fire while they are in use and protect the eyes and skin with suitable clothing; also make sure that the working area is well ventilated as the vapour is harmful if inhaled for prolonged periods. Take care to follow the solvent manufacturer's instructions and to heed any warnings; seek medical advice immediately if an accident should occur.*

33 Engine reassmbly: general

1 Before reassembly of the engine/gearbox box unit is commenced, the various component parts should be cleaned thoroughly and placed on a sheet of clean paper, close to the working area.

2 Make sure all traces of old gaskets have been removed and that the mating surfaces are clean and undamaged. Great care should be taken when removing old gasket compound not to damage the mating surface. Most gasket compounds can be softened using a suitable solvent such as methylated spirits, acetone or cellulose thinner. The type of solvent required will depend on the type of compound used. Gasket compound of the non-hardening type can be removed using a soft brass-wire brush of the type used for cleaning suede shoes. A considerable amount of scrubbing can take place without fear of harming the mating surfaces. Some difficulty may be encountered when attempting to remove gaskets of the self-vulcanising type, the use of which is becoming widespread, particularly as cylinder head and base gaskets. The gasket should be pared from the mating surface using a scalpel or a small chisel with a finely honed edge. Do not, however, resort to scraping with a sharp instrument unless necessary.

3 Gather together all the necessary tools and have available an oil can filled with clean engine oil. Make sure that all new gaskets and oil seals are to hand, also all replacement parts required. Nothing is more frustrating than having to stop in the middle of a reassembly sequence because a vital gasket or replacement has been overlooked. As a general rule each moving engine component should be lubricated thoroughly as it is fitted into position.

4 Make sure that the reassembly area is clean and that there is adequate working space. Refer to the torque and clearance setting wherever they are given. Many of the smaller bolts are easily sheared if overtightened. Always use the correct size screwdriver bit for the cross-head screws and never an ordinary screwdriver or punch. If the existing screws show evidence of maltreatment in the past, it is advisable to renew them as a complete set.

34 Engine reassembly: refitting the lower crankcase half components

1 If removed, refit the selector drum needle roller bearing and secondary shaft left-hand ball bearing. Heat the casting as described in Section 17 of this Chapter and drive the bearing into place using a tubular drift such as a socket spanner that bears only on the bearing outer race. The needle roller bearing must be driven home until it is flush with the surrounding crankcase, but the secondary shaft bearing must be fitted to a depth of 10.7 – 11.3 mm (0.42 – 0.44 in) from the left-hand end of the crankcase bearing boss, ie so that the bearing does not project beyond the inside of the crankcase, and so that the bearing cap fits correctly.

2 Having thoroughly cleaned all oilways, refit the hexagon-headed plugs to each end of the main oil gallery, renewing the sealing O-rings, and refit the three oil feed nozzles, noting that the small black nozzle must be staked to prevent loosening and that the two silver nozzles must be screwed into their housings so that they are sunk to a depth of 1 mm (0.04 in). Apply one or two drops of thread locking compound to the threads of all plugs or nozzles to prevent loosening, but be careful not to use an excessive amount or there is a risk of blocking an oilway.

3 Refit the selector pins to the selector drum left-hand end. Note that one pin is longer than the other five; this must be inserted into the drilling next to the hole drilled at the drum edge, out of line with the other six. See the accompanying illustration. The drum end plate has two recesses drilled in its right-hand face; it must be fitted so that the longer selector pin engages with the recess that is **not** marked by a circle stamped around it. If this is not correct, the neutral indicator lamp will not light at the correct time. Applying a few drops of thread-locking compound to its threads, refit and tighten the end plate counter-sunk retaining screw securely. Do not forget to refit the large circlip (if removed).

4 Fit the selector drum from the left, stopping halfway so that the larger selector fork can be refitted; the extended boss of this fork must face to the left, towards the selector pins. Push the drum fully home, insert the locating dowel in its recess and refit the detent cam,

ensuring that the small projection faces out to the right. Refit the retaining circlip.

5 Turn the casing over and fit the guide bolt in its hole near the drum right-hand end, taking care not to omit the tab washer. Tighten the bolt securely and bend up the locking tab. Moving to the left-hand end of the drum, insert the neutral detent plunger, spring and cap bolt, tightening the latter securely. Rotate the selector drum until the neutral position is found. The detent plunger should be seated in the small cutout located 180° from the cam location pin.

6 The large selector fork guide pin is inserted so that its longer end passes through the fork boss and into the **middle** selector drum track, leaving the split-pin hole at the top. Always fit a new split-pin, passing it through from left to right and spreading its ends securely. Trim any surplus from the pin ends, as shown in the accompanying photograph. The two remaining forks are identical; check that its circlip is securely refitted, then push the selector fork shaft through the crankcases from left to right, stopping to refit the two forks. These are fitted with their extended bosses facing to the left and with their guide pins engaging in the selector drum outside two tracks.

7 Ensuring that each is in its correct position, refit the five main bearing inserts; check that their locating tangs are correctly seated and that all are well lubricated.

34.2 Do not forget to refit oil feed nozzles (if removed) – install as described in text

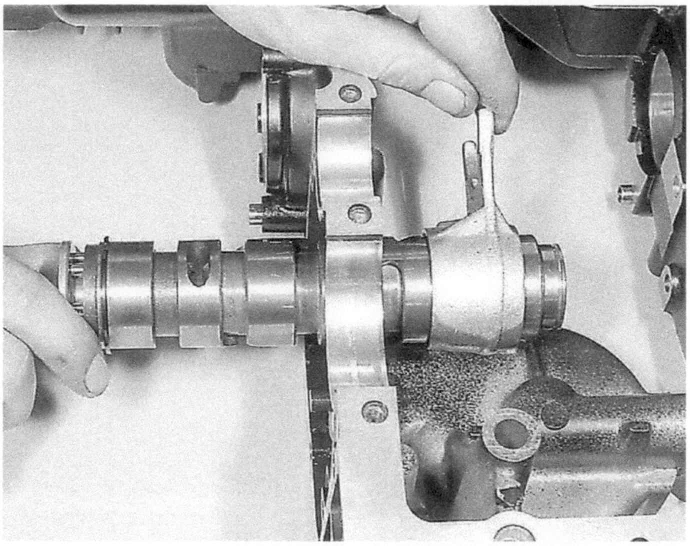
34.4a Do not forget to install larger selector fork when refitting selector drum

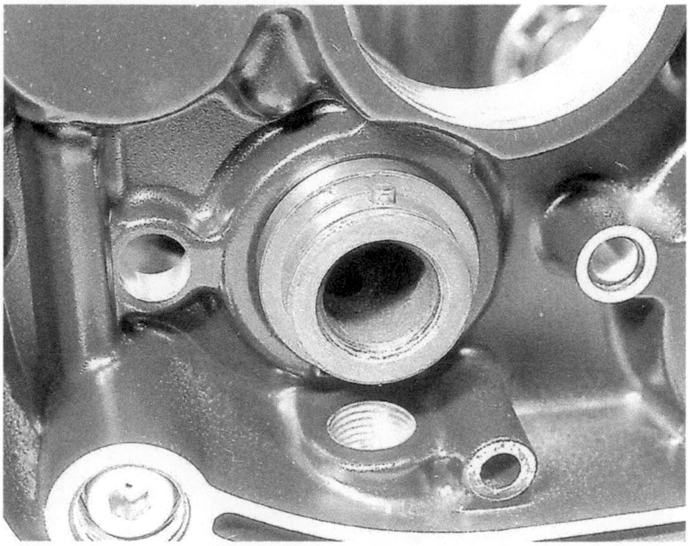
34.4b Insert locating dowel into drum right-hand end ...

34.4c ... and refit detent cam – note small projection

34.4d Detent cam is retained by large circlip

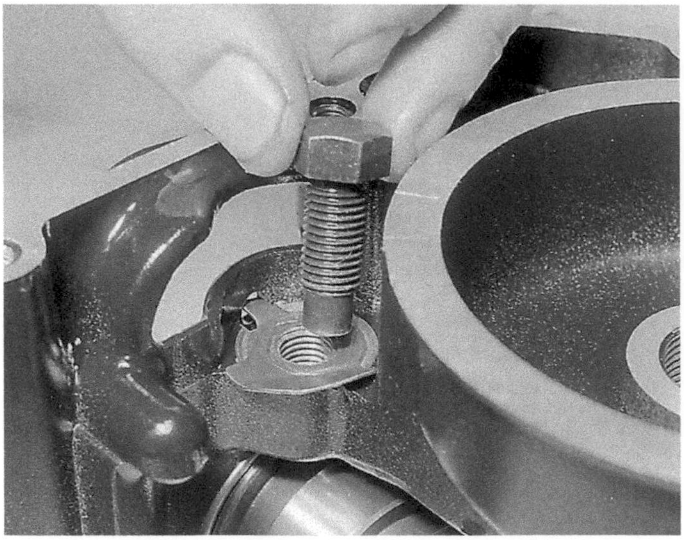
34.5a Refit guide bolt to locate selector drum – note projection on tab washer engaged in hole in crankcase wall

34.5b Secure guide bolt by bending up tab washer as shown

34.5c Neutral detent plunger assembly is fitted into crankcase ...

34.5d ... to engage with detent cam neutral position as shown

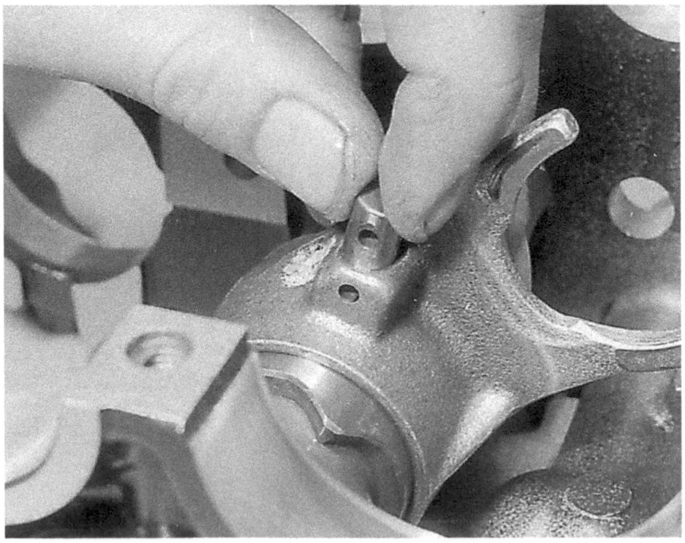
34.6a Ensure split-pin holes align when refitting selector fork guide pin

34.6b Refit split-pin as shown to secure guide pin

34.6c Outer two selector forks are refitted as shown

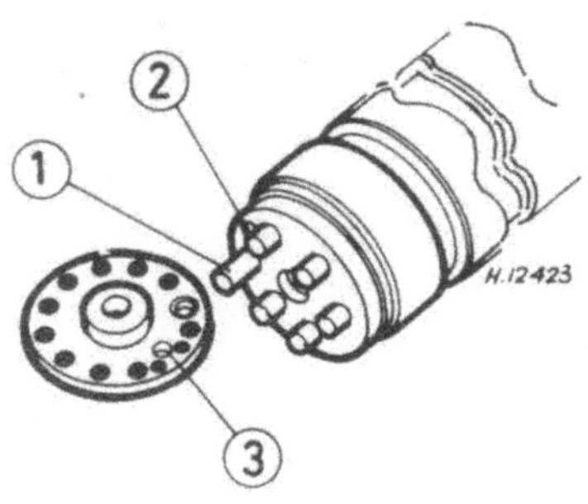

Fig. 1.17 Location of selector pins and selector drum end plate

1 Long pin
2 Hole
3 Non-circle marked hole

35 Engine reassembly: refitting the upper crankcase half components

1 Ensuring that all are in their correct positions, refit the big-end bearing inserts to the connecting rods and their caps, ensuring that the locating tang of each insert fits correctly into its recess. Fit each connecting rod on to its original crankpin only, then refit its cap. Ensure that the bearing surfaces are well lubricated and use an oil can to force some oil into each crankpin oilway. Use the marks made on dismantling or the weight group letter to ensure that the caps are aligned correctly with their respective rods and that the rods are refitted the same way round as before.

2 Refit the cap nuts, tightening them by hand only at first, then tighten them evenly to a torque setting of 2.4 kgf m (17 lbf ft). Check that the rods revolve smoothly and easily about their crankpins; some stiffness is inevitable if new inserts have been fitted, but this must not be excessive.

3 Refit the camshaft and primary drive chains to their respective sprockets and the oil seals to each end of the crankshaft. Note that high melting point grease should be applied to the sealing lips of the seals and that they are designed to operate in the normal direction of crankshaft rotation, arrows to show this are moulded on the outside of each seal. The seal with the arrow pointing clockwise fits on the crankshaft right-hand end, therefore the seal with the anti-clockwise arrow fits on its left-hand end.

4 Ensuring that all are in their correct positions, refit the five main bearing inserts; check that their locating tangs are correctly seated and that all are well lubricated. Use an oil can to prime the oilways of both crankcase halves and of the crankshaft. Ensuring that the upper crankcase half is fully supported, lower the crankshaft into position. The cam drive chain and connecting rods must pass through their respective apertures and the ridges moulded on the circumference of each seal must locate correctly in their grooves in the crankcase. Check that the crankshaft is seated fully and revolving smoothly and easily, then pull the primary drive chain to the rear so that it does not bunch on its sprocket.

5 Refit the two dowel pins in the crankcase gasket surface, if removed, also the half ring retainer and dowel pin which locate the bearing outer races at each end of both gearbox shafts. Do not forget to refit the rubber plug that blanks off the oilway at the input shaft right-hand bearing recess.

6 Their ball bearings can be refitted to the gearbox shafts using a hammer and a long tubular drift that bears only on the bearing inner race. Tap the bearings home until their inner races bear firmly against the locating shoulder or integral gear pinion, as applicable. Refit the spacer over the output shaft left-hand end (chain drive models only) noting that it is identical to the clutch bearing sleeve but has no oil feed drilling.

7 Refit the gearbox input and output shafts ensuring that their ball bearings locate correctly on the half-ring retainers and that their needle bearing outer races locate correctly on the dowels so that the bearings seat fully in their recesses. Oil all gearbox components and check that the shafts rotate easily and smoothly.

8 It is much easier to refit the starter drive idler gear at this stage. Place the gear in position with the smaller pinion on the left-hand side, refit the shaft and secure it with its circlip.

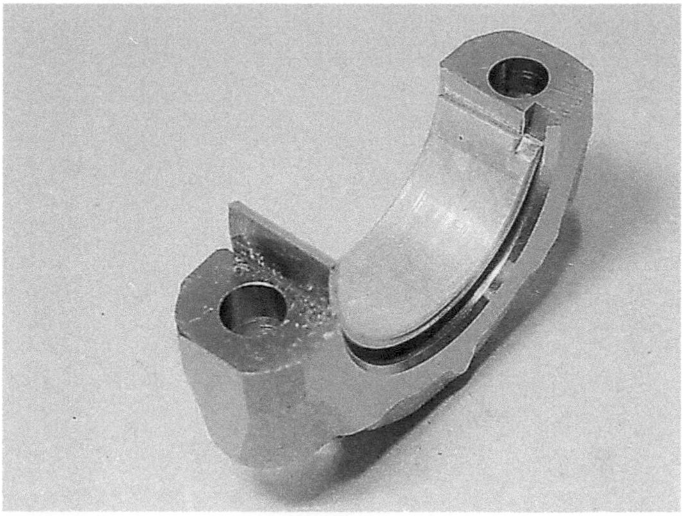

35.1a Tang of each bearing insert must locate in recess provided in bearing cap

35.1b Tighten cap nuts to specified torque setting – use weight group marking to ensure that bearing cap is refitted the original way round

35.4 Ensure all components are correctly installed when refitting crankshaft

35.5a All four gearbox bearings are located either by a dowel pin ...

35.5b ... or by a half-ring retainer – do not omit to refit

35.5c Note rubber ball-shaped plug in oilway of input shaft right-hand bearing – do not omit

35.7 Ensure bearings locate correctly when refitting gearbox shafts

35.8 Refit starter idler gear and secure shaft as shown

36 Engine reassembly: joining the crankcase halves

1 Check that all components are refitted and well lubricated and that they are revolving or sliding smoothly and easily. Use a rag dampened with solvent to wipe over the gasket surfaces removing all traces of oil, then apply a thin film of jointing compound **only** to those areas indicated in the accompanying illustration. Take great care to apply only a thin film and to leave a narrow margin around any oilways so that there is no risk of surplus compound blocking an oilway. Kawasaki recommend the use of a liquid gasket and market their own through their dealers, but one of the many proprietary RTV (Room Temperature Vulcanising) compounds now available would be equally suitable, provided care is taken to follow the manufacturer's instructions.

2 Carefully lower the bottom crankcase half into position, being careful not to dislodge any of the main bearing inserts. Guide the claw ends of the selector forks into the grooves of their respective pinions; the fork on the selector drum engages with the input shaft 3rd/4th gear pinion while of the other two the left-hand one engages with the output shaft 6th gear pinion and the right-hand fork engages with the output shaft 5th gear pinion. With care, this can be done by one person via the sump aperture, but the aid of an assistant may be helpful. Check that the ridges moulded on the circumference of the crankshaft oil seals engage correctly with the grooves in the lower crankcase half.

3 Check that the lower crankcase half seats fully without force, and that all components are free to rotate as necessary. Refit the seven 6 mm bolts and the ten 8 mm bolts, using the cardboard template to ensure that each bolt is refitted in its correct position. Tighten all bolts finger-tight.

4 The tightening order for the 8 mm bolts is indicated by numbers cast on the crankcase next to each bolt. If these are illegible, the sequence is repeated in the accompanying illustration. Working in this sequence, tighten the bolts to an initial setting of 1.5 kgf m (11 lbf ft), then to a final torque setting of 2.5 kgf m (18 lbf ft). Working across the front from the centre outwards, followed by the bolt at the rear, tighten the 6 mm bolts to a torque setting of 1.0 kgf m (7 lbf ft). Turn the unit over, secure the cam chain so that it is engaged correctly on its sprocket and cannot drop into the crankcase, then refit and finger-tighten the upper thirteen 6 mm bolts. In a diagonal sequence from the centre outwards, tighten these to a torque setting of 1.0 kgf m (7 lbf ft).

5 Check that the crankshaft rotates smoothly and easily, also the gearbox shafts. It is impossible to select any gears other than 1st and neutral unless the output shaft can be spun fast enough to disengage the neutral finding mechanism. If there are any signs of undue stiffness, of jerkiness or of any other problems, the fault must be located and rectified before work can proceed.

6 If all is well pack the crankcase mouths and cam chain tunnel with clean rag to prevent the entry of dirt, and carefully wipe away any surplus jointing compound. Renewing both sealing O-rings around the cover joint and bolt head, refit the breather cover to the crankcase top surface, aligning its locating tag with the crankcase boss as shown in the accompanying photograph and tightening the bolt to a torque setting of 0.6 kgf m (4 lbf ft).

36.1 Ensure all components are correctly refitted and well lubricated – note two locating dowels (arrowed)

36.2 Ensure crankshaft oil seals and selector forks locate correctly when refitting crankcase bottom half

36.4 Tighten all crankcase fastening bolts to specified torque settings and in sequence shown

36.6 Breather cover locating tag must align as shown

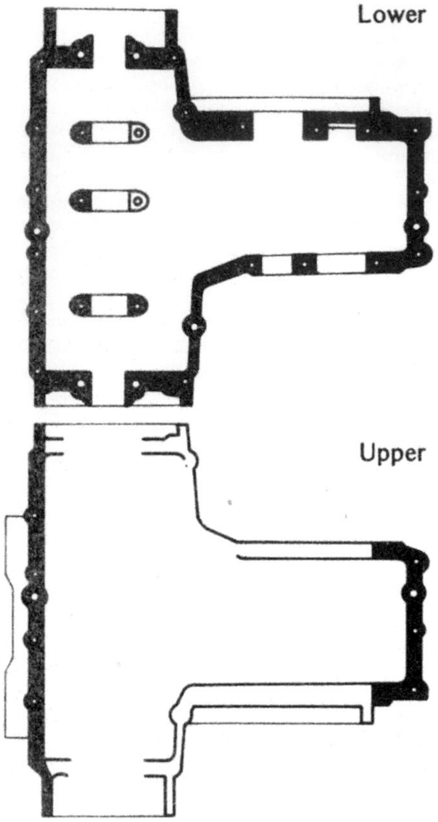

Fig. 1.18 Apply sealant only to the areas shown by the shading

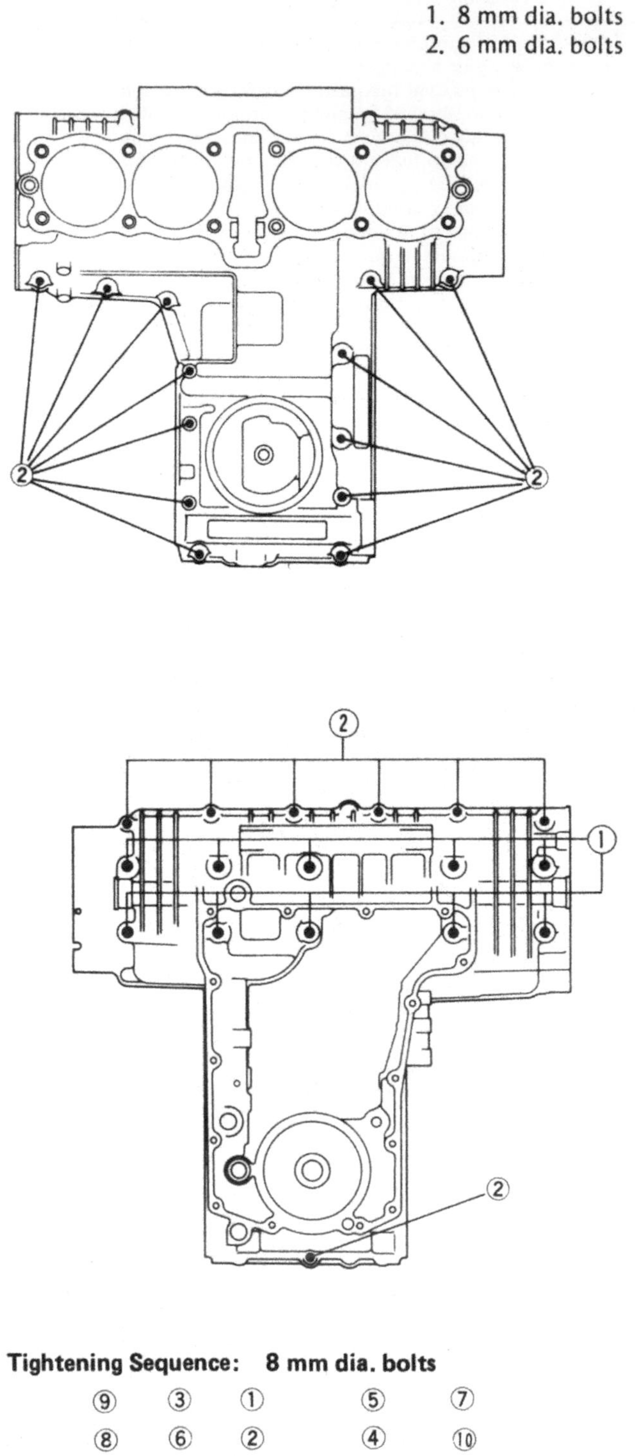

Fig. 1.19 Location and tightening sequence of crankcase fastening bolts

37 Engine reassembly: refitting the external gearchange components

1 Check that the circlip is correctly located in its groove in the end of the selector fork shaft. Offer up the gearchange shaft assembly, having first checked that the locating post for the return spring is secure. If necessary, refit the post using a thread locking compound and tightening it to the correct torque setting. As the assembly is fitted, ensure that the claws are placed on each side of the selector drum end, and make sure that the return spring engages on the locating pin. It is worthwhile temporarily refitting the gearchange pedal to check that gear selection is positive, as far as this is possible.
2 On shaft drive models the front gear case cannot be refitted until the engine/gearbox unit is back in the frame.
3 On all chain drive models, place a new O-ring (where fitted) over the boss on the input shaft end and place a new gasket in position over the two dowel pins. Grease carefully the gearchange shaft and output shaft and offer up the cover. Ease it over the shaft ends to prevent damage to the oil seals, then slide it into place and tighten securely the retaining screws, noting that one secures a wiring clamp. Fit the chainguard and tighten securely its three bolts, using thread-locking compound on their threads. If the engine/gearbox unit is in the frame the clutch push rod, where applicable, can be oiled and refitted in the input shaft; if the engine is still out of the frame plug the shaft end with grease to prevent the entry of dirt.

38 Engine reassembly: refitting the secondary shaft and starter clutch

1 If not already done, refit the starter idler gear as described in Section 35 of this Chapter, and the starter motor, as described in Chapter 6. Tap the secondary shaft right-hand bearing on to the shaft using a hammer and a tubular drift that bears only on the bearing inner race. Tap the bearing down until it butts against the shaft shoulder, then refit the thick spacer and the secondary shaft gear. The gear is fitted with its two threaded holes facing outwards and is tapped into place as described for the bearing. Refit the retaining circlip as soon as its groove is fully exposed.
2 Smear a little oil over the shock absorber rubbers, place them on

the centre and refit the secondary sprocket, ensuring that its vanes engage correctly between the rubbers. Refit the retaining circlip. If it was removed refit the starter clutch body to the shock absorber; apply thread locking compound to the Allen screw threads and tighten them to a torque setting of 3.5 kgf m (25 lbf ft). Refit the springs to the plungers, insert them into their recesses in the body and hold them with a small screwdriver while the rollers are refitted. Place the first thrust washer in the clutch body and refit the starter driven gear, rotating it to allow the rollers to engage with its centre boss. Invert the engine/gearbox unit, fit the assembly into the crankcase and engage the primary drive chain around the sprocket.

3 Place the second thrust washer over the shaft left-hand end then fit the shaft from right to left, aligning its splines with those of the shock absorber. Tap gently the shaft right-hand bearing until it seats against the shoulder machined in its recess; this can be seen via the sump aperture. It will of course be necessary to tap the shaft itself as the bearing is obscured by the gear. If severe resistance is encountered, cease tapping and see below. Once the shaft is fully in place, tap the sleeve into the shaft left-hand bearing.

4 While the above method invariably works well it can put some strain on the bearing if this is a tight fit in the crankcase. If this is the case it will be necessary to withdraw the shaft and to remove from it the circlip, gear, spacer and right-hand bearing as described in Section 13 of this Chapter. The shaft is then refitted as described above, the sleeve fitted into the left-hand bearing and the right-hand bearing tapped into place over the shaft. It will be necessary to find or to make up a thick-walled tubular drift which bears on both inner and outer races at the same time to avoid bearing damage; the correct tool is supplied under Kawasaki Part Number 57001-297 and should be used if available. One possible alternative is to use a $\frac{3}{4}$ inch drive socket spanner that has a suitable outside diameter; this can be reversed so that the face with the square drive hole fits over the shaft and against the bearing, but this is largely a matter for the individual owner's ingenuity. Once the bearing is fitted, the thick spacer, the gear and the circlip may be refitted as described in paragraph 1 above, but care must be taken to support the shaft left-hand end so that the complete shaft assembly is not displaced to the left as the gear is tapped into place; this can be checked by measuring the depth of the shaft left-hand bearing in its recess. See Section 34. Kawasaki avoid this possibility by the use of a special gear 'pusher', Part Number 57001-319.

5 With the secondary shaft correctly fitted and both bearings correctly seated, check that the sleeve is fully in place in the left-hand bearing, then refit the shaft retaining nut. Lock the shaft by the method used on dismantling and tighten the nut to a torque setting of 6.0 kgf m (43 lbf ft).

6 Renewing its sealing O-ring and applying a smear of grease to ease the task, insert the bearing cap into its crankcase recess then refit and tighten securely its two retaining screws, noting that the upper one carries a wiring clamp.

37.1 Arrange gear selector external mechanism as shown

37.3 Do not forget wiring clamp when refitting cover retaining screws

38.1a Insert starter motor boss into crankcase recess ...

38.1b ... and tighten the mounting bolts securely

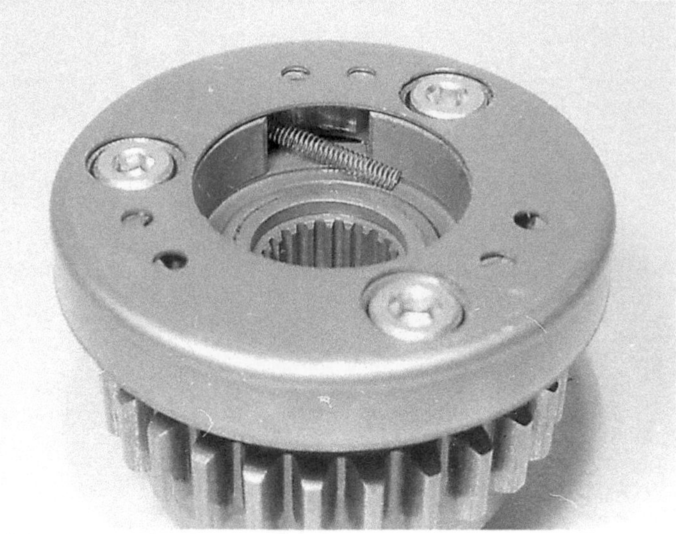

38.2a Insert springs into clutch body ...

38.2b ... followed by plungers

38.2c Press plungers into recess to permit refitting of starter clutch rollers

38.2d Insert starter clutch/shock absorber assembly into crankcase and engage on primary chain

38.3a Install secondary shaft – note second thrust washer

38.3b Tap sleeve into secondary shaft left-hand bearing – note correct bearing location

38.5 Crankshaft must be locked to tighten secondary shaft nut

38.6 Renew O-ring before refitting secondary shaft bearing cap

39 Engine reassembly: refitting the alternator

1 Wipe the crankshaft taper with a rag soaked in solvent to remove all traces of oil or grease, then refit the rotor and its retaining bolt. Lock the rotor by the method used on dismantling and tighten the bolt to a torque setting of 7.0 kgf m (50.5 lbf ft).
2 To refit the stator, press the coil lead wire sealing grommet into its recess in the cover, then gently place the stator in the cover so that its wires are not bent or trapped. With the stator aligned partially by its wires, use the mounting screw holes to position it exactly. Applying thread locking compound to their threads, refit the three Allen screws and tighten them to a torque setting of 1.0 kgf m (7 lbf ft). Refit the wiring guide plate and apply thread locking compound to their threads before refitting and tightening securely the two retaining screws.
3 Refit the two locating dowels and place a new gasket over them, then refit the cover. Refit and tighten securely the retaining screws.

39.1 Note method used to lock crankshaft when tightening rotor retaining bolt

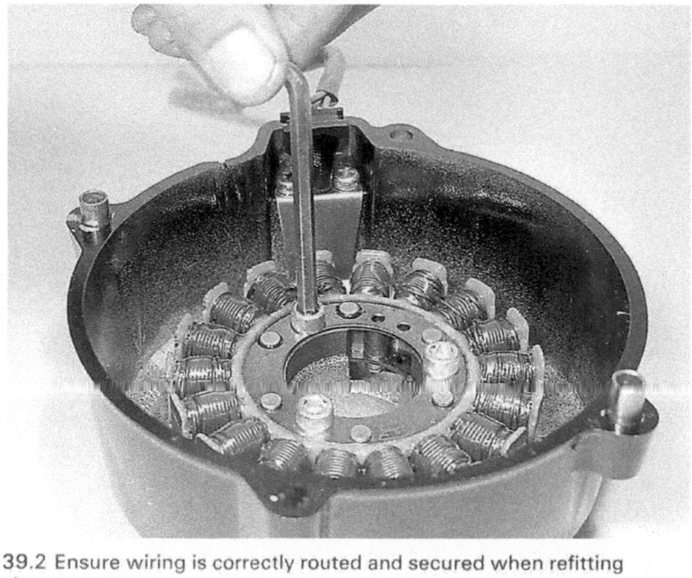
39.2 Ensure wiring is correctly routed and secured when refitting alternator stator

39.3 Always use new gasket when refitting alternator cover – note two locating dowels (arrowed)

40 Engine reassembly: refitting the oil pump and sump

1 Prime the pump by filling it with oil, then position it against the crankcase and push through the crankcase wall from the outside the two long locating dowels to locate it. Check that the pump driven gear meshes correctly with the drive gear on the secondary shaft, then refit and tighten by hand only the retaining bolt.
2 Place the secondary shaft right-hand bearing retaining plate in position, then refit the two countersunk retaining screws. Tighten these securely using an impact driver and stake them against the retaining plate. Tighten securely the pump mounting bolt. Check that the filter screen is clean and refitted securely to the pump body.
3 Place three new O-rings in their recesses around the oilways using a smear of grease to stick them in place; two in the crankcase, one in the oil pump. Note that the flat side of the O-rings must fit in the crankcase or pump recess, leaving the triangular edge facing the sump. Fit a new sump gasket, also using grease to retain it. Place a new O-ring in the groove around the filter chamber and refit the sump. Tighten its retaining bolts in a diagonal sequence from the centre outwards to a torque setting of 1.0 kgf m (7 lbf ft).
4 Refit the oil filter assembly, using a new filter as described in Routine Maintenance and tightening its mounting bolt to a torque setting of 2.0 kgf m (14.5 lbf ft). Refit also the drain plug, renewing its sealing washer if necessary, and tightening it to a torque setting of 3.8 kgf m (27.5 lbf ft).

40.1a Refit two oil pump locating dowels (arrowed) ...

40.1b ... and place oil pump in position

40.2 Oil pump mounting screws must be staked to prevent slackening

40.3a Whenever sump is refitted, always renew gasket, three smaller O-rings (arrowed) ...

40.3b ... and large O-ring around filter chamber

41 Engine reassembly: refitting the clutch

1 Place the special thrust washer over the end of the gearbox input shaft, noting that its chamfered face should be installed towards the bearing. Fit the bearing sleeve and the caged needle roller bearing next, followed by the clutch outer drum and the plain thrust washer. Note that the bearing sleeve can be mistaken for the output shaft sleeve quite easily. Check prior to installation that the component fitted as the bearing sleeve contains an oil hole. Slide the clutch centre into place, followed by the large washer. This is normally of the Belville type, being conical in section. The word 'OUTSIDE' is stamped on its outer face, and it should be fitted accordingly. On some models, however, a plain washer has been fitted as standard, which may be encountered. Fit a **new** clutch centre nut, tightening it to 13.5 kgf m (97.5 lbf ft), holding the clutch centre in the same manner as was used during dismantling.

2 If new clutch plates are to be fitted, apply a coating of oil to their surfaces to prevent seizure. Fit the clutch plates alternately, starting and finishing with a friction plate. On ZX550 A1, A1L and all shaft drive models insert the pull rod into the thrust bearing and press the latter into the rear face of the clutch pressure plate so that all three can be refitted as a single unit. On all other models, insert the $\frac{3}{8}$ in steel ball into the input shaft end, followed by the mushroom-headed pushrod, having applied a coating of molybdenum disulphide grease to both items. Refit the clutch pressure plate, springs, washers and bolts, tightening the latter evenly and progressively to 0.9 kgf m (6.5 lbf ft) of torque.

3 Ensure that the mating surfaces of the clutch cover and crankcase are clean, and that the two locating dowels are in position. Fit a new gasket and offer up the clutch cover. Fit the retaining screws and tighten them securely, noting that if any wiring clamps were fitted they should be refitted and aligned correctly before the final tightening.

42 Engine reassembly: refitting the contact breaker/ignition pickup assembly

1 Lightly oil the ATU (where fitted) with machine oil and check that it is working correctly. If it was removed, refit the locating dowel pin in its crankshaft drilling, then refit the ATU or ignition rotor (as applicable), aligning the notch in its rear face with the pin. Refit the engine turning hexagon, aligning its cutouts with the notches on the ATU (models fitted with ATUs only). Refit the retaining bolt, tightening it to a torque setting of 2.5 kgf m (18 lbf ft), then check again that the ATU (where applicable) is working correctly.

2 Refit the ignition backplate, rotating it so that the marks made on dismantling align exactly, then refit and tighten the retaining screws. Route the wiring lead through any clamps or ties provided for it but do not refit the cover yet as the ignition timing must be checked first.

41.1a Chamfered face of special thrust washer must face inwards, as shown

41.1b Ensure correct sleeve is fitted to input shaft – note oil hole

41.1c Lubricate needle roller bearing before refitting

41.1d Refit clutch outer drum ...

41.1e ... followed by plain thrust washer ...

41.1f ... and clutch centre

41.1g Either plain or Belville washer may be fitted – see text

41.1h Holding tool is essential when rebuilding clutch – always use a new centre nut

41.2a Starting with a friction plate ...

41.2b ... fit all clutch plates alternately

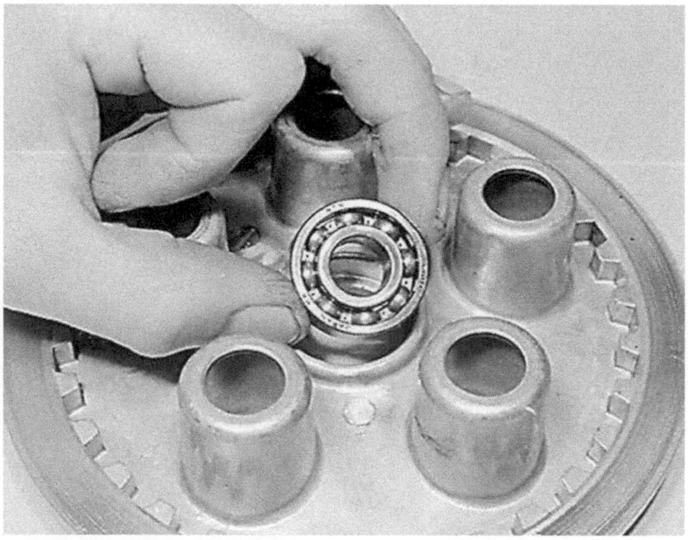

41.2c ZX550 A1 and shaft drive models – lubricate and insert thrust bearing into pressure plate ...

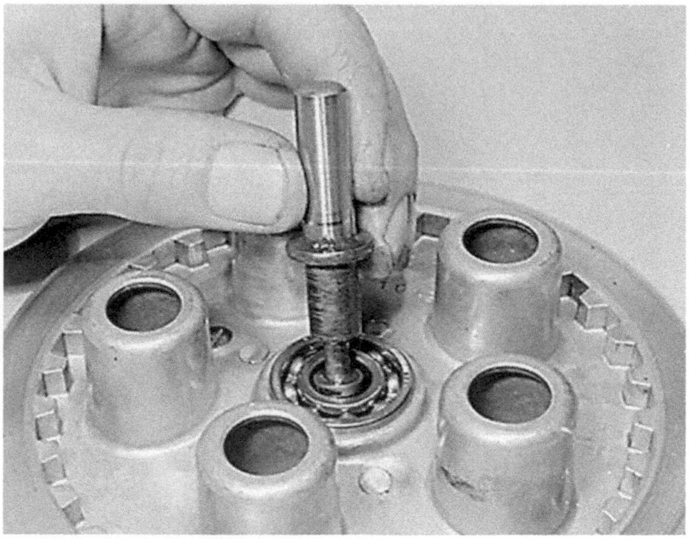

41.2d ... followed by pull rod

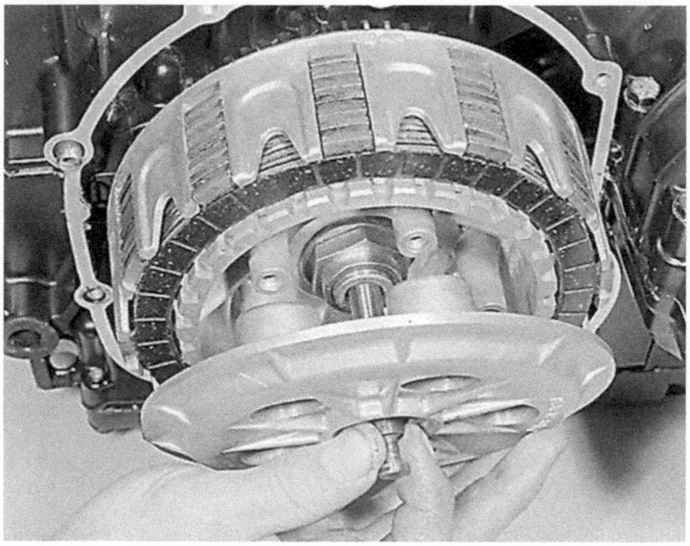

41.2e Do not omit clutch release components when refitting pressure plate

41.2f Refit clutch springs ...

41.2g ... and tighten spring bolts securely

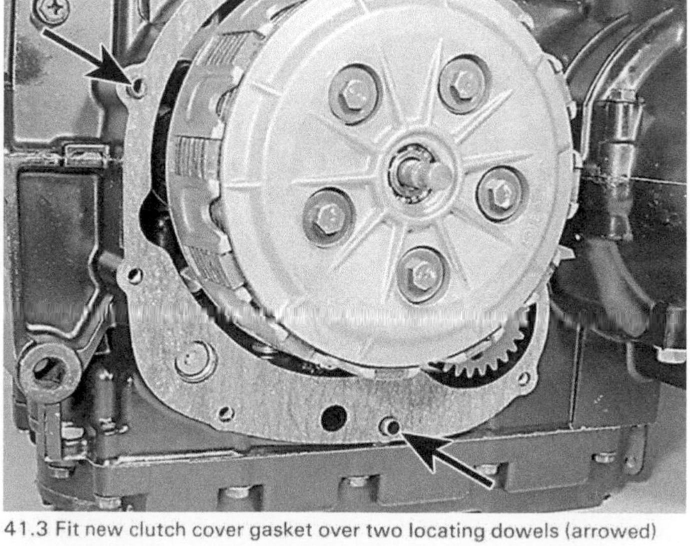

41.3 Fit new clutch cover gasket over two locating dowels (arrowed)

42.1a Ensure notch in rear face of rotor (or ATU) aligns with dowel pin in crankshaft

42.1b Hold rotor as shown to tighten retaining bolt

42.2 Route ignition wiring through all clamps provided

43 Engine reassembly: refitting the camchain tensioner components, the pistons, the cylinder block and the cylinder head

1 Reassemble the tensioner blade assembly and fit it to the crankcase, then apply thread locking compound to the threads, refit and tighten securely the two Allen screws. The front guide blade is inserted into the cylinder block cam chain tunnel with its longer end protruding downwards and is twisted into place. Check that it is securely fixed.

2 Refer to Section 32 of this Chapter for information on curing persistent oil leaks.

3 Fit the oil feed nozzles into their recesses at each end of the crankcase gasket surface, then fit new O-rings around each nozzle and around the two holding down studs on the right-hand side of the cam chain tunnel. Check that both cylinder block and crankcase gasket surfaces are clean and unmarked and wipe them over with a rag moistened in solvent to remove all traces of oil or grease. Fit a new cylinder base gasket over the studs so that the notch cut out of one end is on the left-hand side. Carefully press the gasket on to the crankcase noting that it fits closely around the outer four studs. No jointing compound should be used. Rotate the crankshaft to bring all connecting rod small-ends to approximately the same height and check that the cam chain is correctly engaged on its sprocket. The crankcase mouths should still be packed with clean rag.

4 Refit the rings to the pistons using the method employed on removal and positioning them as follows. One-piece oil scraper rings are fitted with the surface marked with the letter 'T' facing upwards and with the gap to the front of the piston. Three-piece oil scraper rings are fitted expander first, so that its ends butt together at the front of the piston, then the side rails which are fitted with their gaps at least 30° on each side of the expander ends. The side rails when new can be fitted either way up, but if the originals are being reused, they must be fitted the original way up, this being revealed by wear marks. Use the accompanying illustration to identify the compression rings, which have a different cross-section. The second compression ring will have one surface marked near the end gap with the letter 'T'; this surface must face upwards and the gap must be to the rear of the piston. If the top ring has one surface similarly marked, this surface must face upwards; if not it can be refitted either way round, with its gap to the front of the piston.

5 All circlips that were disturbed on dismantling must be renewed; never reuse them. Check that each piston has one circlip fitted and insert the gudgeon pin from the opposite side; if it is a tight fit, the piston should be warmed as described in Section 8 of this Chapter. If the original pistons are being refitted, use the marks made on dismantling to ensure that each is refitted in its original bore, and note that arrow marks cast in the piston crowns show the front face of each piston. Lubricate the gudgeon pins, piston bosses and small-end eyes, lower each piston in turn over its respective connecting rod and push the gudgeon pin through both piston bosses and the small-end eye. Take great care not to cause any damage if a hammer and drift are required. Secure each gudgeon pin with its second, new, circlip, ensuring that the circlip is correctly seated in its groove. Make a final check that all rings, pistons, and circlips are correctly and securely refitted, then smear a small quantity of molybdenum disulphide engine assembly grease over the skirt of each piston and over the surface of all cylinder bores. Failing this, a copious amount of clean engine oil should provide sufficient lubrication as the engine is re-started. It is advisable to enlist the aid of an assistant during block refitting and it helps to support each piston skirt on a clean block of wood placed across the crankcase mouth.

6 The cylinder bores have a generous lead in for the pistons at the bottom, and although it is an advantage on an engine such as this to use the special Kawasaki ring compressor, in the absence of this, it is possible to gently lead the pistons into the bores, working across from one side and gently tapping down the block. Great care has to be taken NOT to put too much pressure on the fitted piston rings. When the pistons have finally engaged remove the rag padding and the wooden blocks, if used, from the crankcase mouths and lower the cylinder

block still further until it seats firmly on the base gasket. Refit the three cap nuts to retain it.

7 Check that the block is seated correctly and that the crankshaft can be rotated smoothly. Hold the block down during this check, and hold the camchain taut to prevent it from 'bunching' around the crankshaft sprocket. Check that the tensioner blade is free to move.

8 Checking that they are of the modified type (see Section 32), insert the two oil feed nozzles, tapered ends upwards, into their recesses at each end of the cylinder block upper gasket surface and fit a new O-ring around each. Refit the two locating dowels around the front outer holding-down studs. Checking that it is of the correct, modified, type (Z400, Z500 models only), fit a new cylinder head gasket so that the 'UP' mark stamped on one surface faces upwards and is at the front of Number 1 cylinder bore. No jointing compound is necessary.

9 Feeding the camchain through its tunnel, lower the cylinder head over the studs and down on to the block. Refit, and tighten by hand only at first, all twelve nuts and five bolts. Working in the order shown in the accompanying illustration, tighten the twelve nuts first to a torque setting of 1.5 kgf m (11 lbf ft), then to their final setting of 2.3 kgf m (16 lbf ft). Tighten the five bolts and three cap nuts evenly to a torque setting of 1.2 kgf m (9 lbf ft).

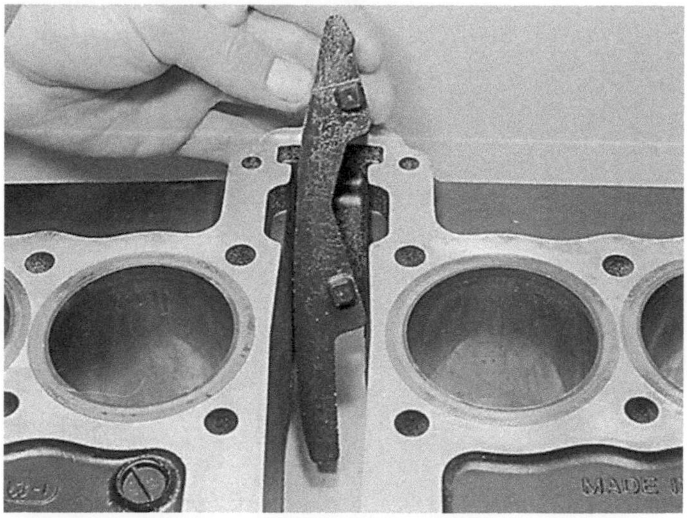

43.1 Refit cam chain front guide blade to cylinder block as shown

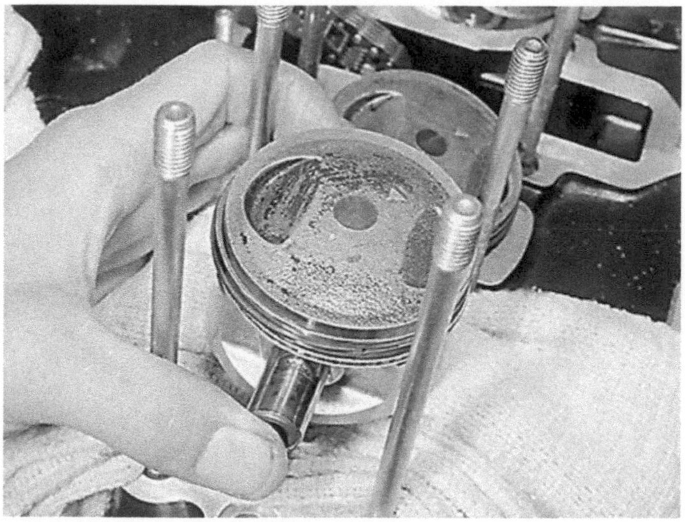

43.3a Arrow marks cast on piston crowns must face forwards on reassembly – note rag packing crankcase mouth

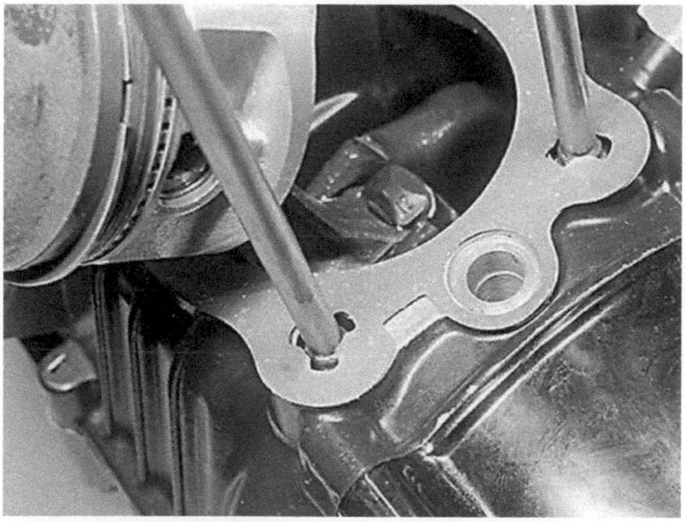

43.3b Cut-out in cylinder base gasket must be on left-hand side, as shown

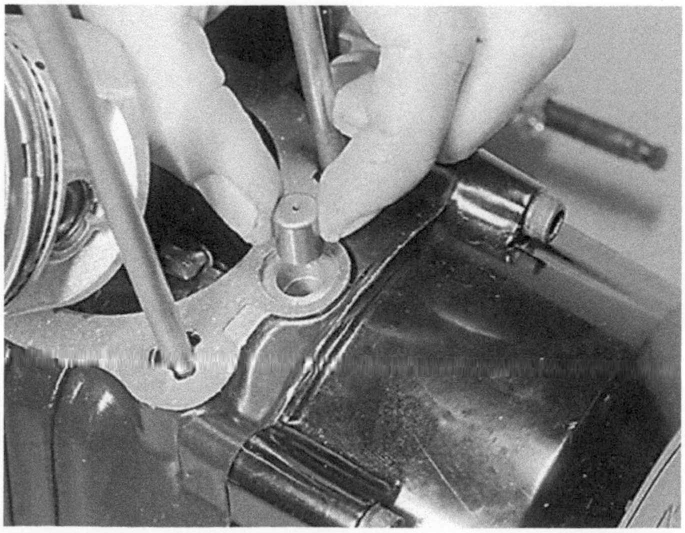

43.3c Refit oil feed nozzles as shown ...

43.3d .. and fit a new O-ring around each one

43.6 Lubricate cylinder bores and pistons before refitting

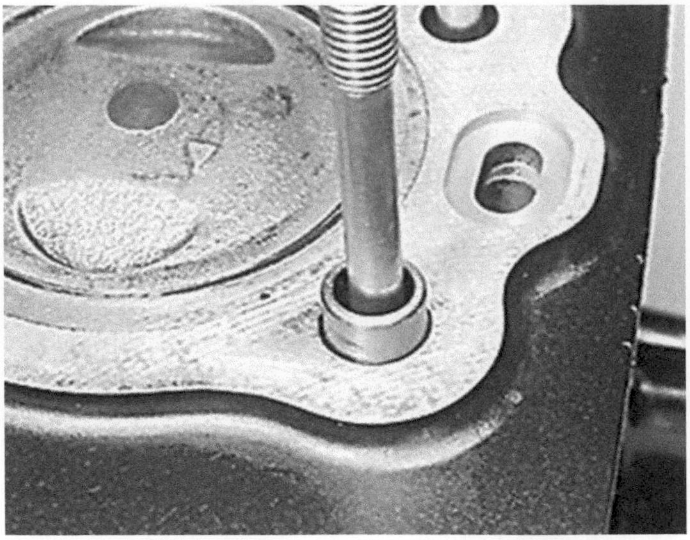

43.8a Note locating dowel around each front outer stud – do not omit

43.8b Always use new O-rings around oil feed nozzles

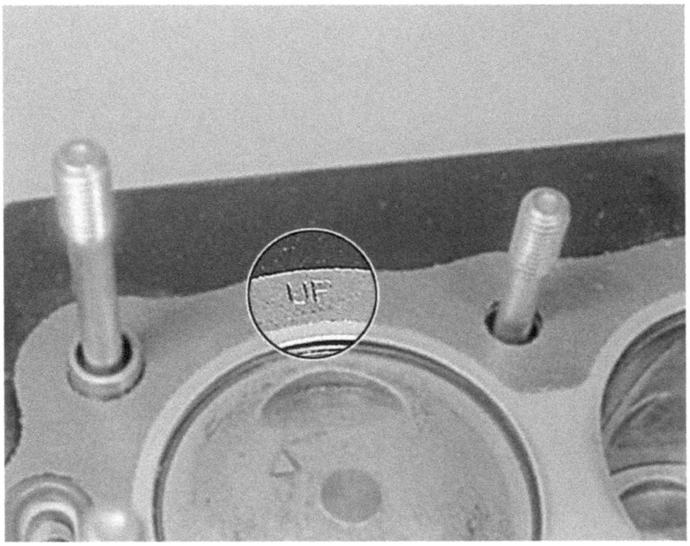

43.8c Cylinder head gasket must be fitted as shown – check that it is of the correct type, where applicable (see text)

43.8d Check that oil feed nozzles are of modified type – all models

43.9a Hold cam chain while refitting cylinder head

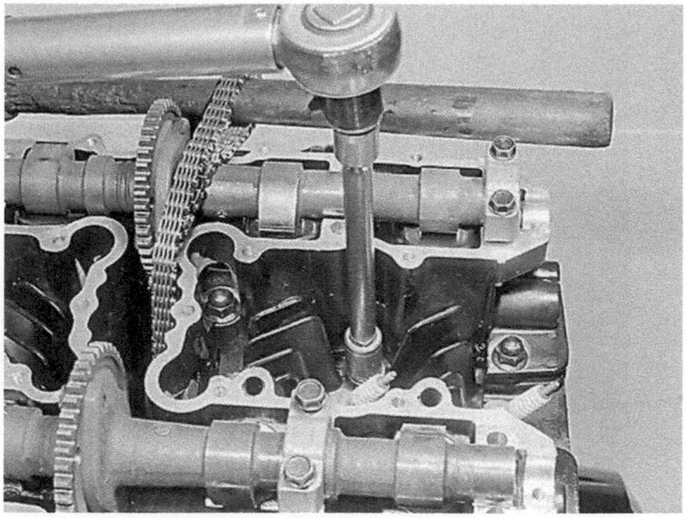

43.9b Tighten cylinder head retaining nuts in sequence to specified torque setting ...

43.9c ... and do not forget five bolts and three cap nuts

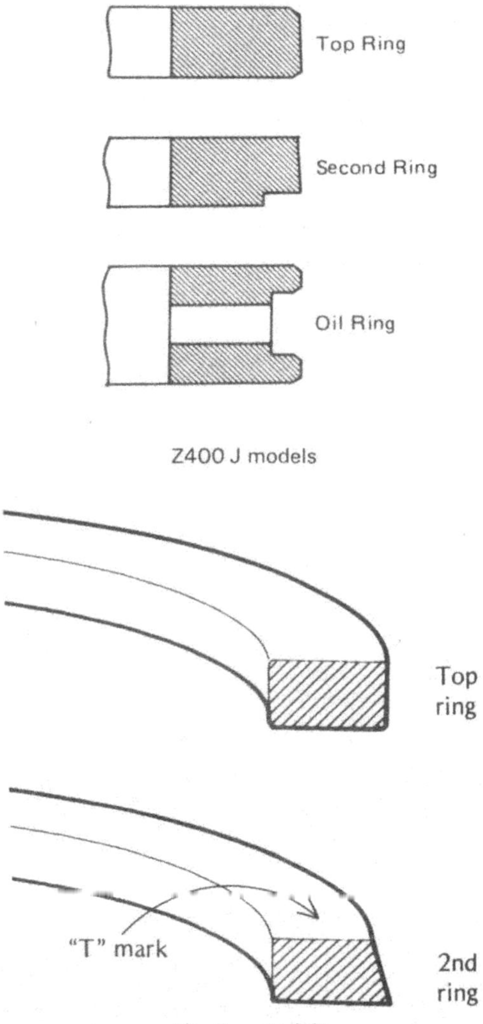

Fig. 1.20 Cross section of piston rings

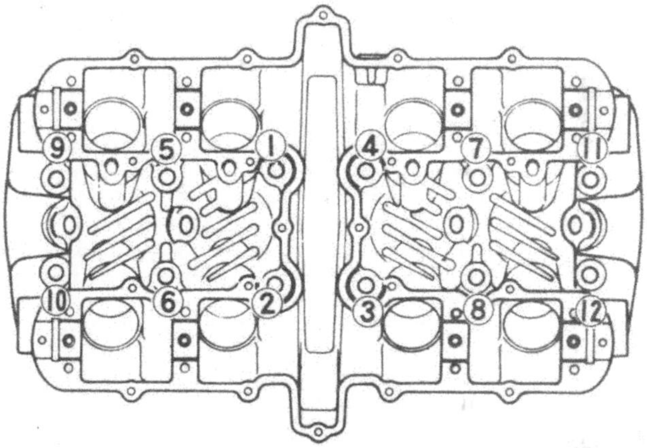

Fig. 1.21 Cylinder head nut tightening sequence

44 Engine reassembly: refitting the camshafts and setting the valve timing

1 Refit the cam chain upper guide blade in the cylinder head cover and remove the wooden dowels (if used) from the camshaft locations. Lay out all the bearing caps with their bolts, keeping separate inlet and exhaust components and remove all traces of dirt from the bearing surfaces and cam followers. Refit all sixteen dowel pins to the cylinder head.

2 If work has been done on the valve gear, making it necessary to check the valve clearances, do not forget to do this at the appropriate time, the work necessary being described in Routine Maintenance.

3 If they were removed, refit the camshaft sprockets over each shaft right-hand end (identified by the square-cut notch) so that its marked surface faces to the right. Rotate the sprocket about the camshaft until the correct pair of sprocket bolt holes is aligned with those in the camshaft flange (refer to the accompanying illustration). Note that all sprockets are identical in appearance the only difference being which pair of bolt holes is drilled for the inlet camshaft mountings. When the correct pair of holes is selected, apply thread locking compound to their threads, then refit the sprocket retaining bolts and tighten them to a torque setting of 1.5 kgf m (11 lbf ft). Refer to Section 7 of this Chapter for notes on camshaft identification.

4 On machines fitted with a mechanically driven tachometer, check that the driven gear and its housing have been removed, as described

in Section 7. **Never** attempt to refit the exhaust camshaft with the tachometer driven gear in place; one or both components will be seriously damaged. On all models, use a rubber sucker or similar to remove each cam follower bucket in turn and check that the adjustment shim is correctly seated with its marked face downwards. These shims are easily displaced and can cause a lot of trouble if not checked at this stage. Holding taut the cam chain to prevent it from snagging, and using only the engine-turning hexagon on the crankshaft right-hand end, rotate the crankshaft forwards, ie, clockwise, until the pistons of 1 and 4 cylinders are at TDC (top dead centre). This is when the '1.4' T mark stamped on the ATU/ignition rotor is aligned exactly with the raised index mark on the crankcase wall. Hold the crankshaft in this position while the valve timing is set and check it at every stage to ensure that it has not moved.

5 The camshaft right-hand ends are identified by the square-cut notches, also by the sprocket markings facing to the right; position each camshaft accordingly when refitting. Smear all bearing surfaces on both cylinder head and camshafts with molybdenum disulphide engine reassembly grease, or failing this, a copious supply of clean engine oil. All bearing caps are located by two dowel pins, which must be refitted in the head before the cap is bolted down; the caps are line-bored with the cylinder head at the factory and so must always be refitted in exactly the same place. To this end arrows cast in each cap top surface indicate the front (exhaust) side of the engine to show which way round a cap fits and the number cast next to the arrow on each cap must match with the number cast on the cylinder head next to each bearing boss to show on which bearing the cap is to be fitted. There is no way of distinguishing between inlet and exhaust cams so these should always be kept separate and marked. Engage the cam chain on the sprocket as described below and place the camshaft in position, followed by the caps and bolts. Partially tighten the left-hand side inner bearing cap bolts to retain the camshaft, then gradually and evenly tighten all the bolts, following the sequence shown in the accompanying illustration, which applies to both camshafts. The final torque setting for these bolts is 1.2 kgf m (9 lbf ft). Check that the cam chain is running correctly in the grooves of its guide blades.

6 To ensure correct valve timing it is essential that the exhaust camshaft is precisely located. With the crankshaft positioned as described above, the exhaust camshaft sprocket must be engaged on the cam chain with the first pin of any one chain link aligned exactly with the sprocket 'EX' mark, as shown in the accompanying photographs and illustrations. Furthermore, it must be arranged so that when the camshaft is bolted down, the cam chain front run is absolutely taut and the 'EX' mark is parallel with the cylinder head top surface. This will require careful work, and may mean that the camshaft must be bolted down and then unbolted several times before the setting is as described. Follow the instructions in paragraph 5 above to fit the camshaft.

7 The positioning of the inlet camshaft is entirely dependent on the exhaust camshaft. Counting the chain pin aligned with the exhaust sprocket 'EX' mark as Number 1, count back along the chain towards the inlet camshaft until the link with pins Number 43 and 44 is reached. The chain must be engaged on the inlet cam sprocket so that this link is astride the 'Z/5 IN' mark on all models except the 400 cc models and ZX550 A1, A1L models where it must be astride the 'Z4 IN' mark. This will mean that pin number 43 is above the mark while pin number 44 is below it, as shown in the accompanying photographs and illustrations. Once the chain is fitted as described, keep the chain top run as taut as possible while the camshaft is bolted down as described in paragraph 6. With the camshaft bolted down fully and the chain top run taut, the relevant inlet sprocket marking should automatically be parallel with the cylinder head top surface; if this is not the case there is no need to worry unless the difference is of the order of one or more chain links, in which case repeat the above procedure to check.

8 When both camshafts are correctly positioned, use a metal rod or similar passed through the tensioner aperture to simulate the action of the chain tensioner by applying (moderate) pressure to the rear of the tensioner blade, and note the positions of the square-cut notches in each camshaft end. Rotate the crankshaft forwards until the '1.4' T mark is again aligned with the fixed index mark and the notches are returned to their previous positions so that 1 and 4 cylinders are again at TDC. (If you go past the mark do not rotate the crankshaft backwards; rotate it forwards again until the marks align.) This will settle all components and place all chain free play on the rear run where the tensioner will compensate for it; thus setting automatically the inlet camshaft position. Stop immediately and find out the cause if any serious resistance is encountered. Check very carefully, referring to the above instructions and to the accompanying photographs and illustrations, that all timing marks are correctly aligned. If not, the procedure must be repeated as necessary until they are correctly aligned.

9 Where applicable, apply molybdenum disulphide engine assembly grease to the tachometer driven gear, insert it into its housing and then fit the two to the cylinder head, securing it with the stop plate and retaining screw.

10 If dismantled, the cam chain tensioner must now be rebuilt. Thoroughly clean all components and apply a smear of molybdenum disulphide grease or engine oil to the push rod and push rod stop. Place the large spring over the push rod end and insert the two into the tensioner body so that the groove cut along the push rod aligns exactly with the hole in the tensioner body extended boss, then insert the

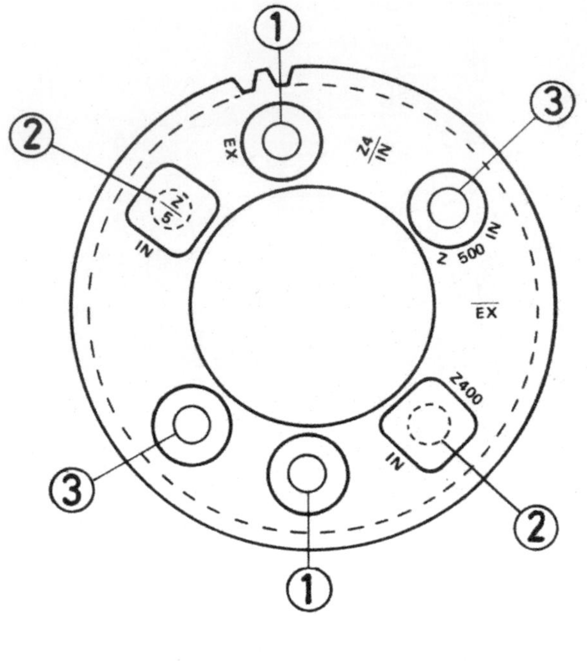

Fig. 1.22 Camshaft sprocket fixing holes

1 Exhaust camshaft
2 Inlet camshaft – all 400 models and ZX550 A1L, A1
3 Inlet camshaft – all other models

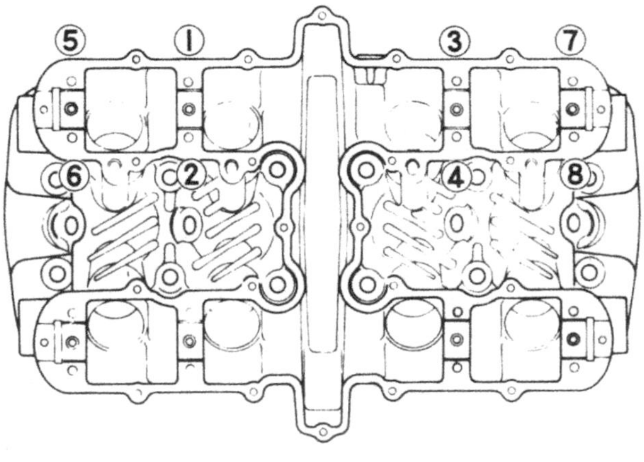

Fig. 1.23 Camshaft cap bolt tightening sequence

small pin to retain the push rod. Check that the push rod moves smoothly in and out. Fit a new sealing O-ring to the tensioner body and refit the assembly to the rear of the cylinder barrel with the retaining pin and push rod stop aperture pointing to the right. Check that the retaining pin is fully inserted; if it drops out there is a possibility of the push rod flying out into the engine while it is running. Tighten securely the tensioner mounting bolts and insert the push rod stop with its tapered surface facing upwards to engage with the underside of the push rod, thus matching its tapered surface.

11 Measure the protruding length of the push rod stop. This should be 11 – 12 mm (0.43 – 0.47 in). If the length is greater than this, the chain free play has not been fully taken up, rotate the crankshaft forwards while maintaining light pressure on the stop end until the setting is correct. Again stop rotating immediately if any resistance is encountered and establish the cause before work can proceed; it may be that the valve timing is incorrect and the valves are contacting the pistons.

12 When the tensioner setting is correct refit the push rod stop spring, the sealing washer and the tensioner cap. Tighten the cap to a torque setting of 2.5 kgf m (18 lbf ft). Having checked that they are clean and correctly gapped as described in Routine Maintenace, refit the spark plugs, tightening them to a torque setting of 1.4 kgf m (10 lbf ft). Wipe the gasket surfaces of the cylinder head and cylinder head cover with a rag moistened in solvent to remove all traces of oil or grease, then check that both faces are absolutely clean and unmarked. Smear a small amount of liquid gasket or RTV over the circumference of each of the four rubber plugs and refit them with the locating shoulder on the inside. Lubricate all bearing surfaces with clean engine oil and fill the pockets around each valve with oil. Fit a new cylinder head cover gasket, using a smear of grease to stick it in place. Do not use jointing compound and note that the gasket will fit correctly only one way.

13 Lower the cylinder head cover into place, noting that the arrow cast in its top surface must face towards the front of the engine. Refit all retaining bolts tightening them by hand only at first. Note that the four longer bolts are fitted at each camshaft end, passing through the rubber end plugs.

14 On US models only, refit the reed valve assemblies and their covers using new gaskets (where fitted).

15 If oil leaks are to be avoided, it is essential that all cylinder head cover retaining bolts are tightened to the correct torque setting. Working in a diagonal sequence from the centre outwards, tighten the bolts progressively and evenly to a final torque setting of 1.0 kgf m (7 lbf ft). This of course includes the reed valve covers on US models, and is essential if distortion of the cylinder head cover is to be avoided.

44.1 Do not forget to refit cam chain upper guide blade to cylinder head cover

44.3 Camshaft right-hand ends are identified by square-cut notches – note locating dowels for bearing caps

44.4 Align crankshaft timing marks exactly, as shown

44.5a Use cast-in markings (arrowed) to ensure each bearing cap is refitted in exactly the correct position

44.5b Lubricate camshaft bearings and tighten bolts in sequence to specified torque setting

44.6 Position camshaft so that 'EX' line is parallel with cylinder head top surface when camshaft is bolted down

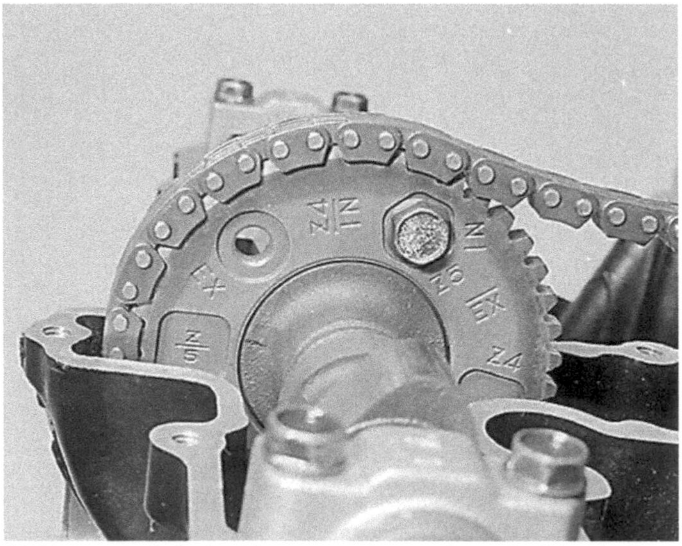

44.7 Inlet camshaft should appear as shown when bolted down

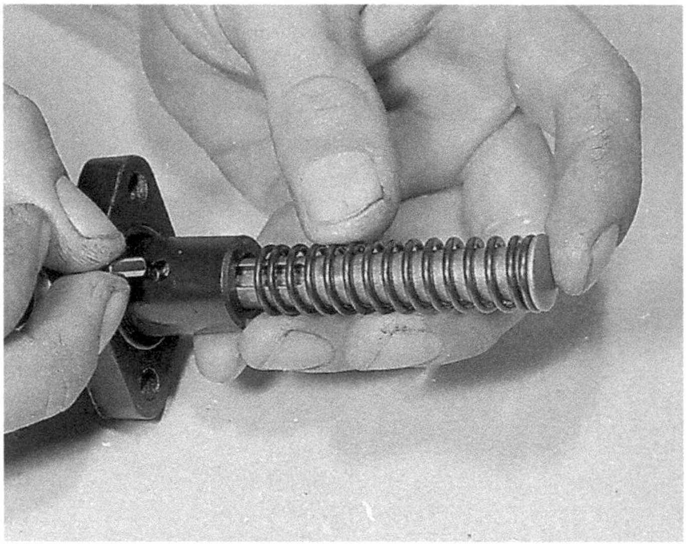

44.10a Refit pushrod and spring to tensioner body and insert pin to retain – note groove in pushrod

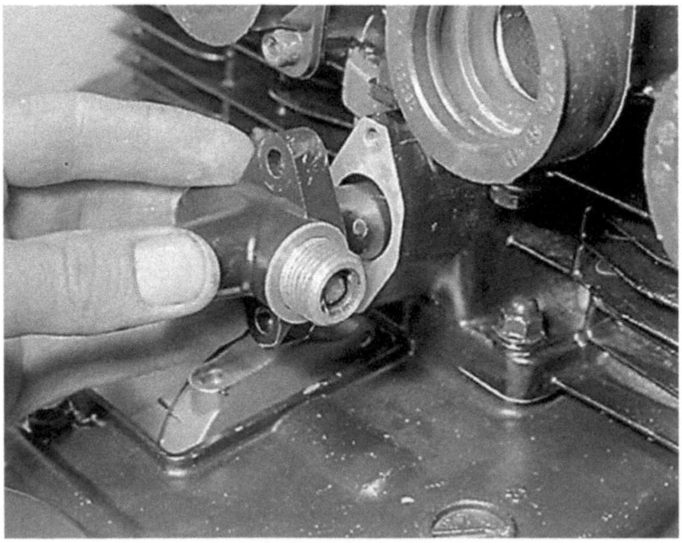

44.10b Refit tensioner assembly to cylinder block ...

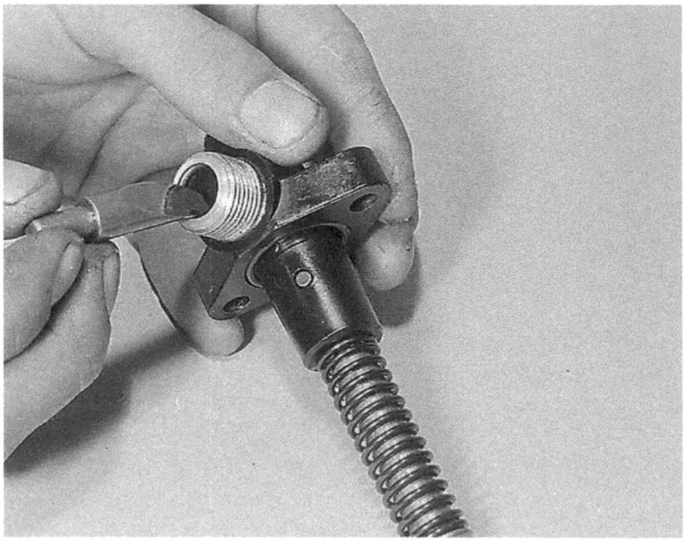

44.10c ... and insert pushrod stop – note alignment of tapered surface

44.11 Measure protruding length of pushrod stop to check tensioner setting

44.12a Refit pushrod stop spring – note flat sealing washer

44.12b Tighten tensioner cap to specified torque setting

44.12c Jointing compound is used only on rubber plugs at camshaft ends – cylinder head cover is refitted dry (with new gasket)

44.13 Arrow mark cast on cylinder head cover must face forwards – tighten bolts only to specified torque setting

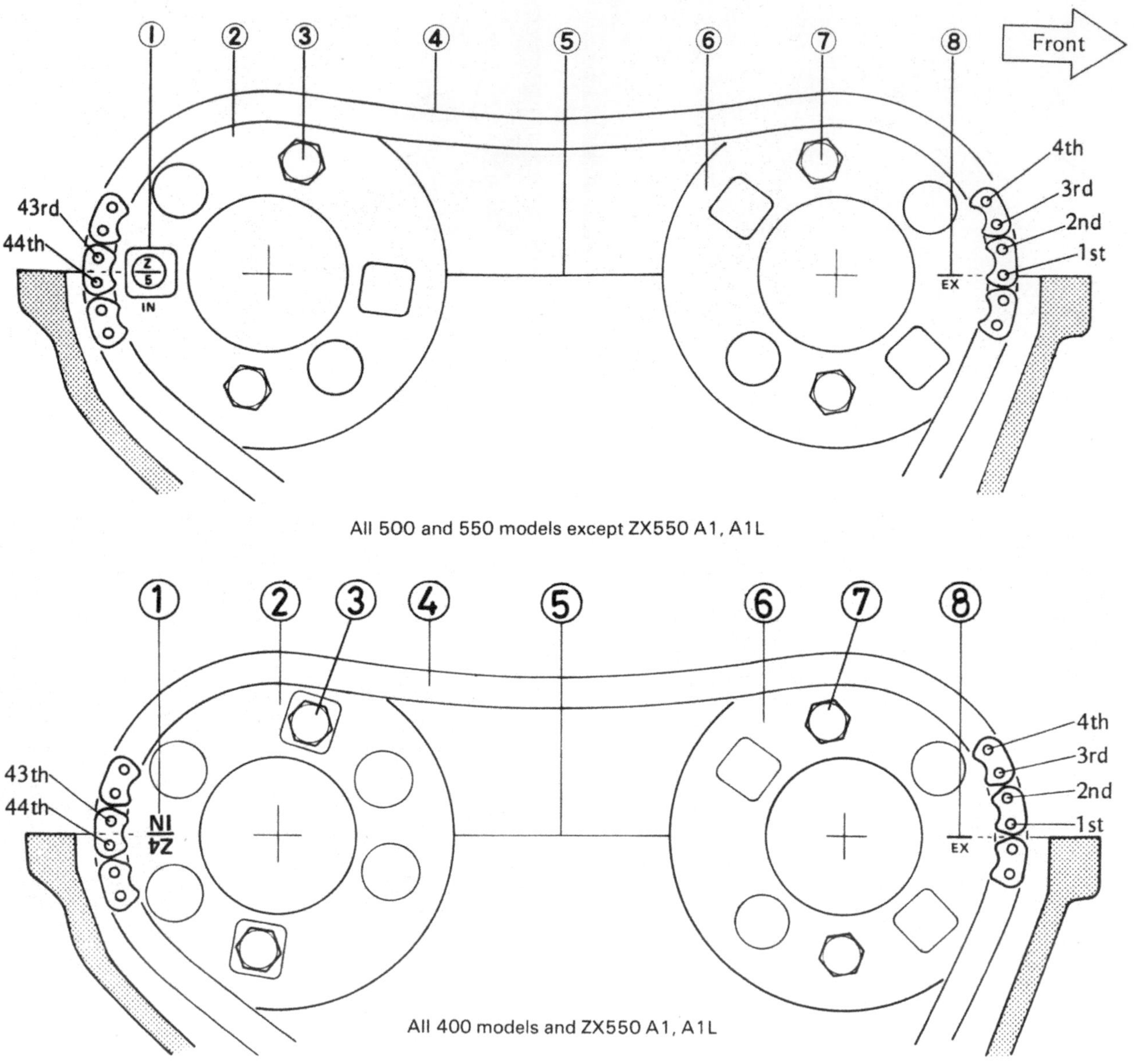

Fig. 1.24 Valve timing marks

1 *Inlet sprocket mark*
2 *Inlet camshaft sprocket*
3 *Sprocket mounting bolt*
4 *Cam chain*
5 *Cylinder head upper surface*
6 *Exhaust camshaft sprocket*
7 *Sprocket mounting bolt*
8 *EX mark*

45 Refitting the engine/gearbox unit in the frame

1 The assistance of at least one, preferably two people, will be necessary to refit the engine/gearbox unit in the frame. Protect the paintwork of the frame tubes with rag or masking tape and check that all mounting components are clean and undamaged. Remove all traces of corrosion from the mounting bolt shanks and smear them with grease, also those crankcase passages through which they must pass, where applicable, to ease the task of refitting.

2 On those models with rubber mounted engines, examine the condition of the rubbers themselves. All can be pressed or drifted from their housings and refitted using soapy water as a lubricant. If cracked, worn or perished the rubbers must be renewed. To assist in identification, the overall length of each rubber bush is given below. Front left-hand mounting bracket, marked '1192' on the inside, bush length 59 mm (2.32 mm). Front right-hand mounting bracket, marked '1193' on the inside, bush length 59 mm (2.32 in). Rear upper right-hand mounting bracket, unmarked on ZX550 A1, A1L models, bush length 42 mm (1.65 in) while on all shaft drive models the bracket is marked '1196' on the inside and is fitted with a bush 50 mm (1.97 in) long. Rear upper left-hand mounting bush (ZX550 A1, A1L only) is 42 mm (1.65 in) long. On all shaft drive models the rear upper mounting bracket is marked '1195' on the inside, the bush length being 50 mm (1.97 in). Finally on shaft drive models only, two bushes 42 mm (1.65 in) long are fitted one at each end of the lower rear engine mounting bolt.

3 Manoeuvre the engine/gearbox unit into the frame from the right-hand side and secure it by refitting the lower rear engine mounting bolt. Take the weight of the unit with a jack or levers as described in Section 5 of this Chapter and refit all remaining engine mountings,

tightening the nuts and bolts by hand only at first. Refer to the illustrations to ensure that all are correctly refitted, then check that the engine unit is seated correctly and without strain on its mountings before tightening the mounting bolts. All 8 mm bolts which secure the mounting brackets to the frame are tightened to a torque setting of 2.4 kgf m (17 lbf ft), while all 10 mm bolts which secure the engine to the frame or to its mounting brackets are tightened either to 3.5 kgf m (25 lbf ft) or to 4.0 kgf m (29 lbf ft) depending on the model. Refer to the Specifications Section of this Chapter for details.

4 Refit all components that were removed, if any, to provide additional working space and are not mentioned specifically in the removal/refitting sequences described.

5 On those models fitted with an oil cooler, refit the assembly to the frame, tightening the mounting bolts to a torque setting of 1.0 kgf m (7 lbf ft). Wipe clean all four mating surfaces, fit new O-rings in the grooves of the pipe mounting flanges and refit the pipes to the crankcase tightening the four mounting bolts to a torque setting of 1.0 kgf m (7 lbf ft). Note that the shape and length of each pipe will clearly show on which crankcase union it is to be mounted. On all models so equipped, connect the tachometer drive cable lower end to the driven gear housing on the cylinder head, checking that the inner cable is correctly engaged.

6 Route the ignition contact breaker/pick-up coil leads along the bottom of the clutch cover and through the frame up to their connector. Secure the wires with any clamps or ties provided and check that the connections are securely fastened. Connect the battery earth lead to the crankcase, tightening securely its retaining screw. On ZX550 A1, A1L and all shaft drive models connect the clutch cable to the release lever and adjust it as described in Routine Maintenance. Note that the angle between release lever and cable should be approximately 80°, and that the release lever should not foul the crankcase top as the clutch is operated.

7 If it was removed, refit the right-hand footrest assembly, then refit the brake pedal, or reset it to its required height below the footrest, as applicable. Refit the stop lamp rear switch and its spring, where these were removed, and connect the switch wires. Moving round to the left-hand side of the machine connect up all the following wires; alternator wires, neutral switch wire, oil pressure/level switch wire and side-stand switch wire (where fitted). Connect the starter motor lead to its terminal, tighten the retaining nut to a torque setting of 1.1 kgf m (8 lbf ft), smear petroleum jelly over the terminal to prevent corrosion and refit the rubber terminal cover. Route all wiring carefully through the clamps, ties or guides provided to secure it out of harm's way.

Shaft drive models

8 If the output shaft shock absorber (damper) cam was removed, this must now be refitted, using a **new** retaining nut. Lock the cam by the method used on dismantling and tighten the nut to 12.0 kgf m (87 lbf ft) and then carefully stake the nut's shouldered section into the slot in the shaft. Take care not to strike the shaft too hard or damage to the gearbox pinions or bearings may result.

9 Offer up the gear case unit, taking great care not to damage or distort the gearchange shaft oil seal as it is slid over the shaft end. It is preferable to use the correct oil seal guide tool, Part Number 57001-264 if it is available, but it is possible to get by without it if care is taken.

10 Refit the gear case mounting bolts noting that they are of different lengths and that care must be taken to refit each in its correct hole. Tighten the bolts evenly to a torque setting of 0.9 kgf m (6.5 lbf ft). Connect the neutral switch lead. Apply a coat of high melting point grease to the output splines and the corresponding internal splines of the driveshaft. Arrange the two components so that the locking pin will align with the hole in the shaft end. If the rubber gaiter (boot) was removed this should be fitted in readiness for the shaft to be reconnected. The shaft can now be pulled forward to engage the front gear case splines. The locating pin should click into place, and this should be checked by pulling hard on the end of the shaft. Once the shaft is secure, refit the rubber gaiter to cover the joint.

11 It is now possible to reassemble the rear wheel end of the drive shaft. Lubricate the shaft splines with high melting point grease. Fit the coil spring over the pinion nut and offer up the final drive casing. Refit the four retaining nuts, having first cleaned the stud threads and applied a locking compound such as Loctite or similar. Tighten the four nuts evenly to 2.3 kgf m (16.5 lbf ft), and refit the rear suspension unit, tightening the nut to a torque setting of 2.5 kgf m (18 lbf ft). Refit the rear wheel as described in Section 4 of Chapter 5.

Chain drive models

12 Engage the sprocket on the chain and refit it to the output shaft splines, followed by the retaining plate. Apply thread locking compound to their threads then refit the sprocket retaining bolts, tightening them to a torque setting of 1.0 kgf m (7 lbf ft). Working as described in Routine Maintenance, check that the chain is correctly adjusted and properly lubricated, and that all disturbed fasteners are correctly tightened.

All models

13 Check that the rear brake is correctly adjusted and that all disturbed fasteners are correctly tightened, also that the stop lamp rear switch is correctly adjusted and working properly.

14 On ZX550 A1, A1L and all shaft drive models, refit the two locating dowel pins in their crankcase recesses, then refit the gearbox sprocket/front gear case cover, tightening the screws securely. On all other models, grease the clutch push rod and insert it into the input shaft until it can be pressed firmly against the steel ball and short push rod. Refit the two locating dowel pins to their crankcase recesses. Check that the clutch release mechanism is correctly fitted and the cable connected, if disturbed, and that the mechanism is well greased. Offer up the cover, ensuring that the push rod engages with the release mechanism and the cover locates on the dowels, then press it into place and refit the retaining screws, tightening them securely. Adjust the clutch as described in Routine Maintenance.

15 Refit the starter motor cover. Where a gearchange lever is fitted, this should be replaced using the marks made on dismantling to ensure that it is correctly aligned. Press the lever on to its shaft and refit the pinch bolt, tightening it securely. On models fitted with a gearchange linkage, check first that the pivot post is secure. If not, apply thread locking compound to its threads and refit it, tightening it to a torque setting of 2.5 kgf m (18 lbf ft). Place one thrust washer over the post; grease it liberally, then refit the linkage front arm followed by the second thrust washer, the circlip and the plastic cap. Arrange the linkage so that the angles formed between the front and rear arms and the adjusting rod are 90°; when the rear arm is replaced on the gearchange shaft the pedal should be automatically at the right height. If adjustment is required, slacken the adjusting rod locknuts; adjustment is a matter of rotating the rod as necessary to lengthen or shorten it, while rotating the rear arm around the gearchange shaft end to preserve the 90° angles. When the correct position is achieved, press the rear arm onto the shaft, refit the pinch bolt to secure it and tighten both the pinch bolt and the locknuts.

16 Thoroughly clean all exhaust system mountings, applying an anti-seize compound such as Copasil to their threads to prevent corrosion. Place new exhaust gaskets in the ports, using grease to stick them in place, then manoeuvre the exhaust system into position under the machine. Lift up the pipes at the front and push them into the ports, then have an assistant hold them in place while the rear is lifted and the rear mountings loosely reassembled. Checking that the gaskets at their lower ends are in good condition, refit Number 2 and 3 exhaust pipes using very gentle blows from a soft-faced mallet to tap them into place. Note that each pipe is identified by its number stamped at its lower end; ensure that each pipe is refitted in its correct place and do not forget the retaining collars. When 2 and 3 pipes are in position, refit the two split collets to each pipe, noting that each collet flanged end is outwards, slide the retaining collars up the pipes and over the studs, then refit the retaining nuts to secure the mountings. Working from the front to the back, tighten securely all exhaust clamp nuts and bolts, and the rear mountings, but be very careful not to overtighten any fastener. It is all too easy to be clumsy with a thread that is stiff and awkward due to corrosion.

17 If it was removed, refit the left-hand footrest assembly. Insert the air filter casing, or casing front half, as applicable, into the frame and refit the breather hose joining it to the crankcase top. Secure the hose at both ends with the wire clips provided. Check that the intake stubs and air filter rubbers are in position. In the case of the latter, make sure that they are located properly and that the spring retainers are rolled back towards the air filter casing. Fit the clamps to the intake stubs with the screws facing outwards. Check that they are slackened off.

18 Manoeuvre the carburettor bank about half way into position, and then reconnect the throttle cable(s). Ease the carburettor bank between the air filter rubbers and intake stubs, and work the carburettor ends into their respective mountings. This stage is rather awkward but may be aided by smearing a very small amount of

suitable lubrication to the rubbers; otherwise it is a matter of care and patience. When the carburettors are finally in place, tighten securely the intake stub clamps, having ensured that the carburettors are pressed fully into the stubs, and roll up the spring retainers over the carburettor rear ends.
19 Secure the air filter casing and refit the filter element(s) following the reverse of the dismantling procedure, then check that the throttle cables are correctly adjusted as described in Routine Maintenance and route carefully the vent hoses.
20 Refit the HT coils to the frame, connect them to the main loom and refit the HT leads to the spark plugs. If in doubt as to the coil connections, refer to the wiring diagrams at the back of this Manual. If the identifying marks have disappeared from the HT leads, the length and shape of each lead should show clearly which lead goes to which spark plug.
21 On US models only, refit the Clean Air System and Evaporative Emission Control System components. The CAS vacuum switch valve is situated over the cylinder head, with one large, short, hose going to each of the cylinder head cover valve housings and one large long hose leading back to the air filter casing. Finally the small diameter hose should be connected to Numbers 1 and 4 carburettor intake stubs. Secure all hose connections with the wire clips, and secure all hoses to the frame with any clamps or ties provided. On ZX550 A1L and KZ550 F2L models only, connect the EECS hoses as follows: the white-marked hose to Number 3 carburettor intake stub, the yellow-marked hose to the carburettor bank and the green marked hose to the air filter front casing. Again, secure all connections with the wire clips and secure the hoses to the frame with any clamps or ties provided. On all US models be careful to check that all emission control hoses and their clamps are routed or placed well clear of all electrical wiring, particularly that of the ignition system. Some cases have occurred of the HT coils being shorted to earth by a hose clamp rubbing against their wires.
22 Refit the battery, remembering to connect the negative (-) terminal last, and securing its retaining strap (where fitted). Refit the fuel tank on its front mountings but keep the rear raised while connecting all hoses and electrical leads (as applicable). Lower the tank and secure its rear mounting, then turn the fuel tap to the 'PRI' position to fill the carburettor float chambers. Refit the sidepanels, tightening their retaining screws, where fitted, and close or refit the seat.
23 Check that the drain plug and oil filter have been refitted and are secured to their correct torque settings, then remove the filler plug and pour in the specified quantity and grade of engine oil. Note that the level will drop as soon as the engine has been started for the first time and oil is beginning to circulate. Be prepared to add more oil to bring the level up to the marks.
24 Working as described in Routine Maintenance, check the contact breaker gaps (Z400 J1, Z500 B1, B2, KZ/Z550 A1 only) are correct. Replace the contact breaker cover.

45.2a Take care that all engine mountings are in good condition ..

45.2b ... that all are refitted in their correct locations ...

45.2c ... and that all are tightened to specified torque settings

45.2d Renew mounting nuts or bolts if damaged or worn

45.5a Refit oil cooler to frame ...

45.5b ... and ensure that unions are correctly fastened – always renew O-rings

45.6 Refit and adjust clutch cable – check clutch operation

45.7 Be careful to route all wiring (and control cables) through clamps provided

45.15a Fit first thrust washer to linkage pivot post (where applicable) ...

45.15b ... grease pivot then refit linkage front arm and second thrust washer ...

45.15c ... secure linkage assembly with circlip and refit cap

45.16a Use grease to stick new exhaust gaskets in place ...

45.16b ... then refit exhaust pipes and position split collets on each ...

45.16c ... before sliding retaining collar into place

45.16d Tighten nuts securely but do not overtighten them

45.16e Do not forget to tighten exhaust clamp bolts

45.16f Exhaust rear mountings should be tightened last

46 Starting and running the rebuilt engine

1 Attempt to start the engine using the usual procedure adopted for a cold engine. Do not be disillusioned if there is no sign of life initially. A certain amount of perseverance may prove necessary to coax the engine into activity even if new parts have not been fitted. Should the engine persist in not starting, check that the spark plugs have not become fouled by the oil used during reassembly. Failing this go through the fault finding charts and work out what the problem is methodically.

2 When the engine does start, keep it running as slowly as possible to allow the oil to circulate. The oil warning light should go out almost immediately the engine has started, although in certain instances a very short delay can occur whilst the oilways fill and the pressure builds up. If the light does not go out the engine should be stopped before damage can occur, and the cause determined. Open the choke as soon as the engine will run without it. During the initial running, a certain amount of smoke may be in evidence due to the oil used in the reassembly sequence being burnt away. The resulting smoke should gradually subside.

3 Check the engine for blowing gaskets and oil leaks. Before using the machine on the road, check that all the gears select properly, and that the controls function correctly. As soon as the engine has warmed up to normal operating temperature, check the oil level as described in Routine Maintenance; it will probably be necessary to add oil to bring the level up to the marks in the level glass, especially where an oil cooler is fitted. Do not forget to check this before and after the machine is taken out on the road for the first time.

4 Working as described in Routine Maintenance, check that the ignition timing is correct. This is essential for all models, whether fitted with contact breaker or electronic ignition.

47 Taking the rebuilt machine on the road

1 Any rebuilt machine will need time to settle down, even if parts have been replaced in their original order. For this reason it is highly advisable to treat the machine gently for the first few miles to ensure oil has circulated throughout the lubrication system and that any new parts fitted have begun to bed down.

2 Even greater care is necessary if the engine has been rebored or if a new crankshaft has been fitted. In the case of a rebore, the engine will have to be run-in again, as if the machine were new. This means greater use of the gearbox and a restraining hand on the throttle until at least 500 miles have been covered. There is no point in keeping to any set speed limit; the main requirement is to keep a light loading on the engine and to gradually work up performance until the 500 mile mark is reached. These recommendations can be lessened to an extent when only a new crankshaft is fitted. Experience is the best guide since it is easy to tell when an engine is running freely.

3 If at any time a lubrication failure is suspected, stop the engine immediately, and investigate the cause. If any engine is run without oil, even for a short period, irreparable engine damage is inevitable.

4 When the engine has cooled down completely after the initial run, check the various settings. During the run most of the engine components will have settled into their normal working locations. Check the various oil levels, particularly that of the engine as it may have dropped slightly now that the various passages and recesses have filled.

Chapter 2 Fuel system and lubrication

For information relating to the 1984 on models, see Chapter 7

Contents

Specifications

Fuel tank capacity	**Litre**	**Imp gal**	**US gal**
KZ/Z550 C1, C2, KZ550 C3, C4:			
Overall	12.4	2.73	3.28
Reserve	1.8	0.40	0.48
KZ550 F1, F2, F2L, M1:			
Overall	13.2	N/App	3.49
Reserve	1.9	N/App	0.50
Z400 J1, J2, Z500 B1, B2, KZ/Z550 A1, A2, D1, KZ550 A3:			
Overall	15.0	3.30	3.96
Reserve	1.5	0.33	0.40
Z400 J3, Z550 A3, KZ550 A4:			
Overall	15.0	3.30	3.96
Reserve	2.0	0.44	0.53
ZX550 A1, A1L:			
Overall	18.0	3.96	4.76
Reserve	3.8	0.84	1.00
ZR400 A1, B1, ZR550 A1, A2, KZ/Z550 H1, H2:			
Overall	18.5	4.07	4.89
Reserve	3.4	0.75	0.90
Z550 G1, G2:			
Overall	21.5	4.73	N/App
Reserve	3.0	0.66	N/App

Fuel grade ... Refer to Chapter 7, Section 5 for details

Carburettors

Slide type:	**Z400 J1,J2, J3**	**Z500 B1,B2**	**Z550 A1,A2, A3,C1,C2, D1**	**KZ550 A1,A2 A3,C1,C2, C3,C4,D1**
Make	Teikei	Teikei	Teikei	Teikei
Model	K21P-2A	K22P-2A	K22P-2D (2F,D1 model)	K22P-2C (2E,D1 model)
Choke size	21 mm	22 mm	22 mm	22 mm
Throttle valve cutaway	2.5	2.5	2.5	2.5
Jet needle	4C91	4C91	4D93	4D92
Clip position – grooves from top	2nd	2nd	2nd	Fixed
Needle jet	V90	V90	V95	V95
Main jet	90	90	92 (94, D1 model)	92 (94, D1 mode
Pilot jet	32	32	32	32
Float valve	1.6	1.6	1.6	1.6
Pilot air screw – turns out	$1\frac{1}{4} \pm \frac{1}{4}$	$1\frac{1}{8} \pm \frac{1}{4}$	$1\frac{3}{8} \pm \frac{1}{4}$	Preset
Fuel level	3.5 ± 1.0 mm	3.5 ± 1.0 mm	3.5 ± 1.0 mm	3.5 ± 1.0 mm
Idle speed – rpm	1150–1250	1000–1100	1000–1100	1000–1100

Early constant depression type:	Z550 H1,H2	KZ550 H1,H2
Make	Teikei	Teikei
Model	K26V-1B	K26V-1A
Choke size	26 mm	26 mm
Jet needle	4 X 30	4 X 31
Clip position – grooves from top	2nd	Fixed
Primary main jet	64	64
Secondary main jet	86	86
Primary main air jet	1.30	1.30
Secondary main air jet	1.00	1.00
Pilot jet	32	32
Pilot air jet	1.30	1.30
Starter jet 1	46	46
Starter jet 2	0.6	0.6
Fuel level	6.0 ± 1.0 mm	6.0 ± 1.0 mm
Float height	27.0 mm	27.0 mm
Pilot screw – turns out	Preset	Preset
Idle speed – rpm	1050 ± 50	1050 ± 50

Late constant depression type:	ZR400 A1,B1	KZ550 A4,F1 M1,F2,F2L	ZR550 A1,A2 Z550 G1,G2	ZX550 A1,A1L
Make	Teikei	Teikei	Teikei	Teikei
Model	K26V	K26V	K26V	K27V
Choke size	26 mm	26 mm	26 mm	27 mm
Jet needle	4A01	4A02	4A00-ZR models 4A01-Z550 G1,G2	4A10
Clip position – grooves from top	4th	Fixed	2nd-ZR models 4th-Z550 G1,G2	Fixed
Main jet	102-ZR400 A1 104-ZR400 B1	110-KZ550 A4 120-All others	100–ZR models 116-Z550 G1,G2	114
Main air jet	N/Av	N/Av	60–ZR models N/Av Z550 G1,G2	1.00
Pilot jet	32	32	32	34
Pilot air jet	N/Av–ZR400 A1 1.40–ZR400 B1	N/Av	1.40–ZR models N/Av–Z550 G1,G2	1.25
Fuel level	7.0 ± 1.0 mm	7.0 ± 1.0 mm	7.0 ± 1.0 mm	7.0 ± 1.0 mm
Float height	27.0 mm	27.0 mm	27.0 mm	27.0 mm
Pilot screw – turns out	2-ZR400 A1 $1\frac{7}{8}$–ZR400 B1	Preset	$2\frac{1}{2}$–ZR models 2–Z550 G1,G2	$2\frac{1}{2}$–UK model Preset–US model
Idle speed – rpm	1200 ± 50	1200 ± 50 (F2L) 1050 ± 50 (all other models)	1000 – 1100	1050 ± 50–UK 1200 ± 50–US

Engine lubrication system

Recommended oil grade	SAE 10W/40, 10W/50, 20W/40 or 20W/50, SE or SF class or equivalent
Capacity:	
At oil change only*	2.6 litre (4.6 Imp pint/2.8 US qt)
At engine rebuild or oil and filter change*	3.0 litre (5.3 Imp pint/3.2 US qt)
Oil pressure – @ 4000 rpm, oil temperature 90°C (194°F)	2.0 – 2.5 kg/cm² (28.5 – 35.6 psi)
Relief valve opening pressure	4.4 – 6.0 kg/cm² (62.6 – 85.3 psi)
Oil pump:	
Type	Trochoidal
Inner rotor/outer rotor maximum clearance	0.30 mm (0.0118 in)
Outer rotor/pump body maximum clearance	0.30 mm (0.0118 in)
Rotor maximum side clearance	0.12 mm (0.0047 in)

**models fitted with an oil cooler will require approximately an additional 0.2 lit (0.4 Imp pt/0.2 US qt) of oil*

Final drive lubrication – Z550 G1, G2, KZ550 M1, F1, F2, F2L

Recommended oil grade	API GL-5 (or GL-6) hypoid gear oil
Viscosity:	
Above 5°C (41°F)	SAE 90
Below 5°C (41°F)	SAE 80
Capacity	190cc (6.9/6.4 Imp/US fl oz)

Torque wrench settings

Component	kgf m	lbf ft
Air suction valve cover bolts – US models only	1.0	7.0
Intake stub:		
Allen screws	1.5	11.0
Hexagon-headed screws	1.2	9.0
Cross-head screws	N/App	N/App
Breather cover bolt	0.6	4.0

Component	kgf m	lbf ft
Sump (oil pan) bolts	1.0	7.0
Engine oil drain plug	3.8	27.5
Oil pressure switch	1.5	11.0
Oil pressure relief valve	1.5	11.0
Oil filter mounting bolt	2.0	14.5
Oil cooler – where fitted:		
Mounting bolts	1.0	7.0
Pipe gland nuts	2.3	16.5
Bottom union mounting bolts	1.0	7.0
Air cleaner housing screws	0.5	3.5

1 General description

The fuel system comprises a fuel tank from which fuel is fed by gravity to the bank of four carburettors via a vacuum operated fuel tap. All later models are fitted with a bank of four Teikei constant depression carburettors. The earlier models make use of Teikei slide carburettors. Air is drawn to the carburettors from a moulded plastic air cleaner casing containing a pleated paper type air filter element. Note that an additional foam filter is fitted to KZ/Z550 H1, H2 and ZR400/550 models.

Engine lubrication is of the wet sump type, the oil being contained in a sump at the bottom of the crankcase.

The gearbox is also lubricated from the same source, the whole engine unit being pressure fed by a mechanical oil pump that is driven off the crankshaft via the secondary shaft.

2 Fuel tank: removal and refitting

1 Where these are secured to the bottom of the tank or will obstruct its removal, remove both sidepanels. Unlock and raise the seat, or unlock it, lift it at the rear and remove it (as applicable). On ZX550 A1, A1L and Z550 G1 and G2 models a latch must be pushed forward to release the seat after it has been unlocked. The tank front mountings consist of two rubbers, one on each side of the frame.

2 The rear mounting on all models except the ZX550 A1, A1L and shaft drive models consists of a prong on the tank underside which engages in a grommet set in the frame. Since the tank is held only by the closed seat, all that is necessary is to lift it at the rear and pull it backwards, once any wiring and the fuel pipes are disconnected. On ZX550 A1 and A1L models the battery retaining plate passes across a tongue projecting from the rear of the tank; remove the two nuts securing the retaining strap to release the tank. Z550 G1 and G2 models are similar except that a single bolt fastens the rear of the tank to the battery retaining strap; remove the three bolts to release the tank and strap. On all US shaft drive machines, remove the two bolts which secure the tank rear mounting.

3 Lift the tank at the rear and disconnect all hoses and electrical leads. All hoses are disconnected by squeezing together the ears of their retaining clips and sliding the clips down the hoses, which can then be worked off the stub with the aid of a small screwdriver. Plug the left-hand (fuel return) union (ZX550 A1L and KZ550 F2L). Electrical leads are disconnected by pulling apart their snap- or multi-pin block connectors. Check that the tank is free, lift it up at the rear and pull it backwards to remove it.

4 Refitting is the reverse of the above, but note that it can be eased by smearing the rubbers with a small amount of lubricant such as soapy water; these are usually a very tight fit. When connecting the hoses, note that the vacuum hose is noticeably smaller in diameter than the fuel hose. On ZX550 A1L and KZ550 F2L models note that the red-marked (fuel return) hose goes to the left-hand union while the blue-marked (vent) hose goes to the right-hand union. Check very carefully for fuel leaks after any tank connections have been disturbed.

5 Fuel tank repairs are for the expert only; any welding or brazing must be preceded by careful flushing out, once the tank has been removed from the machine and stripped of all ancillary components. Dents and scratches due to accident damage can be repaired using the same methods as for minor car bodywork repairs, but note that it is very difficult to obtain an exact match with the original paint finish. It is usually best to have the complete tank resprayed, in which case the painter will usually carry out any repairs and preparation necessary.

3 Fuel tap: removal and overhaul

1 The fuel tap is of the vacuum type and is automatic in operation. The tap lever has three positions marked 'On', 'Res' (reserve) and 'Pri' (prime). In the first two of these settings, fuel flow is controlled by a diaphragm and plunger held closed by a light spring. When the engine is started, the low pressure in the intake tract opens the plunger allowing fuel to flow through the tap to the carburettors. When the tap is set to the 'Pri' position, the diaphragm and plunger are bypassed.

2 In the event of failure, the most likely culprits are the vacuum hose or the diaphragm. If a leak develops in either of these the tap will not operate in anything other than the 'Pri' position. Check the vacuum hose for obvious splits or cracks, and renew it if necessary. If the diaphragm itself is suspect, set the tap lever to 'On' or 'Res' and disconnect the fuel and vacuum hoses at the carburettor. Suck gently on the vacuum hose. If fuel does not flow, remove the tap for inspection as described below.

3 Remove the fuel tank as described in Section 2. If the tank is full or nearly full, drain it into a clean metal container taking great care to avoid any risk of fire. Place the tank on its side on some soft cloth, arranging it so that the tap is near the top. Slacken and remove the two tap mounting bolts and lift the tap away, taking care not to damage the O-ring which seals it.

4 From the front of the tap, remove the two small cross-head screws which secure the tap lever assembly. Withdraw the retainer plate, wave washer, tap lever, O-ring and the tap seal. Examine the tap seal and tap lever O-ring, especially if there has been evidence of leakage. Check that the tap seal has not become damaged and caused the blockage of the outlet hole. Fit a new O-ring and tap seal as required, and reassemble the tap lever assembly by reversing the above sequence.

5 Working from the rear of the tap, remove the four countersunk screws which retain the diaphragm cover, noting the direction in which the vacuum stub faces, and lift it away taking great care not to damage the rather delicate diaphragm. Remove the small return spring. Very carefully dislodge the diaphragm assembly and remove it from the tap body. The diaphragm assembly comprises a plastic diaphragm plate sandwiched between two thin diaphragm membranes. Carried through the centre of the assembly is the fuel plunger which supports a sealing O-ring.

6 Examine the diaphragm closely for signs of splitting or other damage. Carefully remove any dust or grit which may have found its way into the assembly. Check the condition of the O-ring on the end of the plunger. If wear or damage of the above components is discovered, it will be necessary to renew the diaphragm assembly complete. Note that one side of the diaphragm plate has a groove in it, and this must face towards the O-ring on the plunger. When fitting the diaphragm assembly and cover, check that the diaphragm lies absolutely flat, with no creases or folds. Fit the cover with the vacuum stub facing in the correct direction (this varies according to the model). Tighten the securing screws evenly and firmly.

7 Note that some early models were fitted with a plain plug at the base of the tap body. If fitted, this can be used to remove water and other impurities from the tank and tap, as described in Routine Maintenance.

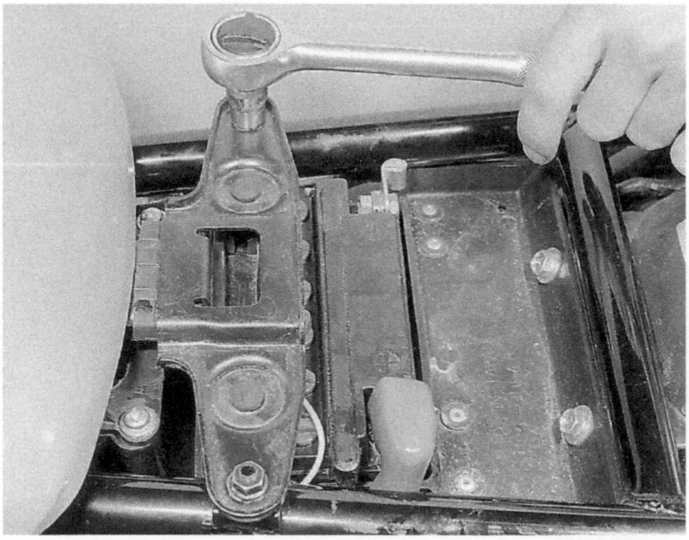

2.2 Where applicable, remove two nuts to release tank rear mounting

2.3 Do not forget to disconnect electrical leads before removing tank

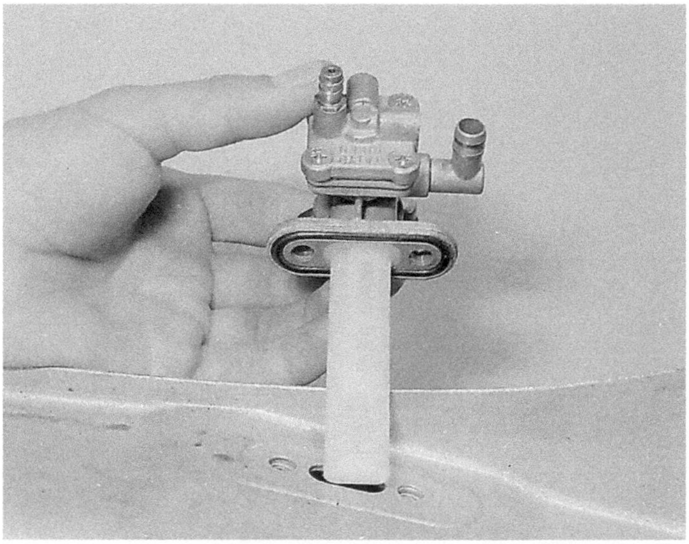

3.3 Renew sealing O-ring if worn or damaged to prevent fuel leaks

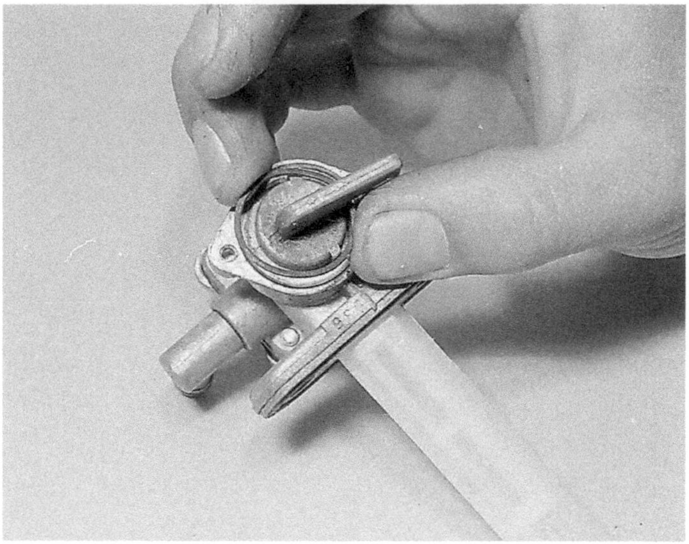

3.4a Withdraw two screws and retainer plate to release wave washer ...

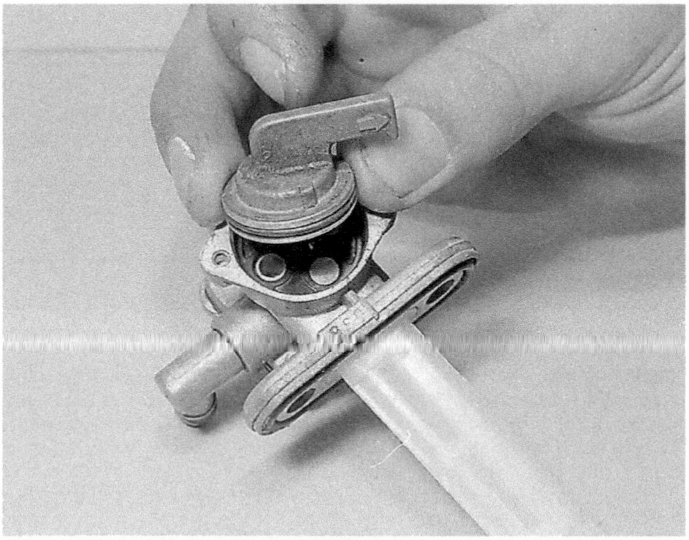

3.4b ... and tap lever – note sealing O-ring

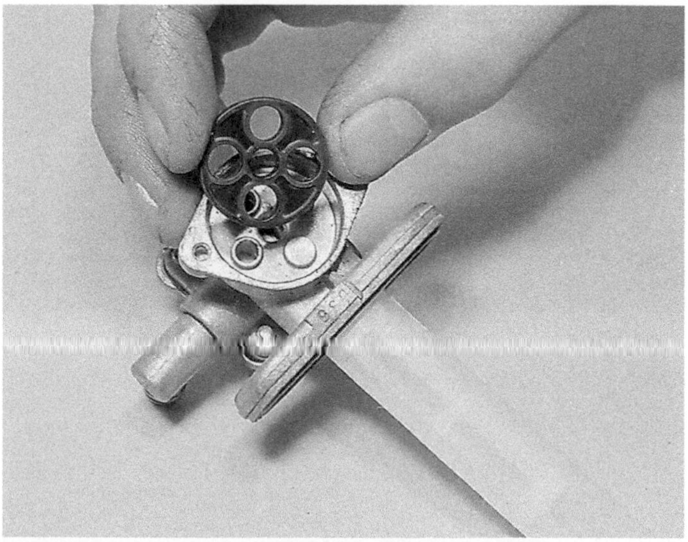

3.4c Check tap seal carefully for signs of wear or damage

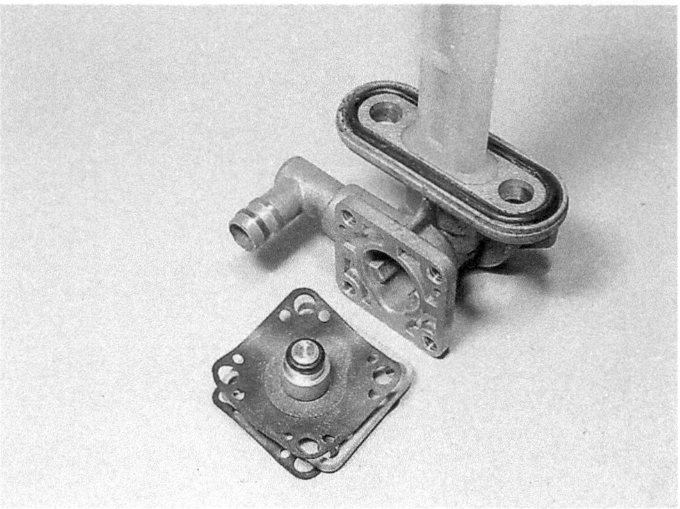

3.5a Diaphragm assembly fits as shown into tap body

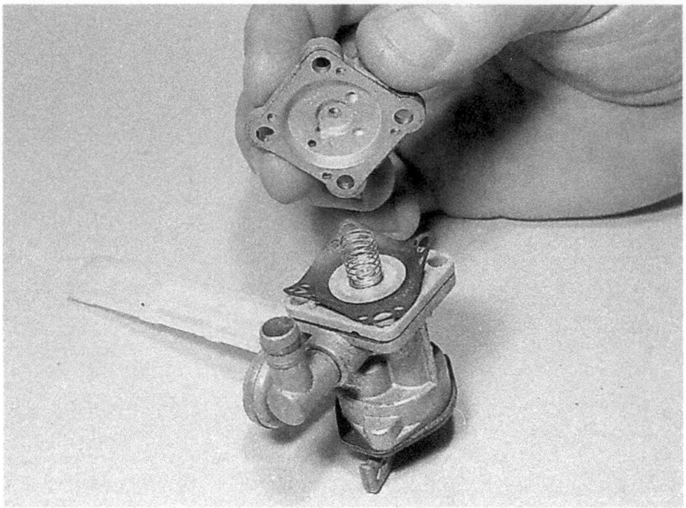

3.5b Do not omit coil spring on reassembly

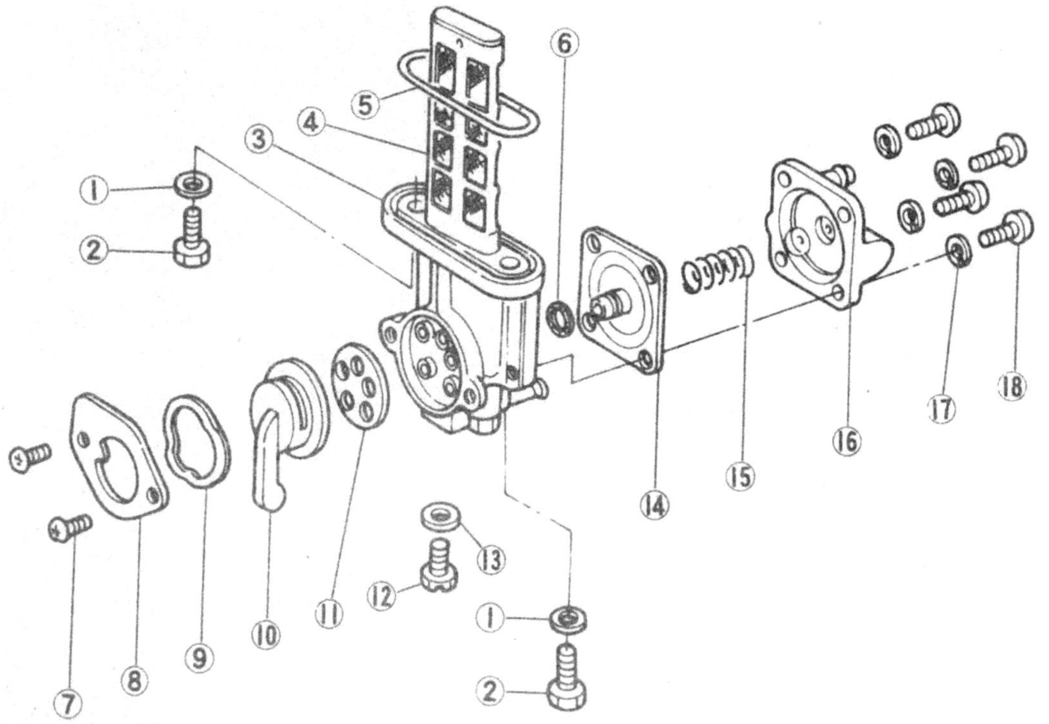

Fig. 2.1 Fuel tap – typical

1 *Gasket*
2 *Bolt*
3 *Fuel tap body*
4 *Filter*
5 *O-ring*
6 *O-ring*
7 *Screw*
8 *Holding plate*
9 *Wave washer*
10 *Tap lever*
11 *Valve gasket*
12 *Drain plug*
13 *Gasket*
14 *Diaphragm*
15 *Spring*
16 *Diaphragm cover*
17 *Lock washer*
18 *Screw*

4 Fuel hoses and other pipes: general

1 Thin-walled synthetic rubber tubing is used for many purposes, whether in the fuel system, emission control system or as drain or breather tubes. All hoses are of the push-on type, being secured by small wire clips. Normally it is necessary to renew hoses only if they become hard or split; it is unlikely that the clips will need frequent renewal as the main seal between hose and union is effected by the interference fit.

2 Check carefully at periodic intervals that the hoses are correctly fitted, undamaged, and secured to the frame by any clamps or ties provided. Check that they are correctly routed and that no drain or breather hoses are long enough to interfere with the final drive chain and gearbox sprocket or with the rear brake or suspension. If the hoses split, it is normally at the end, on or close to the union. In such cases the damaged length can be cut off and the hose refitted.

3 If any hose has to be renewed, use only the genuine Kawasaki replacement parts, particularly on emission control systems. Where hoses are moulded to a particular shape or where they are of an unusual size, this will be necessary anyway. The only exception to this is that it is permissible to use proprietary synthetic rubber or neoprene tubing for vacuum breather and drain hoses and, in emergency, for fuel hoses. Never use natural rubber tubing or clear plastic petrol pipe. Neither of these is suitable for such use.

5 Carburettors: removal and refitting

1 Remove the fuel tank as described in Section 2 of this Chapter, then remove the sidepanels and the electrical cover (where fitted), disconnect and remove the battery and disconnect from the carburettor bank and air filter casing all emission control system hoses (US models only); note carefully all hose connections. Release the carburettor drain and breather hoses and hang them on the right-hand side of the machine.

2 Slacken fully all intake stub clamp screws and roll towards the filter casing the spring retainers which secure the air filter rubbers. Slacken the adjuster locknut(s) and screw in fully the throttle cable twistgrip adjuster(s).

3 Remove the air filter element(s) as described in Routine Maintenance. On Z400 J, Z500 and KZ/Z550 A, C and D models, remove the small baffle plate secured by two bolts to the frame above the filter casing, then remove the two casing rear mounting brackets which are secured to the frame by a single bolt on each side. On ZX550 A1, A1L and all shaft drive models, remove the two bolts and mounting brackets which secure the rear of the air filter element housing, also the two bolts which secure the filter casing front half to the frame, thus separating the two parts of the filter casing. On ZR400, ZR550 and KZ/Z 550 H1 and H2 models, remove both left- and right-hand air filter housings, noting that it will be necessary to detach the battery case to gain access to one of the right-hand side housing retaining screws.

4 Pull the assembly to the rear to clear it from the intake stubs, then twist it carefully to disengage the air filter rubbers and manoeuvre it out to the right until each throttle cable can be disconnected from the pulley, noting that some cables have no lower end adjuster and can be pulled out until the inner wire can be passed through the slot in the bracket. Others are fitted with adjusters which must be slackened before the cable is removed. With the cables removed, pull carefully the carburettor assembly out to the right.

5 The above procedure describes the bare essentials of what is a very awkward and difficult procedure; it is, however, up to the individual owner's ingenuity to discover if there is any easier way. For example, it was found on a Z550 H2 that it was possible to remove and refit the air filter rubbers when the carburettors were still in place; this provided a great deal more working space and aided the task considerably. It is not possible, however, on other models such as the ZX550 A1 featured in the photographs; the filter rubbers project too far inside the casing and cannot be removed. No matter which way the carburettors are removed, great care must be taken to ensure that they cannot be damaged. Exercise care and patience at all times.

6 Refitting is the reverse of the removal procedure. The need for care and patience is just as great, except that a small quantity of light grease may be smeared over the insides of both filter rubbers and intake stubs to ease the task. Do not forget to stop and refit the throttle cable(s). Check that the carburettors are pushed fully forwards, at both ends, into the intake stubs before tightening the clamp screws, then check that the filter rubbers are correctly located before securing them with the spring retainers. Do not forget to secure the filter casing rear mountings. Adjust the throttle cable as described in Routine Maintenance.

7 If the intake stubs are removed at any time, check that they are refitted in exactly the same way so that the vacuum take-off stubs are correctly aligned (where fitted). On UK models the take-off stubs point downwards and to the outside, but on US models, Numbers 1 and 4 cylinder stubs point upwards and to the inside while those for Numbers 2 and 3 cylinders point downwards and to the outside. On later models the take-off stubs project from the carburettor bodies, therefore all intake stubs are the same. Where O-rings are fitted these must be renewed whenever the stubs are disturbed to prevent induction leaks; if no O-rings are fitted, apply a thin film of petrol-proof sealant. Tighten the retaining screws to the recommended torque settings (where given).

6 Carburettors: dismantling and reassembly

Carburettor separation – slide type

1 Do not remove the carburettors from their mounting plate unless absolutely necessary; each carburettor can be dismantled sufficiently for all normal cleaning and adjustment while in place on the mounting plate. If separation is necessary, proceed as follows:

2 Note the relative positions of the pilot air screws and drain plugs, and of the fuel and air vent hoses, these being the only way of identifying individual carburettor bodies. To be safe, mark each carburettor body with the number of its respective cylinder. Unscrew the idle speed adjusting screw and remove its spring and flat washer, prise out the black rubber plugs from each end of the throttle shaft, displace the throttle return spring and remove the cap and gasket from each carburettor top, the caps each being secured by three screws.

3 Remove the screw locking each throttle arm to the shaft, the screws locking the right- and left-hand cable pulleys to the shaft and the screw locking the fast idle arm to the shaft, noting that it may be necessary to slacken its locknut and to remove the fast idle adjusting screw to gain access to this last locking screw. Remove from the mounting plate the shaft set plate which is retained by a screw and lock washer. Push the throttle shaft out to the left, catching the cable pulleys and fast idle arm as they are released. It will be necessary to rotate with a screwdriver the shaft while pressing to assist removal. Withdraw the throttle valve assemblies, marking each one with its respective cylinder number using a felt marker or a tied-on label.

4 Do not disturb the adjusting screw setting at the end of each throttle arm or subsequent carburettor adjustment will be very time-consuming and awkward. Remove the two screws securing each slide to its bracket to release the slide and to gain access to the needle. On reassembly, note that the slide cutaway must be on the opposite side to the throttle arm, and refit each valve assembly in its original carburettor body, taking care that the needle enters correctly into its jet.

5 The fast idle cam is situated between numbers 1 and 2 carburettors; remove the screw locking it to the choke shaft, also the screws locking each carburettor's choke arm to the shaft, then slowly pull out the choke shaft; catching the steel ball and spring from number 1 carburettor body, and collecting the four choke arms and valves and the fast idle cam with its spring and flat washer as they are released. Again, mark all choke assemblies so that they are only refitted in their original carburettor bodies.

6 Finally remove the two mounting screws and withdraw each carburettor in turn from the mounting plate, disengaging the fuel supply and air vent tubes.

7 Reassembly is the reverse of the above procedure, noting the following points. The fuel supply T-piece is fitted between numbers 2 and 3 carburettors, the shorter horizontal end towards number 3; renew the two O-rings sealing each end. Apply thread locking compound to the threads of the carburettor mounting screws and tighten the screws by the very careful use of an impact driver. Support the carburettor bodies while the impact driver is in use. Grease lightly the throttle shaft before refitting and ensure that its notched end is on the right-hand side; note that there is a projection on the set plate which must engage with a recess in the mounting plate so that the set plate is correctly aligned with the throttle shaft groove and will retain the shaft without preventing its free movement.

8 On refitting the choke shaft, refer to the accompanying illustration to ensure that all components are correctly refitted and do not forget to install the coil spring and steel ball in number 1 carburettor body before the shaft is pressed fully into position. When the shaft is fitted and secured, move the lever throughout its full travel and check that each choke valve moves smoothly and easily from the fully closed to the fully open position, closing completely the carburettor bore in the first case, and withdrawing fully above the bore in the second. If necessary, alter the height of each valve by slackening its locking screw and rotating very slightly the choke arm on the shaft.

9 If the throttle valve height adjusting screws were disturbed, they can be reset as follows with sufficient accuracy for the engine to start and tick over. Screw in the idle speed adjusting screw until it just touches the cable pulley, then screw it in another $\frac{1}{4}$ turn from that point. Obtain a metal rod 0.5 – 1.0 mm (0.02 – 0.04 in) in diameter (a sewing needle, for example) and lift number 1 carburettor throttle slide just enough for the rod to be inserted between the bottom of the slide, opposite the cutaway, and the bottom of the bore. Lower the slide to trap the rod then lay the carburettor assembly flat on a clean surface with the engine side downwards.

10 Slacken the locknut and very slowly unscrew the adjusting screw until the rod is just released. Taking care not to move the screw, tighten the locknut lightly. Repeat this procedure on the remaining three carburettors, then unscrew the idle speed adjusting screw until it no longer touches the cable pulley; this last will mean that the engine will not tick over on initial start up but it will prevent the risk of the engine speed rising uncontrollably. Control the speed on the twistgrip until the engine has warmed up sufficiently for a preliminary idle speed to be set. To finish the initial setting procedure, open fully the throttles and check that the front (engine) edge of all slides has risen to just above the top of each carburettor bore. If any slide is protruding into its bore, slacken its locknut and unscrew the throttle stop screw set in the fast idle arm until the setting is correct, then tighten the locknut.

11 Lastly check that there is 1.7 – 1.9 mm (0.07 – 0.08 in) clearance between the bottom front (engine) edge of all slides and the bottom of their respective carburettor bores when the choke lever is fully raised to the closed position, ie choke valves closing off the carburettor bores.

This is easiest to check using a metal rod of the correct diameter in the same way as that described above for initial balancing, but is adjusted using the fast idle adjusting screw. Since this acts on the throttle shaft, all four slides are raised and lowered together; therefore there is no need to repeat the check on the other three carburettors.

12 Remember that it is essential that the carburettor synchronization be finally set using vacuum gauges; this must be done whenever any point of the throttle mechanism is disturbed in any way. Refer to Section 10.

Carburettor separation – constant depression type

13 Never remove the carburettors from their mounting plates unless absolutely necessary; each carburettor can be dismantled sufficiently for all normal cleaning and adjustment while in place on the mounting plates. If separation is necessary, mark each carburettor body with the number of its respective cylinder.

14 Remove the throttle cable bracket, which is retained by two screws, then displace the four circlips from the choke shaft. Slowly pull out the choke shaft collecting the plunger levers and their springs as they are released and catching the steel ball and spring set in number 3 carburettor body. Place the assembly on a clean work surface and remove all sixteen mounting screws, then withdraw both mounting plates. Very carefully disengage each carburettor from its neighbour, freeing the fuel and vent tubes and noting carefully exactly how the throttle linkages engage with each other. Catch the three small coil springs and flat washers that will be released, one from under each adjusting screw, noting that it may be necessary to slacken off the adjusting screws to gain sufficient space to disengage the linkage; if so, note carefully the number of turns required.

15 On reassembly, place the coil spring and flat washer over the pin of the relevant carburettor's throttle linkage, then insert it into the lower part of the next carburettor's linkage and return the adjusting screws to its original position. Very carefully refit the fuel and vent tubes, noting that the fuel pipe T-piece is fitted between numbers 2 and 3 carburettors, while the vent tube T-pieces fit between numbers 1 and 2, 3 and 4 carburettors. Lay the four carburettors on a flat surface and refit both mounting plates, tightening the mounting screws only loosely at first. Use a sheet of plate glass or a straight edge to ensure that all are aligned exactly both horizontally and vertically, then tighten firmly the screws.

16 Applying grease to stick them in place, refit first the coil spring, then the steel ball in number 3 carburettor body. Push the choke shaft slowly through refitting each plunger lever, and spring, as soon as it can be aligned with its respective choke plunger cap. Note that each spring is refitted on the shaft between the two lever edges. Check also that the shaft engages correctly with the spring-loaded ball in number 3 carburettor. When the shaft is refitted, secure it by refitting the four circlips. Check that the choke works properly. With the knob pulled fully out to the left, the ends of numbers 2 and 4 carburettor choke plungers should contact respectively the stops on numbers 1 and 3 carburettor bodies. Push the choke knob slowly in; at about the halfway point the spring-loaded ball should be felt to engage with a groove in the choke shaft, and with the knob fully in the plungers should be fully seated, and the shaft springs should be slightly compressed so that there is clearance between each circlip and the left-hand edge of the plunger lever.

17 If the throttle linkage settings were disturbed, it should be possible to set them accurately enough by eye for the engine to start and to tick over. Check that the clearance between the bottom edge of each butterfly and the bottom of its carburettor bore is the same on carburettors 1 and 2, and on 3 and 4; use the two outer adjusters to make the necessary adjustments. Next check that the clearance is the same between numbers 2 and 3 carburettors, using the centre adjuster to balance both pairs.

Carburettor dismantling and reassembly – both types

18 It is recommended that the carburettor bank be removed from the machine as described in Section 5 before any dismantling work is undertaken; if this is attempted while the carburettors are in place, not only is access very restricted but there is a high risk of damage and of small but vital components being lost.

19 The throttle and choke valve assemblies of slide carburettors are removed as described in paragraphs 1 – 5 above; on constant depression carburettors it is necessary to separate them before the choke plungers can be withdrawn. Unscrew the plunger cap and pull out the plunger.

20 On constant depression carburettors, remove the four screws which secure each carburettor top cover, withdraw the cover and spring, and carefully peel the diaphragm away from its seating groove; do not use any sharp instrument for this. Withdraw the piston, needle and diaphragm taking care not to damage the diaphragm. Displace the needle stop to withdraw the needle and its clip.

21 On both types of carburettor, gain access to the jets by removing the float chamber, each of which is secured by four screws. Press out the float pivot pin and remove the float assembly and float needle. The needle seat is screwed into place on slide carburettors and is pressed into place and retained by a single screw on constant depression carburettors; note the gasket or O-ring may be fitted around the seat. With one or two exceptions, all jets are screwed into their housings in the carburettor body and are removed using a spanner or screwdriver; some are covered by small rubber plugs which must be displaced first. The exceptions are on constant depression carburettors only, where the primary bleed pipe (where fitted) and needle jet are pressed into their seatings and must be displaced by passing a wooden rod down through the vacuum piston bore to push out each jet from above. Note that these carburettors also have a starter jet screwed into the float bowl and air jets screwed into the carburettor upper body, below the diaphragm. In addition to this, the KZ/Z550 H1 and H2 models are fitted with an early type of carburettor which has primary and secondary main and main air jets and a primary bleed pipe; these are shown in the accompanying illustrations and will not be found on other models.

22 The removal of the pilot air (or pilot mixture) screw requires a special approach. On all US models, deform the blanking plug using a punch or scriber, then lever it out. On all models, screw in the pilot screw until it seats lightly, counting the exact number of turns necessary to achieve this, then remove the screw with its O-ring and spring. On reassembly, refit first the O-ring, then the spring to the screw and refit the assembly to the carburettor body. Screw it in until it seats lightly, then unscrew it by the number of turns recorded; where such settings are given in the Specifications Section of this Chapter, the recorded setting should be very close to that recommended. Where the setting is given as 'Preset' it should not be altered except by a qualified mechanic measuring carbon dioxide levels with an exhaust gas analyser. US owners refer to Section 8 of this Chapter. On US models, press a new blanking plug to cover the screw and apply a trace of locking compound to the outside of the plug to secure it. Some UK models are fitted with rubber plugs.

23 On reassembling all carburettors, renew all sealing O-rings and gaskets as a matter of course. Use only close-fitting spanners or screwdrivers to refit the jets and tighten each one by just enough to secure it; it is easy to overtighten and shear a jet. On constant depression carburettors, insert the piston into its bore and lightly push it down until the needle passes into its jet, then press the diaphragm outer edge into its groove aligning the small tongue at the front with the notch in the carburettor body. Check that the diaphragm is fitted correctly without creases, and that the piston moves smoothly up and down its bore before refitting the spring and top cover.

6.15 Constant depression carburettors – ensure throttle linkages are correctly connected on reassembly

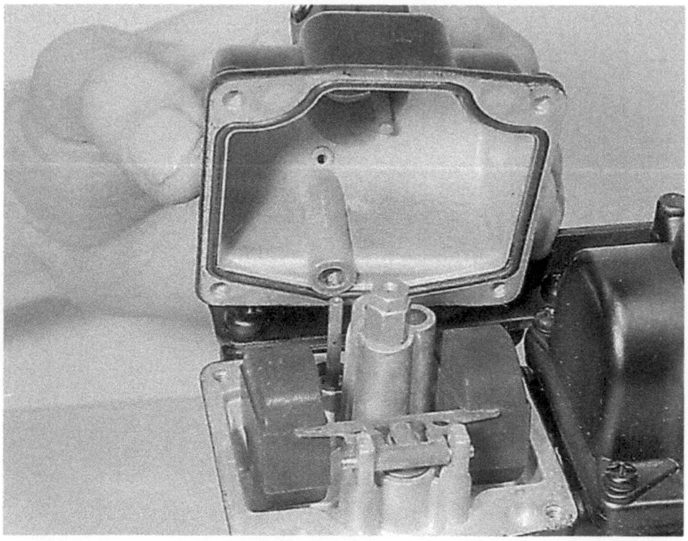

6.21a Remove float bowl to gain access to jets

6.21b Note various air jets in upper body of constant depression carburettors

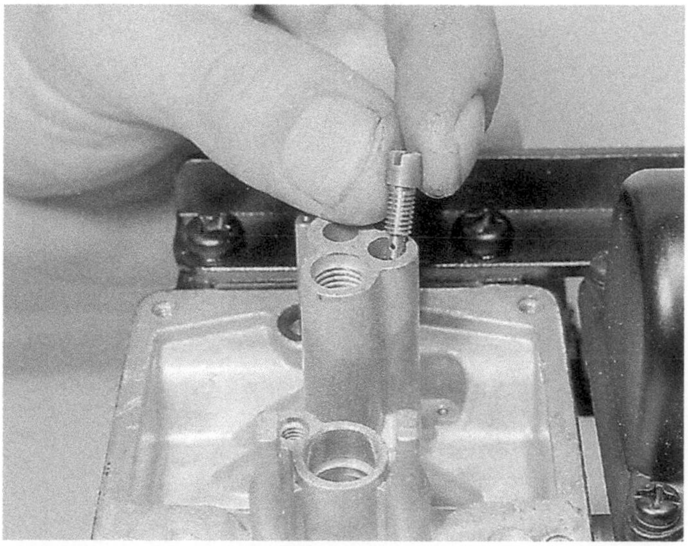

6.22a Pilot jet is screwed into position shown

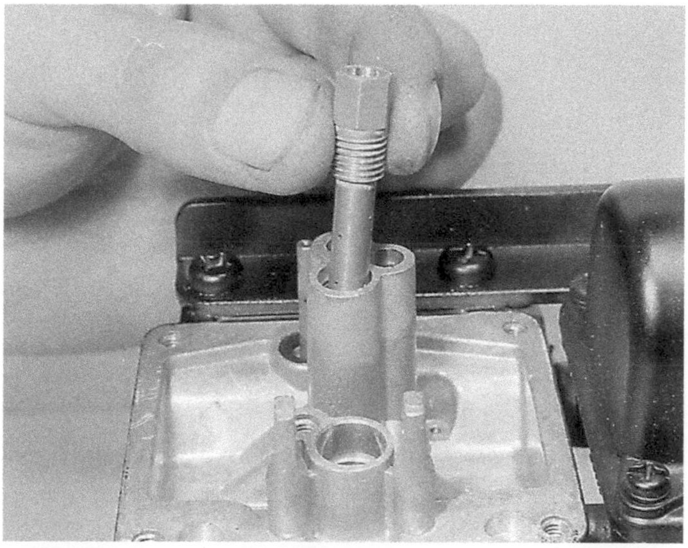

6.22b Needle jet is screwed into central location – ensure bleed holes are clear

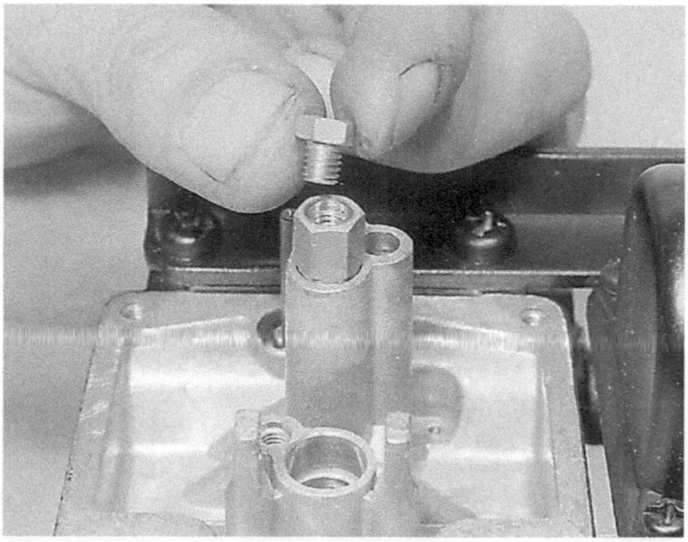

6.22c Do not overtighten main jets on refitting

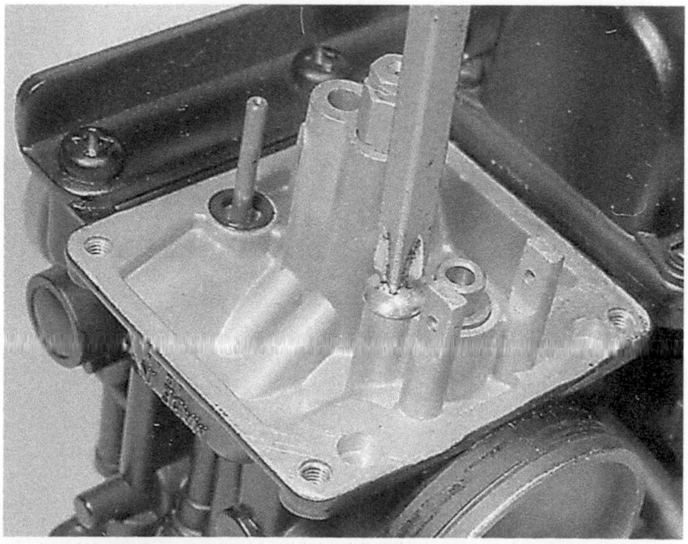

6.22d Float needle seat is retained by a single screw – constant depression carburettors

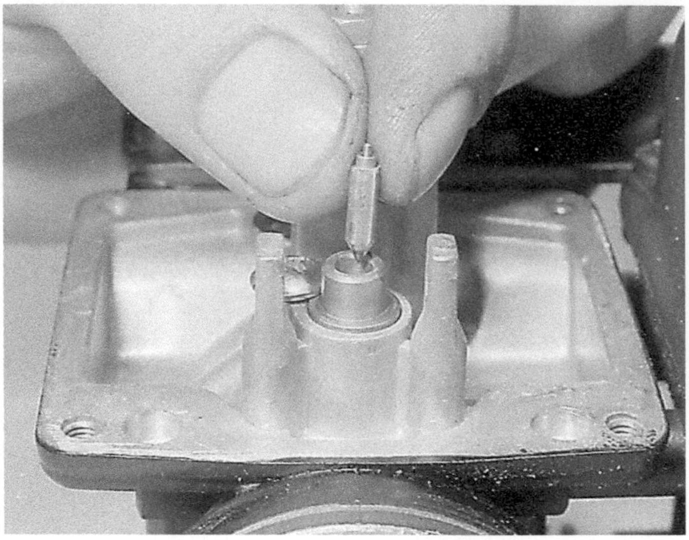

6.22e Do not forget to refit float needle ...

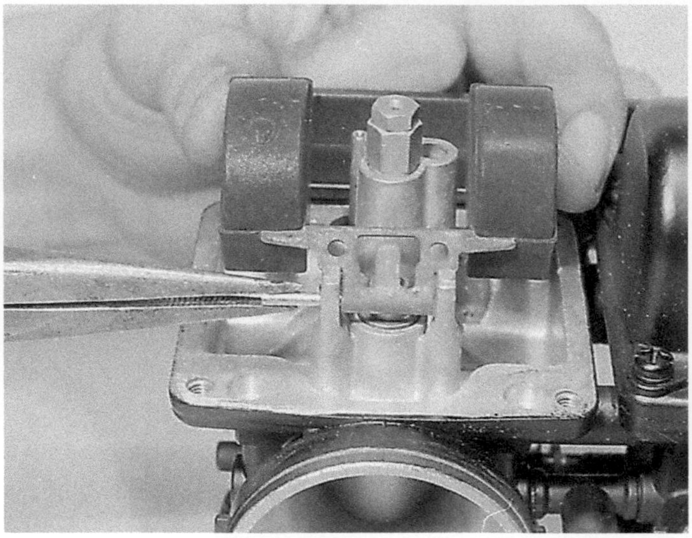

6.22f ... before refitting float assembly and pivot pin

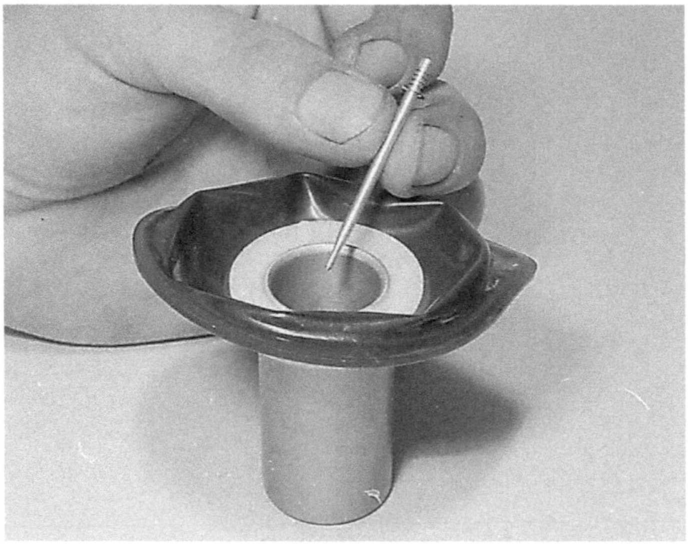

6.23a Constant depression carburettors – insert needle into piston ...

6.23b ... and secure with needle retainer ...

6.23c ... then refit piston assembly as shown

6.23d Ensure diaphragm is correctly positioned before refitting spring and top cover

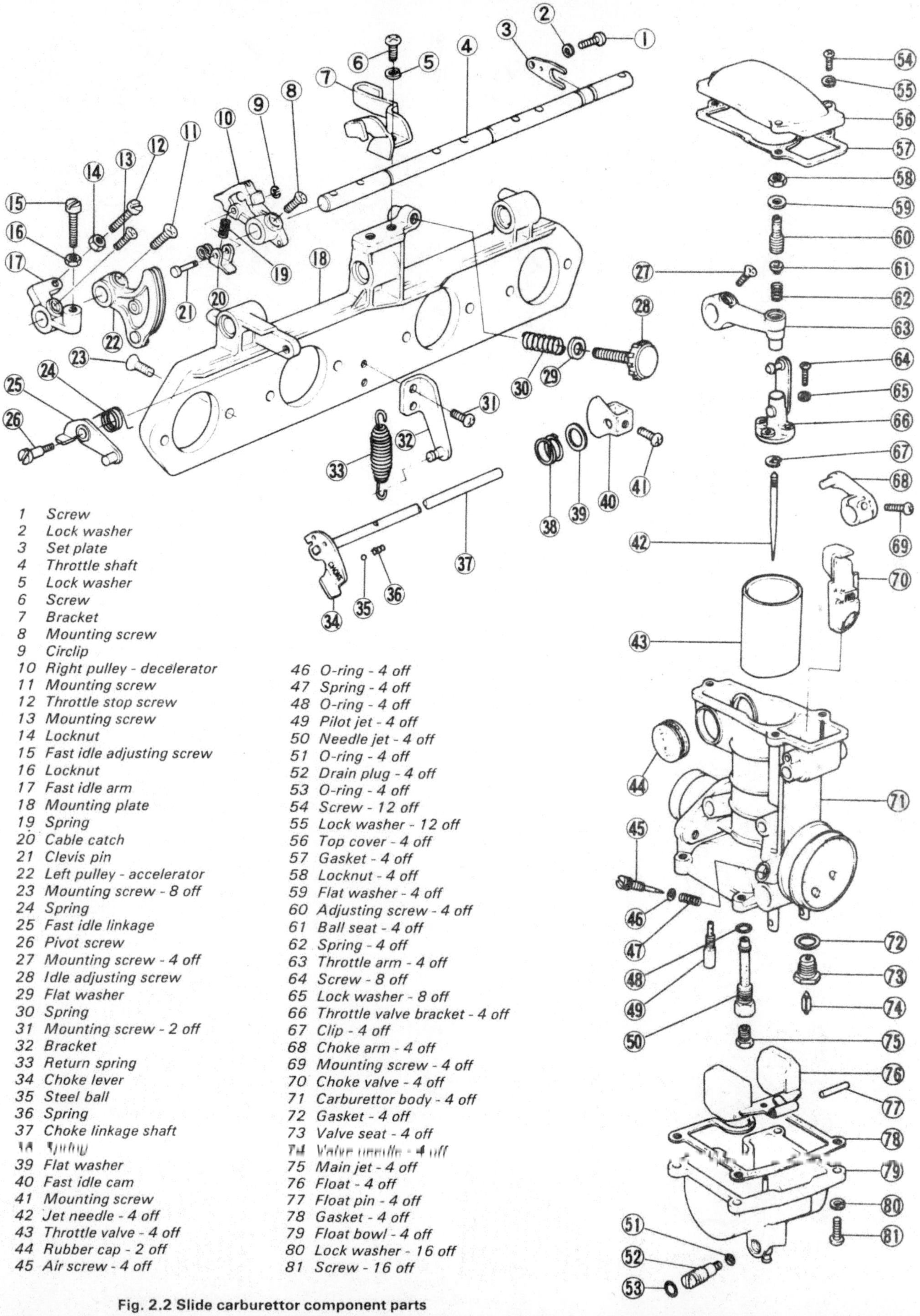

1 Screw
2 Lock washer
3 Set plate
4 Throttle shaft
5 Lock washer
6 Screw
7 Bracket
8 Mounting screw
9 Circlip
10 Right pulley - decelerator
11 Mounting screw
12 Throttle stop screw
13 Mounting screw
14 Locknut
15 Fast idle adjusting screw
16 Locknut
17 Fast idle arm
18 Mounting plate
19 Spring
20 Cable catch
21 Clevis pin
22 Left pulley - accelerator
23 Mounting screw - 8 off
24 Spring
25 Fast idle linkage
26 Pivot screw
27 Mounting screw - 4 off
28 Idle adjusting screw
29 Flat washer
30 Spring
31 Mounting screw - 2 off
32 Bracket
33 Return spring
34 Choke lever
35 Steel ball
36 Spring
37 Choke linkage shaft
38 Spring
39 Flat washer
40 Fast idle cam
41 Mounting screw
42 Jet needle - 4 off
43 Throttle valve - 4 off
44 Rubber cap - 2 off
45 Air screw - 4 off
46 O-ring - 4 off
47 Spring - 4 off
48 O-ring - 4 off
49 Pilot jet - 4 off
50 Needle jet - 4 off
51 O-ring - 4 off
52 Drain plug - 4 off
53 O-ring - 4 off
54 Screw - 12 off
55 Lock washer - 12 off
56 Top cover - 4 off
57 Gasket - 4 off
58 Locknut - 4 off
59 Flat washer - 4 off
60 Adjusting screw - 4 off
61 Ball seat - 4 off
62 Spring - 4 off
63 Throttle arm - 4 off
64 Screw - 8 off
65 Lock washer - 8 off
66 Throttle valve bracket - 4 off
67 Clip - 4 off
68 Choke arm - 4 off
69 Mounting screw - 4 off
70 Choke valve - 4 off
71 Carburettor body - 4 off
72 Gasket - 4 off
73 Valve seat - 4 off
74 Valve needle - 4 off
75 Main jet - 4 off
76 Float - 4 off
77 Float pin - 4 off
78 Gasket - 4 off
79 Float bowl - 4 off
80 Lock washer - 16 off
81 Screw - 16 off

Fig. 2.2 Slide carburettor component parts

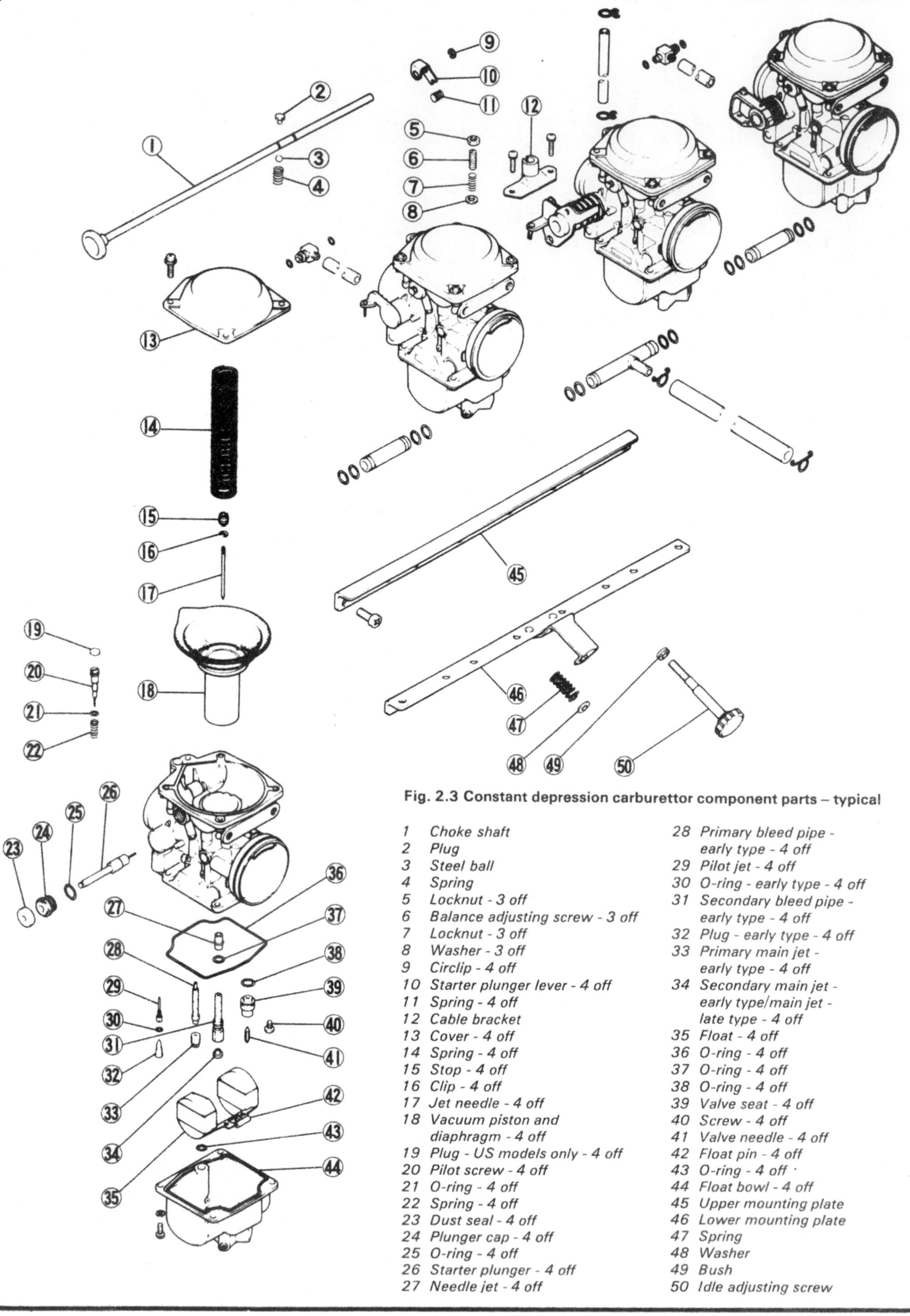

Fig. 2.3 Constant depression carburettor component parts – typical

1 *Choke shaft*
2 *Plug*
3 *Steel ball*
4 *Spring*
5 *Locknut - 3 off*
6 *Balance adjusting screw - 3 off*
7 *Locknut - 3 off*
8 *Washer - 3 off*
9 *Circlip - 4 off*
10 *Starter plunger lever - 4 off*
11 *Spring - 4 off*
12 *Cable bracket*
13 *Cover - 4 off*
14 *Spring - 4 off*
15 *Stop - 4 off*
16 *Clip - 4 off*
17 *Jet needle - 4 off*
18 *Vacuum piston and diaphragm - 4 off*
19 *Plug - US models only - 4 off*
20 *Pilot screw - 4 off*
21 *O-ring - 4 off*
22 *Spring - 4 off*
23 *Dust seal - 4 off*
24 *Plunger cap - 4 off*
25 *O-ring - 4 off*
26 *Starter plunger - 4 off*
27 *Needle jet - 4 off*
28 *Primary bleed pipe - early type - 4 off*
29 *Pilot jet - 4 off*
30 *O-ring - early type - 4 off*
31 *Secondary bleed pipe - early type - 4 off*
32 *Plug - early type - 4 off*
33 *Primary main jet - early type - 4 off*
34 *Secondary main jet - early type/main jet - late type - 4 off*
35 *Float - 4 off*
36 *O-ring - 4 off*
37 *O-ring - 4 off*
38 *O-ring - 4 off*
39 *Valve seat - 4 off*
40 *Screw - 4 off*
41 *Valve needle - 4 off*
42 *Float pin - 4 off*
43 *O-ring - 4 off*
44 *Float bowl - 4 off*
45 *Upper mounting plate*
46 *Lower mounting plate*
47 *Spring*
48 *Washer*
49 *Bush*
50 *Idle adjusting screw*

7 Carburettors: examination and renovation

1 Before any course of action is decided on, check with a good Kawasaki dealer exactly what is available for a particular model; in many cases components are available only as a part of a much larger assembly.
2 Check that the carburettor body, float chamber and mounting plates are free from wear or damage of any sort, and that all gasket faces are flat and unmarked and all bearing surfaces not excessively worn. Reassemble components such as the throttle or choke shafts temporarily to feel for free play; if excessive, the relevant parts must be renewed.
3 Jets should be cleaned using only compressed air; never use wire to poke clear a jet, although in extreme cases it is permissible to use a nylon bristle which will not scratch or enlarge the jet. Use a high-flash point solvent or proprietary carburettor cleaner to remove dirt and gum deposits, but check first that the type chosen will not affect rubber or plastic as there are a number of small seals which cannot be removed. If excessive dirt or water was found in the float chamber on dismantling, flush the fuel tank and clean the fuel tap filter as well before running the machine again.
4 The only components likely to suffer wear are the operating linkages, the throttle valve (slide) or piston, the throttle needle and jet, and the float needle and seating. The first two can be checked by feeling for free play or looking for excessive wear marks, the last two can be examined, using a magnifying glass if necessary. If any ridges or other signs of wear are seen the components concerned must be renewed as a matched pair. Note that a gauze filter is fitted to the float needle seat on some models; check that this is clean before refitting. The throttle needle is easily bent; if in doubt roll it on a sheet of plate glass and renew it, with its jet, if any sign of bending is seen. Do not attempt to straighten a needle; they break very easily.
5 On constant depression carburettors, check that the diaphragm is not split, perished, or otherwise damaged. Holding it up to a strong light source will usually reveal even the smallest hole. The diaphragm must be renewed even if slight damage is found; it is not worthwhile attempting repairs of any sort.
6 Renew the float assembly if damaged, or if petrol can be heard inside when it is shaken.

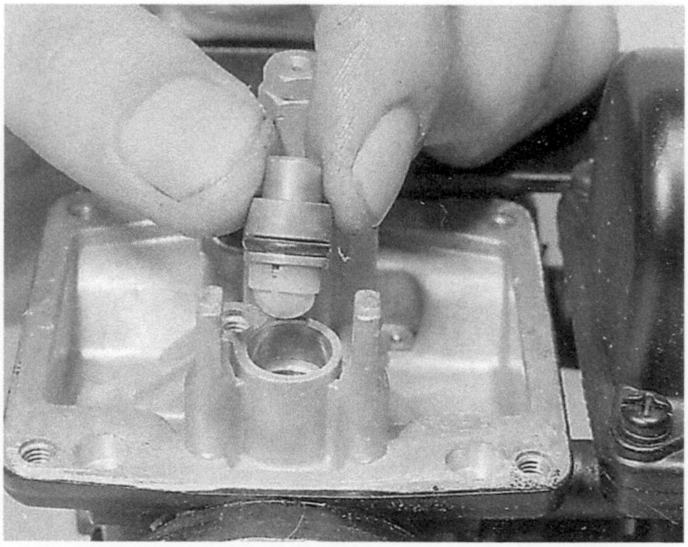

7.4 Gauze filter is fitted to float needle seat on some models – note sealing O-ring

8 Carburettor adjustment and exhaust emissions: general note

1 In some countries legal provision is made for describing and controlling the types and levels of toxic emissions from motor vehicles.
2 In the USA exhaust emission legislation is administered by the Environmental Protection Agency (EPA) which has introduced stringent regulations relating to motor vehicles. The Federal law entitled The Clean Air Act, specifically prohibits the removal (other than temporary) or modification of any component incorporated by the vehicle manufacturer to comply with the requirements of the law. The law extends the prohibition to any tampering which includes the addition of components, use of unsuitable replacement parts or maladjustment of components which allow the exhaust emissions to exceed the prescribed levels. Violations of the provisions of this law may result in penalties of up to $10 000 for each violation. It is strongly recommended that appropriate requirements are determined and understood prior to making any change to or adjustments of components in the fuel, ignition, crankcase breather or exhaust systems.
3 To help ensure compliance with the emission standards some manufacturers have fitted to the relevant systems fixed or pre-set adjustment screws as anti-tamper devices. In most cases this is restricted to plastic or metal limiter caps fitted to the carburettor pilot adjustment screws, which allow normal adjustment only within narrow limits. Occasionally the pilot screw may be recessed and sealed behind a small metal blanking plug, or locked in position with a thread-locking compound, which prevents normal adjustment.
4 It should be understood that none of the various methods of discouraging tampering actually prevents adjustment, nor, in itself, is re-adjustment an infringement of the current regulations. Maladjustment, however, which results in the emission levels exceeding those laid down, is a violation. It follows that no adjustments should be made unless the owner feels confident that he can make those adjustments in such a way that the resulting emissions comply with the limits. For all practical purposes a gas analyser will be required to monitor the exhaust gases during adjustment, together with EPA data of the permissible Hydrocarbon and CO levels. Obviously, the home mechanic is unlikely to have access to this type of equipment or the expertise required for its use, and, therefore, it will be necessary to place the machine in the hands of a competent motorcycle dealer who has the equipment and skill to check the exhaust gas content.
5 For those owners who feel competent to carry out correctly the various adjustments, specific information relating to the anti-tamper components fitted to the machines covered in this manual is given in the relevant Sections of this Chapter.
6 Note that no modification is necessary to any of the US models covered in this Manual to improve their emission control performance when used at altitudes above 4000 feet.

9 Carburettors: checking the settings

1 The various jet sizes, throttle valve cutaway (on slide type carburettors) and needle position are predetermined by the manufacturer and should not require modification. Check with the Specifications list at the beginning of this Chapter if there is any doubt about the types fitted. If a change appears necessary it can often be attributed to a developing engine fault unconnected with the carburettor(s). Although carburettors wear in service, this process occurs slowly over an extended length of time and hence wear of the carburettor is unlikely to cause sudden or extreme malfunction. If a fault does occur, check first other main systems, in which a fault may give similar symptoms, before proceeding with carburettor examination or modification.
2 Where non-standard items, such as exhaust systems, air filters or camshafts have been fitted to a machine, some alterations to carburation may be required. Arriving at the correct settings often requires trial and error, a method which demands skill born of previous experience. In many cases the manufacturer of the non-standard equipment will be able to advise on correct carburation changes.
3 As a rough guide, up to $\frac{1}{8}$ throttle is controlled by the pilot jet, $\frac{1}{8}$ to $\frac{1}{4}$ by the throttle valve cutaway, $\frac{1}{4}$ to $\frac{3}{4}$ throttle by the needle position and from $\frac{3}{4}$ to full by the size of the main jet. These are only approximate divisions, which are by no means clear cut. There is a certain amount of overlap between the various stages. The above remarks apply only in part to constant depression carburettors which utilise a butterfly valve in place of the throttle valve. The first and fourth stages are controlled in a similar manner. The second stage is controlled by the by-pass valve which is uncovered as soon as the throttle valve (piston) is opened. During the third stage the fuel passing through the main jet is metered by the needle jet working in conjunction with the piston needle (jet needle).
4 If alterations to the carburation must be made, always err on the

side of a slightly rich mixture. A weak mixture will cause the engine to overheat which may cause engine seizure. Reference to Routine Maintenance will show how, after some experience has been gained, the condition of the spark plug electrodes can be interpreted as a reliable guide to mixture strength.

5 The basic setting which must be checked if the carburettors are persistently too weak or too rich is the fuel level. This can be checked with the carburettors in place on the machine, provided it is standing upright, but since it is adjusted by altering the float height, it is recommended that the carburettors are removed and mounted absolutely upright on a work surface. This will mean the construction of a suitable stand and that the fuel tank or an alternative fuel supply must be arranged securely above the carburettors. The carburettors must be vertically upright for the check to be accurate.

6 The correct Kawasaki service tool, Part Number 57001-1017 is a clear plastic tube graduated in millimetres; an alternative is to use a length of clear plastic tubing and an accurate ruler. Connect one end of the tube to the float chamber drain outlet (use the drain tube if the Kawasaki tool is being used) and place the tube upper end against the carburettor body as shown in the accompanying illustration. Mark the tube at a point several millimetres above the bottom edge of the carburettor body (use the 'O' mark on the service tool), then unscrew the float chamber drain plug by one or two full turns and switch on the fuel supply (fuel tap to the 'PRI' position). Wait for the fuel level to stabilise then very slowly bring the tube down the carburettor body until the mark is level with its bottom edge. Do not lower the tube beyond this and raise it again or the level will be inaccurate. Measure the distance between the bottom edge of the carburettor body and the top of the fuel in the gauge.

7 The fuel level should be noted and the process repeated on the remaining carburettors. If any level is outside the tolerances given in the Specifications Section of this Chapter the carburettors must be removed from the machine and the setting altered as follows.

8 Remove the float chamber(s) and hold the carburettor assembly vertical with the air filter side upwards, then slowly invert it until each float is just resting on its needle and not compressing it. (See the accompanying illustration.) Measure the distance between the bottom gasket surface of the carburettor body and the bottom of each float of the float assembly. If there is any discrepancy it can be corrected by bending carefully the bridge piece. Note the float heights of all carburettors to be adjusted, then remove the float and bend as necessary the tang which bears on the float needle. Bending the tang up increases the float height and lowers the fuel level, therefore bending it down decreases the float height and raises the fuel level. Be very careful when bending the tang; only the smallest movement is necessary to effect a major change in float height. Note that the standard float height is 27 mm (1.063 in) for all constant depression carburettors.

9 When adjustment is complete, reassemble the carburettors and recheck the fuel level. Make the adjustments again, if necessary, but note that if serious difficulties are encountered, the float assembly, float needle and seat must be removed and checked very carefully for wear. Refit the carburettors to the machine when all carburettor fuel levels are at the correct setting, or at least within tolerances.

10 Where these are given as 'Preset' in the Specifications Section of this Chapter, the pilot screw settings should be regarded as fixed and should not be altered except by an experienced and qualified mechanic using the necessary diagnostic equipment. This is beyond the scope of most private owners. It is recommended that the same attitude be applied to all other models, the factory setting is usually best for all normal use and while badly-adjusted pilot screws will have a serious effect on engine performance, setting them accurately is by no means easy for the inexperienced. The object is to find the setting at which the engine runs fastest and smoothest when warmed up to normal operating temperature.

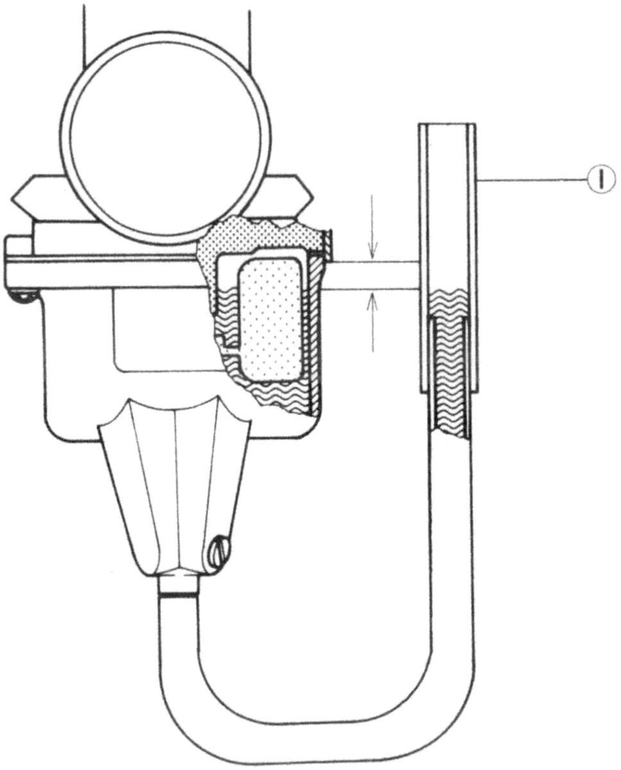

Fig. 2.4 Measuring the fuel level

1 Fuel level gauge

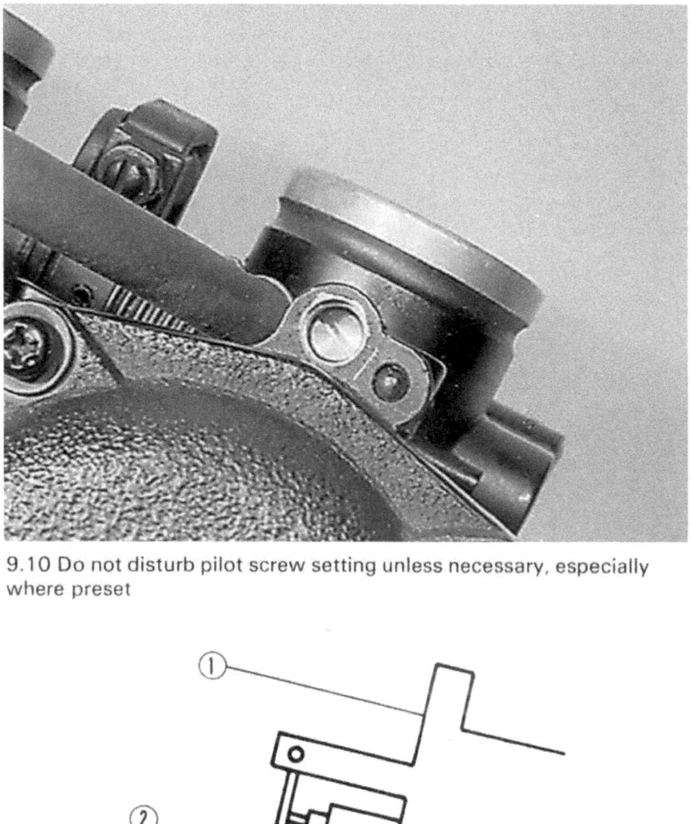

9.10 Do not disturb pilot screw setting unless necessary, especially where preset

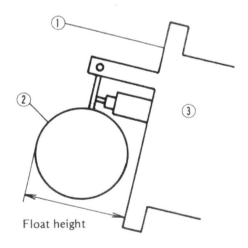

Fig. 2.5 Measuring the float height

1 Float bowl mating surface
2 Float
3 Carburettor body

10 Carburettors: synchronisation

1 Carburettor synchronisation must be checked at the interval specified in Routine Maintenance, and whenever the carburettors have been disturbed or if the engine is running roughly. Always check valve clearances and ignition timing before starting work. A set of accurate vacuum gauges is essential for the synchronisation, and if these are not available the job should be entrusted to a Kawasaki dealer. On no account attempt to adjust synchronisation by 'feel'. It will almost always make things worse.

2 Remove the fuel tank and arrange a temporary fuel supply, either by using a small temporary tank or by using extra long fuel pipes to the now remote fuel tank on a nearby workbench. **Note**: if the vacuum hose is bypassed, it is important to plug its open end before attempting the check. Connect the vacuum gauge hoses to the four vacuum take-off points, having first disconnected the relevant hose(s) and cap(s); the vacuum take-off points are to be found on the top or bottom of the intake stubs, or on the top of each carburettor body, where it enters the intake stub. Start the engine and allow it to warm up to normal operating temperature. If the gauges are fitted with damping adjustment, set this so that needle flutter is just eliminated but so that they can still respond to small changes in pressure.

3 Running the engine at idle speed, check that all needles produce the same reading. A tolerance of up to 2 cm Hg between cylinders is permissible but it is better to have all cylinders adjusted to the same reading; this is by no means as difficult as it would appear, requiring only a little care and patience. Note that it does not matter what the reading is; only that it is the same for all cylinders. Stop the engine and allow it to cool down if it overheats.

4 To adjust the setting on slide carburettors, remove the carburettor top covers (each retained by three screws) slacken the adjuster locknut and rotate the adjusting screw set in the end of the throttle arm. All carburettors are linked to the throttle shaft in the same way, and all are fitted with adjusting screws; therefore there is no particular order of adjustment. On constant depression carburettors, three adjusting screws and locknuts are fitted in the throttle linkage between the carburettors at the front. First check that each outer pair of carburettors is producing the same reading, using the outer adjusters, then balance both pairs against each other using the centre adjuster.

5 On both types when making adjustments, do not press down on the adjusting screw while rotating it, hold the screw in exactly the same position while fastening the locknut and do not overtighten the locknut. Open and close the throttle quickly to settle the linkage after each adjustment is made, wait for the gauge reading to stabilise and note the effect of the adjustment before proceeding.

6 When the carburettors are correctly synchronised, stop the engine, disconnect the gauges and refit all disturbed components.

10.2 It may be necessary to remove vacuum and balance hoses to connect vacuum gauges

11 Air filter: general

The care and maintenance of the air filter element(s) is described in Routine Maintenance. If it is necessary to remove the casing for any reason, this can be done after the carburettors have been removed. Never run the engine with the air filter disconnected or the element removed. Apart from the risk of increased engine wear due to unfiltered air being allowed to enter, the carburettors are jetted to compensate for the presence of the filter and a dangerously weak mixture will result if the filter is omitted.

US owners should note that the air filter is subject to the anti-tampering legislation currently in force, which means the machine must never be run with the filter element removed or rendered inoperative, or with the assembly altered in any way. Furthermore, only genuine Kawasaki replacement parts may be used if the renewal of any component is necessary.

12 Exhaust system: general

1 All models are fitted with an exhaust system consisting of two exhaust pipe/silencer assemblies that are joined by a connector pipe under the rear of the engine/gearbox unit. Numbers 2 and 3 cylinder exhaust pipes are separate, being fitted into stubs projecting from each main assembly.

2 No maintenance is required save to ensure that all mountings are secure. Renew all gaskets to prevent leaks whenever the system is disturbed.

3 Corrosion, both from inside and outside, is the most serious problem. Take care to keep the system as clean as possible at all times and to protect its finish using wax polish or similar products. Apply liberal quantities of penetrating fluid before attempting to dismantle the system; it is easiest to remove it as an assembly as described in Chapter 1, Section 5 and then to separate both main assemblies. Refitting is described in Section 45.

4 Note that if the renewal is necessary of any part of the exhaust system only genuine Kawasaki replacement parts should be used so that the machine complies with all noise and pollution regulations in force. Under (Federal) US law it is an offence to replace any part of the exhaust with a component not EPA-approved, and to modify in any way the system if the modification results in increased noise levels.

12.2 At regular intervals, check that all exhaust mountings and clamps are secure

13 Clean air system: description and renovation – US models only

1 The US models incorporate an air injection system designed to enhance the burning of hydrocarbons in the exhaust gases, thus reducing toxic emissions. The system employs a modified cylinder head and cover, in which air is drawn through a reed valve arrangement into the exhaust ports.

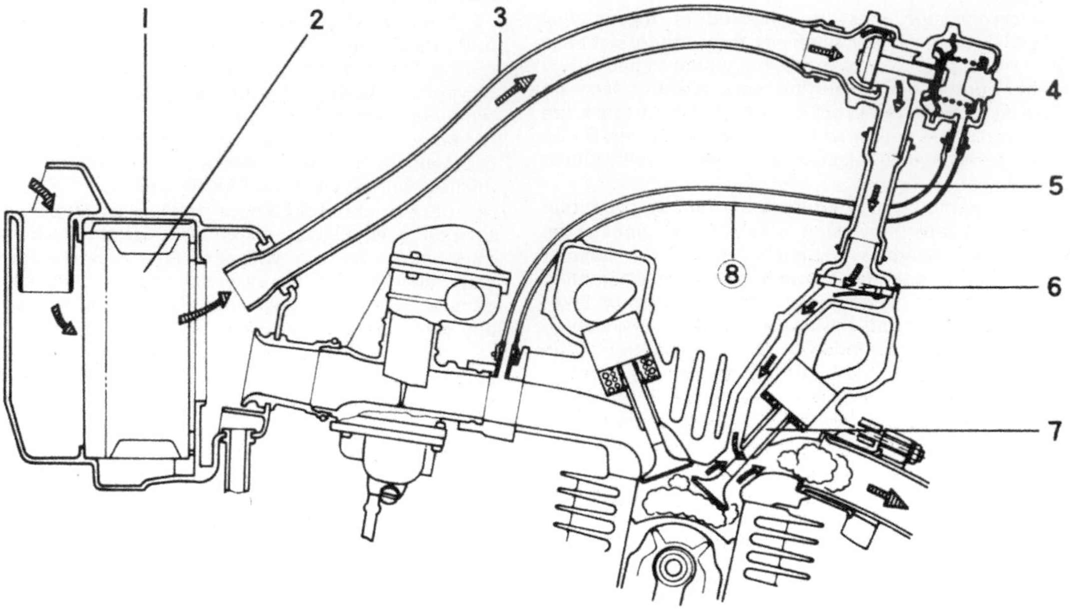

Fig. 2.6 Kawasaki Clean Air System – typical

1 Air filter housing
2 Air filter element
3 Connecting hose
4 Vacuum switch valve
5 Connecting hose
6 Air suction valve
7 Exhaust valve
8 Vacuum hose

2 The clean air system is automatic in operation and should not normally require attention. The most likely fault is that unfiltered air may be drawn into the system through a damaged air filter element or leaking hose making the tickover unstable and reducing engine power. Backfiring or other unusual noises may be apparent.

Air suction valve (reed valve)

3 The reed valves may be removed for examination after releasing the covers which house them. Check each valve for signs of deterioration, specifically examining the reeds for signs of delamination, cracking or scoring. Wash off any contaminants with a suitable solvent, and guard against scraping or scoring the sealing faces. The reeds and stopper plates may be removed if necessary, noting that Loctite or a similar compound must be applied to the screw threads prior to reassembly.

Vacuum switch valve

4 Regular inspection of the vacuum switch valve is unnecessary, and should be avoided unless a fault has been indicated. A vacuum gauge and a syringe-type vacuum pump are required to check that the valve closes at 38 – 48 cmHg and opens at 37 cmHg. Since few owners are likely to have access to this type of equipment, it is suggested that the check is done by a dealer, or by temporary substitution of a sound valve. Note that if the test is to be performed, no more than 50 cmHg should be applied to the valve. Note also that Kawasaki state that a faulty valve **must** be renewed and that 'adjustment is not permitted' despite the adjuster screw and locknut fitted to the valve.

Hoses

5 Damaged or perished hoses are probably the most likely cause of trouble in the system, and are fortunately the cheapest problem to remedy. Remember that air leaks will cause erratic running, and if located between the vacuum switch valve and reed valves, will allow unfiltered air to enter the reed valves.

General

6 The removal and refitting of the system components is described in Sections 5 and 45 of Chapter 1, and that of the air suction valves in Sections 7 and 44 of the same Chapter. The system is subject to the anti-tampering legislation currently in force in the US which means that the machine must never be used with any part of the system disconnected, missing, rendered inoperative or altered in any way. Use only genuine Kawasaki replacement parts if renewal of any component is necessary.

14 Evaporative emission control system: description and renovation – ZX550 A1L and KZ550 F2L only

1 To comply with legislation in force in that state and applying to all machines sold in California from 1984 onwards, these two models are fitted with equipment which prevents the escape into the atmosphere of any vapours produced by evaporation in any part of the fuel system. The equipment consists of a modified fuel tank, a canister of activated charcoal and a separator/pump unit, in addition to the connecting hoses and fittings.

2 Whilst the engine is stopped, vapour emitted by the evaporation of fuel in the tank passes through a blue-marked vent hose to the top of the separator, where some of it condenses and passes through the separator into the pump unit, but the majority passes into the canister. Vapour emitted from the carburettor float chambers passes through a yellow-marked vent hose directly to the canister. From there it can only escape to the atmosphere by passing through the activated charcoal which traps the vapour completely. This works equally well in reverse when the engine is running, air passing via the air filter and the green-marked purge hose to the canister and backwards through the system into the fuel tank and carburettors to compensate for the fuel consumed.

3 As soon as the engine is started, the system is purged using the partial vacuum created at various points as the engine is running. All vapour remaining in the canister is drawn via the green-marked purge hose into the air filter casing where it passes into the engine in the usual way. At the same time a simple pump, operated by the vacuum transmitted from number 3 carburettor inlet tract by the white-marked hose, returns all liquid fuel in the separator to the fuel tank; both these purging operations being completed within moments of starting the engine. Since the pump runs automatically whenever the engine is running, it maintains the system at a pressure below that in the air filter casing, allowing air to enter the system so that the fuel system components can 'breathe' as described above.

Fuel tank

4 This is fitted with a sealed cap in addition to the vent and fuel return pipes, and requires no maintenance other than to ensure that the cap seal, gasket and mounting screw O-rings are in good condition at all times. Renew the seals immediately if there is any doubt about their condition. On a general note, always plug the left-hand (fuel return) hose union whenever the tank is removed with fuel still inside it, to prevent the loss of fuel, and never fill the tank to above the bottom of the filler neck. If fuel rises under expansion into the filler neck, it may enter the system via the vent hose and flood it, which would produce hard starting and indifferent engine performance due to an over-rich mixture. Use compressed air to clear the cap vent and tank pipes if they are blocked.

Liquid/vapour separator and pump unit

5 Test the unit by removing the blue-marked vent hose from its top, then add about 20cc of gasoline via the hose fitting. Disconnect the fuel return hose from the tank union and place the hose open end in a container, level with the tank top. Start the engine and allow it to idle; all the gasoline should be ejected into the container almost immediately. If not, the unit must be renewed. Keep it upright at all times to prevent surplus fuel from escaping.

Charcoal canister

6 This should last the life of the machine in normal use and will not require attention of any sort other than to check its mountings and connections, but it should be noted that if fuel, solvent, water or any other liquid is allowed into the canister, its absorbing ability will be reduced to the point where it must be renewed. Check closely, therefore, for cracks or other damage.

Hoses

7 These should be examined and renewed if necessary as described in Section 4 of this Chapter. Use only genuine Kawasaki replacement parts and ensure that the hoses are connected following the instructions given in the accompanying illustration whenever they are disturbed.

8 Check that the hoses are not pinched or trapped, or unusual symptoms may arise. If the engine performs erratically at high speeds or if it stops with apparent signs of fuel starvation, or if the tank sides bulge out because of excessive internal pressure, check the blue-marked vent hose. If the engine is difficult to start and hesitates due to an over-rich mixture (ie produces clouds of black smoke), this is due to the canister being flooded because liquid fuel cannot return to the tank; check the red-marked fuel return hose and the white-marked vacuum hose, but note that similar symptoms may be caused by the presence of excess fuel vapour due to a blocked or pinched green-marked purge hose. If the engine is difficult to start and hesitates due to a weak mixture, or stops with apparent signs of fuel starvation, check the yellow-marked vent hose.

General

9 The system is subject to the anti-tampering legislation currently in force in the US which means that the machine must never be used with any part of the system disconnected, missing, rendered inoperative or altered in any way. Use only genuine Kawasaki replacement parts if renewal of any component is necessary.

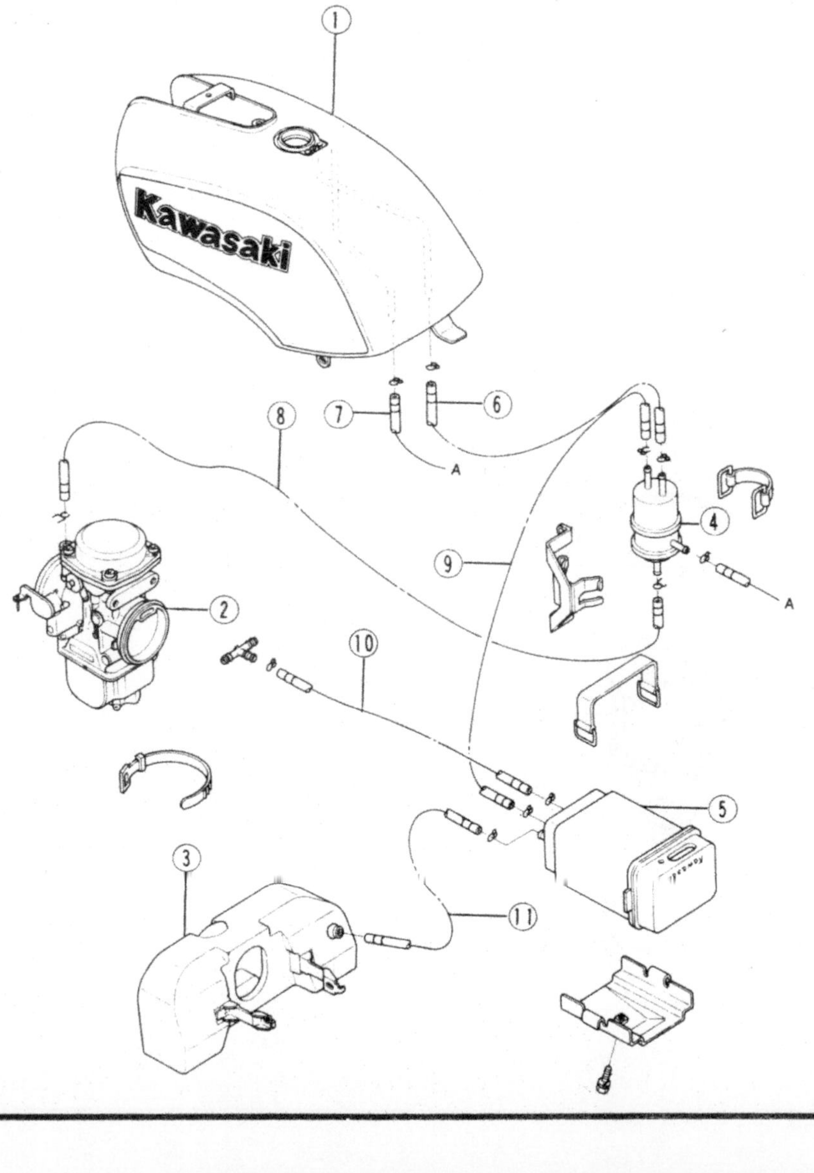

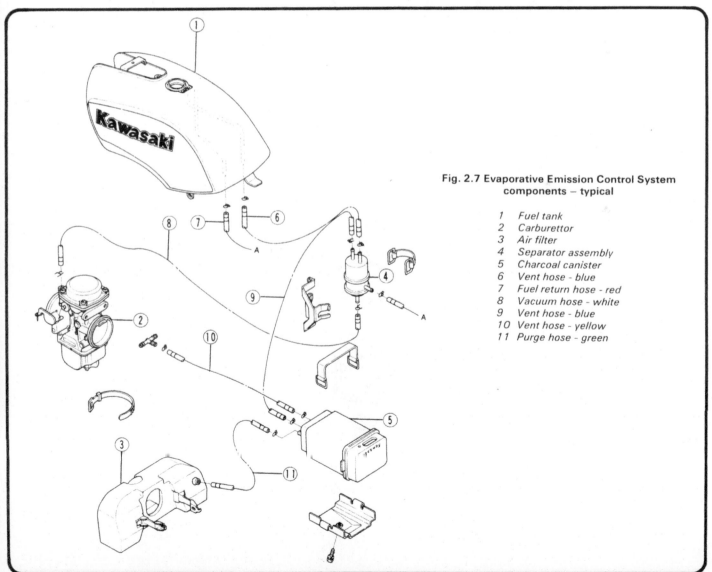

Fig. 2.7 Evaporative Emission Control System components – typical

1 *Fuel tank*
2 *Carburettor*
3 *Air filter*
4 *Separator assembly*
5 *Charcoal canister*
6 *Vent hose - blue*
7 *Fuel return hose - red*
8 *Vacuum hose - white*
9 *Vent hose - blue*
10 *Vent hose - yellow*
11 *Purge hose - green*

15 Lubrication system: checking the oil pressure

1 The efficiency of the lubrication system is dependent on the oil pump delivering oil at the correct pressure. This can be checked by fitting an oil pressure gauge to the right-hand oil passage plug, which is located immediately below the ignition pickup housing. Note that the correct threaded adaptor must be obtained or fabricated for this purpose. The best course of action is to obtain the correct Kawasaki pressure gauge and adaptor, Part Numbers 57001-164 and 57001-403/1188 respectively.

2 Remove the end plug and fit the adaptor and gauge into position. Start by checking the pressure with the engine cold. Note that if the engine is warm, **hot oil may be expelled when the end plug is removed.** Start the engine and note the pressure reading at various engine speeds. If the system is working normally, a reading of 4.4 – 6.0 kg/cm^2 (63 – 85 psi) should be maintained. If it exceeds the higher figure by a significant amount it is likely that the relief valve is stuck closed. Conversely, an abnormally low reading indicates that the valve is stuck open or the engine very badly worn. The test should now be repeated after the engine has warmed up. The correct pressure at 4000 rpm and 90°C (194°F) should be 2.0 – 2.5 kg/cm^2 (28 – 36 psi). If the oil pressure is significantly below this figure, and no obvious oil leakage is apparent, the oil pump should be removed for examination. On no account should the machine be used with low oil pressure, as plain bearing engines in particular rely on oil pressure as much as volume for effective lubrication.

3 It is likely that the normal oil pressure will be slightly above the specified pressure, but if it proves to be abnormally high, it is likely to be due to the oil pressure relief valve being jammed or damaged. The latter component is fitted to the inside of the sump. Refer to Section 17 of this Chapter.

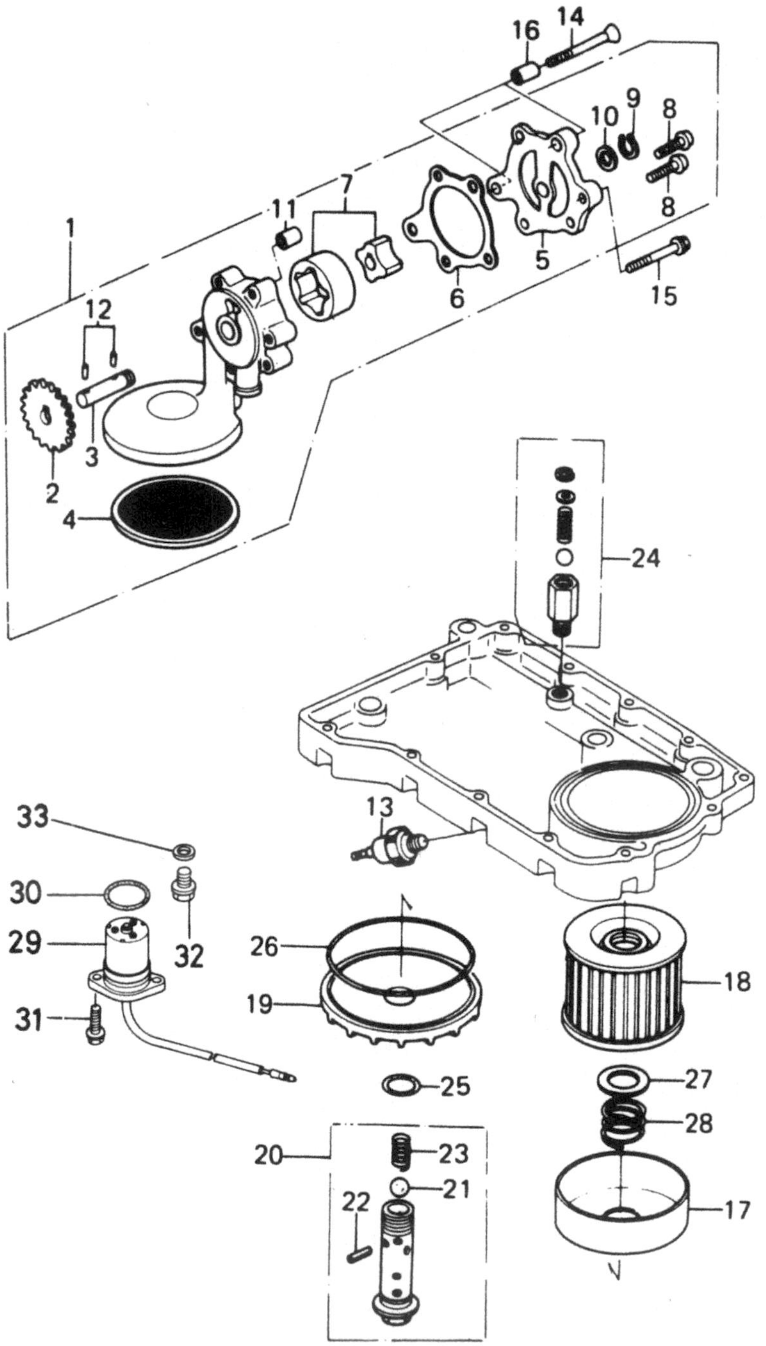

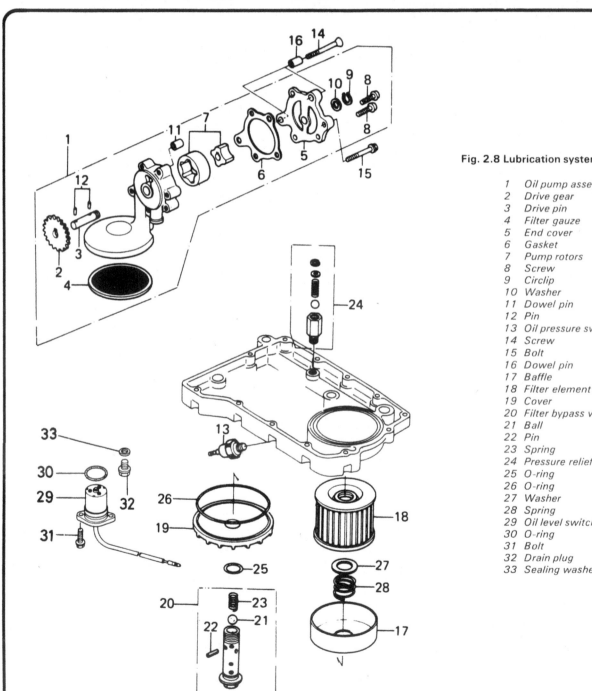

Fig. 2.8 Lubrication system components

1 Oil pump assembly
2 Drive gear
3 Drive pin
4 Filter gauze
5 End cover
6 Gasket
7 Pump rotors
8 Screw
9 Circlip
10 Washer
11 Dowel pin
12 Pin
13 Oil pressure switch
14 Screw
15 Bolt
16 Dowel pin
17 Baffle
18 Filter element
19 Cover
20 Filter bypass valve
21 Ball
22 Pin
23 Spring
24 Pressure relief valve
25 O-ring
26 O-ring
27 Washer
28 Spring
29 Oil level switch
30 O-ring
31 Bolt
32 Drain plug
33 Sealing washer

16 Oil pump: removal, examination and renovation

1 The oil pump is removed as described in Section 11 of Chapter 1 and is refitted as described in Section 40 of that chapter.
2 Remove the three oil pump cover screws and the circlip and thrust washer on the pump spindle. Lift the cover and gasket away. The inner and outer rotors can be shaken out of the pump body, the driving pin displaced, and the pump spindle withdrawn, noting the second pin locating the pump driven gear.
3 Wash all the pump components with petrol and allow them to dry. Check the pump casing casting for breakage or fracture, or scoring on the inside perimeter.
4 Reassemble the pump rotors and measure the clearance between the outer rotor and the pump body, using a feeler gauge. If the measurement exceeds the service limit of 0.30 mm (0.012 in) the rotor or the body must be renewed, whichever is worn. Measure the clearance between the outer rotor and the inner rotor, using a feeler gauge. If the clearance exceeds 0.30 mm (0.012 in) the rotors must be renewed as a set. With the pump rotors installed in the pump body, lay a straight edge across the mating surface of the pump body. Again with a feeler gauge measure the clearance between the rotor faces and the straight edge. If the clearance exceeds 0.12 mm (0.005 in) the rotors should be replaced as a set.
5 Examine the rotors and the pump body for signs of scoring, chipping or other surface damage which will occur if metallic particles find their way into the oil pump assembly. Renewal of the affected parts is the only remedy under these circumstances, bearing in mind that the rotors must always be replaced as a matched pair.
6 Reassemble the pump components by reversing the dismantling procedure. The component parts must be ABSOLUTELY clean or damage to the pump will result. Always renew the pump gasket. Replace the rotors and lubricate them thoroughly before refitting the cover.
7 Check that the pump turns smoothly, then refit it to the casing. Before refitting the sump, remove and examine the pressure relief valve as described in Section 17. Do not omit the large 'O' ring which must be fitted to the oil pump before the sump is refitted, and do not forget to clean the pump pick-up filter gauze. This can be done easily using solvent and a soft-bristled brush and should be carried out whenever the sump is removed. Ensure that the gauze is seated securely in the pump recess before refitting the sump.

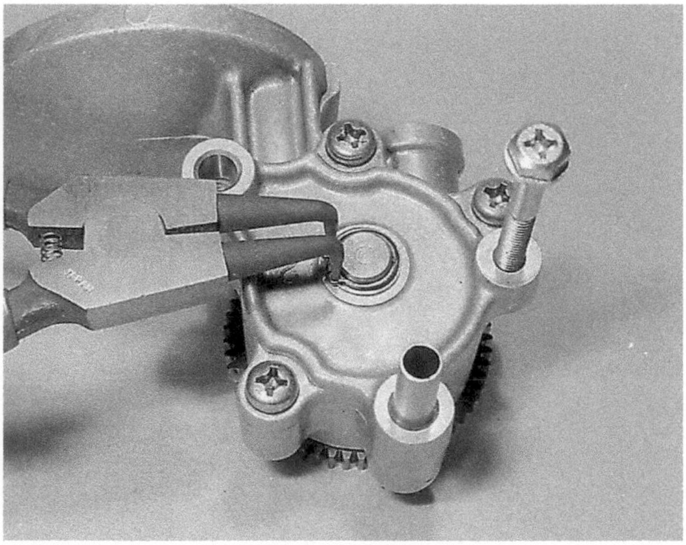

16.2a Remove from pump spindle the circlip ...

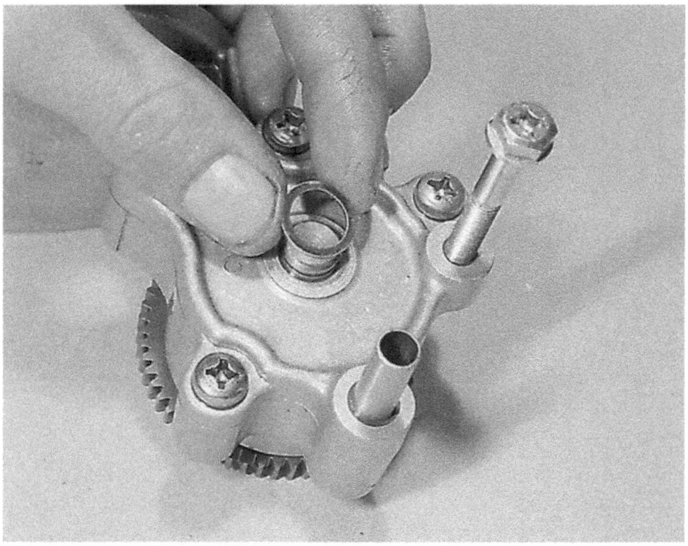

16.2b ... and thrust washer ...

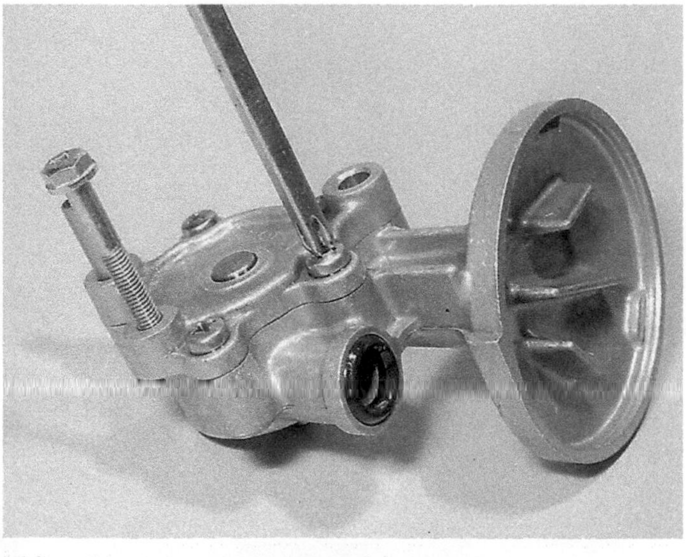

16.2c ... then remove pump cover retaining screws ...

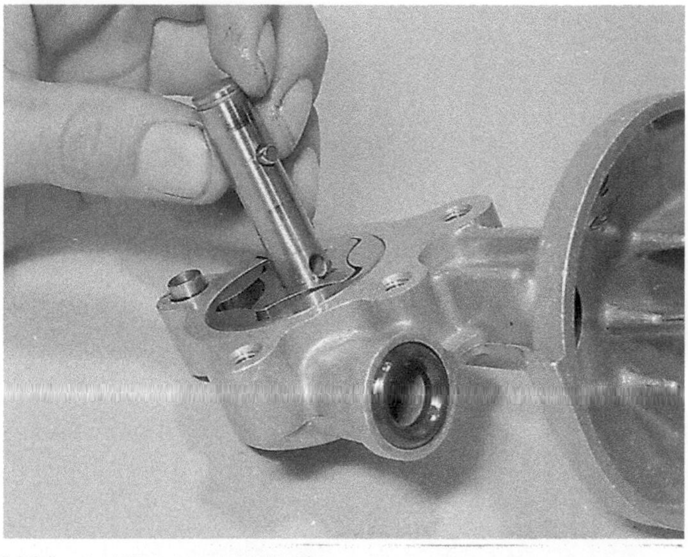

16.2d ... and lift out pump spindle ...

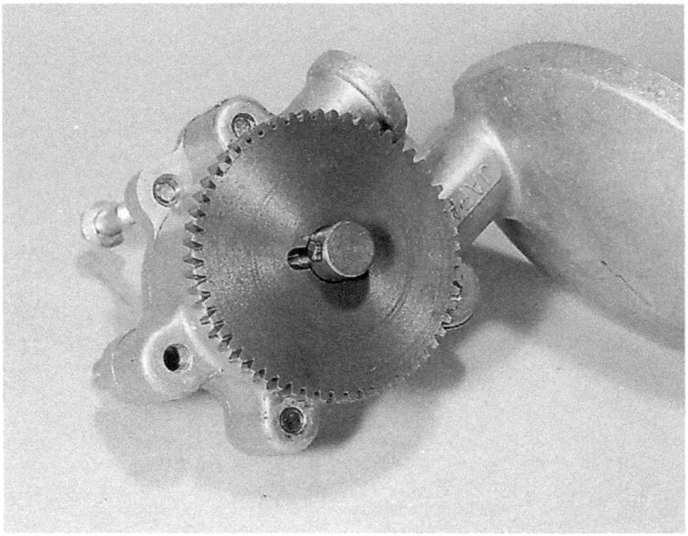

16.2e ... noting the second driving pin

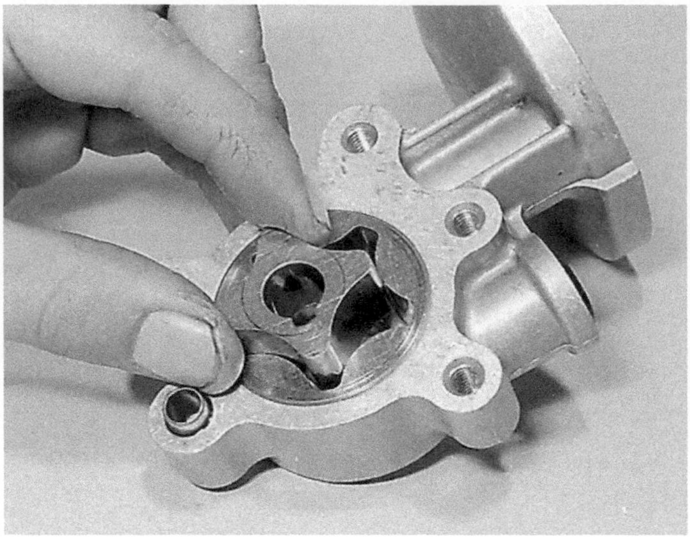

16.2f The pump rotors can then be withdrawn

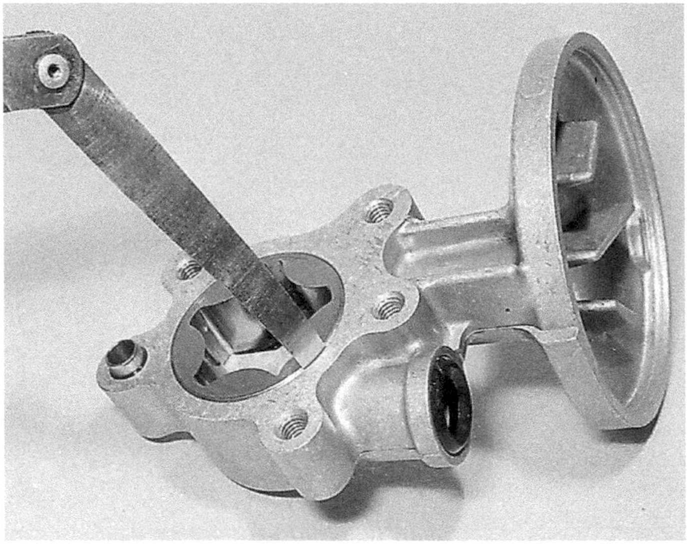

16.4a Measuring outer rotor/pump body clearance

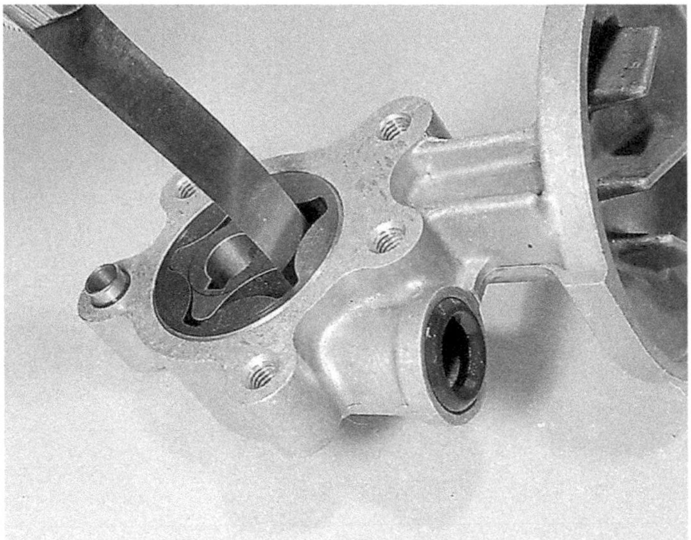

16.4b Measuring inner rotor/outer rotor clearance

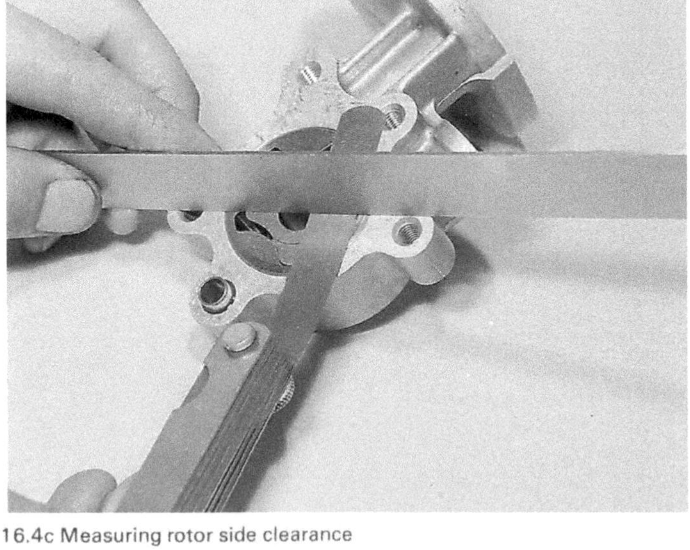

16.4c Measuring rotor side clearance

16.6 Always renew pump gasket (and O-rings) on reassembly – note dowel pin

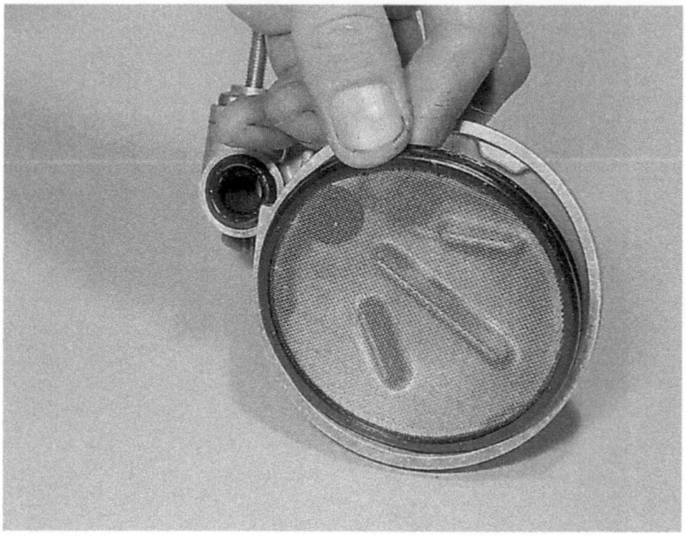

16.7 Do not forget to clean pick-up filter gauze whenever sump is removed

17 Oil pressure relief valve: removal and testing

1 The pressure relief valve is screwed into the sump (oil pan) from above; it is therefore necessary to remove the sump to gain access to it as described in Section 11 of Chapter 1. On reassembly, apply thread locking compound to its threads and tighten the valve to a torque setting of 1.5 kgf m (11 lbf ft), then refit the sump as described in Section 40 of Chapter 1.

2 Kawasaki caution against dismantling the valve, because it is felt that doing so may itself upset the valve assembly and cause inaccuracy. Using a wooden dowel or plastic rod, push the ball off its seat against spring pressure, noting that the ball should move smoothly, with no rough spots. If any hesitation is noted, wash the valve assembly thoroughly in a high flash point solvent, and blow it dry with compressed air. If no improvement is noted, renew the valve assembly as a unit, there being no provision for obtaining the parts individually.

18 Oil filter bypass valve: examination and renovation

1 The filter bypass valve is situated in the centre bolt of the oil filter assembly and is therefore removed and refitted with the oil filter as described in Routine Maintenance. Note that if the engine oil and filter are renewed at the specified intervals, it is unlikely that the bypass valve will ever come into operation or give trouble of any sort.

2 Its function is to ensure that the engine always receives a supply of oil (even if it is unfiltered) if the filter itself is too clogged to pass oil in sufficient quantities. If the filter does become this clogged, the oil is diverted underneath it and into the centre bolt via the bottom pair of holes which are kept open by the presence of the coil spring and flat washer. The bypass valve ball is forced off its seat when subjected to this extra pressure so that the oil can pass on into the engine.

3 It will be evident that diagnosis of a bypass valve fault is very difficult, but as the valve is so simple and so rarely used it is not likely to give trouble. It should be washed in high-flash point solvent whenever it is removed, and care should be taken that the filter is refitted correctly so that the valve can operate correctly if the need arises. If dismantling is necessary, place a wad of rag over the centre bolt open end, press out the retaining pin and tip out the spring and ball. All components should be examined for signs of wear, which should be evident, except for the spring which can only be compared with a new component, and renewed if necessary. Check that dirt is not present in the centre bolt then refit the ball, the spring and compress the spring while inserting the retaining pin.

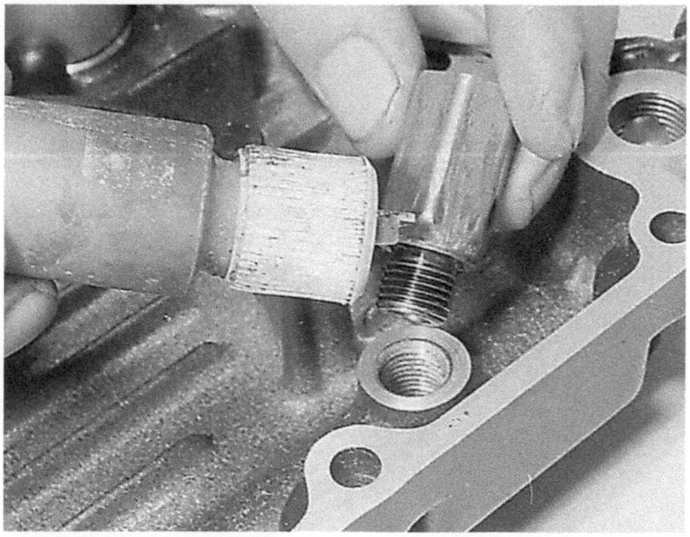

17.1a Apply thread locking compound to its threads on refitting pressure relief valve

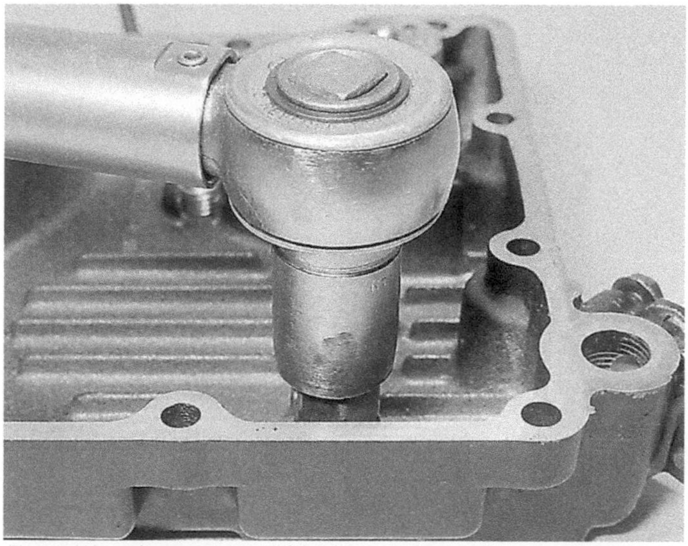

17.1b Do not overtighten – always use recommended torque setting

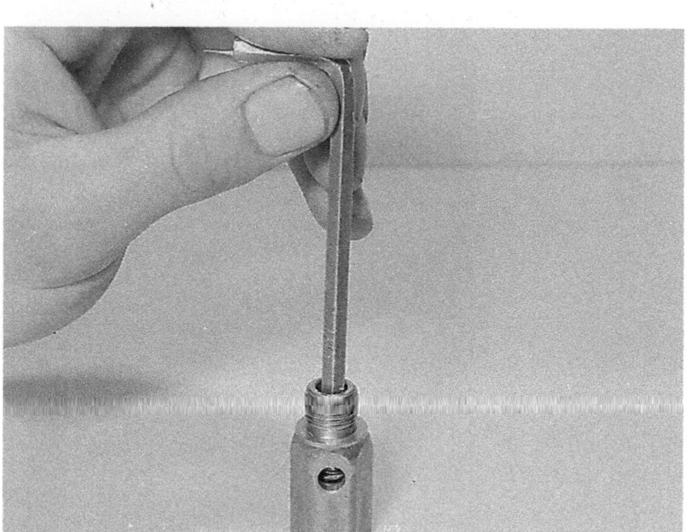

17.2 Test valve operation as shown – do not atttempt to dismantle valve

19 Oil pressure/level switch: removal and refitting

1 All models are fitted with a warning indicator which operates either when the oil pressure is too low or when the oil level is too low. The sender units are mounted on the outside of the oil pan (sump) and can be removed after the engine oil has been drained and their electrical leads disconnected.

2 The pressure switch is screwed into a recess next to the filter housing and is removed and refitted using a ring or socket spanner. On refitting wipe clean its threads and the surrounding area, then tighten the switch to a torque setting of 1.5 kgf m (11 lbf ft).

3 The oil level switch is secured by two bolts to the underside of the sump left-hand front corner. On reassembly, always renew its sealing O-ring and tighten securely its two mounting bolts.

4 To test the pressure switch, disconnect its lead, switch on the ignition and earth the lead against the crankcase. If the lamp lights, the switch is proved defective and must be renewed; if it does not light, check carefully the bulb and wiring. The level switch is tested as described in Chapter 6.

Chapter 3 Ignition system

For information relating to the 1984 on models, see Chapter 7

Contents

Specifications

Ignition system

Type:	
Z400 J1, Z500 B1, B2, KZ/Z550 A1	Battery and coil, contact breaker
All other models	Transistor-controlled electronic
Advance type:	
Z550 G2, KZ550 F2, F2L, ZX550 A1, A1L	Electronically controlled
All other models	Mechanically controlled
Advance curve – mechanically controlled advance/retard only:	
Advance begins at	1400 – 1600 rpm
Full advance at	3000 – 3400 rpm

Ignition timing

400 models:	
Initial	15° BTDC @ 1200 rpm
Full advance	35° BTDC @ 3200 rpm (40° – ZR400 A1, B1)
Z550 G2, ZX550 A1, A1L, KZ550 F2, F2L:	
Initial	12.5° BTDC at idle speed (7.5° KZ550 F2L)
Full advance	40° BTDC @ 10 000 rpm (32.5° KZ550 F2L)
All other models:	
Initial	10° BTDC @ 1050 rpm
Full advance	35 ° BTDC @ 3200 rpm

Contact breaker

Gap	0.35 mm (0.014 in)
Tolerance	0.30 – 0.40 mm (0.012 – 0.016 in)
Dwell angle	192.5 ± 7.5° (53.5 ± 2.5%)
Condenser capacity	0.24 ± 0.02 microfarad

Ignition HT coil

Minimum spark gap	6 mm (0.24 in)	
Winding resistances:	Contact breaker ignition	Electronic ignition
Primary	3.2 – 4.8 ohm	1.8 – 2.8 ohm
Secondary	10 – 16 K ohm	10 – 16 K ohm

Pick-up coil resistance

Z550 G2, ZX550 A1, A1L, KZ550 F2, F2L	376 – 564 ohm (Bk to Y cyl 1 and 4, BK/W to Bl cyl 2 and 3)
All other models	360 – 540 ohm (Bk to Bl cyl 1 and 4, Y to R cyl 2 and 3)

Spark plugs

Type:	NGK	ND
ZX550 A1L, A1 – US models	DP9EA-9	X27EP-U9
ZR400 B1, ZX550 A1 – UK model	DPR9EA-9	X27EPR-U9
Z400 J3, ZR400 A1, Z550 A3, A4, H1, H2, G1, G2, ZR550 A1, A2	DR8ES	X24ESR-U
All other models	D8EA	X24ES-U
Gap:		
ZR400 B1, ZX550 A1, A1L	0.8 – 0.9 mm ([illegible] in)	
All other models	0.6 – 0.7 mm (0.024 – 0.028 in)	

Torque wrench settings

Component	kgf m	lbf ft
Spark plugs	1.4	10.0
Advance/retard unit or ignition rotor mounting bolt	2.5	18.0
Ignitor unit mounting bolts – 1982 models on, units fitted with rubber mounting bushes	0.65	4.5

1 General description

Two basic types of ignition system are fitted. The Z400 J1, Z500 B1, B2 and KZ/Z550 A1 models are fitted with conventional battery and coil, contact breaker triggered ignition, while all other models are fitted with a magnetically-triggered, transistor-controlled system. Ignition advance is by a mechanical automatic timing unit (ATU) on all models except Z550 G2, KZ550 F2, F2L and ZX550 A1/A1L where it is controlled by separate circuits built into the ignitor unit.

On all models the system comprises two identical circuits, each circuit firing two of the four cylinders; numbers 1 and 4 cylinders being grouped in one circuit, while numbers 2 and 3 are in the other. For any given cylinder the plug is fired twice for every engine cycle, but one of the sparks occurs on the exhaust stroke and thus performs no useful function. The arrangement is usually known as a 'spare' or 'wasted' spark system. Note that since the spark plugs are connected in series, a fault occurring at one plug will also stop its fellow from firing.

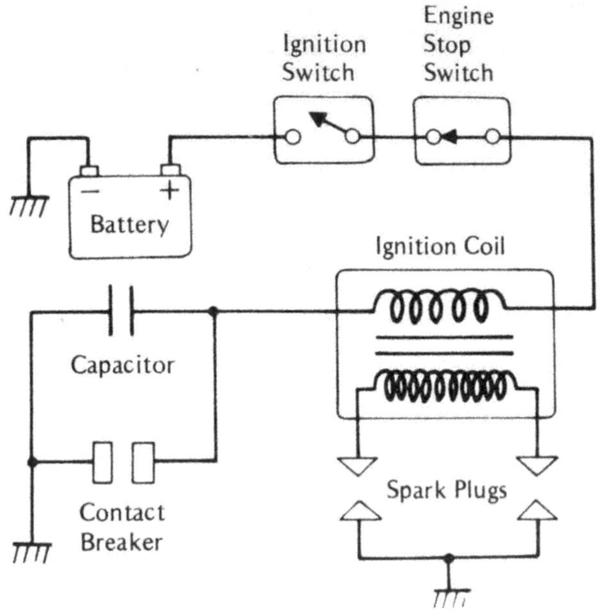

Fig. 3.1 Contact breaker ignition circuit – typical

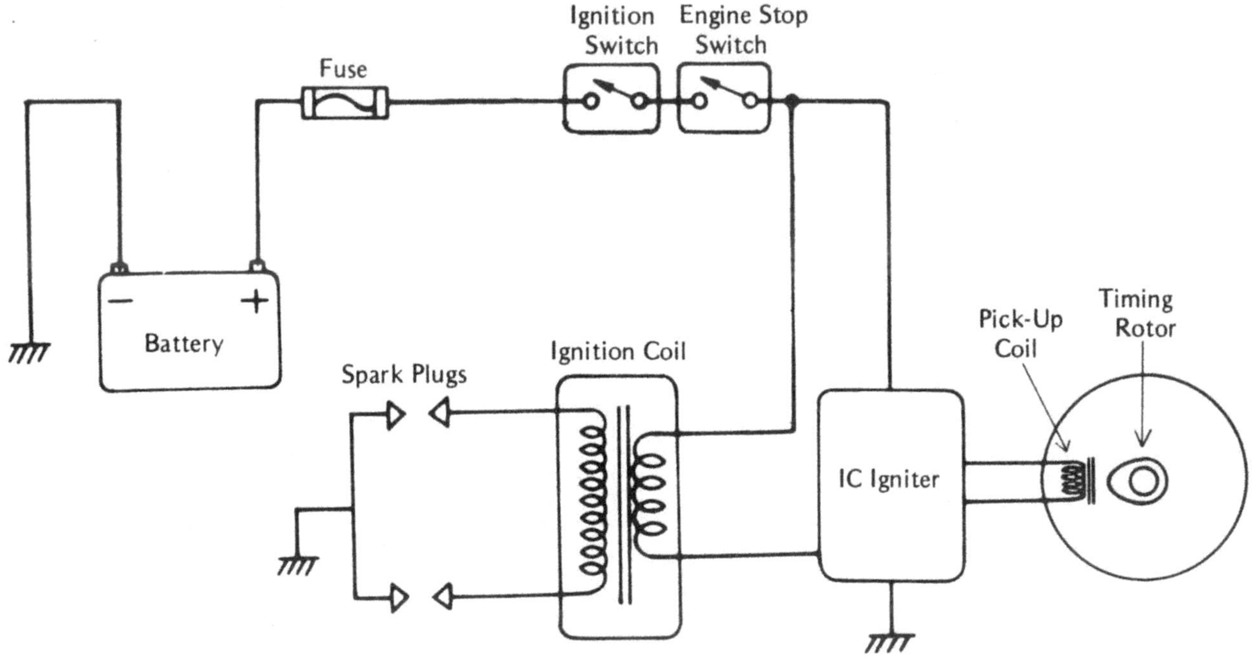

Fig. 3.2 Electronic ignition circuit – typical

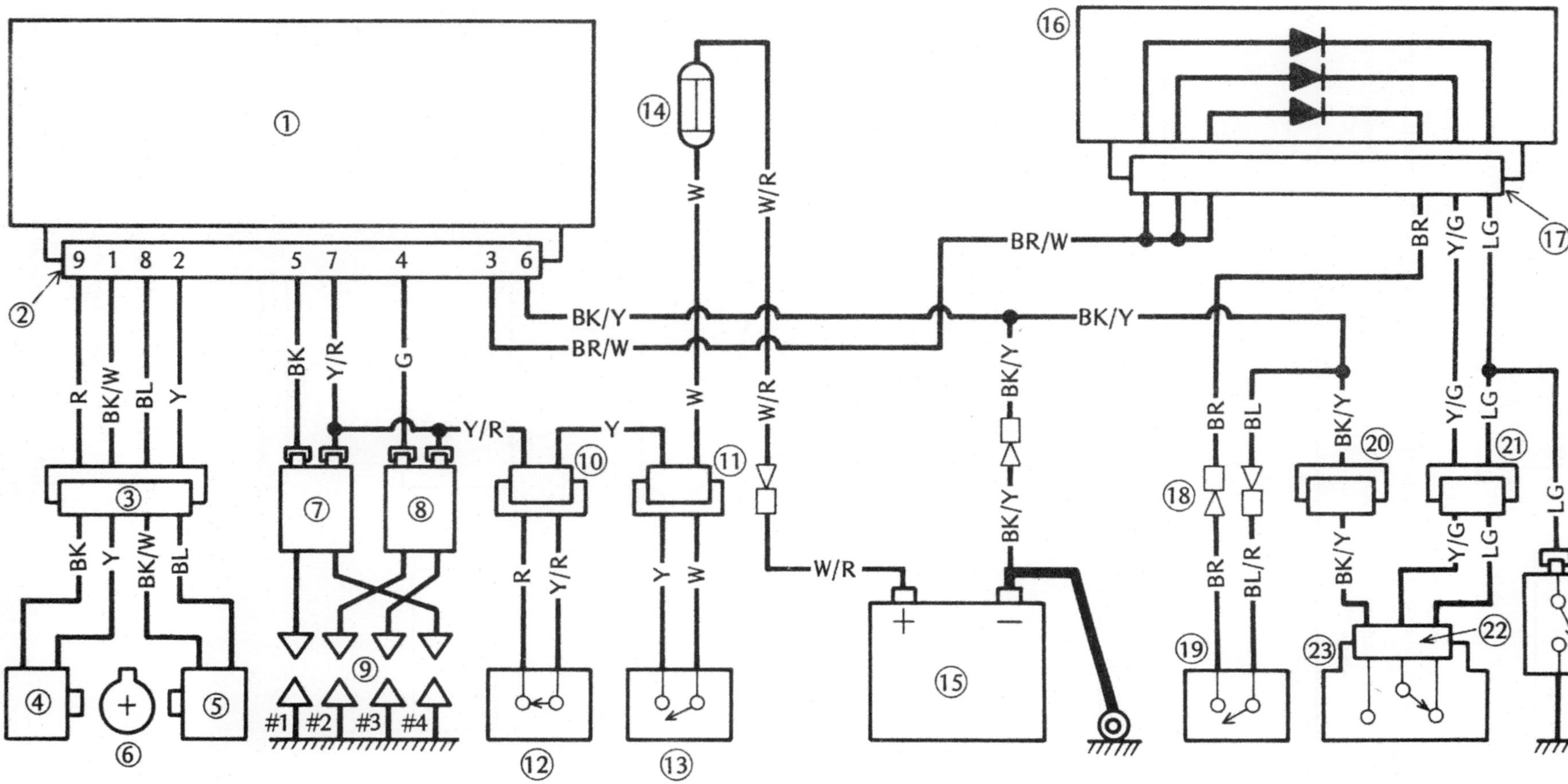

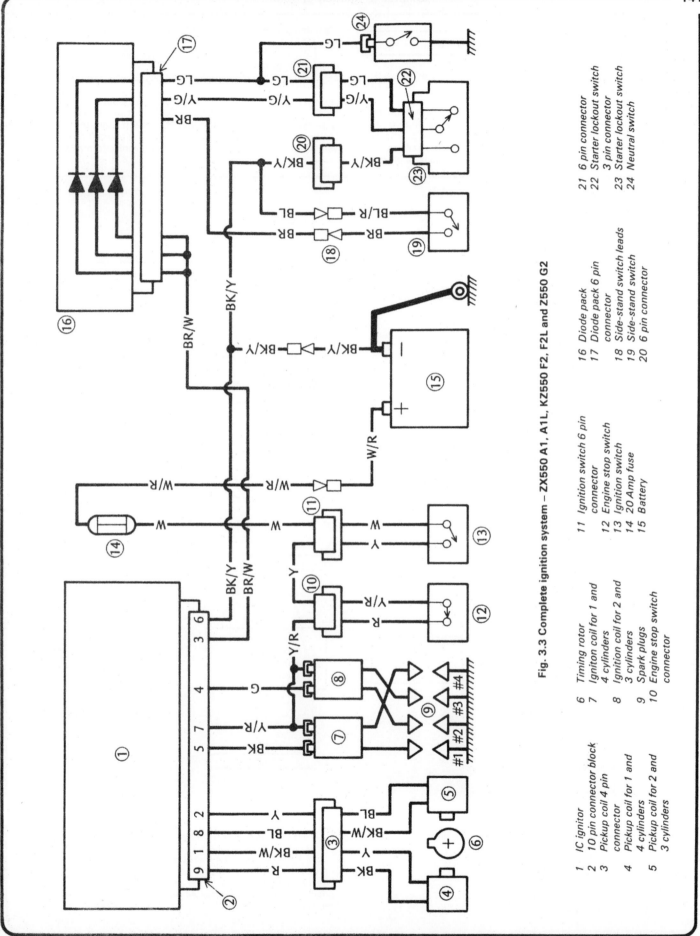

Fig. 3.3 Complete ignition system – ZX550 A1, A1L, KZ550 F2, F2L and Z550 G2

1 *IC ignitor*
2 *10 pin connector block*
3 *Pickup coil 4 pin connector*
4 *Pickup coil for 1 and 4 cylinders*
5 *Pickup coil for 2 and 3 cylinders*
6 *Timing rotor*
7 *Igniton coil for 1 and 4 cylinders*
8 *Ignition coil for 2 and 3 cylinders*
9 *Spark plugs*
10 *Engine stop switch connector*
11 *Ignition switch 6 pin connector*
12 *Engine stop switch*
13 *Ignition switch*
14 *20 Amp fuse*
15 *Battery*
16 *Diode pack*
17 *Diode pack 6 pin connector*
18 *Side-stand switch leads*
19 *Side-stand switch*
20 *6 pin connector*
21 *6 pin connector*
22 *Starter lockout switch 3 pin connector*
23 *Starter lockout switch*
24 *Neutral switch*

2 Ignition system: locating and identifying faults

Contact breaker ignition system

1 The contact breaker ignition system fitted to the early models is extremely simple in operation and is usually reliable. It has the advantage of being easy to put right if a fault occurs.

2 If a fault should occur, use the Fault Diagnosis section at the front of this Manual to trace it, working in a logical sequence. Note that as a general rule the spark plugs and suppressor caps should be checked first, followed by the contact breakers, the wiring and finally the HT coils and condensers. Refer to the appropriate Sections of Routine Maintenance or of this Chapter for information on the components concerned.

Electronic ignition system

3 Although equally simple in principle and requiring much less maintenance than contact breaker systems, this type of electronic ignition requires a more sophisticated approach when tracing faults. For this reason the accompanying flow chart should be used as a guide to working methodically through the system until the fault is found. The numbers next to the major fault-finding operations indicate which Section of this Chapter contains the relevant information.

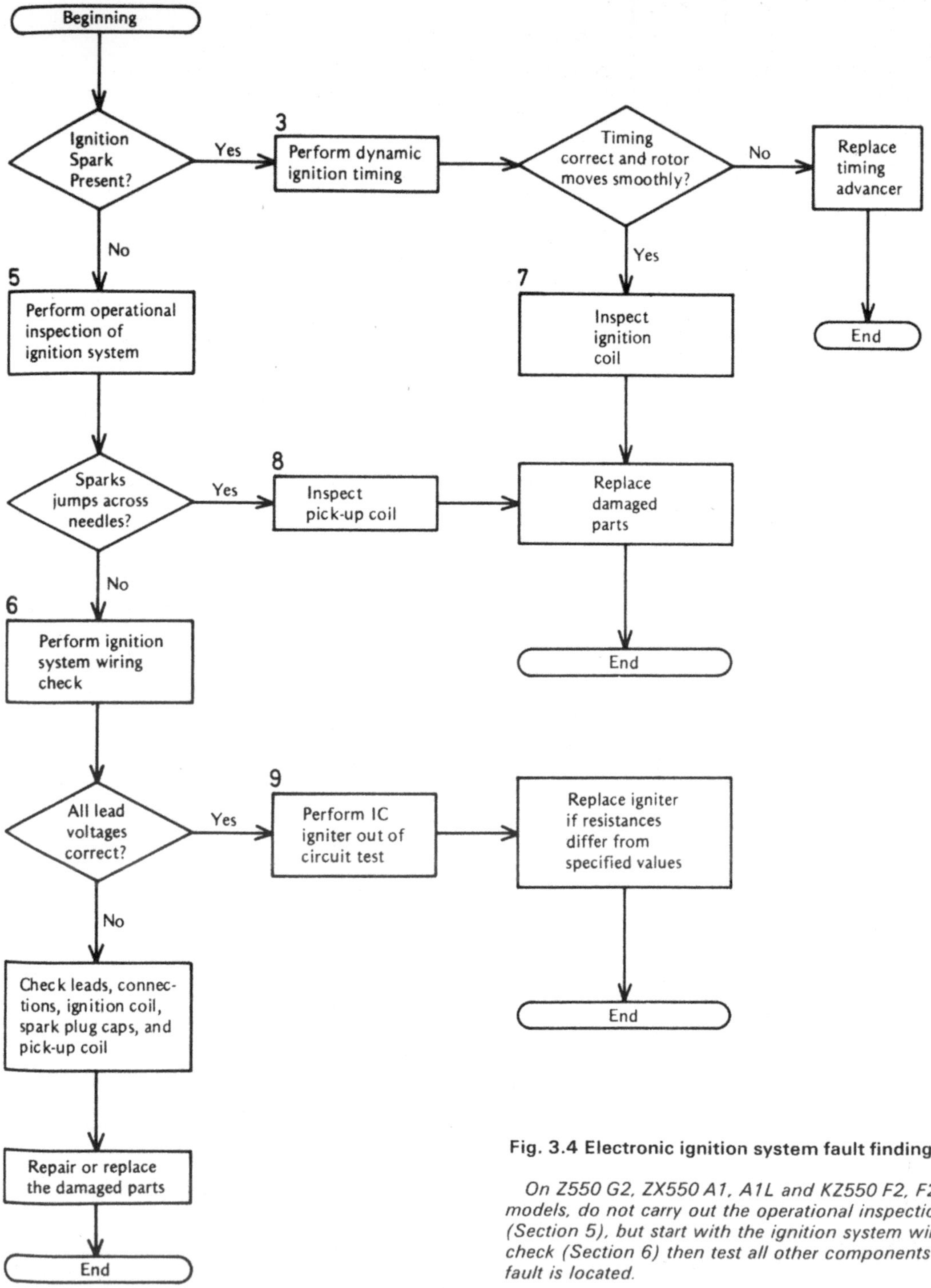

Fig. 3.4 Electronic ignition system fault finding flow chart

On Z550 G2, ZX550 A1, A1L and KZ550 F2, F2L models, do not carry out the operational inspection (Section 5), but start with the ignition system wiring check (Section 6) then test all other components until the fault is located.

3 Ignition timing: checking

Refer to the relevant Section of Routine Maintenance for details of this operation.

4 Automatic timing unit (ATU): general

Refer to the relevant Section of Routine Maintenance for details of this component, and its examination and maintenance.

5 Operational inspection of the ignition system: electronic ignition – except Z550 G2, ZX550 A1, A1L, KZ550 F2, F2L

1 This test requires the use of a separate 6 or 12 volt motorcycle battery in addition to that fitted to the machine. Kawasaki recommend the use of an electrotester to provide a pre-set 5 – 8 mm electrode gap, the triggering being by rapidly connecting and disconnecting an ohmmeter powered by a 1.5 – 12V battery. Do not use a DC voltage source (battery or meter) with more than 12 volts or the ignitor unit will be damaged. In view of the fact that this equipment will probably not be available it is recommended that a similar piece of equipment be fabricated using an insulated stand of wood, and four electrodes of stiff wire or fashioned from nails, the test to be triggered by a separate battery. The opposing ends of the electrodes, across which the spark will jump, should be sharpened, and the outer ends of the two electrodes to which the HT voltage is to be applied should be ground down so that the sparking plug caps are a push fit. The accompanying figure shows the general arrangement.

2 Remove the right-hand side panel, then trace and separate the four-pin connector between the IC ignitor and the pick-up coil leads. The power source must be connected to the IC ignitor as shown in Fig. 3.6. Note that the blue lead is connected to the negative (-) terminal and the black lead to the positive (+) terminal; if an ohmmeter is used, or a multimeter set to the resistance scale, the normal polarity is reversed, ie red probe will be negative (-), black probe positive (+). The switch shown in the circuit is not vital; in practice the test circuit can be made and broken by joining and separating the black lead to the positive terminal.

3 Connect the test apparatus so that the plug caps are connected to one side of the electrodes, and connect the other side of the electrodes to a good earth point on the engine. Check that the electrode gaps do not exceed 8 mm. If intermittent earthing occurs or if the spark gap is too great to allow the spark to jump (in normal operation) damage to the electronic ignition may result. Turn the ignition switch on.

4 The test is conducted by applying voltage from the power source to the IC ignitor for a few seconds, then disconnecting the supply. This energises the low tension windings in the ignition coil, and a spark should be produced at both plugs when the supply is disconnected. If no spark occurs, investigate the system wiring as described in Section 6. If the plugs spark properly, the fault must lie in the pick-up coil, which can be checked as described in Section 8. **Important note:** do not connect the power source for more than 30 seconds, otherwise the ignition coil and ignitor may overheat and be damaged. If the test shows the circuit for cylinders 1 and 4 to be operating normally, repeat the test on the cylinders 2 and 3 circuit, connecting the appropriate high tension leads to the test apparatus, and noting that the power source should be connected to the yellow lead (positive terminal) and the red lead (negative terminal) at the 4-pin connector.

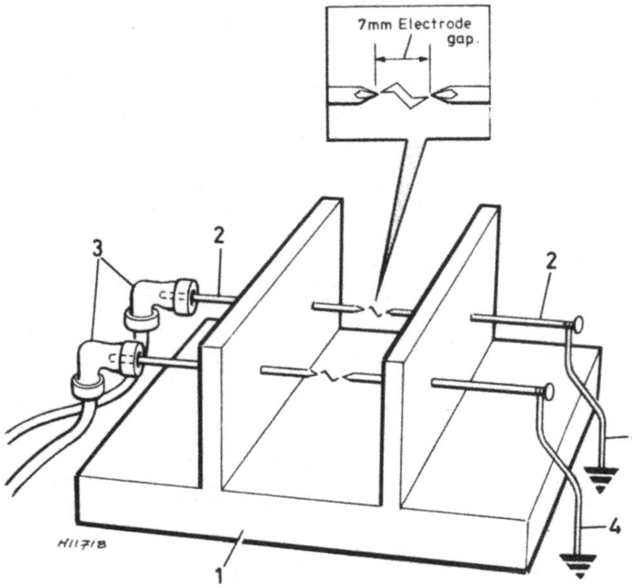

Fig. 3.5 Fabricated electrode gap tester

1 Wood stand — *3 HT leads*
2 Nails — *4 Earth wires*

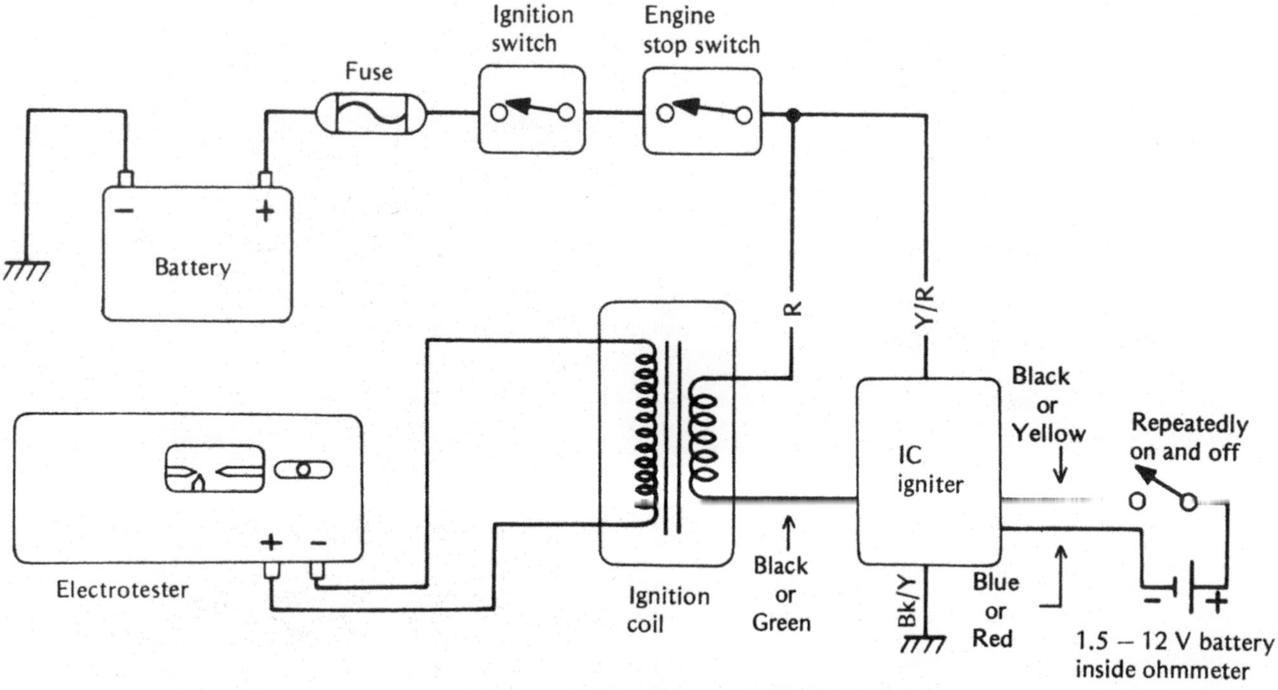

Fig. 3.6 Operational inspection of electronic ignition system

6 Ignition system wiring check: electronic ignition

1 A general wiring test can be performed using a multimeter set on the 0 – 25 volts dc range. The system should be intact, all connectors attached normally, and the meter probe should be pushed into the back of the IC ignitor terminals so that the connections remain intact. The test is performed with the engine stopped but with the ignition switched on and the kill switch in the 'Run' position. Results should be as follows for all except ZX550 A1, A1L, KZ550 F2, F2L and Z550 G2 models. With the meter negative (-) probe attached to a good frame earth point, a reading of full battery voltage should be obtained when its positive (+) probe is connected to each in turn of the three wires leading to the ignition HT coils, ie yellow/red, black or green. Similarly, with the meter negative (-) probe attached to a good frame earth point, a reading of 0.5 – 1.0 volt should be obtained when its positive (+) probe is connected to each in turn of the four wires leading to the pick-up coils, ie black, blue, yellow or red. On ZX550 A1, A1L, KZ550 F2, F2L and Z550 G2 models, attach the meter positive (+) probe to the yellow/red wire terminal and its negative (-) probe to the black/yellow wire terminal; a reading of full battery voltage should be obtained.

2 On all models, if the results of the above test indicate that power is not reaching the IC ignitor unit, check the various connector blocks for corrosion or moisture. Should these be discovered, a water dispersing fluid such as WD40 can be used as a temporary remedy in most instances. For a more permanent cure, clean off as much of the corrosion as possible to ensure a sound connection, and pack the connector halves with silicone grease to prevent further trouble. The engine 'kill' switch and the main ignition switch can suffer from similar problems and should be dealt with in a similar manner.

3 Referring to the relevant wiring diagram at the back of this Manual, check that all components related to the ignition system are in good working order. Using a meter check that there is continuity between their terminals when the kill switch is in the 'Run' position, when the sidestand is raised (sidestand switch), when the clutch lever is pulled in (starter lockout switch) and between the neutral switch terminal and earth when neutral gear is selected. Not all these components will be connected to the ignition circuit on all models; the relevant wiring diagram will give precise details of the components and their connections.

4 Check the main fuse, battery terminals and all other wires and connectors. On ZX550 A1, A1L, KZ550 F2, F2L and Z550 G2 models, check also the diode pack. This is a small rectangular black plastic unit with six wires connected to it via a multi-pin connector and is to be found behind the left-hand sidepanel. Using an ohmmeter or a multimeter set to the appropriate resistance scale, test each of the three diodes by measuring the resistance found first in one direction, then in the other. Reverse the meter probes to take the second reading as shown in the accompanying illustration. If a diode is in good condition the resistance will be low in one direction and at least ten times as much in the reverse direction. If any diode is found to have high or low readings in both directions, it is faulty and the complete pack must be renewed.

7 Ignition HT coils: checking – all types

1 The ignition HT coils are bolted to the frame immediately above the cylinder head and can be removed after the seat and fuel tank have been withdrawn. The wiring diagrams show the connections for each coil. The coils are of the same type whether fitted to contact breaker or electronic ignition systems and are tested in the same way; the most accurate test is with a spark-gap tester or electrotester, Part Number 57001-980/1242.

2 Connect the coil to the tester when the unit is switched on, and open out the adjusting screw on the tester to 6 mm (0.24 inch). The spark at this point should bridge the gap continuously. If the spark starts to break down or is intermittent, the coil is faulty and should be renewed.

3 In the absence of a coil tester, the winding may be checked for broken or shorted windings using a multimeter, noting that the test will not reveal insulation breakdown which may only be evident under high voltage.

4 The primary winding resistance can be measured by connecting the meter probes between the two low tension terminals. Set the meter on the ohm x 1 range and note the reading, which should be 1.8 – 2.8 ohm on electronic ignition coils, 3.2 – 4.8 ohm on contact breaker units. To check the secondary windings, remove the plug cap from the high tension (plug) lead, and connect the meter probes to the high tension leads. With the meter set on the ohm x 1000 (k ohms) scale, a reading of 10 – 16 k ohm is to be expected.

5 Finally check for continuity between low tension lead and the coil core, followed by the high tension lead, repeating the test with the second low tension lead. If anything other than infinite resistance (insulation) is shown, the coil must be renewed.

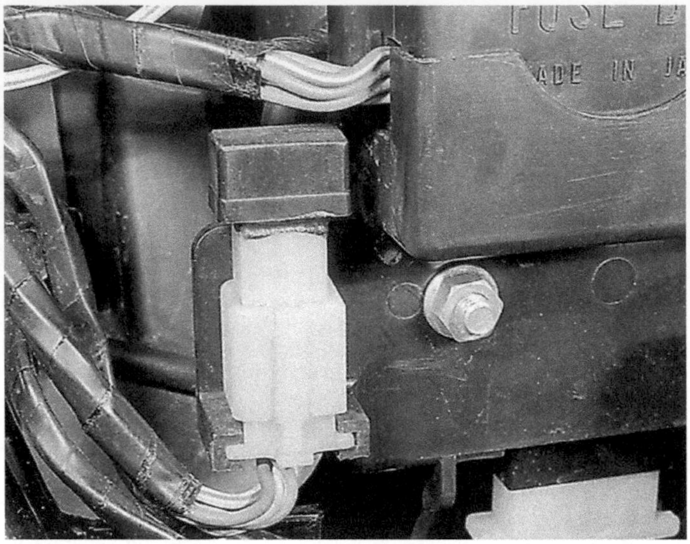

6.4 Location of diode pack – ZX550 A1

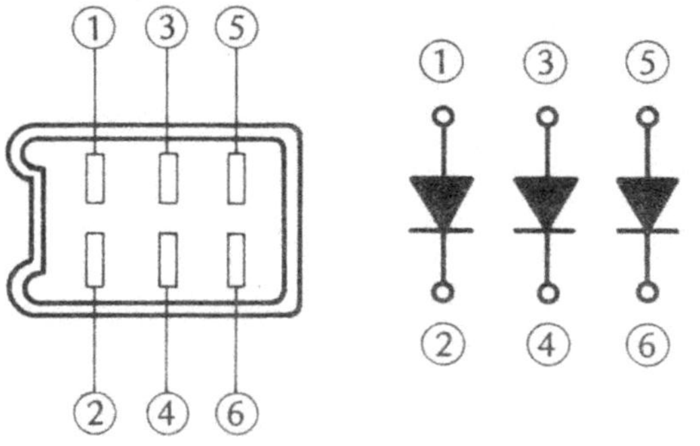

Fig. 3.7 Testing the diode pack – ZX550 A1, A1L, KZ550 F2, F2L and Z550 G2

7.1 Ensure HT coil connections and mountings are clean and securely fastened

8 Ignition pick-up coils: checking

1 The electronic ignition system pick-up coils are mounted on a backplate at the front right-hand end of the crankcase and are removed and refitted as described in Sections 9 and 42 of Chapter 1. On ZX550 A1, A1L, KZ550 F2, F2L and Z550 G2 models, note that each pick-up coil must be positioned so that its air gap (the clearance between the coil core and the tip of the rotor projection) is between 0.4 – 0.6 mm (0.016 – 0.024 in).
2 If the operational test described in Section 5 (where applicable) has indicated a fault in the pick-up assembly, the coil resistances should be checked. Disconnect the pick-up coil lead at the IC ignitor 4-pin connector, set a multimeter to the ohm x 100 scale and measure the resistance between each pair of wires given in the Specifications Section of this Chapter; note that the first pair of wires in each case is for the pick-up coil for cylinders 1 and 4, the second pair is for the coil of cylinders 2 and 3. Switch the meter to its highest resistance range and measure the resistance between each of the four wires in turn and a good earth point: no continuity (ie infinite resistance) should be found in each case.
3 If the reading obtained is significantly different from that specified, or if the second test reveals a short circuit the coil is faulty and the complete assembly must be renewed. Note however that it is worth checking first whether the fault is due to damaged wires which might easily be repaired.

9 IC ignitor unit: out of circuit test

1 The ignitor unit is a flat, square sealed unit mounted behind the right-hand sidepanel which has either eight wires connected to it via two multi-pin block connectors or ten wires via a single block connector.
2 On early models the ignitor was bolted directly to a mounting bracket but experience revealed that carelessness while removing the battery could damage the ignitor, and all units are now isolated by rubber mountings. The modification is available as a kit, but only applies to the new units; however an enterprising owner could easily fabricate a set of rubber mounts if desired. When refitting a rubber-mounted ignitor, tighten the bolts only to a torque setting of 0.65 kgf m (4.5 lbf ft). The problem does not arise if care is taken when disconnecting or connecting the battery.
3 **Note**: on models fitted with 'Unitrak' rear suspension, check very carefully that the wiring is routed properly and secured well clear of the suspension components.
4 If the system wiring test described in Section 6 indicates that all voltages are correct, the IC ignitor unit should be tested out of circuit by measuring its internal resistances. Check that the ignition switch is off, then pull off the connector at the ignitor. Using the relevant table shown in the accompanying illustrations, connect the meter probes to the appropriate terminals and note the resistance reading obtained. If the figures obtained differ markedly from those shown the unit may require renewal. Note that different meters may give slightly different readings from the Kawasaki tester, Part Number 57001-983, so confirm the diagnosis by taking the suspect unit to a Kawasaki dealer before buying a new one.
5 If the ignitor is one of the non rubber-mounted units fitted to an early model and damage is thought to be due to careless disconnection of the battery, check for continuity between the black/yellow lead of the larger block connector and the ignitor body; if any resistance is measured the internal earth wire has been damaged. In such a case a separate earth lead can be made up to run between the ignitor body (attached by one of the mounting points) and either the battery negative (-) terminal or a good frame earth. This will bypass the damaged earth circuit and make possible the continued use of the ignitor. **Note**: damage of this nature will be avoided completely if care is taken when connecting and disconnecting the battery as described in the relevant Sections of Routine Maintenance and Chapter 6.

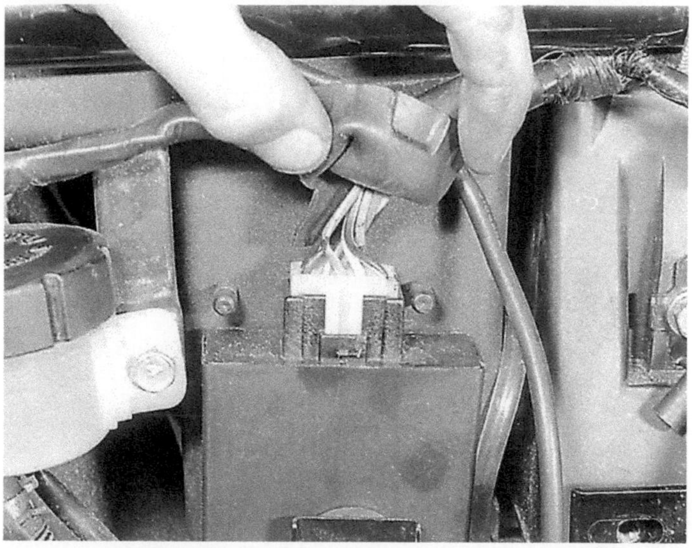

9.1 Location of IC ignitor unit – ZX550 A1

Meter Range	Lead Location	Connections	Reading
x 1 kΩ	Male ignition coil connector (disconnected)	Meter (+) → Black/Yellow Meter (–) → Black, Green	∞
x 100 Ω		Meter (+) → Black, Green Meter (–) → Black/Yellow	200 – 500 Ω
		Meter (+) → Yellow/Red Meter (–) → Black/Yellow	200 – 600 Ω
		Meter (+) → Black/Yellow Meter (–) → Yellow/Red	300 – 1,100 Ω
x 1 kΩ	Male pickup coil connector (disconnected)	Meter (+) → Blue, Red Meter (–) → Black, Yellow	20 – 40 kΩ
		Meter (+) → Black, Yellow Meter (–) → Blue, Red	25 – 45 kΩ

Fig. 3.8 IC ignitor test table – all models except ZX550 A1, A1L, KZ550 F2, F2L and Z550 G2

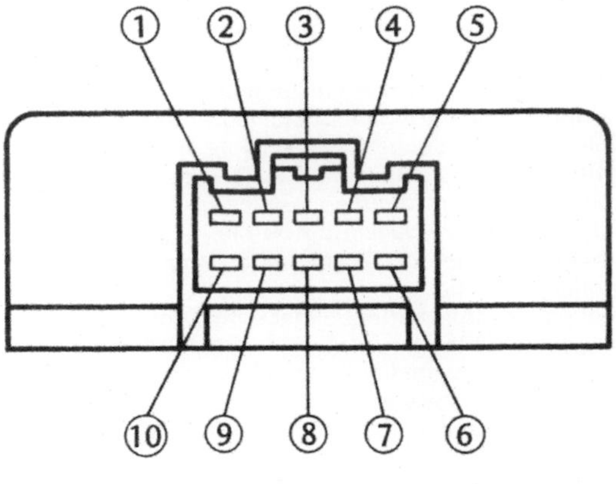

Value (kΩ)	
O	Zero
A	0.3 – 4.2
B	6.6 – 21.4
C	25 – 75
D	125 – 375
∞	Infinity

Tester (–) Lead Connection: Terminal Number	Tester (+) Lead Connection: 1	2	3	4	5	6	7	8	9	10
1		D	D	D	D	D	D	D	D	∞
2	D		D	D	D	D	D	D	D	∞
3	C	C		B	B	B	B	B	B	∞
4	∞	∞	∞		∞	∞	∞	∞	∞	∞
5	∞	∞	∞	∞		∞	∞	∞	∞	∞
6	C	C	B	A	A		A	O	O	∞
7	C	C	B	A	A	A		A	A	∞
8	C	C	B	A	A	O	A		O	∞
9	C	C	B	A	A	O	A	O		∞
10	∞	∞	∞	∞	∞	∞	∞	∞	∞	

Fig. 3.9 IC ignitor test connections and resistances – ZX550 A1, A1L, KZ550 F2, F2L and Z550 G2

10 Condensers (capacitors): removal, testing and refitting

1 There are two condensers contained in the ignition system, each one wired in parallel with a set of points. If a fault develops in a condenser, ignition failure is likely to occur.

2 If the engine proves difficult to start, or misfiring occurs, especially when the engine is hot, it is possible that a condenser is at fault. To check, separate the contact points by hand when the ignition is switched on. If a spark occurs across the points as they are separated by hand and they have a burnt or blackened appearance, the condenser connected to that set of points can be regarded as unserviceable.

3 Test the condenser on a coil and condenser tester unit or alternatively fit a new replacement. In view of the small cost involved it is preferable to fit a new condenser, and observe the effect on engine performance as a result of the substitution.

4 Check that the screws that hold the condensers to the contact breaker plate are tight, and also form a good earth connection.

11 Spark plugs: checking and resetting the gaps

1 The spark plugs recommended by the manufacturer as being most suitable for all normal use are given in the Specifications Section of this Chapter; alternatives are available to allow for varying altitudes, climatic conditions and the use to which the machine is put. If a spark plug is suspected of being faulty it can be tested only by the substitution of a brand new (not second-hand) plug of the correct make, type, and heat range; always carry a spare on the machine.

2 Note that the advice of a competent Kawasaki Service Agent or similar expert should be sought before the plug heat range is altered from standard. The use of too cold, or hard, a grade of plug will result in fouling and the use of too hot, or soft, a grade of plug will result in engine damage due to the excess heat being generated. If the correct grade of plug is fitted, however, it will be possible to use the condition of the spark plug electrodes to diagnose a fault in the engine or to decide whether the engine is operating efficiently or not. The series of colour photographs in Routine Maintenance will show this clearly.

3 It is advisable to carry a new spare sparking plug on the machine, having first set the electrodes to the correct gap. Whilst spark plugs do not fail often, a new replacement is well worth having if a breakdown does occur. Ensure that the spare is of the correct heat range and type.

4 The gap can be assessed using feeler gauges. If necessary, alter the gap by bending the outer electrode, preferably using a proper electrode tool. **Never** bend the centre electrode, otherwise the porcelain insulator will crack, and may cause damage to the engine if particles break away whilst the engine is running.

5 Before refitting a spark plug into the cylinder head, coat the threads sparingly with a graphited grease to aid future removal. Use the correct size spanner when tightening a plug, otherwise the spanner may slip and damage the ceramic insulator. The plug should be tightened by hand only at first and then secured with a quarter turn of the spanner so that it seats firmly on its sealing ring. If a torque wrench is available, tighten the plug to a torque setting of 1.4 kgf m (10 lbf ft).

6 Never overtighten a spark plug otherwise there is risk of stripping the threads from the cylinder head, especially as it is cast in light alloy. A stripped thread can be repaired without having to scrap the cylinder head by using a 'Helicoil' thread insert. This is a low-cost service operated by a number of dealers.

7 Note that some UK models are fitted with resistor plugs, these being denoted by the letter 'R' in the plug code. This indicates that the resistor necessary to suppress the interference of the HT pulse with radio and TV reception is built into the plug, not the cap. The ignition system is designed to suit such plugs and will be damaged if a non-resistor type is fitted.

12 Spark plug (HT) lead and suppressor cap: examination

1 Erratic running faults and problems with the engine suddenly cutting out in wet weather can be attributed to leakage from the high tension lead and spark plug cap. If this fault is present, it will often be possible to see tiny sparks around the lead and cap at night. One cause of this problem is the accumulation of mud and road grime around the lead, and the first thing to check is that the lead and cap are clean. It is often possible to cure the problem by cleaning the components and sealing them with an aerosol ignition sealer, which will leave an insulating coating on both components.

2 Water dispersant sprays are also highly recommended where the system has become swamped with water. Both these products are easily obtainable at most garages and accessory shops. Occasionally the suppressor cap or the lead itself may break down internally. If this is suspected, the components should be renewed.

3 Where resistor spark plugs are used, make sure that the new 'suppressor' caps are compatible. If an old-style cap and resistor plug are used together, the combination of heavy resistances can damage the ignition system. Combination of a non-suppressed cap and non-resistor plug will have the same effect.

Chapter 4 Frame and forks

For information relating to the 1984 on models, see Chapter 7

Contents

Specifications

Frame

Type	Welded tubular steel

Front forks

Type:			
Z400 J1, Z500 B1, B2, KZ/Z550 A1	Hydraulically damped telescopic		
All other models	Hydraulically damped, air assisted telescopic		
Wheel travel:			
Z550 G1, G2, ZX550 A1, A1L, KZ550 F1, M1, F2, F2L	160 mm (6.3 in)		
ZR400 A1, B1, ZR550 A1, A2	175 mm (6.9 in)		
All other models	180 mm (7.1 in)		
Air pressure:	**Standard**	**Adjustment range**	**Maximum**
Z400 J1, Z500 B1, B2, KZ/Z550 A1	N/App	N/App	N/App
KZ/Z550 C1, C2, KZ550 C3, C4, F1, M1, F2, F2L	0.6 kg/cm² (8.5 psi)	0.5 – 0.7 kg/cm² (7.0 – 10.0 psi)	2.5 kg/cm² (35.5 psi)
All other models	0.7 kg/cm² (10.0 psi)	0.6 – 0.8 kg/cm² (8.5 – 11.0 psi)	2.5 kg/cm² (35.5 psi)
Fork spring minimum free length:			
Z400 J1, Z500 B1, B2, KZ/Z550 A1	543.0 mm (21.38 in)		
KZ/Z550 C1, C2, KZ550 C3, C4	605.0 mm (23.82 in)		
Z400 J2, J3, KZ/Z550 A2, A3, D1, KZ550 A4	562.0 mm (22.13 in)		
ZR400 A1, B1, ZR550 A1, A2	549.0 mm (21.61 in)		
KZ/Z550 H1, H2	584.0 mm (22.99 in)		
ZX550 A1, A1L	500.0 mm (19.69 in)		
Z550 G1, G2	530.0 mm (20.87 in)		
KZ550 F1, M1, F2, F2L	504.0 mm (19.84 in)		
Recommended fork oil:			
Z550 G1, G2, KZ550 F1, M1, F2, F2L	SAE 10W		
All other models	SAE 5W20		
Fork oil capacity per leg – oil change:*	cc	Imp fl oz	US fl oz
Z400 J1, ZR400 A1, B1, Z500 B1, B2, KZ/Z550 A1, ZR550 A1, A2, ZX550 A1, A1L	200.0	7.04	6.76
KZ/Z550 C1, C2, KZ550 C3, C4	270.0	9.50	9.13
Z400 J2, J3, KZ/Z550 A2, A3, D1, KZ550 A4	225.0	7.92	7.61
KZ/Z550 H1, H2	215.0	7.57	7.27
Z550 G1, G2	255.0	8.98	8.62
KZ550 F1, M1, F2, F2L	280.0	9.86	9.47

*Quantities approximate only – add sufficient oil to achieve correct oil level

Fork oil capacity per leg – on reassembly:			
Z400 J1, Z500 B1, B2, KZ/Z550 A1	220.0 ± 2.5	7.74 ± 0.09	7.44 ± 0.08
KZ/Z550 C1, C2, KZ550 C3, C4	290.0 ± 2.5	10.21 ± 0.09	9.81 ± 0.08
Z400 J2, J3, KZ/Z550 A2, A3, KZ550 A4	245.0 ± 2.5	8.62 ± 0.09	8.28 ± 0.08
KZ/Z550 D1	244.0 ± 2.5	8.59 ± 0.09	8.25 ± 0.08
ZR400 A1, B1, ZR550 A1, A2	236.0 ± 2.5	8.31 ± 0.09	7.98 ± 0.08
KZ/Z550 H1, H2	234.0 ± 2.5	8.24 ± 0.09	7.91 ± 0.08
ZX550 A1, A1L	229.0 ± 2.5	8.06 ± 0.09	7.74 ± 0.08
Z550 G1, G2	299.0 ± 4.0	10.53 ± 0.14	10.11 ± 0.13
KZ550 F1, M1, F2, F2L	327.0 ± 4.0	11.51 ± 0.14	11.06 ± 0.13

Fork oil level:*	
Z400 J1, Z500 B1, B2, KZ/Z550 A1	505.0 ± 4.0 mm (19.88 ± 0.16 in)
KZ/Z550 C1, C2, KZ550 C3, C4	356.0 ± 4.0 mm (14.02 ± 0.16 in)
Z400 J2, J3, KZ/Z550 A2, A3, KZ550 A4	520.0 ± 4.0 mm (20.47 ± 0.16 in)
KZ/Z550 D1	517.0 ± 4.0 mm (20.35 ± 0.16 in)
ZR400 A1, B1, ZR550 A1, A2	495.0 ± 2.0 mm (19.49 ± 0.08 in)
KZ/Z550 H1, H2	489.0 ± 2.0 mm (19.25 ± 0.08 in)
ZX550 A1, A1L	472.0 ± 2.0 mm (18.58 ± 0.08 in)
Z550 G1, G2, KZ550 F1, M1, F2, F2L	416.0 ± 2.0 mm (16.38 ± 0.08 in)

*Oil level measured from top of stanchion, top plug and fork spring removed, forks fully extended

Rear suspension

Type:	
ZR400 A1, B1, ZR550 A1, A2, KZ/Z550 H1, H2	'Unitrak' rising-rate system, single full-floating suspension unit
ZX550 A1, A1L	Pivoted fork 'Unitrak' system, single suspension unit
Z550 G1, G2, KZ550 F1, M1, F2, F2L	Pivoted fork incorporating drive shaft, two suspension units
All other models	Pivoted fork, two coil sprung hydraulically damped suspension units
Wheel travel:	
Z550 G1, G2	105 mm (4.1 in)
KZ550 F1, M1, F2, F2L	110 mm (4.3 in)
ZR400 A1, B1, ZR550 A1, A2, KZ/Z550 H1, H2, ZX550 A1, A1L	140 mm (5.5 in)
All other models	120 mm (4.7 in)
Swinging arm pivot – except shaft drive and 'Unitrak' models:	
Pivot sleeve minimum OD	21.96 mm (0.8646 in)
Pivot bolt maximum runout	0.14 mm (0.0055 in)
Suspension linkage bearings – ZR400 A1, B1, ZR550 A1, A2, KZ/Z550 H1, H2:	
Sub-frame pivot sleeve minimum OD	24.96 mm (0.9827 in)
Rocker arm pivot sleeve minimum OD	21.96 mm (0.8646 in)
Tie-rod spherical bearing maximum wear	0.30 mm (0.0118 in)
Swinging arm pivot – ZX550 A1, A1L:	
Pivot sleeve minimum OD	21.96 mm (0.8646 in)
Swinging arm left-hand pivot lug/frame clearance – Z550 G1, G2, KZ550 F1, M1, F2, F2L	1.4 – 1.6 mm (0.0551 – 0.0630 in)

Rear suspension units – Z550 G1, G2, KZ550 F1, F2, F2L

Type	Air-sprung, hydraulically damped
Air pressure:	
Standard	0.8 kg/cm² (11.0 psi)
Adjustment range	0.8 – 1.5 kg/cm² (11.0 – 21.0 psi)
Maximum	5.0 kg/cm² (71.0 psi)

	Z550 G1	**Z550 G2**	**KZ550 F1**	**KZ550 F2, F2L**
Air chamber capacity per unit	180cc	167cc	200cc	184cc
Approximate oil capacity per unit	225cc	223cc	205cc	204cc
Recommended oil (fork oil)	SAE 5W	SAE 5W	SAE 5W	SAE 5W

Torque wrench settings

Component	**kgf m**	**lbf ft**
Steering stem head bolt:		
Models with cup and cone steering head bearings	4.5	32.5
Models with taper roller steering head bearings	4.3	31.0
Steering stem head clamp bolt – only models with cup and cone steering head bearings	1.8	13.0
Steering head bearing adjusting nut – models with cup and cone steering head bearings	3.0	22.0
Steering head bearing adjusting nut – models with taper roller steering head bearings:		
Initial setting on assembly	4.0	29.0
Normal setting	Nil	Nil
Handlebar clamp bolts:		
Z550 G1, G2, ZX550 A1, A1L	2.5	18.0
ZR400 A1, B1, ZR550 A1, A2, KZ/Z550 H1, H2	3.5	25.0
All other models	1.8	13.0
Handlebar holder bolts:		
ZR400 A1, B1, ZR550 A1, A2, Z550 G1, G2, KZ/Z550 H1, H2	10.0	72.0

Component	kgf m	lbf ft
ZX550 A1, A1L	7.5	54.0
All other models	N/App	N/App
Front master cylinder clamp bolts	0.9	6.5
Front fork air valves – Z400 J2, J3, KZ/Z550 A2, A3, C1, C2, D1, KZ550 A4, C3, C4	1.2	9.0
Front fork top plugs – Z550 G1, G2, KZ550 F1, F2, F2L M1	2.3	16.5
Top yoke pinch bolts:		
Z550 G1, G2, KZ550 F1, F2, F2L, M1	2.1	15.0
All other models	1.8	13.0
Bottom yoke pinch bolts:		
Z550 G1, G2, KZ550 F1, F2, F2L, M1	2.8	20.0
All other models	1.8	13.0
Front and rear brake caliper mounting bolts:		
Front caliper – ZX550 A1, A1L	2.5	18.0
All others	3.0	22.0
Anti-dive assembly – ZX550 A1, A1L:		
Plunger/valve assembly Allen screws	0.4	3.0
Valve assembly mounting Allen screws	0.75	5.5
Brake hose union mounting bolts – models with twin front disc brakes	0.9	6.5
Front fork drain plugs:		
Z550 G1, G2, KZ550 F1, F2, F2L, M1	0.15	1.0
All other models	0.8	6.0
Front fork bottom Allen bolts:		
Z550 G1, G2, KZ550 F1, F2, F2L, M1	2.0	14.5
All other models	1.8	13.0
Front wheel spindle nut:		
ZR400 A1, B1, ZR550 A1, A2, Z550 G1, G2, KZ550 F1, F2, F2L, M1	6.5	47.0
All other models	8.0	58.0
Front wheel spindle clamp bolt:		
ZR400 A1, B1, ZR550 A1, A2, ZX550 A1, A1L	1.4	10.0
All other models	2.0	14.5
Engine mounting bolts – 10 mm – engine/frame or mounting bracket:		
KZ/Z550 H1, H2, ZR400 A1, B1, ZR550 A1, A2, Z550 G1, G2, KZ550 F1, M1, F2, F2L	3.5	25.0
All other models	4.0	29.0
Engine mounting bolts – 8 mm – frame/mounting bracket	2.4	17.0
Oil cooler mounting bolts – where fitted	1.0	7.0
Air cleaner mounting screws	0.5	3.5
Turn signal assembly mounting nuts:		
Standard	1.3	9.5
Maximum permissible	1.5	11.0
Rear brake pedal cap nut – Z400 J1, J2, Z500 B1, B2 KZ/Z550 A1, A2	2.0	14.5
Rear suspension pivot shaft or pivot bolt retaining nut:		
Z550 G1, G2, KZ550 F1, F2, F2L, M1	1.3	9.5
ZX550 A1, A1L	9.0	65.0
All other models	8.0	58.0
Rear suspension unit fasteners:		
Z550 G1, G2, KZ550 F1, F2, F2L, M1	2.5	18.0
ZX550 A1, A1L	5.0	36.0
ZR400 A1, B1, ZR550 A1, A2, KZ/Z550 H1, H2	7.0	50.5
All other models	3.0	22.0
Rear suspension unit connecting hose – Z550 G1, G2, KZ550 F1, F2, F2L:		
Hose/unit unions	1.2	9.0
Gland nut	2.0	14.5
Unitrak linkage – ZR400 A1, B1, ZR550 A1, A2, KZ/Z550 H1, H2, ZX550 A1, A1L:		
Rocker arm pivot shaft – except ZX550 A1, A1L	9.0	65.0
Rocker arm pivot shaft – ZX550 A1, A1L	5.0	36.0
Tie rod bolts – except ZX550 A1, A1L	3.8	27.5
Tie rod bolts – ZX550 A1, A1L	5.0	36.0
Final drive rear gear case/swinging arm mounting nuts – shaft drive models	2.3	16.5
Rear wheel spindle nut:		
Z550 G1, G2, KZ550 F1, F2, F2L, M1	7.5	54.0
ZX550 A1, A1L	9.5	69.0
All other models	8.0	58.0
Rear brake torque link nuts or bolts	3.0	22.0
Rear brake caliper holder clamp bolt – ZX550 A1, A1L	0.65	4.5

1 General description

All models are fitted with front forks of the conventional coil-spring hydraulically-damped telescopic type, but later models use air pressure to provide a greater range of adjustment to suit the riders' needs. In addition, ZX550 A1 and A1L models are fitted with a system whereby front brake hydraulic pressure activates a system of valves to stiffen the hydraulic damping on the compression stroke, thus eliminating most of the fork dive on braking. The valves are mounted on the front of the fork lower leg and can be adjusted for effect through three positions.

The frame is a full-cradle design, of welded tubular steel and on most models the rear suspension is by swinging arm (pivoted fork) acting on two suspension units. All GPz and ZR400/550 models employ the Unitrak suspension system whereby a swinging arm (pivoted fork) of tubular steel (square-section aluminium alloy tubing on ZX550 A1 and A1L models) acts on a single suspension unit via a linkage which is designed to vary the rate at which the suspension unit controls wheel movement.

2 Front forks: removal and refitting

1 Place the machine on its centre stand on level ground and remove the fuel tank or at least cover it with a thick layer of padding to protect its paintwork. Remove the front wheel and the front mudguard, then remove the brake caliper(s) from the fork lower legs. Place a piece of wood between the brake pads to prevent their being displaced and tie each caliper loosely to the frame to keep it out of the way. On ZX550 A1 and A1L models remove the two Allen screws securing each anti-dive plunger assembly to the top of its valve unit and the single bolt or two Allen bolts securing each junction block to the fork lower legs, then withdraw the plunger assemblies and junction blocks to permit the calipers to be moved. On all GPz models the fairing can be removed, if desired, to gain additional working space. On models with conventional chromed-tube handlebars remove the rubber or plastic cap from the top of each fork leg.

2 On models with four-piece handlebars, remove the two securing screws and lift away the ignition switch/fork top yoke cover, prise off the rubber or plastic cap or plug from the top of each leg and remove the bolt securing each handlebar locating plate. Remove the plates, noting the direction of their arrow marks. Remove the large threaded bolt or plug securing each handlebar half to the top yoke; these are fastened to a high torque setting and care must be taken to hold the forks steady while they are slackened. Ensuring that all control cables, brake hose and wiring are not stretched, kinked or trapped, allow the handlebar halves to hang down on each side of the forks.

3 Where applicable, remove the valve cap(s) and depress the valve core(s) to release the air pressure in the fork legs. Make a note of the position of the fork stanchion upper ends relative to the top yoke upper surface. Slacken the pinch bolts in both top and bottom yokes and pull each fork leg downwards out of the yokes, twisting the stanchion while pulling to make the task easier. If the stanchions are stuck in the yokes, apply penetrating fluid, allow time for it to work, and try again. In extreme cases it is permissible to remove fully the pinch bolts and to ease apart the yoke clamps by wedging a screwdriver blade between them, but great care must be taken when doing this or the clamps may be broken, necessitating the renewal of the yoke. Where applicable, the headlamp brackets should remain in place between the yokes, thus retaining the various metal and rubber washers and spacers. **On ZX550 A1 and A1L models** note that the air union must be held firmly against the top yoke, as each leg is withdrawn, to protect it from damage. A wire circlip is fitted around each stanchion to retain the air union; use a small screwdriver to displace the circlip from its groove, then lift it off the top of the stanchion before attempting to pull the stanchion through the bottom yoke. When both fork legs are removed, the air union will be released; note the presence of the rubber spacer and metal washer around each stanchion above the union.

4 The fork legs should be refitted when completely reassembled. Smear grease over the stanchion upper end to help its passage through the yokes and, where applicable, apply grease to the air union sealing O-rings to prevent damage to them and to ensure that they are not displaced. Ensuring that the air union (where fitted), headlamp brackets (where fitted) and all rubber and metal spacers and washers are refitted and correctly located, work each stanchion up through the yokes. On all models with conventional handlebars and the Z550 G1 and G2 models the stanchion upper end must be exactly flush with the upper surface of the top yoke; on GPz and all ZR400/550 models the stanchion upper end must be a little way below the top yoke upper surface. On all models with chrome-tube handlebars tighten the top yoke pinch bolts to the recommended torque setting.

5 On all models with four-piece handlebars, position each handlebar half on the top yoke and secure it with the large threaded bolt or plug, then refit the locating plates with their arrow marks pointing to the rear. Rotate the handlebars to ensure that the plates locate correctly. Tighten the large bolts to a torque setting of 10.0 kgf m (72 lbf ft) on all models except the ZX550 A1, A1L where the plug is tightened to a torque setting of 7.5 kgf m (54 lbf ft). With the stanchions now pulled into their correct positions, tighten the top yoke pinch bolts to the recommended torque setting, then tighten the locating plate Allen screws securely. Refit the rubber or plastic cap or plug to the top of each leg, followed by the ignition switch/top yoke cover.

6 On ZX550 A1 and A1L models release the calipers, refit the junction blocks to the fork lower legs and the plunger assemblies to the anti-dive units. On all models refit the front mudguard, not forgetting the brake hose clamps (where fitted). Refit the front wheel but tighten the spindle nut and clamp bolt only lightly at this stage. Refit the brake calipers, tightening their mounting bolts to the recommended torque setting, then apply the brake repeatedly to bring the pads back into firm contact with the discs. Push the machine off its stand, apply the front brake and pump the forks up and down to ensure that all disturbed components are correctly aligned. When normal fork movement is restored, put the machine back on its stand.

7 Tighten the bottom yoke pinch bolts to the recommended torque setting, followed by tightening the mudguard mounting bolts securely. The spindle nut and clamp bolt should be tightened last, to their recommended torque settings, as described in Section 3 of Chapter 5, to ensure that the fork legs are clamped in their correct working positions. Where applicable, fill the forks with air to the desired pressure, as described in Routine Maintenance, and check carefully for air leaks before refitting the valve cap(s) and rubber or plastic cap at the top of each fork leg.

8 Check that the front brakes are working properly with full lever pressure restored, and that all disturbed components are correctly refitted and securely fastened before taking the machine out on the road. On all models fitted with fork gaiters, secure the gaiters with the breather holes facing to the rear.

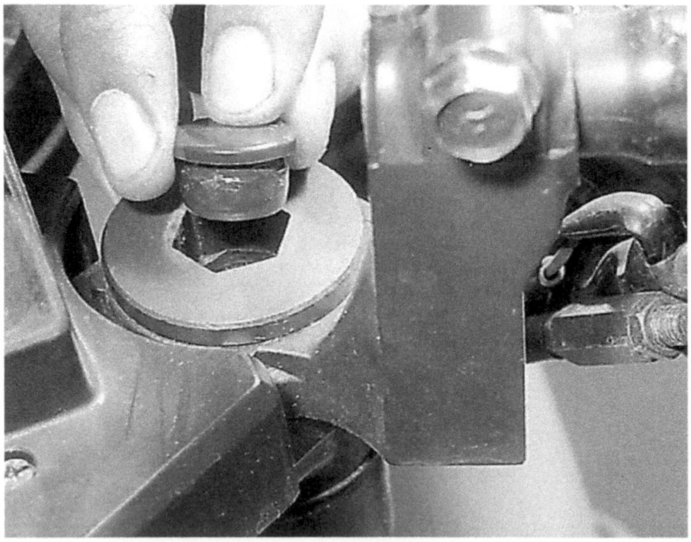

2.2a Remove cap or plug from top of fork leg

2.2b Remove top bolt or plug – models with four-piece handlebars

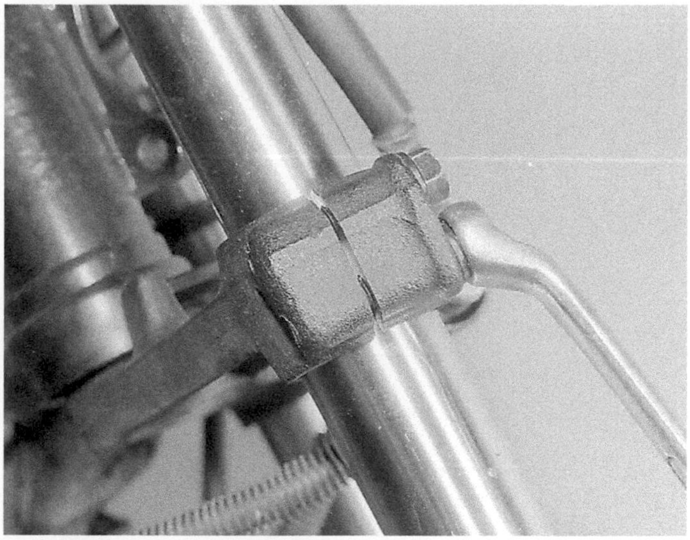
2.3a Slacken fork yoke pinch bolts ...

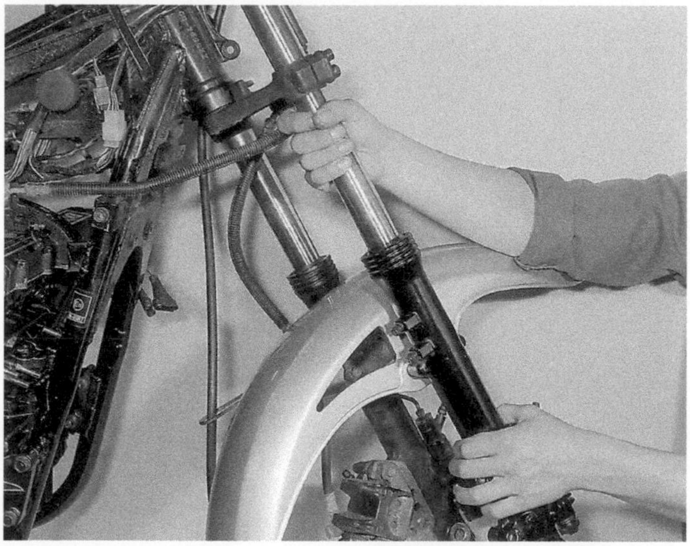
2.3b ... and withdraw fork leg from yokes

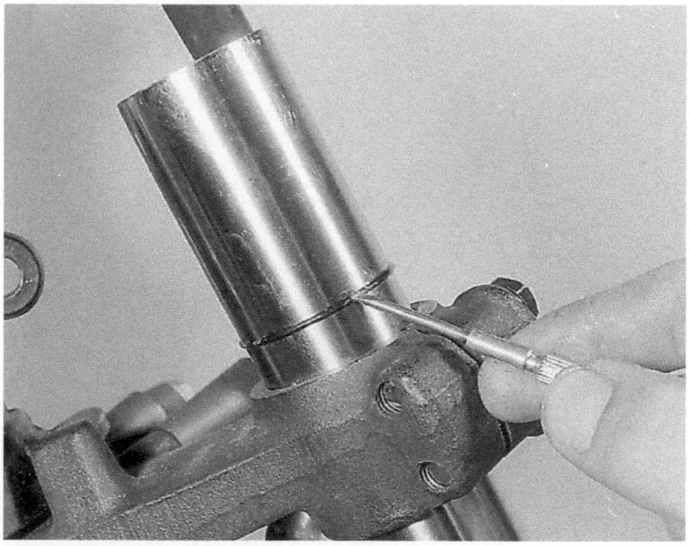
2.3c ZX550 A1, A1L – note air union retaining circlip which must be removed before fork leg can be withdrawn

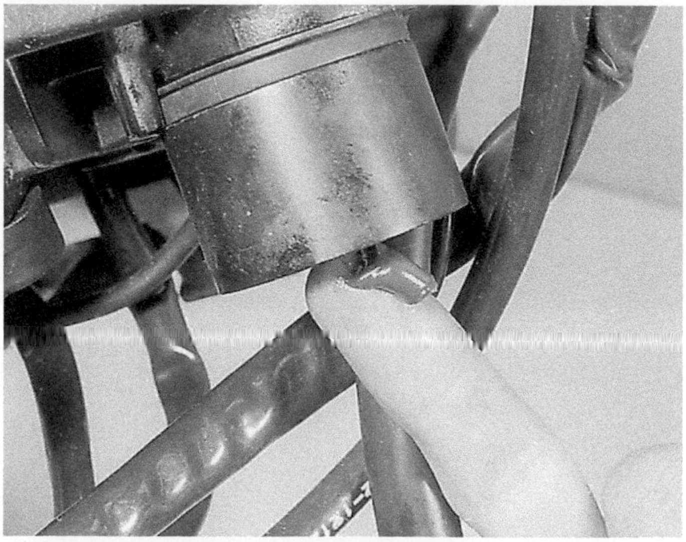
2.4 Apply grease to air union O-rings (where fitted) on refitting

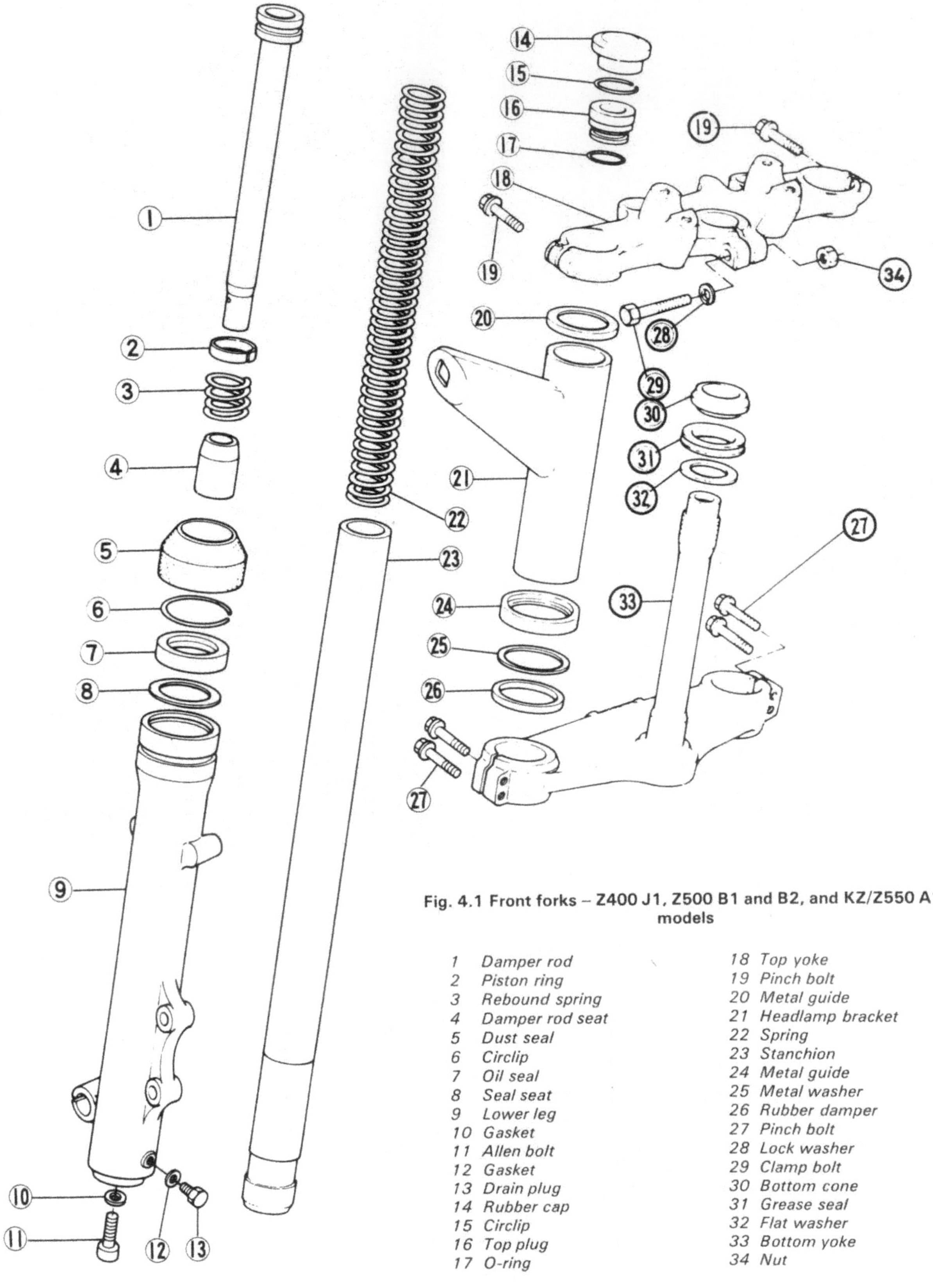

Fig. 4.1 Front forks – Z400 J1, Z500 B1 and B2, and KZ/Z550 A1 models

1 Damper rod
2 Piston ring
3 Rebound spring
4 Damper rod seat
5 Dust seal
6 Circlip
7 Oil seal
8 Seal seat
9 Lower leg
10 Gasket
11 Allen bolt
12 Gasket
13 Drain plug
14 Rubber cap
15 Circlip
16 Top plug
17 O-ring
18 Top yoke
19 Pinch bolt
20 Metal guide
21 Headlamp bracket
22 Spring
23 Stanchion
24 Metal guide
25 Metal washer
26 Rubber damper
27 Pinch bolt
28 Lock washer
29 Clamp bolt
30 Bottom cone
31 Grease seal
32 Flat washer
33 Bottom yoke
34 Nut

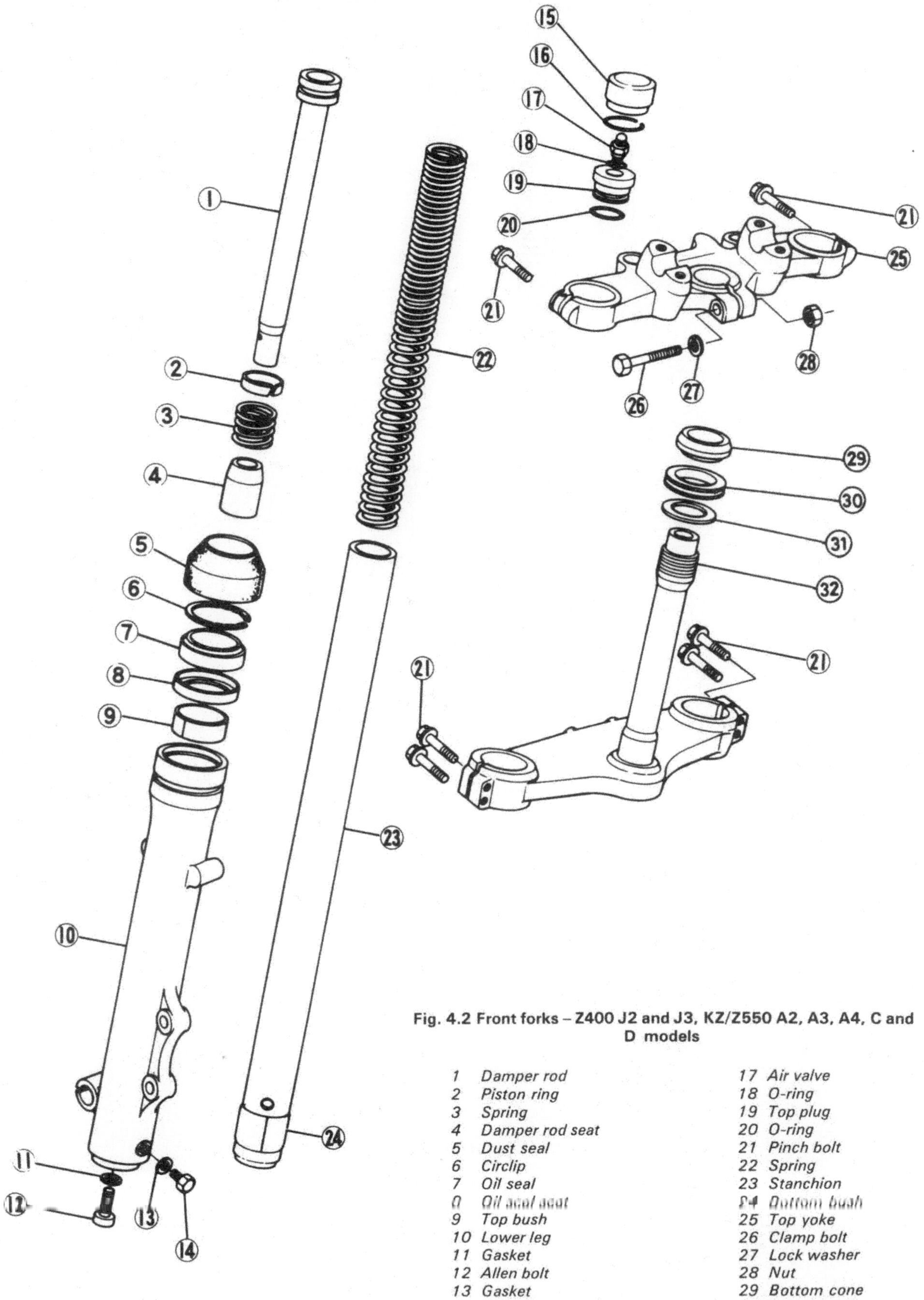

Fig. 4.2 Front forks – Z400 J2 and J3, KZ/Z550 A2, A3, A4, C and D models

1 *Damper rod*
2 *Piston ring*
3 *Spring*
4 *Damper rod seat*
5 *Dust seal*
6 *Circlip*
7 *Oil seal*
8 *Oil seal seat*
9 *Top bush*
10 *Lower leg*
11 *Gasket*
12 *Allen bolt*
13 *Gasket*
14 *Drain plug*
15 *Rubber cap*
16 *Circlip*
17 *Air valve*
18 *O-ring*
19 *Top plug*
20 *O-ring*
21 *Pinch bolt*
22 *Spring*
23 *Stanchion*
24 *Bottom bush*
25 *Top yoke*
26 *Clamp bolt*
27 *Lock washer*
28 *Nut*
29 *Bottom cone*
30 *Grease seal*
31 *Flat washer*
32 *Bottom yoke*

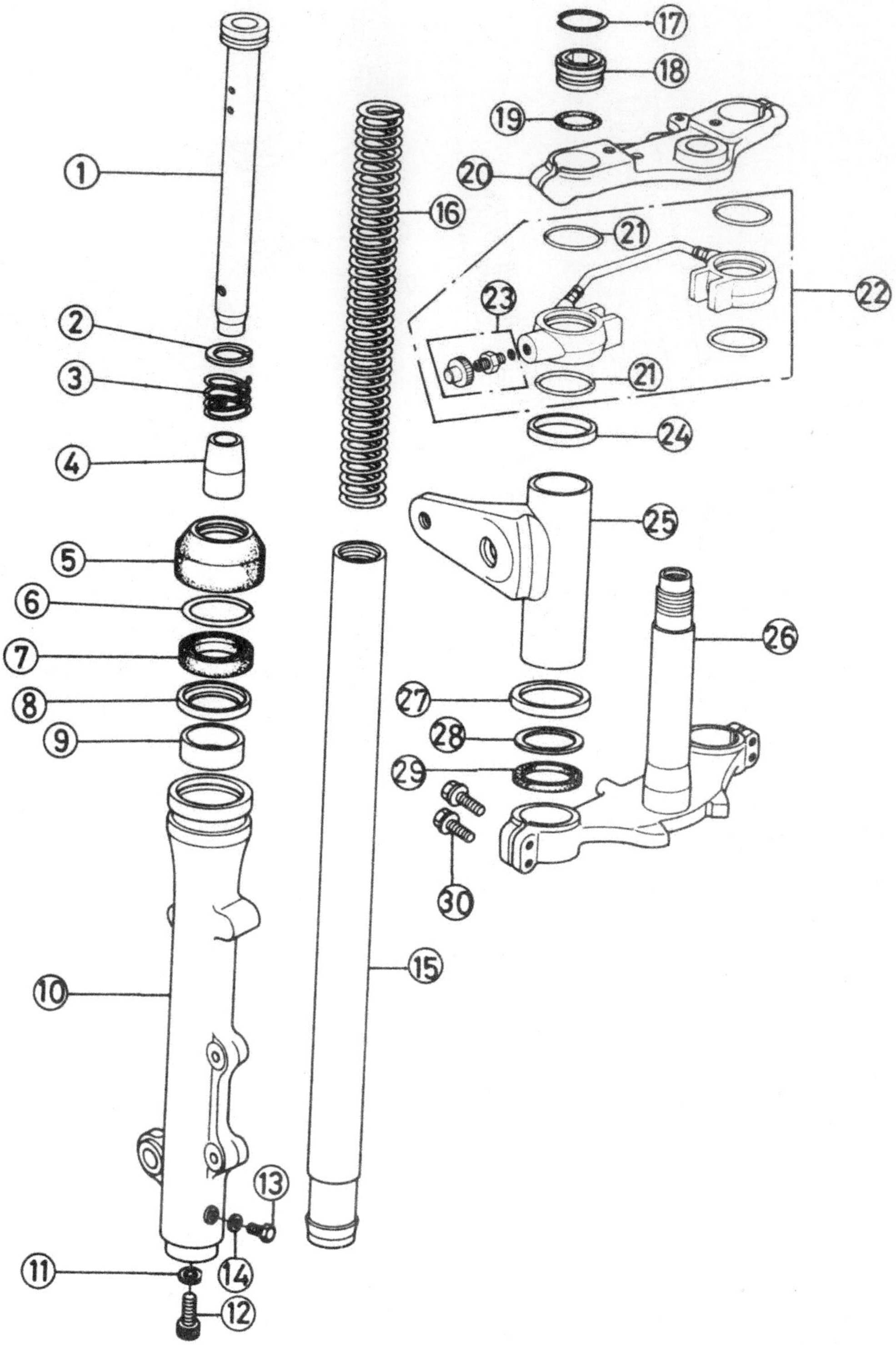

Fig. 4.3 Front forks – KZ/Z550 H1, H2 and all ZR400/550 models

1 Damper rod
2 Piston ring
3 Rebound spring
4 Damper rod seat
5 Dust seal
6 Circlip
7 Oil seal
8 Seal seat
9 Top bush
10 Lower leg
11 Gasket
12 Allen bolt
13 Drain plug
14 Gasket
15 Stanchion
16 Spring
17 Circlip
18 Top plug
19 O-ring
20 Top yoke
21 O-ring
22 Air union
23 Air valve
24 Guide
25 Headlamp bracket
26 Bottom yoke
27 Guide
28 Metal washer
29 Rubber spacer
30 Pinch bolt

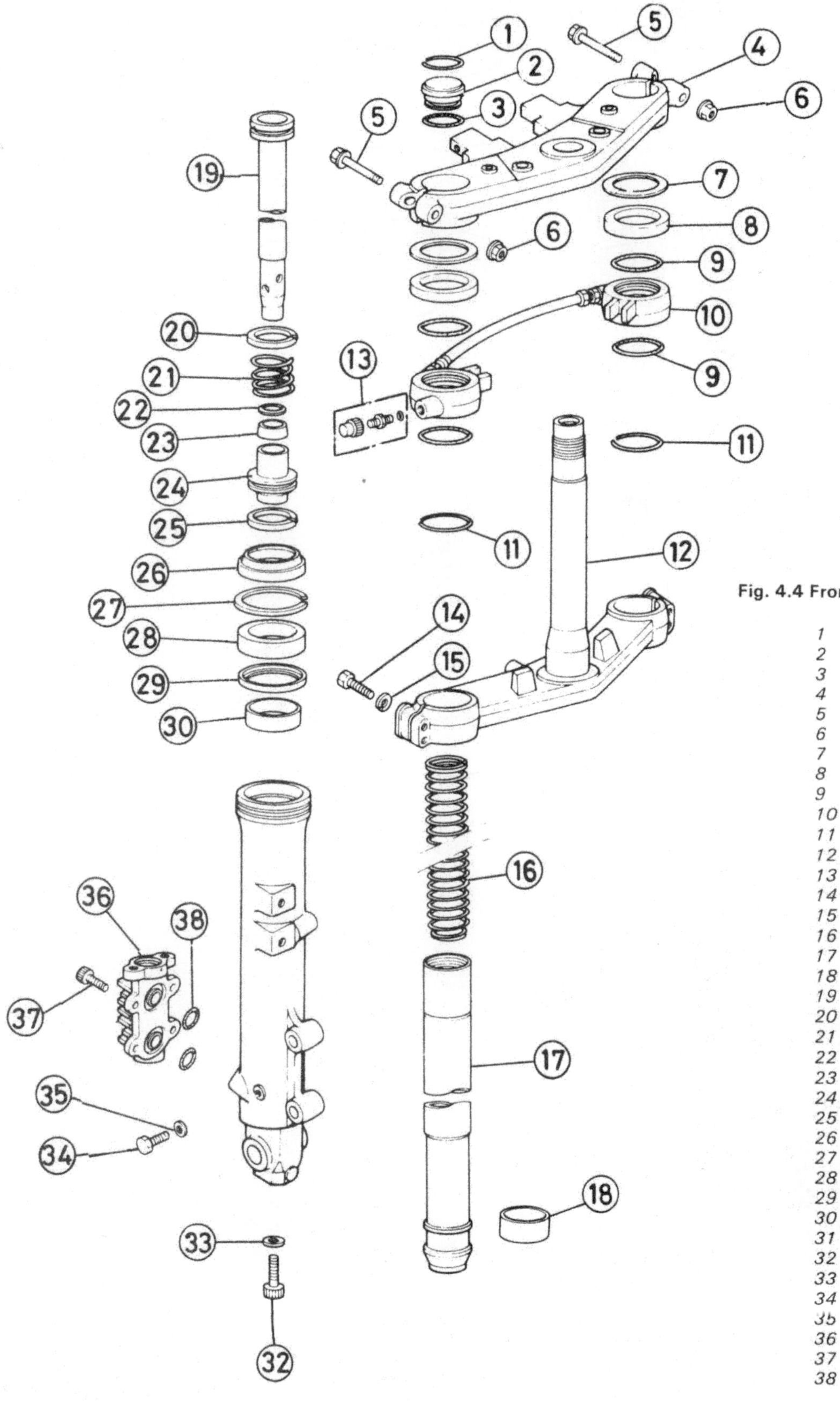

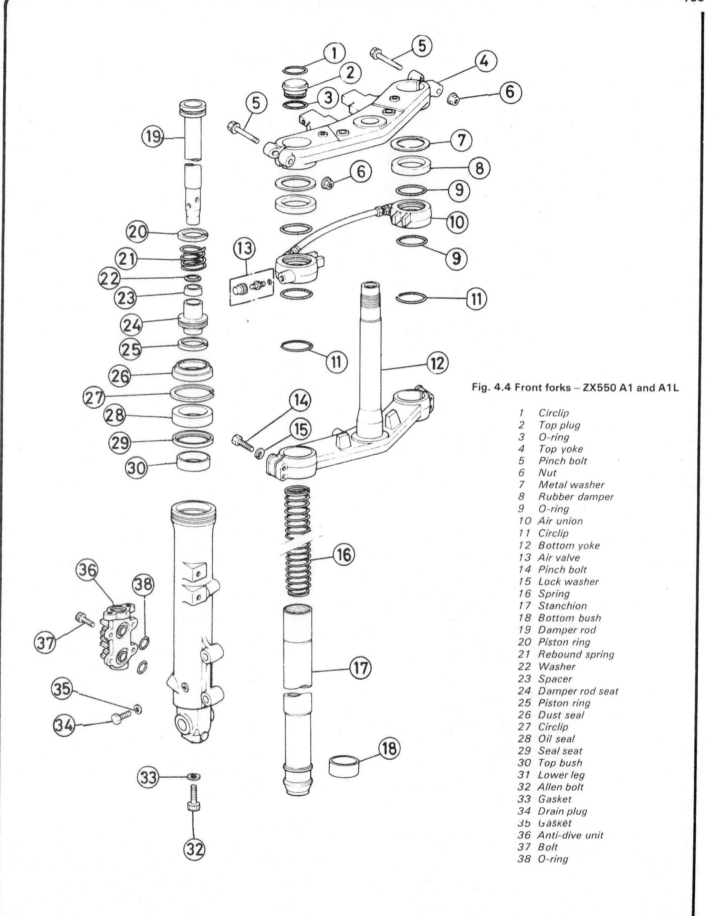

Fig. 4.4 Front forks – ZX550 A1 and A1L

1 *Circlip*
2 *Top plug*
3 *O-ring*
4 *Top yoke*
5 *Pinch bolt*
6 *Nut*
7 *Metal washer*
8 *Rubber damper*
9 *O-ring*
10 *Air union*
11 *Circlip*
12 *Bottom yoke*
13 *Air valve*
14 *Pinch bolt*
15 *Lock washer*
16 *Spring*
17 *Stanchion*
18 *Bottom bush*
19 *Damper rod*
20 *Piston ring*
21 *Rebound spring*
22 *Washer*
23 *Spacer*
24 *Damper rod seat*
25 *Piston ring*
26 *Dust seal*
27 *Circlip*
28 *Oil seal*
29 *Seal seat*
30 *Top bush*
31 *Lower leg*
32 *Allen bolt*
33 *Gasket*
34 *Drain plug*
35 *Gasket*
36 *Anti-dive unit*
37 *Bolt*
38 *O-ring*

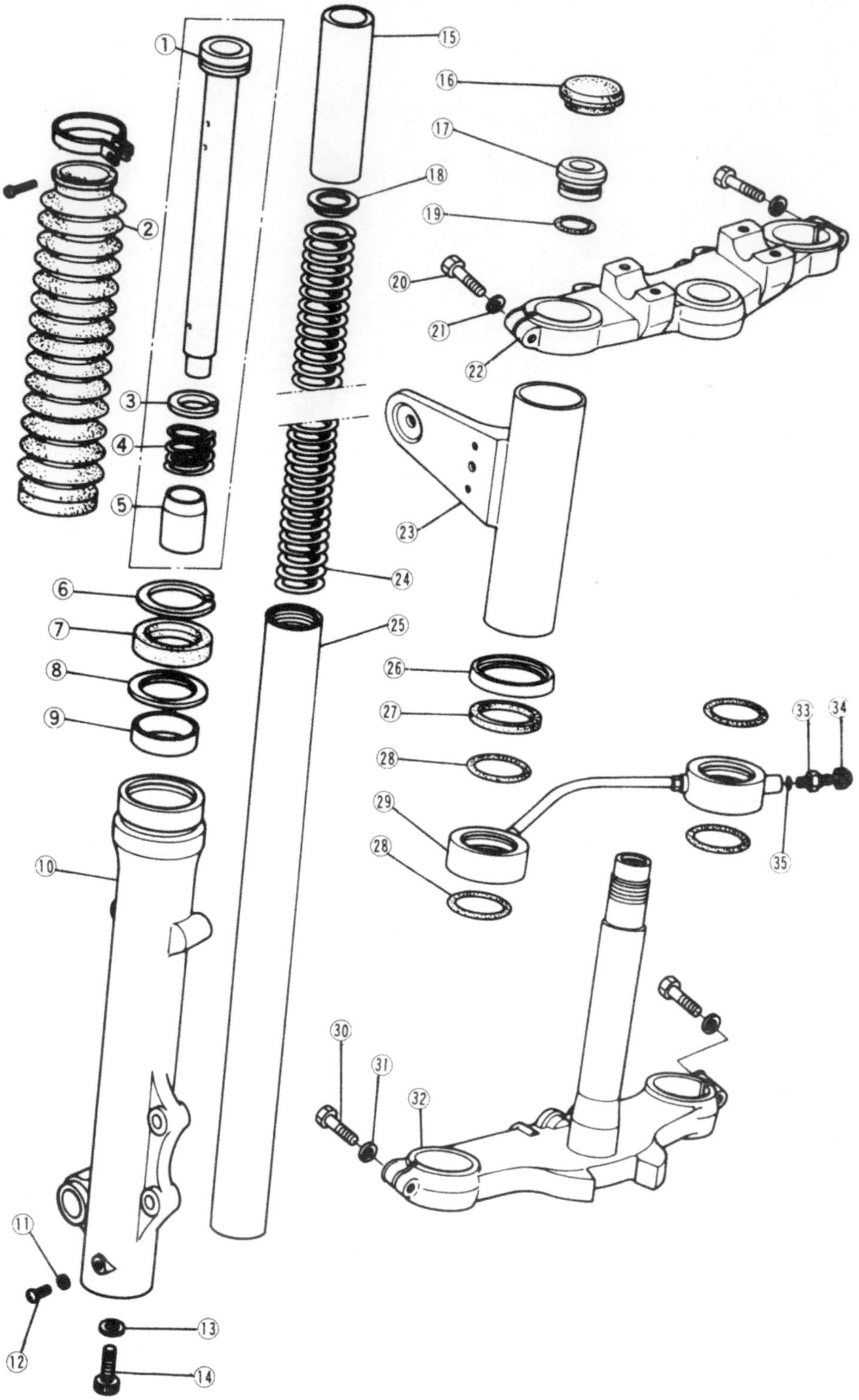

Fig. 4.5 Front forks – shaft drive models

1 Damper rod
2 Gaiter
3 Piston ring
4 Spring
5 Damper rod seat
6 Circlip
7 Oil seal
8 Seal seat
9 Top bush
10 Lower leg
11 Gasket
12 Drain plug
13 Gasket
14 Allen bolt
15 Spacer
16 Rubber cap – US models only
17 Top plug
18 Washer
19 O-ring
20 Pinch bolt
21 Lock washer
22 Top yoke
23 Headlamp bracket
24 Spring
25 Stanchion
26 Guide
27 Rubber damper
28 O-ring
29 Air union
30 Pinch bolt
31 Lock washer
32 Bottom yoke
33 Air valve
34 Cap
35 O-ring

3 Front forks: dismantling and reassembly

1 Remove the fork legs from the machine as described in the previous Section of this Chapter, then dismantle each leg in turn, taking care that the various components are kept separate at all times.

2 On all shaft drive models use an Allen key to unscrew the threaded top plug, noting its sealing O-ring, then remove the spacer, thick washer and the fork spring. Depress the stanchion far enough to reach the fork spring and lift it out, noting which way round it is fitted. On chain drive models an assistant will be required. Placing a suitable implement in the top plug central recess or around the air valve, press down on the top plug while the wire circlip is removed from its groove in the stanchion, then slowly allow fork spring pressure to drive out the plug, noting its sealing O-ring. On KZ/Z550 D1 models only remove the thick washer above the fork spring. On all models withdraw the fork spring, noting which way round it is fitted.

3 Invert the fork leg over a suitable container to tip out the fork oil, then pump the stanchion in and out to expel as much as possible of the surplus. Where gaiters are fitted, slacken the top clamp, pull the gaiter off the top of the fork lower leg and withdraw it; on all models prise the dust seal off the top of the fork lower leg and withdraw it. Remove the circlip from the top of the fork lower leg, above the fork seal. Push the stanchion fully into the lower leg.

4 Using a vice fitted with soft alloy or wooden jaw covers, clamp the fork leg by the wheel spindle lug so that it is horizontal. Unscrew the Allen bolt at the bottom of the fork lower leg to release the damping mechanism; if one is lucky this will unscrew easily. If the screw merely breaks free and rotates without unscrewing, the damping mechanism is rotating with the screw and must be held. The Kawasaki tools for this task consist of an adaptor Part Number 57001-1011 (57001-1057 on shaft drive models) which is fitted to a long T-handle Part Number 57001-183 and passed down inside the stanchion to engage with the head of the damper rod. A usable substitute can be made by grinding a coarse taper on a wooden rod of sufficient length. With an assistant holding the protruding end of the rod with a self-locking wrench and applying pressure to the damper rod head, the Allen screw can be removed. The stanchion and lower leg can now be separated.

5 On Z400 J1, Z500 B1, B2 and KZ/Z550 A1 models, pull the stanchion out of the lower leg, then invert it to tip out the damper rod and rebound spring. Invert the lower leg to tip out the damper rod seat then lever the fork seal out of the top of the lower leg and tip out the seal seat. If the seal is a tight fit, pour boiling water over the upper end of the fork leg, taking great care to avoid personal injury; this will expand the alloy of the lower leg sufficiently to loosen its grip on the seal. Use a lever with rounded edges to remove the seal and take care not to damage the seal housing or to damage the fork leg in any way. Note that heavy leverage should not be required.

6 On all other models, check that the lower leg is clamped securely in the vice and pull the stanchion sharply out as far as possible, then push it fully in and repeat the process until the slide hammer action of the bushes against the seal seat drives the seal out of the lower leg. Pull the stanchion out of the lower leg and pull off the fork seal, the seal seat and the top bush, then invert the stanchion to tip out the damper rod and rebound spring. Remove the bottom bush by inserting a screwdriver blade into the vertical split and springing apart the two ends by just enough to allow the bush to slide down the stanchion. Invert the lower leg to tip out the damper rod seat; on ZX550 A1 and A1L models note that there are two wave washers and a cone-shaped spacer around the damper rod above the seat, also that the seat has a close-fitting piston ring fitted which may make removal difficult.

7 When all components have been cleaned, dried and checked for wear or renewed, they should be reassembled in the reverse of the dismantling sequence.

8 On Z400 J1, Z500 B1, B2 and KZ/Z550 A1 models, place the seal seat in its recess in the fork lower leg, then refit the oil seal. Using a tubular drift such as a socket spanner which bears only on the outer edge of the oil seal, to prevent seal distortion or damage to the sealing lips, tap the oil seal into the lower leg keeping the seal square to its housing, until the circlip groove is exposed, then refit the circlip. Refit the piston ring and rebound spring to the damper rod and drop this into the stanchion so that it protrudes from the stanchion lower end; insert the wooden dowel used on dismantling or the fork spring into the stanchion to retain the damper rod, smear the stanchion surface and seal lips with clean fork oil, and refit the damper rod seat over the lower end of the damping rod, using a smear of grease to stick it in place. Insert the stanchion assembly into the lower leg, taking care not to damage the oil seal lips, and press it down until both stanchion and damper rod are pressed firmly against the bottom of the fork lower leg. Applying liquid gasket to its sealing washer and thread locking compound to its threads, refit the Allen bolt, tightening it to a torque setting of 1.8 kgf m (13 lbf ft). Pack grease around the fork seal and refit the dust seal.

9 On all other models, refit the top and bottom bushes, followed by the seal seat and the oil seal. Smear oil over the stanchion to protect the seal as it is slid into place. Refit the piston ring and rebound spring to the damper rod and drop the rod assembly into the stanchion so that it protrudes from the stanchion lower end. Use the wooden dowel to retain the rod while the damper rod seat is refitted. On ZX550 A1 and A1L models, refit the two wave washers and the cone-shaped spacer, refit the piston ring to the damper rod seat and place the seat over the damper rod end. On all models, smear clean fork oil liberally over the stanchion and bushes, then insert the assembly into the lower leg until it seats firmly in the bottom of the lower leg. Applying liquid gasket to its sealing washer and thread locking compound to its threads, refit the Allen bolt, tightening it to the recommended torque setting. Press the top bush firmly into the lower leg and the seal seat down to the bottom of its recess, then use a length of tubing as a drift to press or tap the seal into place until the circlip groove is exposed. Ensure that the tube has no sharp or raised edges which might damage the seal and that it bears only on the hard outer edge of the seal. Refit the circlip. On models fitted with gaiters, pack grease around the seal and lightly smear the stanchion with grease to prevent corrosion before refitting the gaiter to the fork lower leg. On all other models, pack grease around the oil seal before refitting the dust seal.

10 Fill each fork leg with the amount of oil given for each machine on reassembly then slowly pump the stanchion up and down to distribute the oil around the damping mechanism. When the level has settled with the stanchion pulled out of the lower leg as far as possible, measure the oil level and correct it if necessary as described in Routine Maintenance. It is essential that the oil level is correct, especially on models with air-assisted front forks.

11 Refit the fork springs with the tapered ends downwards so that the close-pitched coils are uppermost, then refit the spacer and/or thick washer, as applicable. On shaft drive models refit the top plug, tightening it securely by hand only using the correct size Allen key. On chain drive models, press the plug down into the stanchion until the circlip can be refitted in its groove. Release the plug and check that it is securely retained. Refit the forks to the machine. **Note**: always stand the forks upright or oil may leak out, especially on models fitted with linked air-assisted forks. This would necessitate resetting the oil level.

3.2a Depress top plug to release wire circlip ...

3.2b ... and withdraw top plug – chain drive models

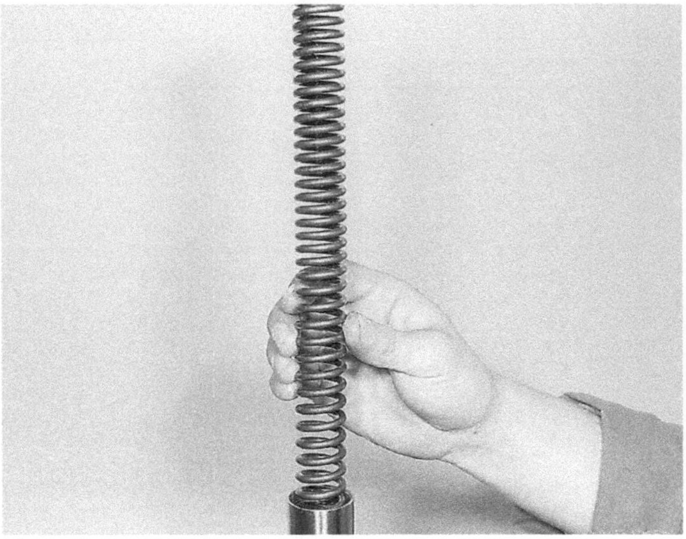
3.2c Note that close-pitched spring coils are uppermost

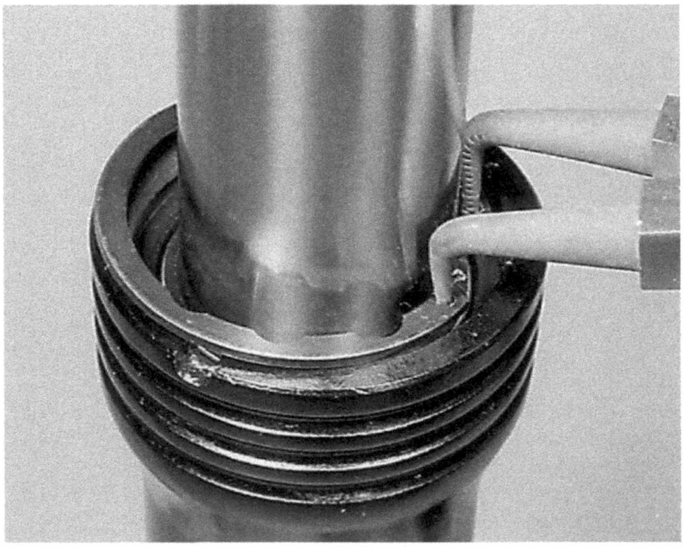
3.3 Remove circlip to release fork oil seal

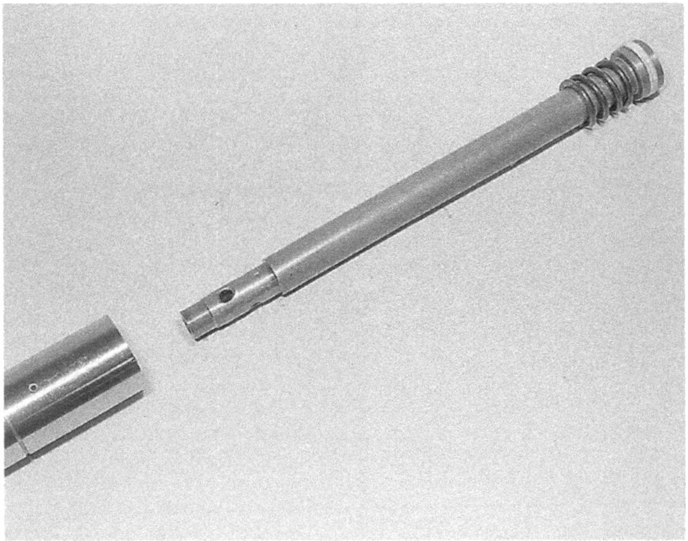
3.9a Refit rebound spring and piston ring to damper rod ...

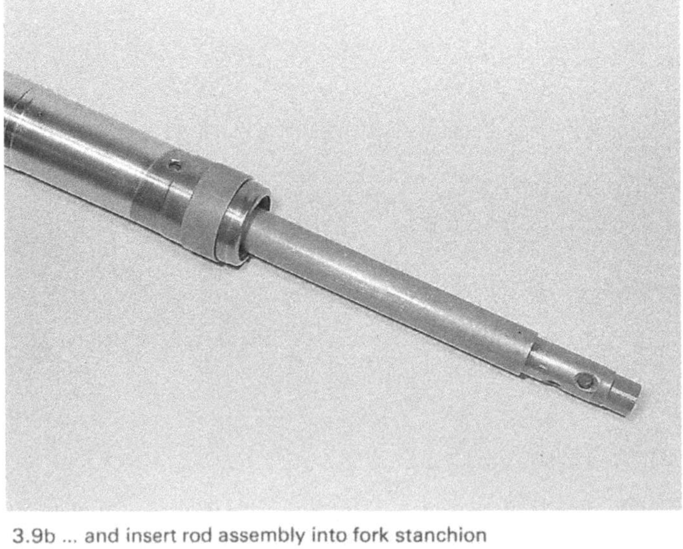
3.9b ... and insert rod assembly into fork stanchion

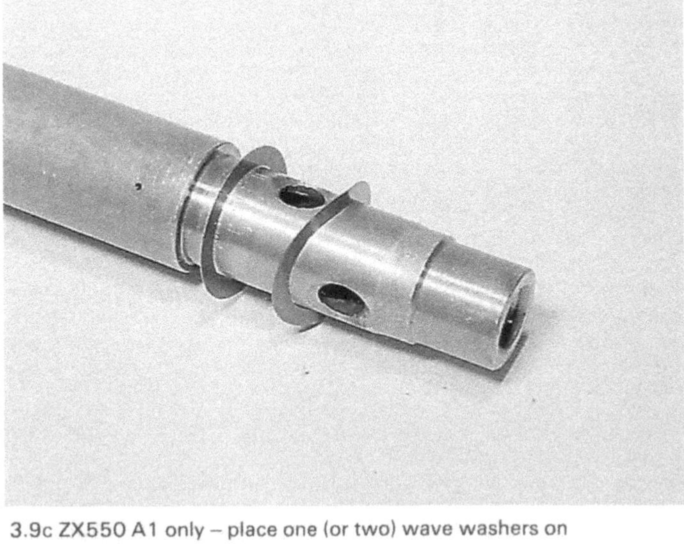
3.9c ZX550 A1 only – place one (or two) wave washers on damper rod ...

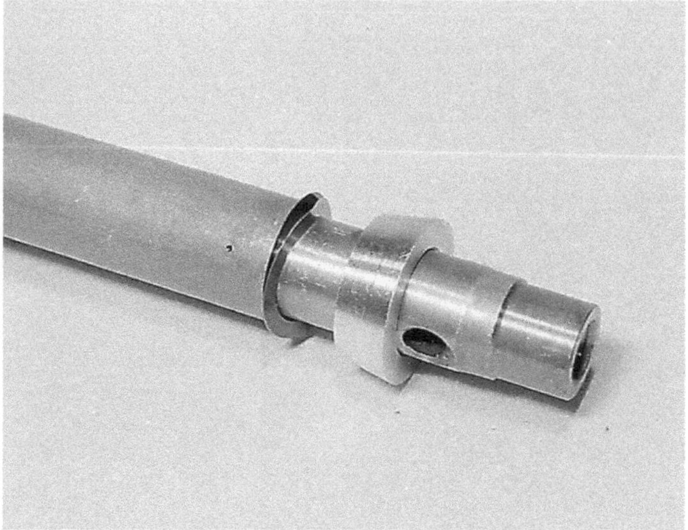

3.9d ... followed by conical spacer ...

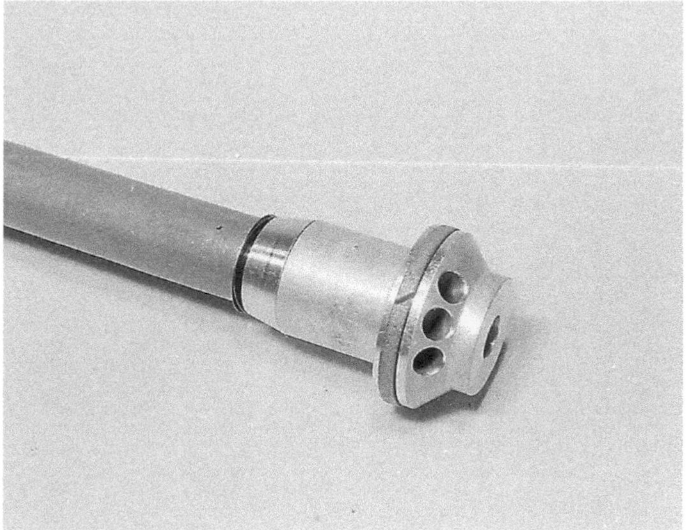

3.9e ... and damper rod seat – renew piston ring if necessary

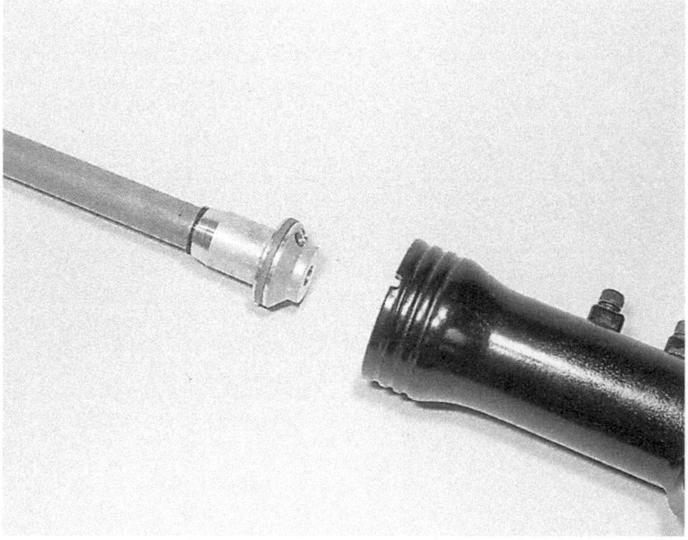

3.9f Insert stanchion assembly into fork lower leg ...

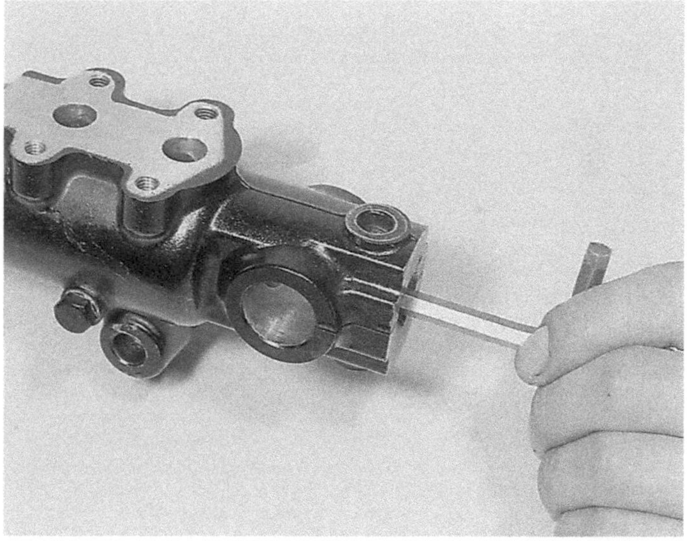

3.9g ... and secure Allen bolt – use torque setting if possible

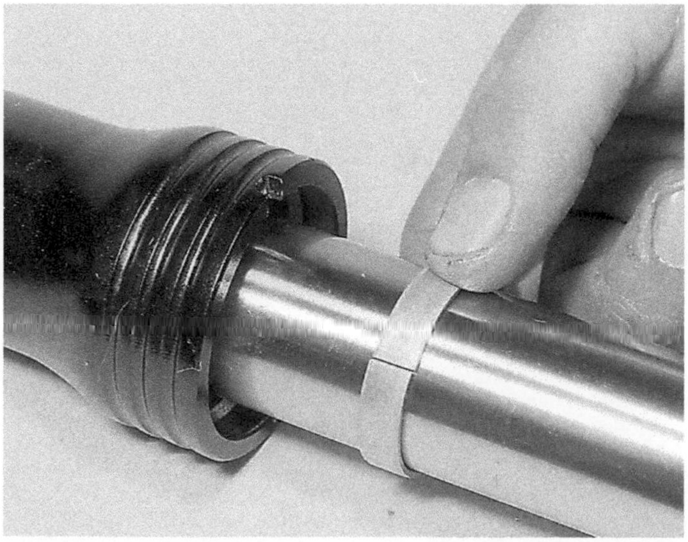

3.9h Slide top bush into place ...

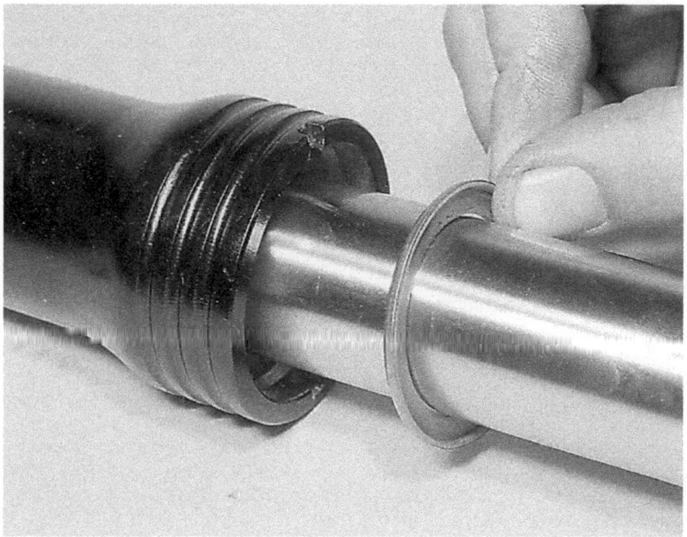

3.9i ... followed by oil seal seat, as shown ...

3.9j ... then refit fork oil seal

3.9k Secure seal with circlip then refit dust seal

4 Front forks: examination and renovation

1 Wash all fork components thoroughly to remove all traces of old oil. Check particularly carefully that all dirt has been removed from the bottom of the fork lower leg and from the passageways in the damping mechanism. On ZX550 A1 and A1L models it is advisable to remove the anti-dive valve unit so that one can be certain of cleaning it thoroughly and drying it completely. Refer to Section 5 of this Chapter.
2 Discard the fork oil seal; this should be renewed whenever it is disturbed. Similarly check the gaiters or dust seals for splits, cracks or other damage, and all sealing O-rings and washers. Renew any component that is found to be worn or damaged. On models so equipped, do not forget the seals around the air valves and, where applicable, in the air union. Also check the piston ring around the head of the damper rod and, on ZX550 A1 and A1L models only, around the damper rod seat. These rings must fit closely and seal tightly in their respective bores if the damping mechanism is to function efficiently.
3 Apart from the above, the only other components likely to wear are the bearing surfaces of the stanchion and fork lower leg. Insert the stanchion into the lower leg, complete with bushes where fitted, and feel for free play between the two at all points from full compression to full extension. No specifications are given, therefore if free play appears to be excessive, the worn components should be renewed. Excessive wear is normally revealed by score marks on one or both surfaces; if such signs are found the component concerned must be renewed. Take the components to a good Kawasaki dealer for an expert opinion if in doubt. On those models fitted with separate bushes the top bush can be renewed, if worn, but this would not appear to be true of the bottom bush in some cases; check with a local Kawasaki dealer for advice in this respect.
4 The stanchions can be checked for straightness by rolling them on a flat surface such as a sheet of plate glass; any bending or distortion should immediately be evident. It is usually possible to straighten slightly bent stanchions provided that the work is undertaken only by an expert; any local motorcycle dealer should be able to recommend such a person. However if the stanchion is bent so much that the tubing has creased or even split, it must be renewed; straightening, even if possible, would induce severe stresses resulting in a fatigue failure at a later date.
5 Check that the stanchions upper surfaces are clean and free from chips, dents or corrosion which might weaken the tubing or cause oil seal failure. Use fine emery paper to polish off any corrosion; but chips or dents, if minor, can be filled with Araldite or similar and rubber down to restore the original shape when the filling compound has set. UK owners should note that such damage will cause the machine to fail its DOT test certificate. If in doubt about the stanchion's strength, renew it in the interests of safety.
6 The fork spring will take a permanent set after considerable usage and will need renewal if the fork action becomes spongy. The service limit for the total free length of each spring is given in Specifications.
7 Make a careful check of all other fork leg components, checking for cracks in castings, damaged threads, defective air valves, and any other signs of wear or damage, renewing any faulty components.

5 Anti-dive valve assembly: testing and renewal – ZX550 A1 and A1L models

1 The valve assembly is mounted on the front of each fork lower leg, being retained by four Allen screws. The oil passages are each sealed by O-rings. Note that no replacement parts are available with which the assembly can be reconditioned. If it becomes worn out or is damaged in an accident, each assembly can only be renewed as a single unit. Note that actual wear is unlikely; the unit is most likely to fail due to dirt jamming a valve.
2 Note that the anti-dive is hydraulically activated; refer to the relevant Sections of Chapter 5 for information on the plunger assembly and hydraulic system. To test the valve assembly, remove the fork legs from the machine as described in Section 2 of this Chapter, then remove the fork top plug and the fork spring and tape up the hole drilled in the stanchion to allow the passage of air via the air union. Clamp the fork leg vertically in the vice, ensuring that the vice has soft alloy or wooden jaw covers that bear only on the wheel spindle lug, then set the anti-dive to the softest (Number 1) setting ie fully anti-clockwise.
3 Ensuring that no oil is spilled, pump the stanchion gently and smoothly up and down through its full travel, feeling the amount of damping present on both compression and rebound strokes when the anti-dive is not operating. Insert a rod such as an Allen key into the hole in the top of the valve assembly and press firmly downwards; do not apply heavy pressure as it is not necessary and may damage the valves. The hydraulic plunger extends by only 2 mm. Repeat the pumping action and compare the difference when the rod is released with when it is depressed, then repeat the test with the anti-dive on each of the stronger settings.
4 The amount of effort required to compress the fork should increase noticeably as soon as the rod is depressed and should return to normal as soon as the rod is released. As the setting is increased to its stiffest the effort required to compress the fork should increase in proportion and should still return to normal when the rod is released. If compression damping is heavy when no pressure is applied to the rod, or if it does not return to normal when the rod is released, also if there is no difference in damping with pressure applied to the rod, the anti-dive valve unit is faulty and must be renewed. As a safety measure the manufacturer recommends that the assembly should not be dismantled with a view to repair and does not supply replacement parts as a result.
5 The assemblies can, however, be dismantled for cleaning. Each

can be removed from its fork leg after the oil has been drained and after the plunger assembly has been withdrawn. Remove the four Allen screws and withdraw the unit, noting the two sealing O-rings. Remove the single retaining screw and displace the adjusting knob, then unscrew the bottom plug and remove the heavy coil spring. Displace the circlip from the top of the unit and withdraw the upper valve assembly followed by the light coil spring, then pull the valve plunger out of the valve body. Although it is possible to remove the valve centre from the unit body once a wire circlip has been displaced there is no point in doing so unless the valve is jammed with dirt. Similarly, the valve plunger can be dismantled further, but there is no point in this.

6 Thoroughly clean and dry all components and remove all traces of dirt or corrosion. If any O-rings are found to be damaged or worn the assembly must be renewed unless replacements of exactly the correct size can be found. Similarly, if any component is found to be severely worn or damaged, the assembly must be renewed.

7 Reassembly is the reverse of the dismantling procedure. Apply clean fork oil to all valve components and ensure that they are securely fastened. On refitting the assembly to the fork leg always renew the sealing O-rings and tighten the Allen screws to a torque setting of 0.75 kgf m (5.5 lbf ft). Repeat the test described above; if no improvement is found, the assembly is obviously faulty and must be renewed.

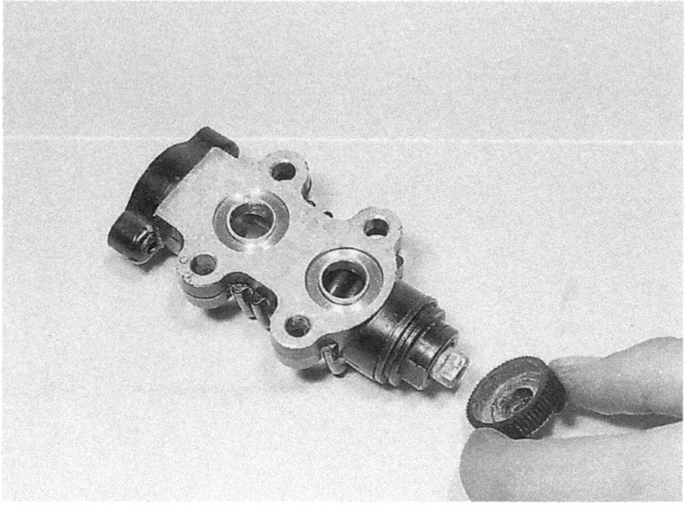

5.5a Remove retaining screw and withdraw adjusting knob ...

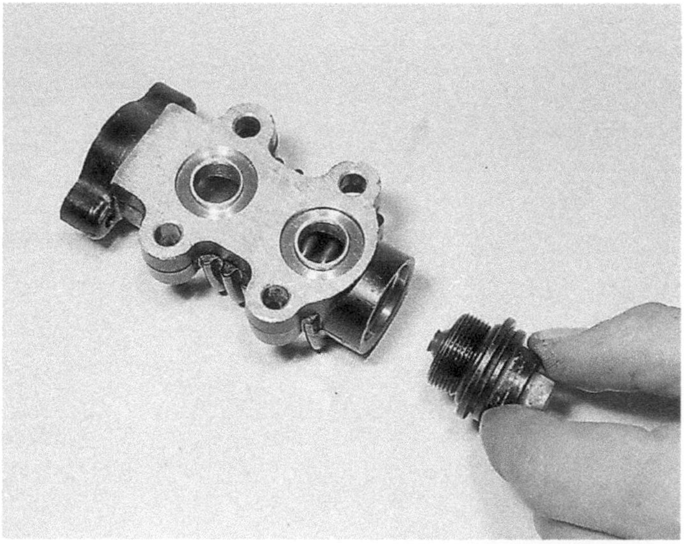

5.5b ... then unscrew bottom plug assembly – note sealing O-ring ...

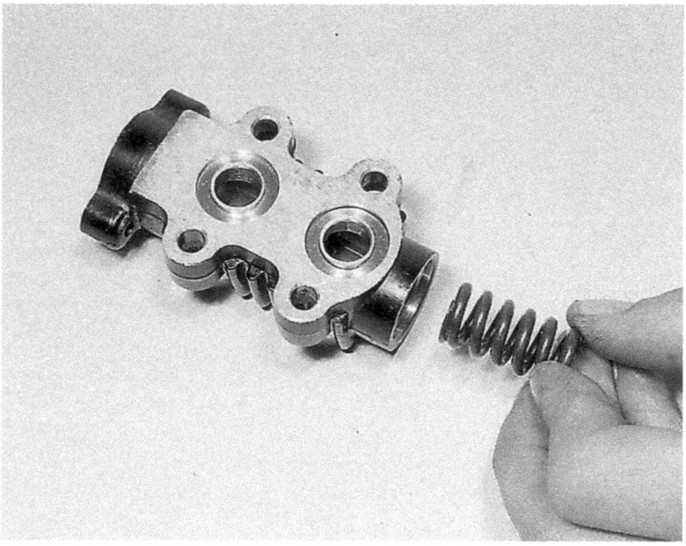

5.5c ... and withdraw heavy coil spring

5.5d Remove circlip from top of unit ...

5.5e ... and withdraw upper valve assembly – note sealing O-ring ...

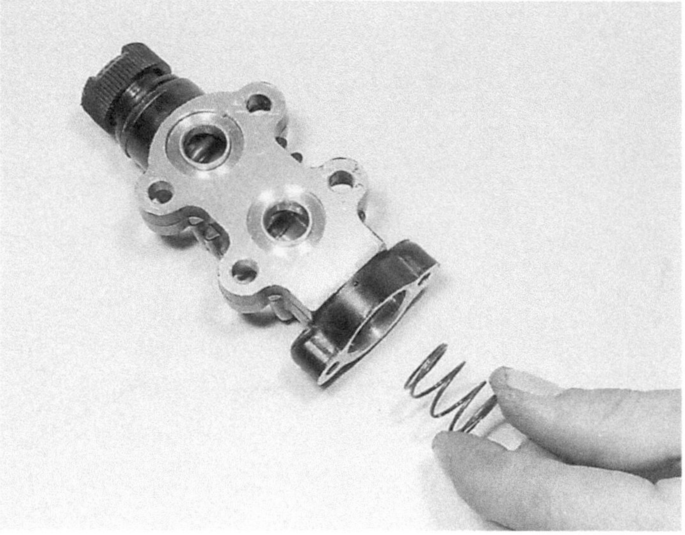

5.5f ... followed by light coil spring

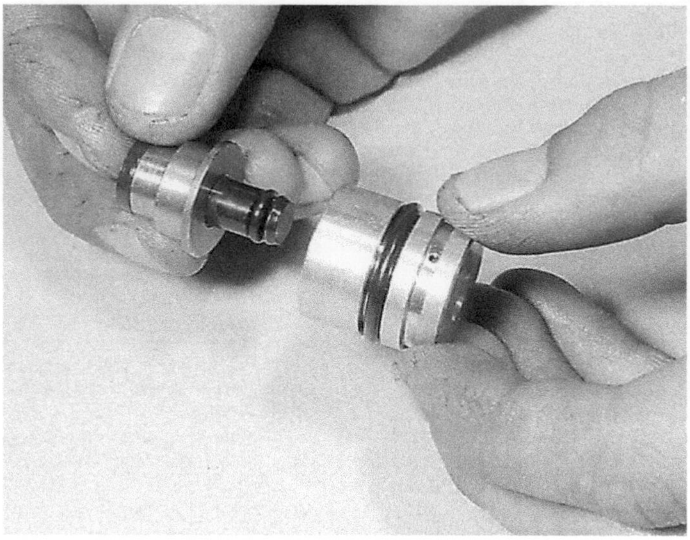

5.5g Plunger can be separated from valve assembly but ...

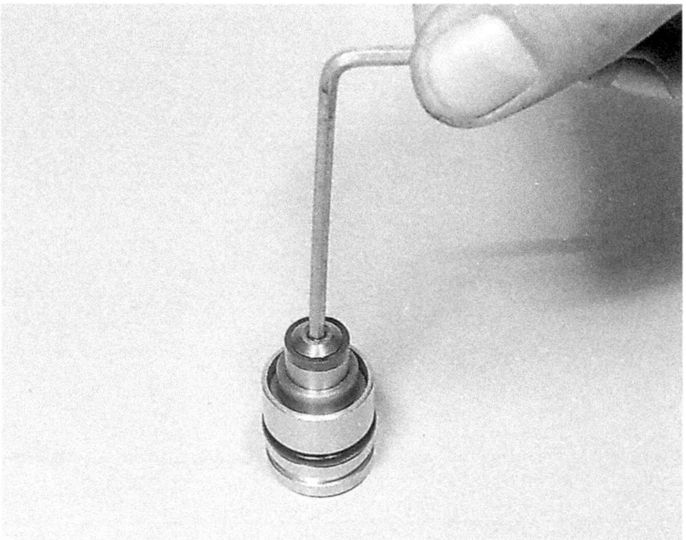

5.5h ... further dismantling is not recommended

5.5i Do not disturb valve centre unless absolutely necessary

5.7a Renew sealing O-rings on refitting ...

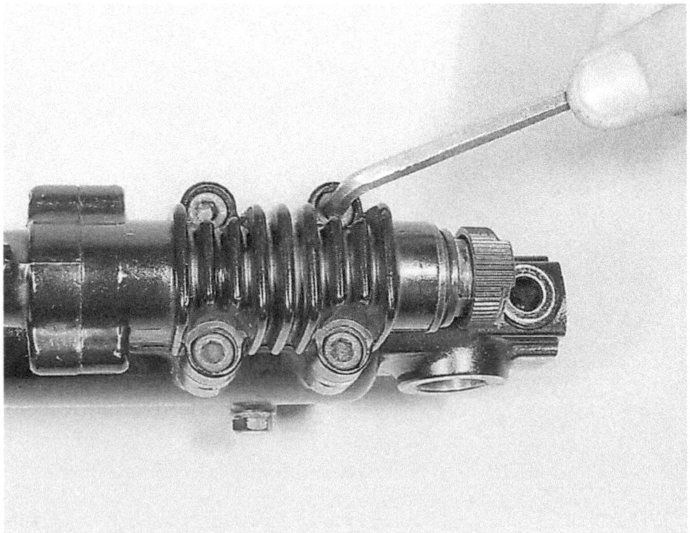

5.7b ... and tighten Allen screws securely

6 Steering head: removal and refitting

1 Remove the seat and fuel tank as described in Section 2 of Chapter 2, remove the fairing (where fitted) and remove both front fork legs as described in Section 2 of this Chapter.

2 Remove the headlamp unit and separate those wiring connectors which lead to the instrument panel and to the turn signals, then lift the headlamp shell and brackets away from the steering head, noting carefully the correct positions of the various metal and rubber guides, washers and spacers. Remove also, where applicable, the fork air union. Remove the cover, where fitted, from the bottom yoke and release the brake hose from its clamps. On models with twin disc front brakes remove the brake hose junction mounting bolts.

3 The handlebar assembly, on models with chrome tube handlebars is removed next, noting that two approaches are possible. The official method is to systematically remove the clutch cable, left-hand switch cluster and wiring clips, right-hand switch cluster and wiring clips and the starter lockout switch. The four handlebar clamp bolts are now removed and the handlebar lifted away. An easier method is to remove the handlebar clamps and lift the entire handlebar assembly rearwards, resting across the frame with the various control cables and electrical leads attached. This avoids a great deal of dismantling work, but does mean that it will be necessary to work around the cables and leads during steering head removal. Disengage all cables and wiring from the guides on the fork top yoke.

4 On models with cup and cone steering head bearings (see Routine Maintenance) slacken the steering stem head clamp bolt and remove the large top bolt, the plain washer and lock washer. Using a soft-faced mallet, tap the underside of the top yoke until it comes free of the steering stem. Lift the yoke clear, together with the instruments, and place the assembly to one side, noting that the instruments should be kept upright.

5 Arrange any remaining cables, leads, etc, so that they do not foul the bottom yoke. Obtain a small box or tin in which the steering head balls can be placed safely. Note that as the bottom yoke and stem are released, the bottom race balls will drop free, and some provision must be made to catch them. As a precaution place a large piece of rag below the headstock to catch any displaced balls. Slacken the slotted adjusting nut, using a C-spanner, and lower the steering head stem and bottom yoke assembly clear of the frame. Note that there are 20 balls in the bottom race and 19 in the upper one, all being of the same size, namely $\frac{1}{4}$ in diameter.

6 Before reassembly, examine and clean the bearing races, then stick the 20 bottom race balls into position using high melting point grease. Reassemble the fork unit, following the dismantling sequence.

7 On models fitted with taper roller steering head bearings, remove the steering stem top bolt and plain washer. Remove the top yoke and instrument panel. If necessary, tap the underside of the yoke to free it from the steering stem.

8 Support the bottom yoke, then using a C-spanner, slacken and remove the two slotted nuts which retain the steering stem. Lower the bottom yoke and steering stem and displace the upper bearing inner race.

9 When refitting the yokes and bearings, grease the latter prior to installation, and coat the steering stem with grease. Offer up the bottom yoke and fit the slotted adjuster nut finger tight. The raised collar should face downward. It is important to bed the bearings in on assembly as described below.

10 To set up the bearings initially, a C-spanner, preferably the correct Kawasaki item, Part Number 57001-1100, will be required. If using any other spanner, note that it will be necessary to make some

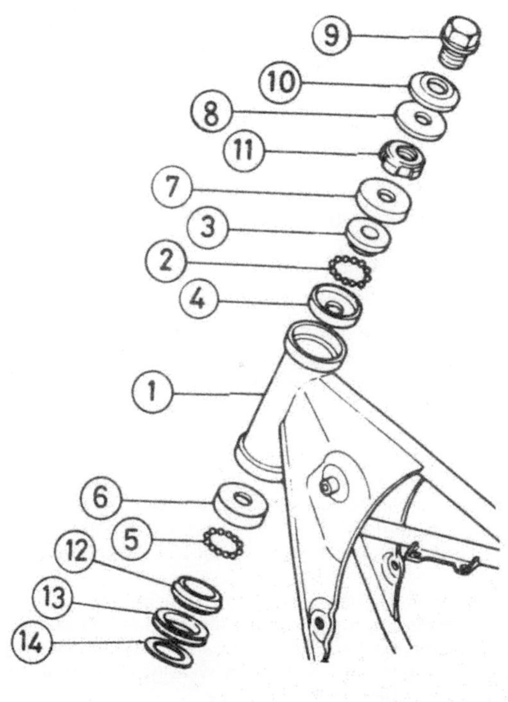

Fig. 4.6 Steering head bearings – all Z400 J, Z500 and KZ/Z550 A, C and D models

1 Frame
2 Ball – 19 off
3 Top cone
4 Top cup
5 Ball – 20 off
6 Bottom cup
7 Dust cover
8 Wave washer
9 Bolt
10 Washer
11 Adjusting nut
12 Bottom cone
13 Rubber seal
14 Metal washer

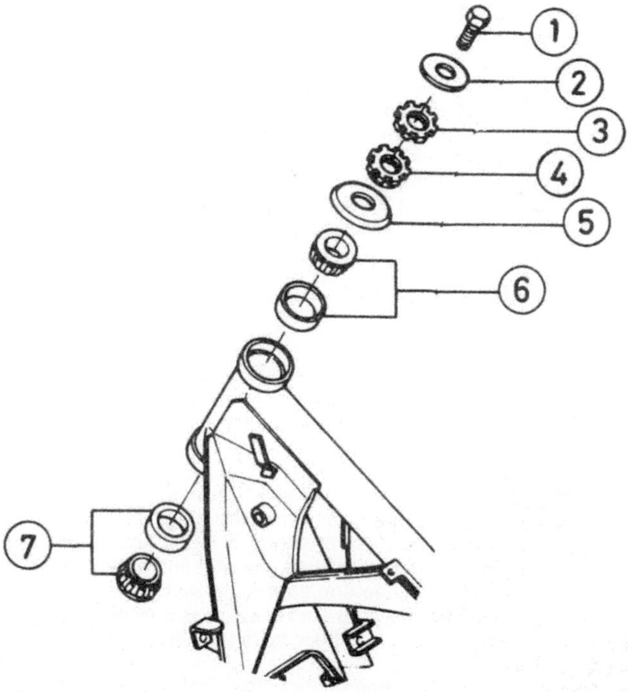

Fig. 4.7 Steering head bearings – all other models

1 Bolt
2 Washer
3 Nut
4 Nut
5 Dust cover
6 Bearing
7 Bearing

provision for a spring balance to be attached at a point 180 mm (7.1 in) from the centre of the steering stem. To this end, extend the spanner as required and drill a hole in the handle at the correct distance. A spring balance capable of reading above 22.2 kg (48.94 lb) will also be required.
11 Fit the C-spanner and apply 4.0 kgf m (29 lbf ft) to the adjuster nut. This is achieved by hooking the spring balance to the hole in the spanner and pulling on it until a reading of 22.2 kg (48.94 lb) is shown. Check that the steering head assembly turns smoothly with no evidence of play or tightness.
12 Slacken the nut slightly until pressure is just released, then turn it slowly clockwise until resistance is **just** evident. Take great care not to apply excessive pressure because this will cause premature failure of the bearings. The object is to set the adjuster so that the bearings are under a **very light** loading, just enough to remove any free play. Once set correctly, run the slotted locknut into place and tighten it firmly whilst holding the adjuster nut in the correct position. On reassembly, refit the top yoke bolt but do not fasten the top bolt until the fork stanchions are refitted and correctly positioned in the yoke (see Section 2). Use a torque wrench to fasten all nuts and bolts to their recommended torque settings.
13 Referring to the relevant Section of Routine Maintenance check that the steering head bearings are correctly adjusted as soon as the forks and front wheel are refitted, but before the handlebars are replaced.
14 Continue assembly by reversing the dismantling sequence. Check that all electrical cables and control cables are routed so that they do not impede steering movement. When refitting the wiring connectors in the headlamp shell, check that the wiring colour codes match up. When assembly has been completed check the operation of the front brake, throttle and clutch and adjust the headlamp alignment and rear view mirror setting.

7 Steering head bearings: examination and renovation

Cup and cone bearings

1 The ball bearing tracks of the respective cup and cone bearings should be polished and free from indentations, cracks or pitting. If signs of wear are evident, the cups and cones must be renewed. For straight line steering to be consistently good, the steering head bearings must be absolutely perfect. Even the smallest amount of wear on the cups and cones may cause steering wobble at high speeds and judder during heavy front wheel braking. The cups and cones are an interference fit on their respective seatings, and can be tapped from position with a suitable drift.
2 Ball bearings are relatively cheap. If the originals are marked or discoloured they **must** be renewed. To hold the steel balls in place during reassembly of the fork yokes, pack the bearings with grease. The upper and lower races contain 19 and 20 $\frac{1}{4}$ in steel balls respectively. Although a small gap will remain when the balls have been fitted, on no account must an extra ball be inserted, as the gap is intended to prevent the balls from skidding against each other and wearing quickly.
3 The bottom cone can be levered off its seat on the steering stem, followed by the rubber seal and metal washer; these last should be renewed if worn or damaged in any way. On reassembly, refit first the metal washer, then the seal and use a length of metal tubing, only slightly larger in its internal diameter than the steering stem, as a tubular drift to tap the bottom cone firmly on to its seat. The bearing cups can be tapped into place using a socket spanner or similar as a a drift; the drift must bear only on the cup outer edge, not on its bearing surface.
4 On reassembly, pack grease around the bottom cone and stick the balls to it as described in Section 6. Grease the inside of the headstock and the steering stem to prevent corrosion. Pack grease into the top cup, refit the balls and place the top cone in position. Remember to hold the cone steady while the bottom yoke is refitted.

Taper roller bearings

5 The inner races are easily checked after all traces of old grease have been removed by washing in a suitable solvent. Turn each race slowly, checking for marks or discolouration of the roller faces.
6 Clean the outer races, and examine the bearing surface for wear or damage. If any wear is discovered, Kawasaki recommend that **both** bearings, including outer races, should be renewed.
7 If renewal is necessary, removal of the old bearing outer races and installation of the new outer races may be accomplished using the correct Kawasaki service tools. Failing this, proceed as described below.
8 The outer races are a fairly tight fit in the steering head tube. Most universal slide-hammer type bearing extractors will work here, and these can often be hired from tool shops. Alternatively, a long drift can be passed through one race and used to drive out the opposite item. Tap firmly and evenly around the race to ensure that it drives out squarely. It may prove advantageous to curve the end of the drift slightly to improve access. Note that with this method there is a real risk of damage unless care is taken. If the race refuses to move, **stop**. Leave the job until a proper extractor can be obtained.
9 The lower inner race can be levered off the steering stem, using screwdrivers on opposite sides to work it free. To fit the new item, find a length of tubing slightly larger in its internal diameter than the steering stem. This will suffice as a tubular drift. Grease the bearing thoroughly and wipe a trace of grease around the steering stem. Drive the bearing home evenly and fully.
10 The new outer races can be installed using a home-made version of the drawbolt arrangement shown in the accompanying illustration.

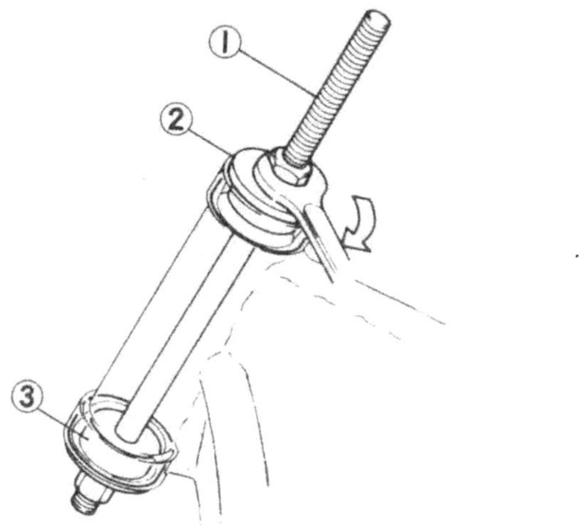

Fig. 4.8 Drawbolt arrangement for fitting bearing outer races

1 High tensile bolt
2 Heavy washer
3 Guide

8 Frame: examination and renovation

1 The frame is unlikely to require attention unless accident damage has occurred. In some cases, renewal of the frame is the only satisfactory remedy if the frame is badly out of alignment. Only a few frame specialists have the jigs and mandrels necessary for resetting the frame to the required standard of accuracy, and even then there is no easy means of assessing to what extent the frame may have been overstressed.
2 After the machine has covered a considerable mileage, it is advisable to examine the frame closely for signs of cracking or splitting at the welded joints. Rust corrosion can also cause weakness at these joints. Minor damage can be repaired by welding or brazing, depending on the extent and nature of the damage.
3 Remember that a frame which is out of alignment will cause handling problems and may even promote 'speed wobbles'. If misalignment is suspected, as a result of an accident, it will be necessary to strip the machine completely so that the frame can be checked, and if necessary, renewed.

9 Rear suspension: removal and refitting

1 Remove the seat and side panels, then remove the rear wheel as described in Section 4 of Chapter 5. The remainder of the procedure is given under the relevant sub-heading.

Chain drive models with swinging arm suspension

2 Remove the chainguard, if this has not been done and release the brake hose from its clamps on the swinging arm, on Z500 B1, B2 and KZ/Z550 D1 models. Remove the rear suspension unit lower mounting bolts and lock washers. Remove the swinging arm pivot shaft nut. Displace the shaft and withdraw the swinging arm clear of the frame.

3 On all models, grease the bearings thoroughly, then grease and fit the pivot sleeve. Refit the swinging arm assembly remembering to pass the left-hand fork end through the chain, manoeuvre the arm into place, taking care not to displace any components then grease and refit the pivot bolt. Tighten its retaining nut to a torque setting of 8.0 kgf m (58 lbf ft). Check that the arm moves smoothly and easily with no trace of free play, then refit the rear suspension units, tightening their mounting nuts and bolts to a torque setting of 3.0 kgf m (22 lbf ft).

Shaft drive models

4 Slacken the suspension unit top mountings and remove the bottom mounting nuts and washers, and pull both units off their studs. Remove its four retaining nuts and withdraw the final drive casing from the swinging arm, taking care not to lose the coil spring. Slacken its clamp screw and pull back the gaiter from the front gear case unit. Rotate the shaft until a small hole is located in the shaft end, then insert a metal rod into the hole to depress the locking pin inside. The shaft can then be pulled backwards off its splines. Remove the dust caps, slacken their locknuts and unscrew the two pivot shafts to allow the swinging arm to be manoeuvred clear of the frame.

5 On reassembly, grease the bearings and those shaft components described in Routine Maintenance, then manoeuvre the swinging arm into the frame and screw in the pivot shafts to retain it. Slacken the locknut fully and rotate the shafts until the clearance between the swinging arm left-hand pivot stub (bearing housing) and the frame gusset is between 1.4 – 1.6 mm (0.0551 – 0.0630 in), as measured with feeler gauges.

6 To settle the bearings, tighten each pivot shaft as hard as possible, using hand pressure only on an Allen key of the correct size; this will approximate the correct torque setting. Never try to increase leverage by extending the key as this may damage the bearings or pivot shaft thread. Unscrew both pivot shafts until pressure is just released, check that the clearance is correct on the left-hand side, then very carefully tighten each shaft to a torque setting of 1.3 kgf m (9.5 lbf ft). Check that the swinging arm moves smoothly and easily up and down with no sign of resistance in the bearings, and yet with no trace of free play. Hold steady each shaft while its locknut is firmly tightened, then refit the dust caps.

7 Check that the locking pin and its hole are aligned and press the shaft on to its splines; the pin should be heard to click into place. Pull backwards on the shaft to check that it is secured, then refit the gaiter and tighten its clamp. Ensuring that the coil spring is correctly positioned, refit the final drive casing and tighten its mounting nuts to a torque setting of 2.3 kgf m (16.5 lbf ft). Refit the suspension units and tighten their mountings to a torque setting of 2.5 kgf m (18 lbf ft). Refit the rear wheel as described in Section 4 of Chapter 5.
setting of 2.5 kgf m (18 lbf ft). Refit the rear wheel as described in Section 4 of Chapter 5.

KZ/Z550 H1, H2 and all ZR400/550 models

8 Remove the left-hand air filter housing, the battery and its tray, and the gearbox sprocket cover. Remove the gearbox sprocket and hang the chain over the swinging arm pivot. Remove the rocker arm pivot shaft nut and tap out the shaft, then disengage the brake hose from the swinging arm clamps, where applicable. Remove the two pivot shafts and withdraw the rear suspension as a complete assembly, with the final drive chain. Note that the pivot shaft S-shaped nuts cannot be rotated to disengage them from their retaining lugs until the engine/gearbox is removed. Do not attempt to remove the nuts with the engine in place, by bending the retaining ears, as refitting is almost impossible if the nuts are to be kept in a useable condition.

9 On reassembly, grease the bearings thoroughly and check that all components, including the drive chain, are in place. Manoeuvre the assembly into place, refit the pivot shafts to retain it and tighten them to a torque setting of 8.0 kgf m (58 lbf ft). Check that the swinging arm moves easily and smoothly with no trace of free play, then align the rocker arm pivot with the frame gusset, grease and refit the pivot shaft and tighten its retaining nut to a torque setting of 9.0 kgf m (65 lbf ft). Complete reassembly by reversing the dismantling procedure. Be very careful to route all wiring clear of the suspension components.

ZX550 A1 and A1L models

10 Lift the machine on to a strong wooden box or similar support so that it is held securely with the centre stand raised. This is to permit access to the rocker arm front pivot bolt, which is also obscured by the exhaust system. Although not absolutely necessary it is recommended that the exhaust system be removed as this will improve access a great deal and will not prolong the operation unduly; a great deal of time and care is required to extract and refit the pivot bolt with the exhaust in place.

11 Disengage the brake hose from the swinging arm clamps and the remote preload adjuster from its frame clamp. Remove their retaining nuts and drive out first the suspension unit top mounting bolt, then the rocker arm front pivot bolt and finally the swinging arm pivot bolt. Remove the rear suspension as a single unit, taking care that the preload adjuster is not damaged and that the chain is disengaged.

12 On reassembly, grease the bearings thoroughly and check that all components are in place, then manoeuvre the assembly into place remembering to pass the swinging arm left-hand fork end through the chain and to pass the preload adjuster out between the frame downtubes underneath the IC ignitor unit. Refit the swinging arm pivot bolt and tighten its retaining nut to a torque setting of 9.0 kgf m (65 lbf ft). Check that it moves smoothly and easily, with no trace of free play, then align the rocker arm front end with its mounting and refit the pivot bolt, followed by refitting the suspension unit top mounting bolt. Tighten both pivot bolt retaining nuts to a torque setting of 5.0 kgf m (36 lbf ft).

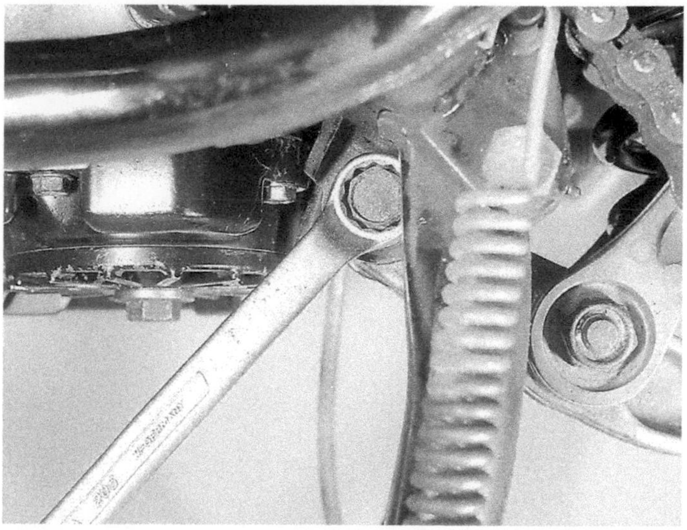

9.10 ZX550 A1 – centre stand must be raised to release rocker front pivot bolt

9.11a Remove suspension unit top mounting bolt

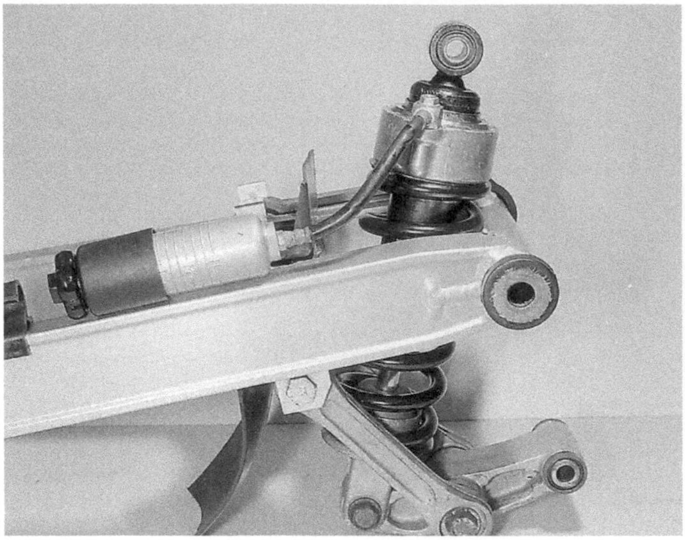

9.11b Rear suspension is best removed as a single unit – Unitrak models

9.12 Always use recommended torque settings on reassembly

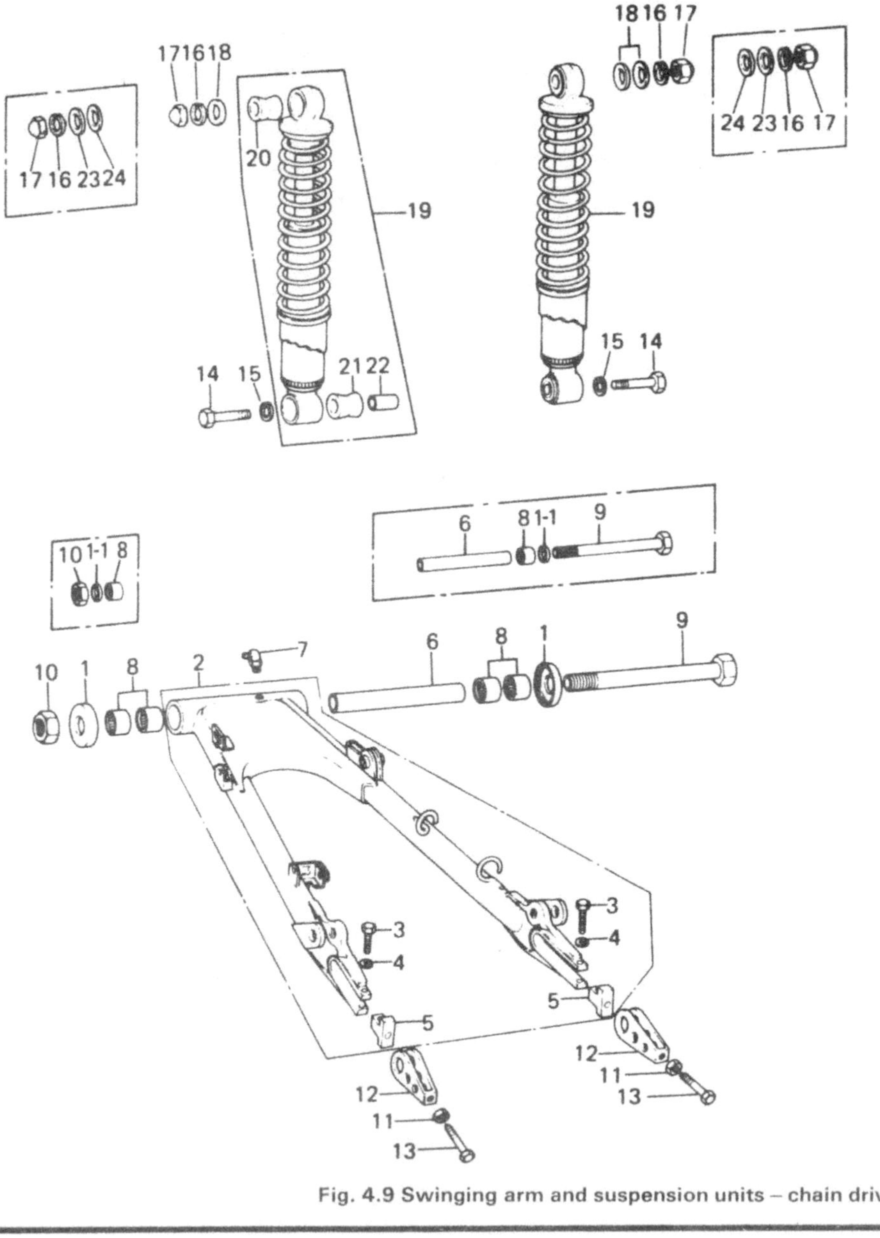

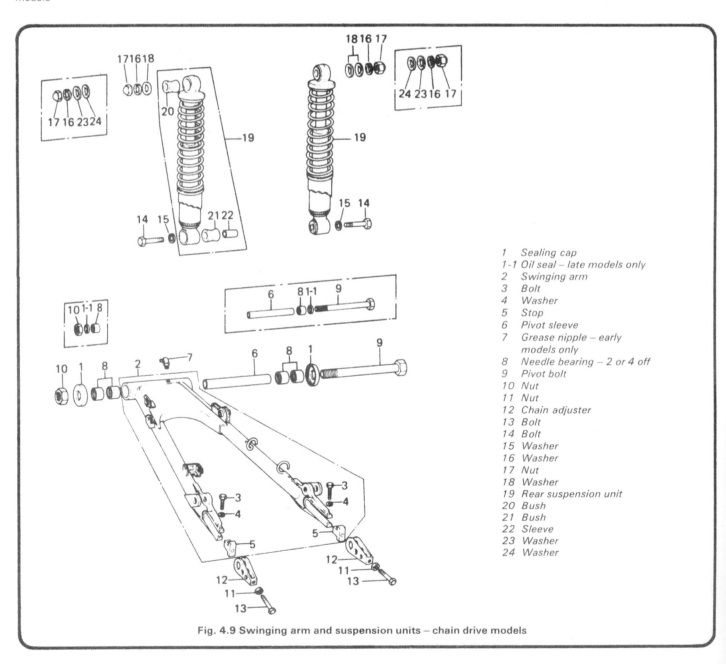

1 Sealing cap
1-1 Oil seal – late models only
2 Swinging arm
3 Bolt
4 Washer
5 Stop
6 Pivot sleeve
7 Grease nipple – early models only
8 Needle bearing – 2 or 4 off
9 Pivot bolt
10 Nut
11 Nut
12 Chain adjuster
13 Bolt
14 Bolt
15 Washer
16 Washer
17 Nut
18 Washer
19 Rear suspension unit
20 Bush
21 Bush
22 Sleeve
23 Washer
24 Washer

Fig. 4.9 Swinging arm and suspension units – chain drive models

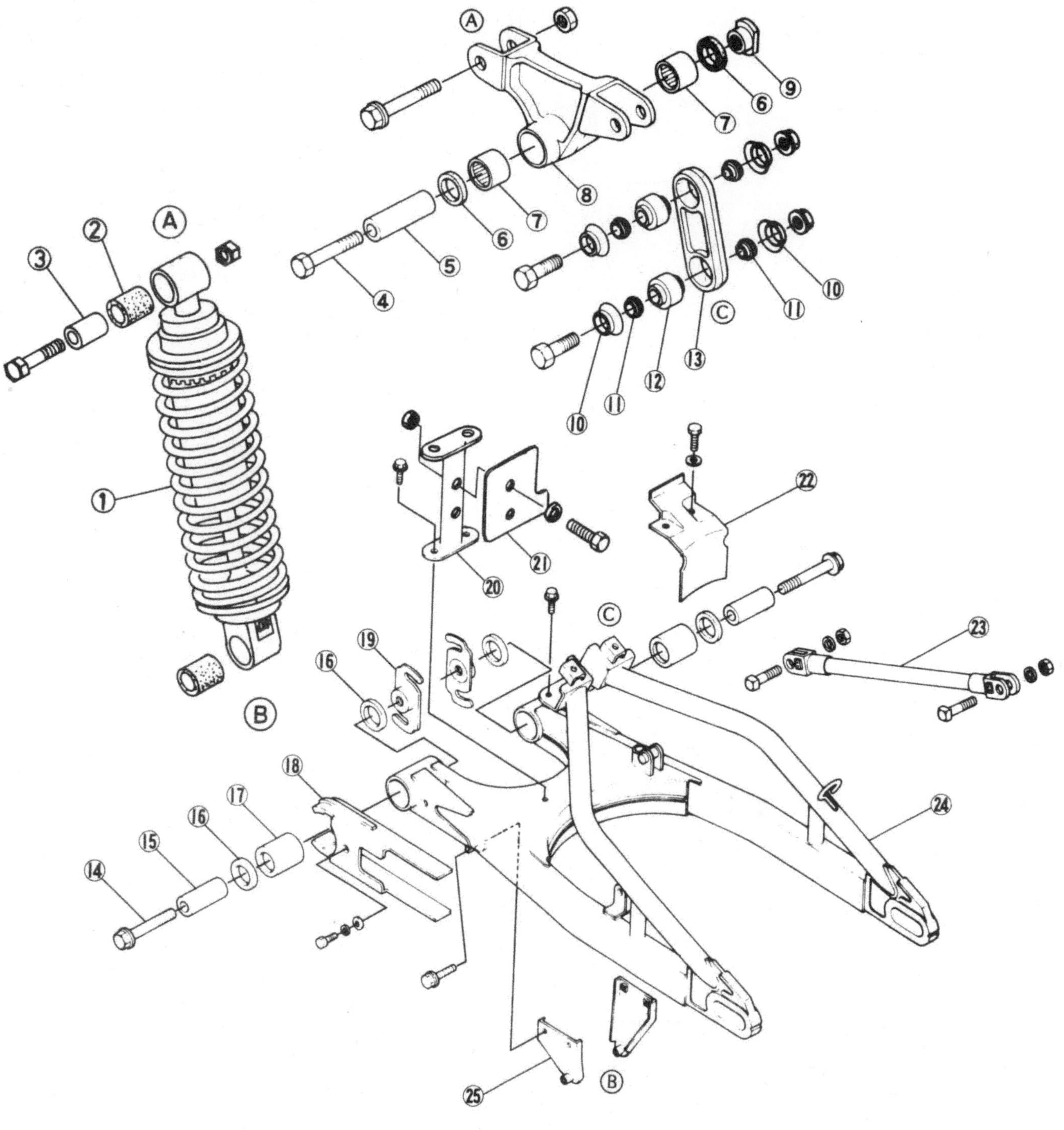

Fig. 4.10 Rear suspension – KZ/Z550 H1, H2 and all ZR400/550 models

1 *Suspension unit*
2 *Rubber bush*
3 *Sleeve*
4 *Pivot shaft*
5 *Sleeve*
6 *Grease seal*
7 *Needle bearing*
8 *Rocker arm*
9 *Nut*
10 *Grease seal*
11 *Collar*
12 *Spherical bearing*
13 *Tie-rod*
14 *Pivot shaft*
15 *Sleeve*
16 *Grease seal*
17 *Needle bearing*
18 *Chain guide*
19 *S-nut*
20 *Bracket*
21 *Mudguard*
22 *Mudguard*
23 *Torque link*
24 *Swinging arm*
25 *Bracket*

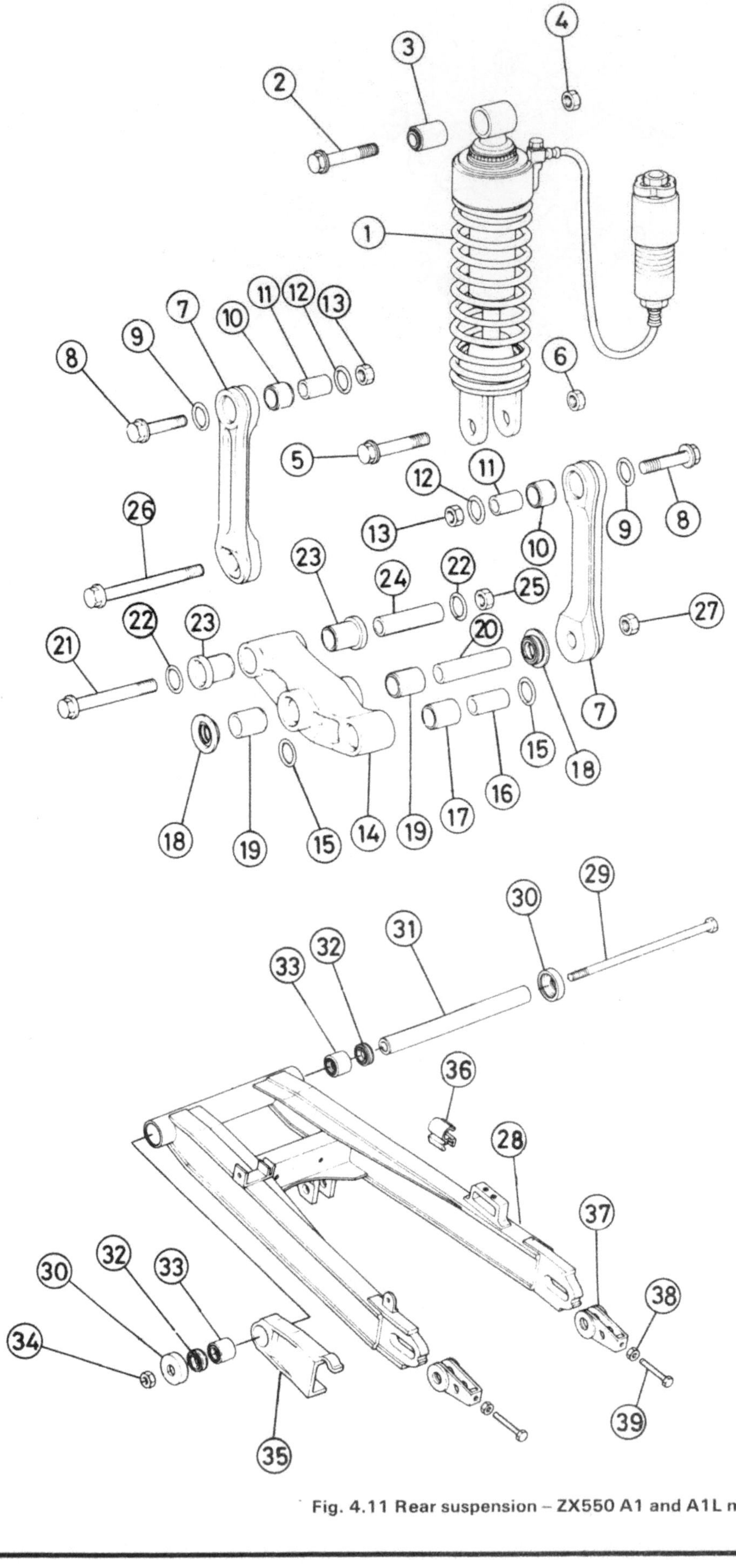

1 Suspension unit
2 Bolt
3 Top mounting bush
4 Nut
5 Bolt
6 Nut
7 Tie rod
8 Bolt
9 O-ring
10 Pivot bush
11 Inner sleeve
12 O-ring
13 Nut
14 Rocker arm
15 O-ring
16 Inner sleeve
17 Bottom bush
18 Seal
19 Pivot bush
20 Inner sleeve
21 Bolt
22 O-ring
23 Pivot bush
24 Inner sleeve
25 Nut
26 Bolt
27 Nut
28 Swinging arm
29 Pivot bolt
30 Sealing cap
31 Inner sleeve
32 Seal
33 Pivot bearing
34 Nut
35 Chain guide
36 Brake hose clamp
37 Chain adjuster
38 Nut
39 Bolt

Fig. 4.11 Rear suspension – ZX550 A1 and A1L models

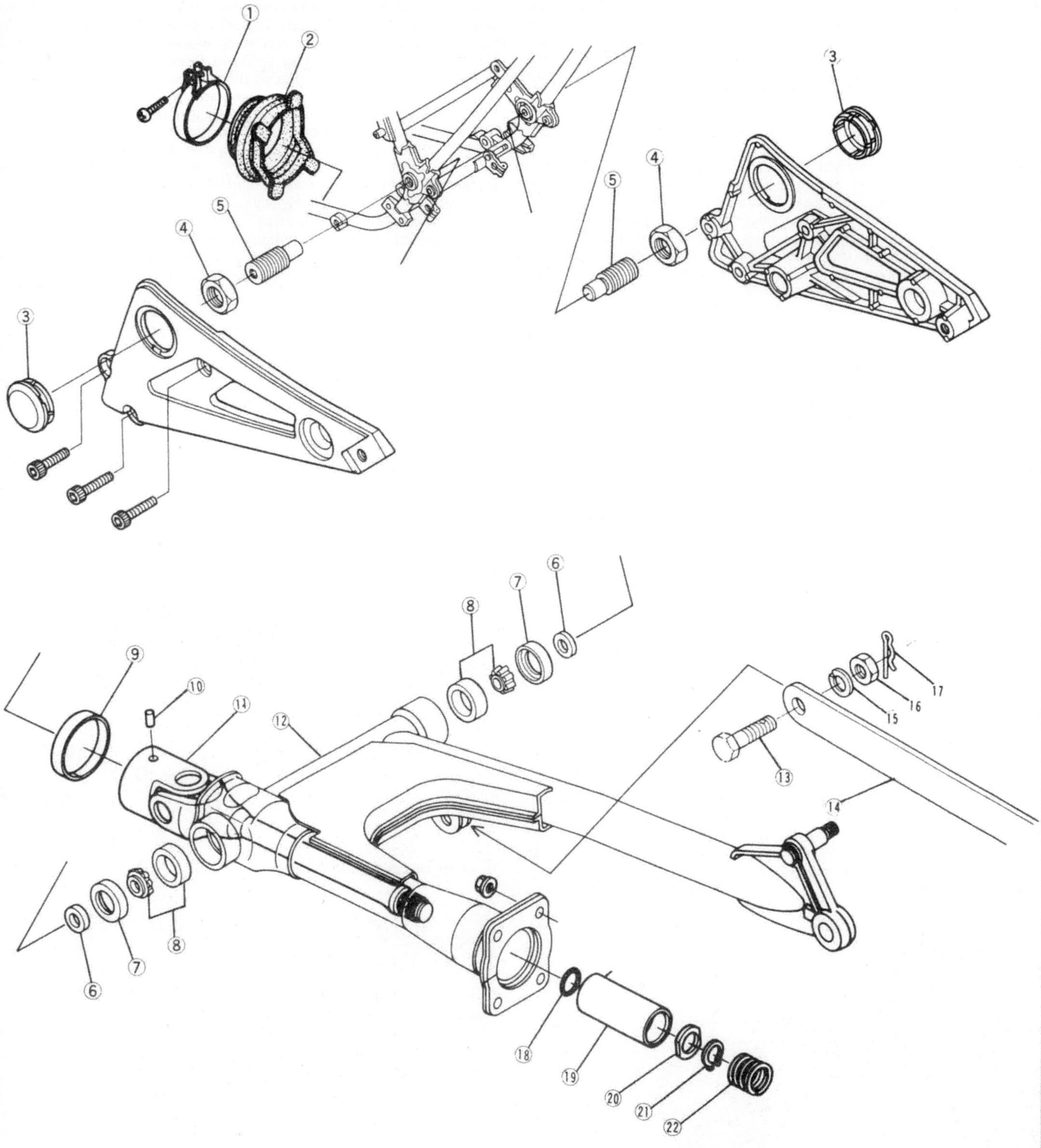

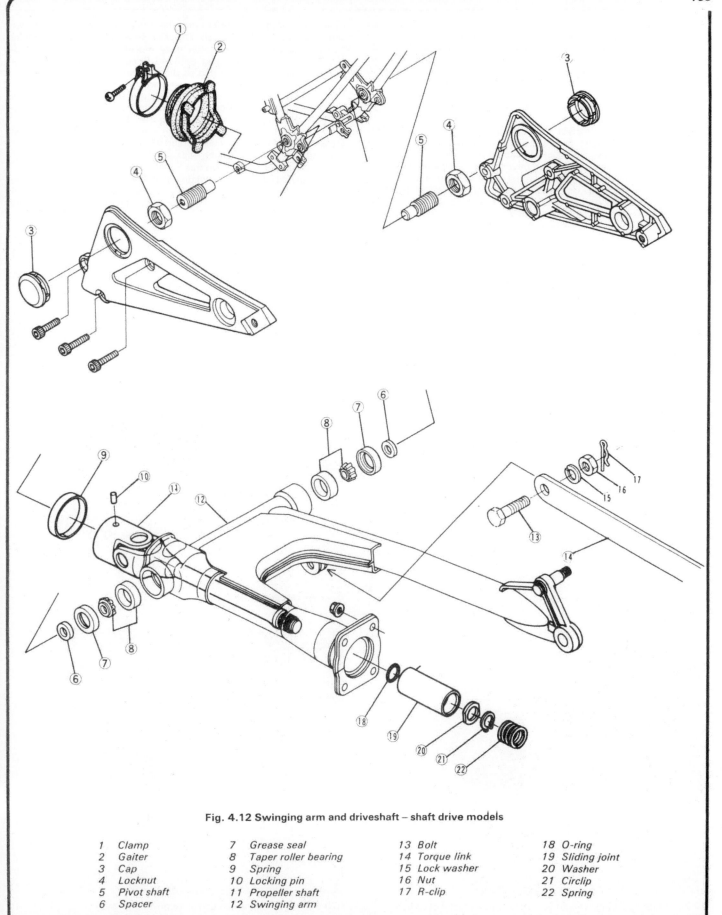

Fig. 4.12 Swinging arm and driveshaft – shaft drive models

1	*Clamp*	7	*Grease seal*	13	*Bolt*	18	*O-ring*
2	*Gaiter*	8	*Taper roller bearing*	14	*Torque link*	19	*Sliding joint*
3	*Cap*	9	*Spring*	15	*Lock washer*	20	*Washer*
4	*Locknut*	10	*Locking pin*	16	*Nut*	21	*Circlip*
5	*Pivot shaft*	11	*Propeller shaft*	17	*R-clip*	22	*Spring*
6	*Spacer*	12	*Swinging arm*				

10 Rear suspension: examination and renovation

1 On models fitted with Unitrak rear suspension, dismantle the linkage components referring to the accompanying illustrations.
2 On all models, remove the seals, where fitted, by levering them out of their housings. Take care not to damage the seal housings. Discard all seals and O-rings; these should never be reused and must be renewed on reassembly.
3 Press or tap out the pivot bearing inner sleeves, clean them and check for obvious signs of wear such as heavy scoring. Where measurements are given the sleeves should be checked for wear by measurement and renewed if worn beyond the service limits given.
4 Check that all pivot bolts are straight and unworn and that their threads are undamaged. Use emery paper to polish off all traces of corrosion and smear grease over them on reassembly.
5 Needle bearings should be washed while in place to remove all old grease and checked for wear by refitting the inner sleeve and feeling for free play; if any is found the bearings must be renewed. Always have new components ready, as removal will destroy the original components. Pass a drift through the pivot tube or bearing housing and tap out the bearings from the inside. On Unitrak linkage components, a drawbolt arrangement can be made up as shown in the accompanying illustration, the dimensions of the drawbolt components will vary according to the size of the bearing being removed. This method can also be used to displace the tie-rod spherical bearings (KZ/Z550 H1, H2 and all ZR400/550 models) or the linkage plain bushes (ZX550 A1 and A1L models).
6 On shaft drive models check that the bearing rollers are smooth and unmarked with no signs of pitting or discolouration, also the outer races. If the bearings are found to be worn, they can be removed by drifting out the outer races as described above. On refitting, use a large socket as a drift, ensuring that the races enter the bores squarely. The oil seals can be refitted in a similar manner, once the bearing rollers have been packed with grease and refitted.
7 On models fitted with Unitrak suspension, check very carefully all components of the linkage. Remove all traces of grease, dirt and corrosion and check for wear by reassembling temporarily the components concerned. If any free play can be felt, or if the bearing surfaces are scored, the worn components must be renewed. On KZ/Z550 H1 and H2 and all ZR400/550 models, check that the spherical bearings do not have excessive free play in their normal direction of operation; if free play is 0.3 mm (0.0118 in) or more the bearings must be renewed. Note that Kawasaki actually recommend that the complete tie-rod assembly be renewed, as it is a very highly-stressed component.
8 On reassembly, refit all needle roller bearings, plain bushes and spherical bearings, as applicable (chain drive models only) using a version of the drawbolt arrangement shown in the accompanying illustration. Do not attempt to drift these bearings into place; they will almost certainly be damaged. On chain drive models with swinging arm rear suspension, note that early models (those with a grease nipple fitted to the swinging arm) are fitted with four needle roller bearings; these must be fitted so that the outer bearing on each side is flush with the end of the swinging arm pivot lug. On later models each bearing (there are only two) must be seated to a depth of 5 mm (0.20 in) inside the end of the pivot lug to make room for the grease seal, which is tapped into place using a socket as a drift.
9 Pack each bearing liberally with molybdenum disulphide-based grease then refit the inner sleeve and where applicable, the sealing O-rings, grease seals or sealing caps.
10 On models with Unitrak rear suspension refer to the accompanying illustrations when rebuilding the linkage and ensure that all components are correctly located. Tighten all pivot bolt retaining nuts to the torque settings given in the Specifications Section of this Chapter.
11 Note that shims are available to remove endfloat from the swinging arm pivot on all chain drive models. If endfloat is felt with the arm installed and the pivot bolt tightened to the correct torque setting, measure the gap with feeler gauges, remove the swinging arm and add the necessary number of shims. These are 0.5 mm (0.020 in) thick and are available under Part Number 92025-1228.

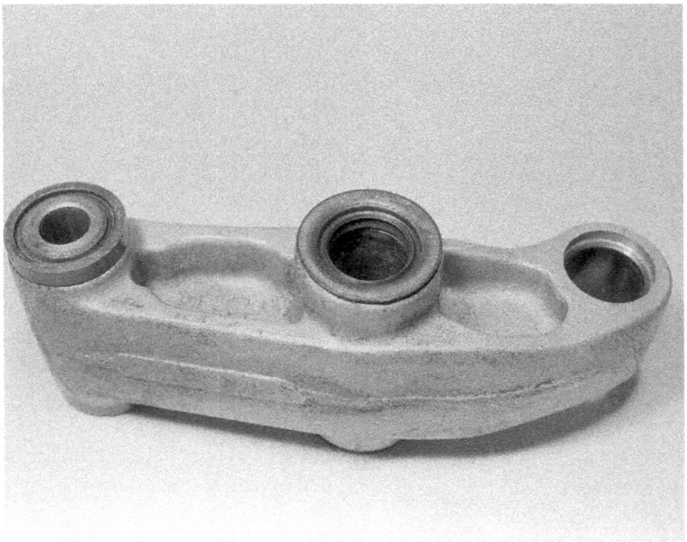

10.2a All rear suspension seals and ...

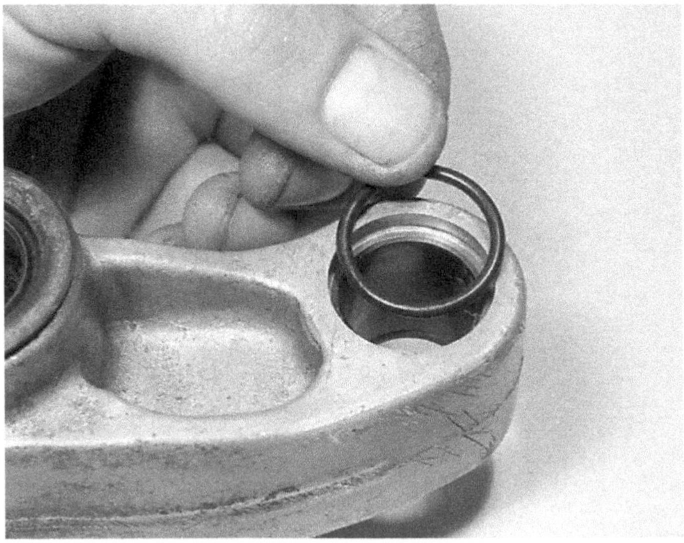

10.2b ... sealing O-rings should be renewed as a matter of course

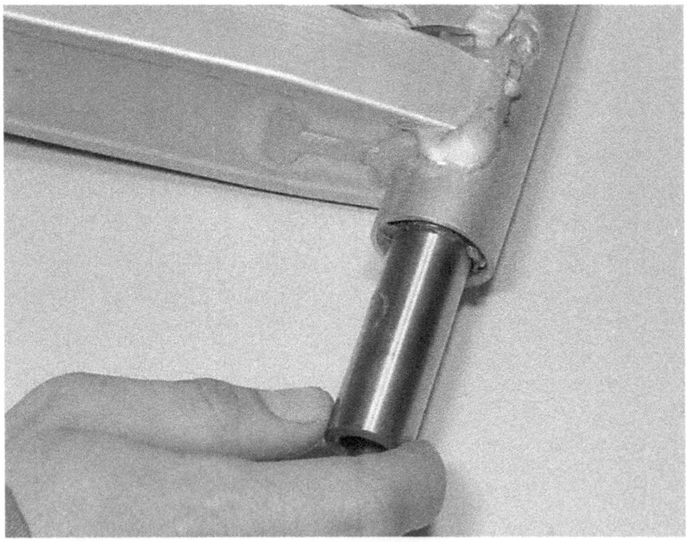

10.3 Bearing inner sleeves can be removed by hand

10.5a Needle roller bearings should be washed clean before checking for wear

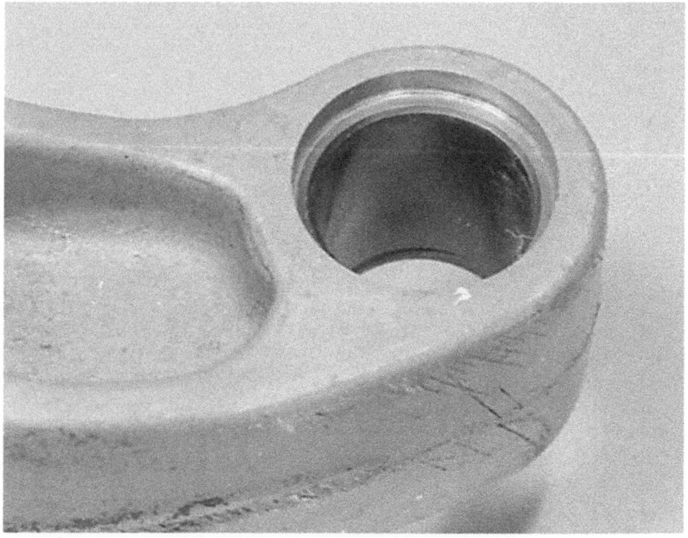

10.5b Use drawbolt arrangement to replace worn bushes, where fitted

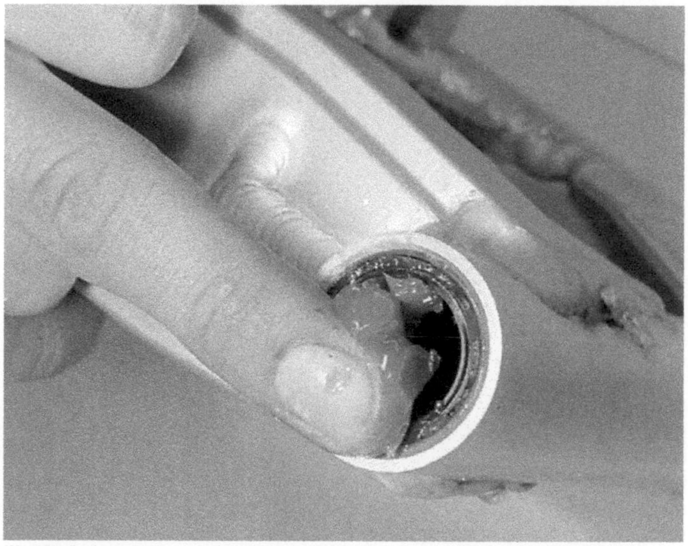

10.9a Pack all bearings with the recommended grease on reassembly ...

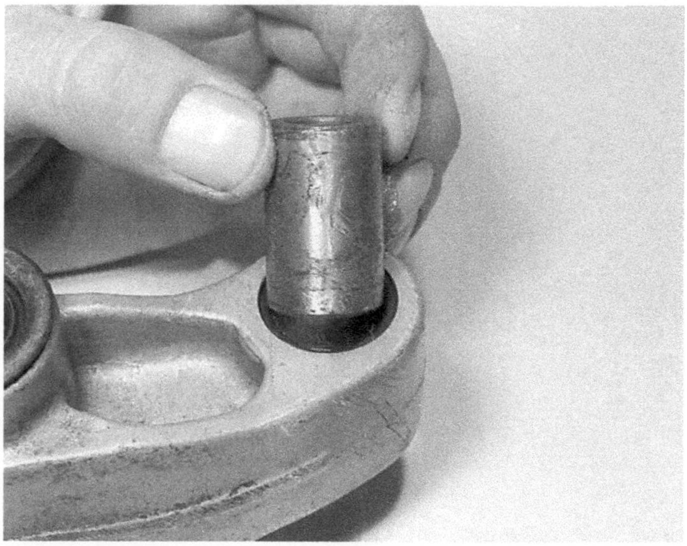

10.9b ... and smear grease over inner sleeves before refitting

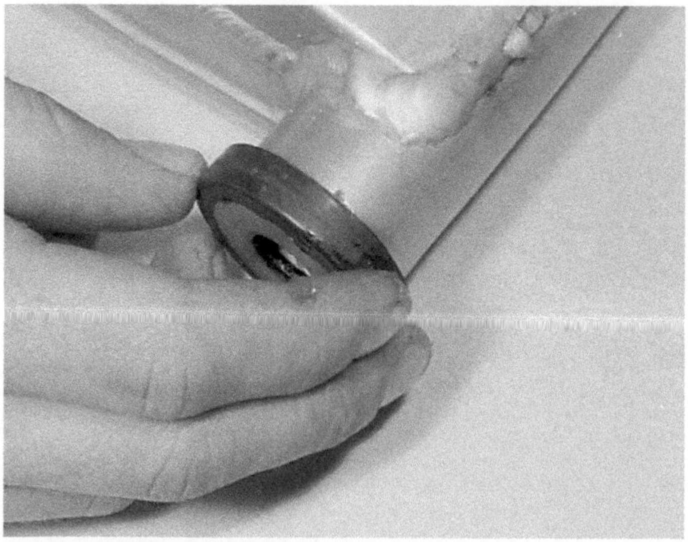

10.9c Do not forget to refit sealing caps, where fitted

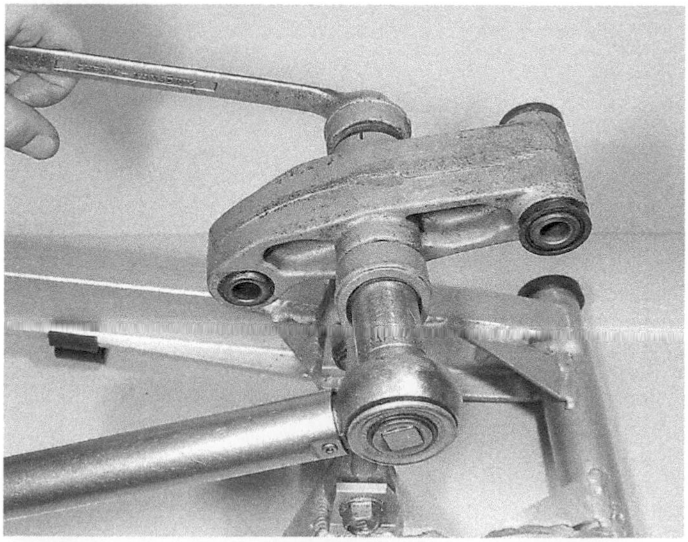

10.10 Use recommended torque settings when rebuilding Unitrak suspension

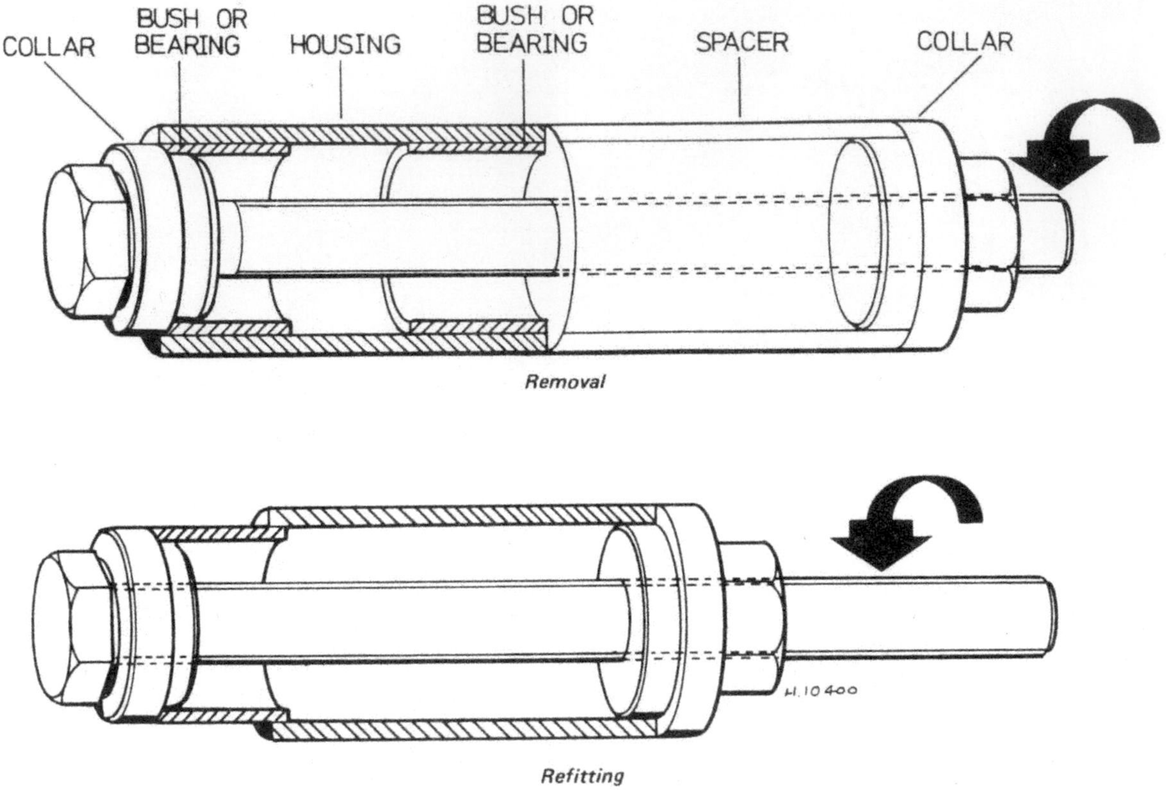

Fig. 4.13 Drawbolt tool for removing and refitting suspension linkage and swinging arm bearings

11 Swinging arm pivot bearings: adjustment – shaft drive models

1 Place the machine on its centre stand and check that there is no free play at the swinging arm pivot, by pulling and pushing the rear wheel from side to side, and that there is no endfloat. If free play or endfloat are found, the pivot bearings can be adjusted.
2 Remove the dust caps and slacken the pivot shaft locknuts, then screw in the pivot shafts until they seat lightly. Check that the clearance between the swinging arm left-hand pivot stub (bearing housing) and the frame gusset is between 1.4 – 1.6 mm (0.0551 – 0.0630 in), as measured with feeler gauges. Screw in or out the pivot shafts until the clearance is correct, then tighten both shafts to a torque setting of 1.3 kgf m (9.5 lbf ft). Recheck the clearance. Hold the shafts while tightening the locknuts securely, then refit the dust caps.
3 If this fails to eliminate free play, or if there is any other sign that the bearings are worn or damaged, they must be removed and checked.

12 Rear suspension units: general – except Z550 G and KZ550 F models

1 The suspension units are sealed for life and can only be renewed if found to be faulty. Check the suspension as described in Routine Maintenance. If movement is found to be jerky and uncontrolled with little evidence of damping, or if oil is seen to be leaking from any unit it must be renewed. Where two suspension units are fitted, they should always be renewed as a matched pair.
2 On models with swinging arm suspension, place the machine on its centre stand and remove the pillion grab rail or lifting handle as appropriate, then remove the suspension unit mounting nuts, bolts and washers and withdraw the units. On reassembly, tighten all mounting nuts and bolts to the specified torque settings.
3 On models with Unitrak suspension, the rear suspension should be removed complete, as described in Section 9 of this Chapter, so that the unit can be removed from the linkage. In view of the difficulty of gaining access to the unit mounting bolts this is easier and does not take much longer than attempting to remove the unit alone. Again, use the recommended torque settings on reassembly.

13 Rear suspension units: oil changing – Z550 G and KZ550 F models

1 The suspension units fitted to these models are of the air spring type in which the quality and smoothness of ride can be adjusted by varying air pressure and oil quantity and viscosity, as well as the adjustable damping facility. Note, however, that the units are sealed; if changing the oil or altering settings fails to correct deteriorating suspension performance, or if oil leaks are found, the units must be renewed as a matched pair.
2 To remove the units, remove the valve cap and depress the core to expel the air, then disconnect the connecting hose at the gland nut which is usually found on the left-hand unit. With the machine on its centre stand, remove the pillion grab rail (where applicable) and remove the mounting nuts and washers, then pull each unit off its studs. On reassembly, tighten the mounting nuts to a torque setting of 2.5 kgf m (18 lbf ft). **Note:** The units rely entirely on the correct oil level for efficient performance; never compress one unit alone when the hose is connected as oil may flow from one unit to the other, producing a marked disparity in oil levels with a resultant decrease in the machine's handling and stability. Similarly, whenever the units are removed from the machine, they should be supported in the upright (vertical) position at all times with the air valves uppermost so that oil cannot leak out. Never hold them upside down or with one unit higher than the other.

3 Check the condition of the mounting rubbers and gaiters; these can be renewed separately and must be renewed if found to be perished, split, cracked or otherwise damaged.

4 To change the oil, remove the air hose unions from each unit, set the damping knob to the softest setting, invert the unit over a suitable container and pump it to expel as much oil as possible. Leave the units in the inverted position to drain.

5 When as much oil as possible has been removed, turn the unit right way up and check that it is fully extended. Holding it upright with the air hose orifice at the highest point, fill the unit completely with SAE 5W fork oil. Stop halfway and pump the unit slowly several times to distribute fresh oil around the damping mechanism and to expel any air bubbles. Each unit should take approximately 400 cc of oil. When the unit is completely filled and at full extension, tip out into an accurate measuring vessel an amount of oil equal to the specified air volume.

6 When the oil quantity (and therefore the level) is correct, refit each unit to the machine, connect the air hose unions and pressurise the units to the required setting. Do not overtighten the air hose unions; use the recommended torque settings where possible. Be very careful not to spill any oil.

7 Note that the oil should be changed at regular intervals as it deteriorates in service. See Routine Maintenance.

14 Propeller shaft and final drive gear case: general – shaft drive models

1 Refer to Routine Maintenance for details of the only maintenance tasks necessary on shaft drive models, ie changing the gear case oil and checking its level, and greasing the propeller shaft splines. If excessive backlash and noise from the drive train indicate that serious wear is present the machine should be taken to a competent Kawasaki dealer for expert attention.

2 The propeller shaft can be checked for wear at any time that work is being done on the drivetrain, but if wear is found it can only be renewed after the swinging arm has been removed; the universal joint is a sealed part of the assembly.

3 The gear case is very similar in design to the front gear case unit described in Chapter 1. While it is simple enough in layout and not beyond the ability of most owners to dismantle or rebuild, the task of setting it up requires a number of specialist tools, a large range of shims, preload collars and thrust washers, as well as a great deal of skill. For this reason it is recommended that the ordinary private owner takes the complete machine to a Kawasaki dealer for attention.

4 Owners of KZ550 F2 and F2L machines with frame numbers between 002402 – 002822 should note the following. The left-hand side of the swinging arm is 1 mm too short, causing the gear case to

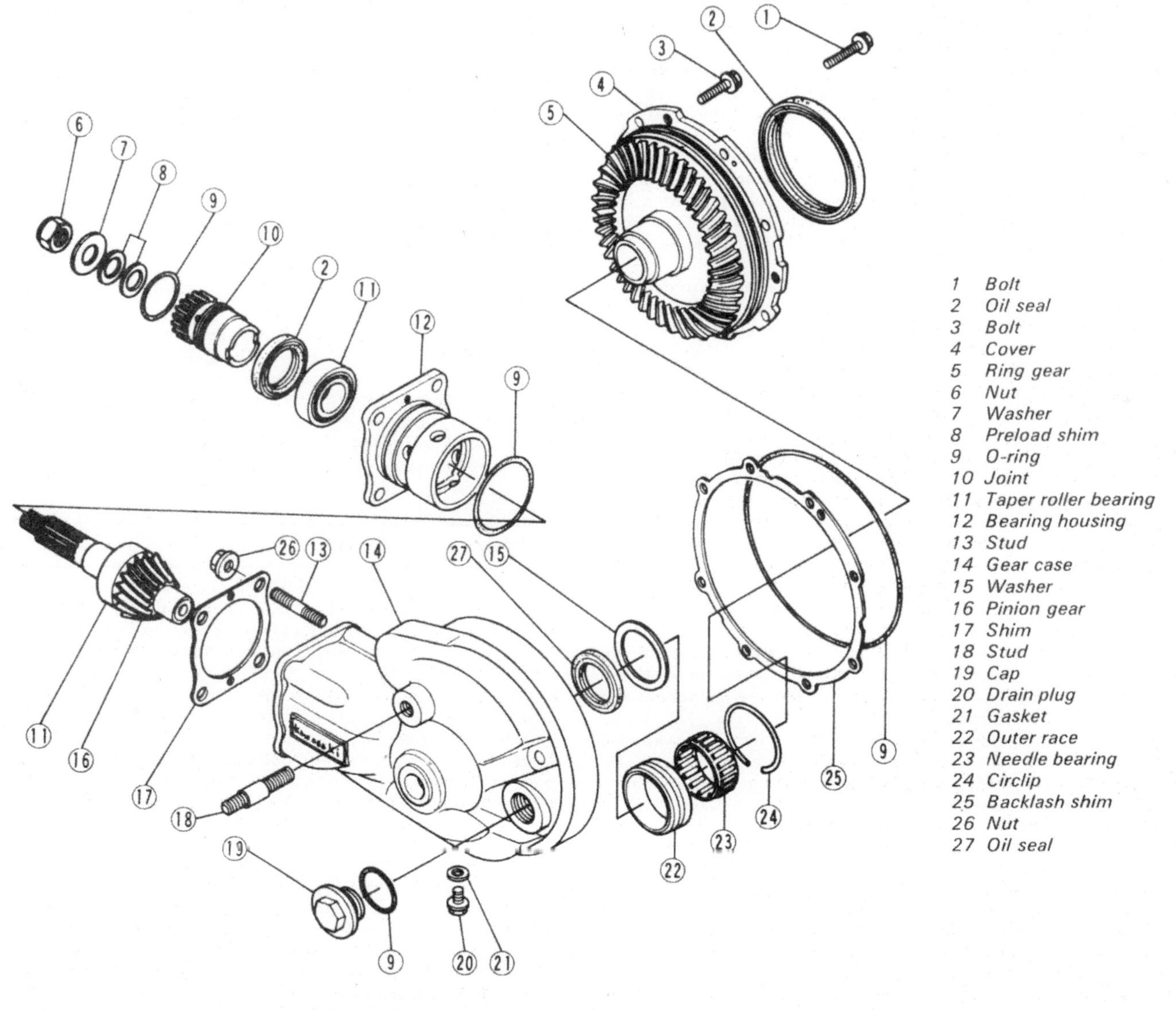

Fig. 4.14 Rear gear case components – shaft drive models

be incorrectly seated and the rear wheel to run out of alignment. All defective machines should have been corrected by the insertion of a shim of appropriate thickness (Part Number 92025-1318) between the final gear case and the end of the swinging arm. This can be checked easily; the shim should be visible from the outside with no dismantling being necessary. If in doubt, check with the Kawasaki dealer who originally sold the machine; the shims were supplied with each machine and should have been fitted before the machine was delivered. If such a machine is being worked on, always refit the shim whenever the gear case is disturbed.
5 Finally, note that the accompanying illustration refers only to the final drive case of Z550 G1 and KZ550 F1 and M1 models; all subsequent models are fitted with a unit which differs in detail but uses the same basic layout. The most noticeable difference is that a ball bearing is fitted around the ring gear/wheel spindle boss instead of the needle roller bearing shown.

15 Instruments and drive components: general

1 Except for the warning panels on some models, the instruments are grouped in a console bolted to the fork top yoke and can be removed as a single unit once the drive cables, electrical wiring and mounting nuts or bolts have been removed. Refer to Section 6 of this Chapter. Once removed the instruments can be separated from their mounting brackets after the bottom cover has been withdrawn, the cover(s) being held by cap nuts or self-tapping screws. Refer to Chapter 6 for details of electrical instruments, but for mechanical instruments proceed as follows.
2 These instruments must be carefully handled at all times and never dropped or held upside down. Dirt, oil, grease and water all have an equally adverse effect on them, and so a clean working area must be provided if they are to be removed.
3 The instrument heads are very delicate and should not be dismantled at home. In the event of a fault developing, the instrument should be entrusted to a specialist repairer or a new unit fitted. If a replacement unit is required, it is well worth trying to obtain a good secondhand item from a motorcycle breaker in view of the high cost of a new replacement.
4 Remember that a speedometer in correct working order is a statutory requirement in the UK. Apart from this legal necessity, reference to the odometer readings is the most satisfactory means of keeping pace with the maintenance schedules.
5 It is advisable to detach the speedometer and tachometer drive cables from time to time in order to check whether they are adequately lubricated and whether the outer cables are compressed or damaged at any point along their run. A jerky or sluggish movement at the instrument head can often be attributed to a cable fault.
6 To grease the cable, uncouple both ends and withdraw the inner cable. After removing any old grease, clean the inner cable with a petrol soaked rag and examine the cable for broken strands or other damage. Do not check the cable for broken strands by passing it through the fingers or palm of the hand, this may well cause a painful injury if a broken strand snags the skin. It is best to wrap a piece of rag around the cable and pull the cable through it, any broken strands will snag the rag.
7 Regrease the cable with high melting point grease, taking care not to grease the last six inches closest to the instrument head. If this precaution is not observed, grease will work into the instrument and immobilise the sensitive movement.
8 The cables on all models are secured at both ends by large knurled rings which must be tightened or slackened using a pair of pliers. Do not overtighten the knurled rings, or they will crack necessitating renewal of the complete cable.
9 When refitting drive cables, always ensure that they have smooth easy runs to minimise wear, and check that the cables are secured by any clamps or ties provided for the purpose of keeping the cables away from any hot or moving parts.
10 Where fitted, the tachometer drive is by a worm gear off the exhaust camshaft. Removal and refitting is described in Chapter 1. If the gear teeth are found to be worn or damaged at any time the components concerned must be renewed. No maintenance is required.
11 The speedometer drive is by a separate gearbox mounted on the left-hand side of the front wheel. Apart from packing it with grease whenever the front wheel is removed, the unit requires no maintenance. If it proves to be defective it can be dismantled. Two different types are fitted, both being shown in the illustrations accompanying Chapter 5. Before starting work, note that the manufacturer recommends that the unit be renewed rather than attempt repairs; replacement parts, although listed, may be difficult to obtain.
12 On the early type of drive, remove the circlip and lift off the thrust washer, drive dog and speedometer gear. The grease seal can be levered out and renewed if necessary. If the speedometer pinion requires attention the roll pin must be removed. Passing a 1 mm drill bit up through the pin, drill through the gearbox housing, then use this as a pilot hole to drill a hole (using a 2 mm drill bit) through to the rear of the pin which can then be tapped out, using a punch. The cable bush, thrust washers and pinion can then be extracted. On reassembly, always use a new pin to retain the cable bush, and stake it in place to prevent it from dropping out.
13 The later type of drive is dismantled in exactly the same way except that the circlip, thrust washer and drive dog are now in the wheel hub. It is only necessary to lever out the grease seal to displace the speedometer gear.

16 Footrests, stands and controls: examination and renovation

1 At regular intervals all footrests and stands and the brake pedal and gearchange lever should be checked and lubricated. Check that all mounting nuts and bolts are securely fastened, using the recommended torque wrench settings where these are given. Check that any securing split-pins are correctly fitted.
2 Check that the bearing surfaces at all pivot points are well greased and unworn, renewing any component that is excessively worn. If lubrication is required, dismantle the assembly to ensure that grease can be packed fully into the bearing surface. Return springs, where fitted, must be in good condition with no traces of fatigue and must be securely mounted.
3 If accident damage is to be repaired, check that the damaged component is not cracked or broken. Such damage may be repaired by welding, if the pieces are taken to an expert, but since this will destroy the finish, renewal is usually the most satisfactory course of action. If a component is merely bent it can be straightened after the affected area has been heated to a full cherry red, using a blowlamp or welding torch. Again the finish will be destroyed, but painted surfaces can be repainted easily, while chromed or plated surfaces can only be replated, if the cost is justified. **Note**: this only applies to components constructed from steel plate or tubing. In some cases, particularly footrest mounting plates, components are made of cast or forged aluminium alloy which requires a very much more careful approach.
4 It is not usually possible to straighten bent alloy components, and this should in any case be attempted only by an expert. Similarly, alloy welding requires the skill and equipment of such an expert.

17 Fairing: general – GPz models

1 The fairing fitted to KZ/Z550 D1, H1 and H2 models is a simple plastic moulding mounted at four points and fitted with a separate windscreen secured to the fairing lower half by a number of screws threading into special sleeved nuts. A rubber strip is fitted between windscreen and fairing.
2 The fairing is removed by unscrewing the two cap nuts which secure it to brackets attached to the top yoke and the two bolts which pass through rubber bushes and into the bottom yoke. On reassembly, tighten securely the nuts and bolts but be careful not to overtighten them or to distort the fairing in any way as it is refitted.
3 The fairing fitted to ZX550 A1 and A1L models is a much larger assembly mounted on a separate subframe and including a separate dash panel as well as the windscreen. It also serves as the mounting point for the headlamp, rear view mirrors and turn signal lamps.
4 To remove the fairing, disconnect the single multi-pin block connector connecting the three lamps to the main loom, then remove the cap nut from the bracket on the front right-hand side of the steering head, then remove the two bolts securing the subframe lower mountings to the frame downtubes and withdraw the fairing. This latter task is best done with an assistant so that there is no danger of damage to paintwork. Manoeuvre the fairing around the fork legs and withdraw it from the machine. Refitting is the reverse of this procedure.

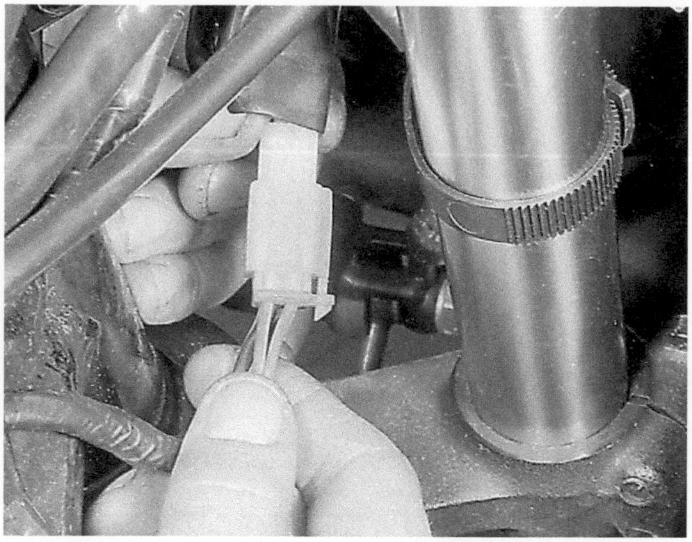

17.4a ZX550 A1 – disconnect wiring at connector ...

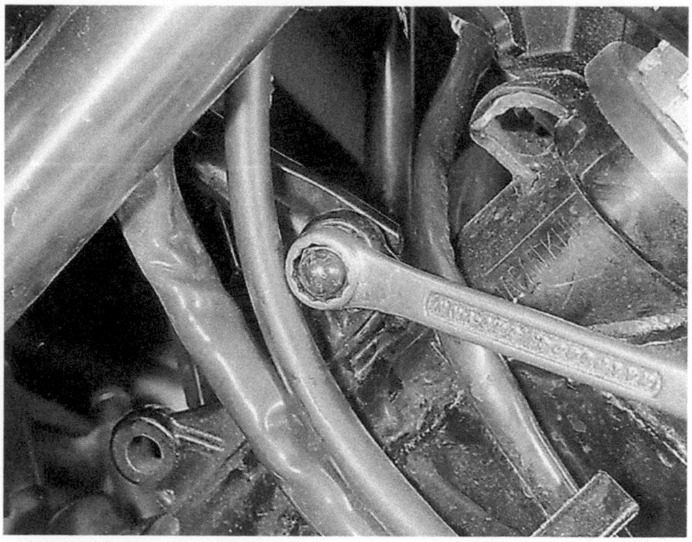

17.4b ... then remove single cap nut at steering head ...

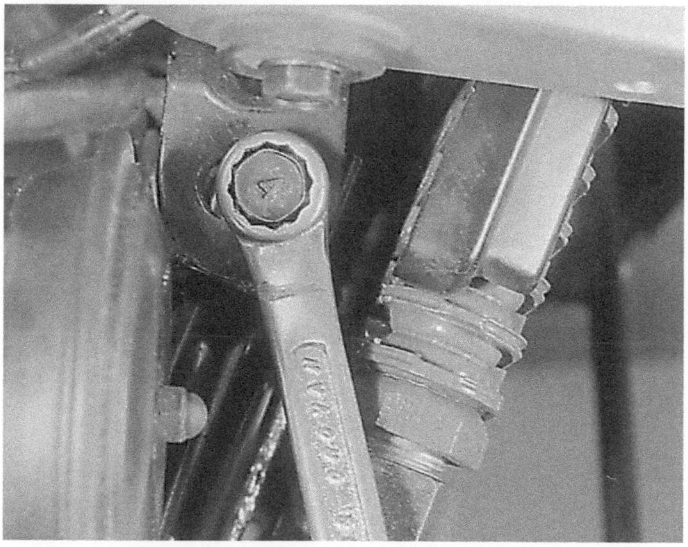

17.4c ... and remove mounting bolt on each frame downtube to release fairing

Chapter 5 Wheels, brakes and tyres

For information relating to the 1984 on models, see Chapter 7

Contents

Specifications

Wheels

Type:	
ZR400 A1, B1, ZR550 A1, A2, KZ/Z550 H1, H2	5-spoke cast alloy
ZX550 A1, A1L	6-spoke cast alloy
All other models	7-spoke cast alloy
Rim maximum runout:	
Radial	0.8 mm (0.0315 in)
Axial	0.5 mm (0.0197 in)
Spindle maximum runout	0.2/100 mm (0.008/3.937 in)

Brakes

Type – front:	
All UK models, KZ550 D1, H1, H2, ZX550 A1, A1L	Twin hydraulic disc
All other US models	Single hydraulic disc
Type – rear:	
Z500 B1, B2, KZ/Z550 D1, H1, H2, ZX550 A1, A1L	Single hydraulic disc
All other models	Single leading shoe drum, rod-operated

Brake disc

Minimum thickness – models with twin front discs:	
ZR400 A1, B1, ZR550 A1, A2, Z550 G1, G2	3.5 mm (0.1378 in)
KZ/Z550 H1, H2	4.0 mm (0.1575 in)
All other models	4.5 mm (0.1772 in)
Minimum thickness – models with single front disc:	
KZ550F1, F2, F2L, M1	4.5 mm (0.1772 in)
All other models	6.0 mm (0.2362 in)
Minimum thickness – rear disc:	
Z500B1, B2	6.0 mm (0.2362 in)
All other models	5.5 mm (0.2165 in)
Disc maximum runout	0.3 mm (0.0118 in)
Brake pad friction material minimum thickness	1.0 mm (0.0394 in)

Front master cylinder – except ZR400A1, B1, ZR550A1, A2, Z550 G1, G2, KZ550 F1, F2, F2L, M1

Cylinder maximum ID:	
Z400 J1, J2, J3, Z500 B1, B2, Z550 A1, A2, A3, C1, C2, KZ/Z550 D1, H1, H2, ZX550 A1, A1L	15.95 mm (0.6280 in)
KZ550 A1	14.08 mm (0.5543 in)
KZ550 A2, A3, A4, C1, C2, C3, C4	12.78 mm (0.5032 in)

Piston minimum OD:	
Z400 J1, J2, Z500 B1, B2, Z550 A1, A2, C1, C2	15.80 mm (0.6221 in)
KZ550 A1	13.90 mm (0.5472 in)
KZ550 A2, C1, C2	12.63 mm (0.4972 in)
KZ550 A3, A4, C3, C4	12.50 mm (0.4921 in)
Z400 J3, Z550 A3, KZ/Z550 D1, H1, H2, ZX550 A1, A1L	15.67 mm (0.6169 in)
Primary cup minimum OD:	
Z400 J1, J2, J3, Z500 B1, B2, Z550 A1, A2, A3, C1, C2, KZ/Z550 D1, H1, H2, ZX550 A1, A1L	16.00 mm (0.6299 in)
KZ550 A1	14.10 mm (0.5551 in)
KZ550 A2, A3, A4, C1, C2, C3, C4	12.80 mm (0.5039 in)
Secondary cup minimum OD:	
Z400 J1, J2, J3, Z500 B1, B2, Z550 A1, A2, A3, C1, C2, KZ/Z550 D1, H1, H2, ZX550 A1, A1L	16.40 mm (0.6457 in)
KZ550 A1	14.50 mm (0.5709 in)
KZ550 A2, A3, A4, C1, C2, C3, C4	13.10 mm (0.5158 in)
Spring minimum free length:	
Z400 J1, J2, Z500 B1, B2, Z550 A1, A2, C1, C2	34.70 mm (1.3661 in)
KZ550 A1	40.50 mm (1.5945 in)
KZ550 A2, A3, A4, C1, C2, C3, C4,	47.20 mm (1.8583 in)
Z400 J3, Z550 A3, KZ/Z550 D1, H1, H2, ZX550 A1, A1L	34.80 mm (1.3701 in)

Front brake caliper – except ZR400 A1, B1, ZR550 A1, A2, Z550 G1, G2, ZX550 A1, A1L

Caliper bore maximum ID:	
Z400 J1, Z500 B1, B2, Z550 A1	38.25 mm (1.5059 in)
All other models	42.92 mm (1.6898 in)
Piston minimum OD:	
Z400 J1, Z500 B1, B2, Z550 A1	38.08 mm (1.4992 in)
All other models	42.75 mm (1.6831 in)

Rear master cylinder – except ZX550 A1, A1L

Cylinder maximum ID:	
Z500 B1, B2	15.95 mm (0.6280 in)
KZ/Z550 D1, H1, H2	14.08 mm (0.5543 in)
Piston minimum OD:	
Z500 B1, B2	15.80 mm (0.6221 in)
KZ/Z550 D1, H1, H2	13.80 mm (0.5433 in)
Primary cup minimum OD:	
Z500 B1, B2	16.30 mm (0.6417 in)
KZ/Z550 D1, H1, H2	14.00 mm (0.5512 in)
Secondary cup minimum OD:	
Z500 B1, B2	16.30 mm (0.6417 in)
KZ/Z550 D1, H1, H2	14.60 mm (0.5748 in)
Spring minimum free length:	
Z500 B1, B2	37.20 mm (1.4646 in)
KZ/Z550 D1, H1, H2	32.40 mm (1.2756 in)

Rear brake caliper – except ZX550 A1, A1L

Caliper bore maximum ID	42.92 mm (1.6898 in)
Piston minimum OD	42.75 mm (1.6831 in)

Rear drum brake

Drum ID:	
ZR400 A1, B1, ZR550 A1, A2	160.00 – 160.16 mm (6.2992 – 6.3055 in)
Service limit	160.75 mm (6.3287 in)
All other models	180.00 – 180.16 mm (7.0866 – 7.0929 in)
Service limit	180.75 mm (7.1161 in)
Brake shoe friction material minimum thickness:	
ZR400 A1, B1, ZR550 A1, A2	1.8 mm (0.0709 in)
All other models	2.5 mm (0.0984 in)
Brake shoe return spring maximum free length:	
ZR400 A1, B1, ZR550 A1, A2	N/Av
All other models	69.0 mm (2.7165 in)
Operating camshaft bearing surface minimum OD:	
ZR400 A1, B1, ZR550 A1, A2	N/Av
All other models	16.83 mm (0.6626 in)
Backplate passage maximum ID:	
ZR400 A1, B1, ZR550 A1, A2	N/Av
All other models	17.22 mm (0.6780 in)

Tyres

Type:	
Z400 J1, Z500 B1, B2, KZ/Z550 A1	Tubed
All other models	Tubeless

Size:	Front	Rear
Z400 J1, Z500 B1, B2, KZ/Z550 A1, A2, A3, D1, KZ550 A4	3.25H – 19 4PR	3.75H – 18 4PR
Z400 J2, J3	3.25S – 19 4PR	3.75S – 18 4PR
ZR400 A1, B1	90/90 – 19 52S	110/90 – 18 61S
KZ/Z550 C1, C2, KZ550 C3, C4	3.25S – 19 4PR	130/90 – 16 67S
KZ550 F1, F2, F2L, M1	100/90 – 19 57S 4PR	130/90 – 16 67S
ZR550 A1, A2, KZ/Z550 H1, H2	3.25H – 19 4PR	4.00H – 18 4PR
Z550 G1, G2	100/90 – 19 57H	120/90 – 18 65H
ZX550 A1, A1L	100/90 – 18 56H	120/80 – 18 62H

Tyre pressures – cold

Note: loads given are total weight of rider, passenger and any accessories or luggage. Pressures apply to original equipment tyres and may be different for other makes or types – check with tyre manufacturer or importer for correct settings if different tyres are fitted.

	Front	Rear – at specified load
KZ/Z550 C1, C2, KZ550 C3, C4, F1, F2, F2L, M1	1.75 kg/cm² (25 psi)	1.50 kg/cm² (21 psi) @ 0.97.5 kg (0 – 215lb)
	1.75 kg/cm² (25 psi)	2.00 kg/cm² (28 psi) @ 97.5 – 155.0 kg (215 – 342 lb)
Z400 J1, J3, Z500 B1, B2, KZ/Z550 A1	2.00 kg/cm² (28 psi)	2.50 kg/cm² (36 psi) @ 0 – 97.5 kg (0 – 215 lb)
	2.00 kg/cm² (28 psi)	2.80 kg/cm² (40 psi) @ 97.5 – 165.0 kg (215 – 364 lb)
Z400 J2, ZR550 A1, A2	2.00 kg/cm² (28 psi)	2.25 kg/cm² (32 psi) @ 0 – 97.5 kg (0 – 215 lb)
	2.00 kg/cm² (28 psi)	2.50 kg/cm² (36 psi) @ 97.5 – 165.0 kg (215 – 364 lb)
ZR400 A1, B1	2.00 kg/cm² (28 psi)	2.25 kg/cm² (32 psi) @ 0 – 97.5 kg (0 – 215 lb)
	2.00 kg/cm² (28 psi)	2.50 kg/cm² (36 psi) @ 97.5 –150.0 kg (215 – 331 lb)
KZ/Z550 A2, A3, D1, KZ550 A4	2.00 kg/cm² (28 psi)	2.25 kg/cm² (32 psi) @ 0 – 97.5 kg (0 – 215 lb)
	2.00 kg/cm² (28 psi)	2.80 kg/cm² (40 psi) @ 97.5 – 165.0 kg (215 – 364 lb)
Z550 H1, H2	2.00 kg/cm² (28 psi)	2.25 kg/cm² (32 psi) @ 0 – 165.0 kg (0 – 364 lb)
	2.00 kg/cm² (28 psi)	2.50 kg/cm² (36 psi) @ 165.0 – 202.0 kg (364 – 445 lb)
KZ550 H1, H2	2.00 kg/cm² (28 psi)	2.25 kg/cm² (32 psi) @ 0 – 97.5 kg (0 – 215 lb)
	2.00 kg/cm² (28 psi)	2.25 kg/cm² (32 psi) @ 97.5 – 165.0 kg (215 – 364 lb)
Z550 G1, G2	2.00 kg/cm² (28 psi)	2.00 kg/cm² (28 psi) @ 0 – 97.5 kg (0 – 215 lb)
	2.00 kg/cm² (28 psi)	2.50 kg/cm² (36 psi) @ 97.5 – 180.0 kg (215 – 397 lb)
ZX550 A1 – UK	2.00 kg/cm² (28 psi)	2.25 kg/cm² (32 psi) @ 0 – 150.0 kg (0 – 331 lb)
	2.00 kg/cm² (28 psi)	2.50 kg/cm² (36 psi) @ 150.0 – 186.0 kg (331 – 410 lb)
ZX550 A1 – US, A1L	2.00 kg/cm² (28 psi)	2.25 kg/cm² (32 psi) @ 0 – 97.5 kg (0 – 215 lb)
	2.00 kg/cm² (28 psi)	2.50 kg/cm² (36 psi) @ 97.5 – 186.0 kg (215 – 410 lb)

Torque wrench settings

Component	kgf m	lbf ft
Front wheel spindle nut:		
ZR400 A1, B1, ZR550 A1, A2, Z550 G1, G2, KZ550 F1, F2, F2L, M1	6.5	47.0
All other models	8.0	58.0
Front wheel spindle clamp bolt:		
ZR400 A1, B1, ZR550 A1, A2, ZX550 A1, A1L	1.4	10.0
All other models	2.0	14.5
Rear wheel spindle nut:		
Z550 G1, G2, KZ550 F1, F2, F2L, M1	7.5	54.0
ZX550 A1, A1L	9.5	69.0
All other models	8.0	58.0
Brake disc mounting bolts	2.3	16.5
Rear sprocket mounting nuts:		
Z400 J3, ZR400 A1, B1, ZR550 A1, A2, KZ/Z550 A3, H1, H2, KZ550 A4, C3, C4	3.5	25.0
All other models	4.0	29.0
Front brake lever pivot bolt – Z400 J1, J2, J3, Z500 B1, B2, KZ/Z550 A1, A2, A3, C1, C2, D1, KZ550 A4, C3, C4	0.3	2.0

Component	kgf m	lbf ft
Front brake lever pivot bolt locknut	0.6	4.0
Rear brake pedal cap nut – Z400 J1, J2, Z500 B1, B2, KZ/Z550 A1, A2	2.0	14.5
Front master cylinder clamp bolts	0.9	6.5
Rear master cylinder/reservoir hose clamp screws – KZ/Z550 D1, H1, H2, ZX550 A1, A1L	0.1	0.5
Brake hose banjo bolts	3.0	22.0
Brake hose union mounting bolts – models with twin front disc brakes	0.9	6.5
Bleed nipples	0.8	6.0
Front and rear brake caliper mounting bolts:		
Front caliper – ZX550 A1, A1L	2.5	18.0
All others	3.0	22.0
Rear brake torque link nuts or bolts	3.0	22.0
Rear brake caliper holder clamp bolt – ZX550 A1, A1L	0.65	4.5
Front brake caliper/mounting bracket axle bolt retaining nuts – Z400 J1, Z500 B1, B2, KZ/Z550 A1	2.6	19.0
Front or rear brake caliper/mounting bracket axle bolts – except Z400 J1, Z500 B1, B2, KZ/Z550 A1	1.8	13.0
Rear caliper Allen bolts – Z500 B1, B2	3.0	22.0
Front brake fixed pad mounting screws – Z400 J1, Z500 B1, B2, KZ/Z550 A1	0.3	2.0
Anti-dive assembly – ZX550 A1, A1L:		
Plunger/valve assembly Allen screws	0.4	3.0
Valve assembly mounting Allen screws	0.75	5.5
Anti-dive assembly brake pipe gland nuts	1.5	11.0
Tyre air valve mounting nut and locknut – tubeless tyres only	0.15	1.0

1 General description

All models are fitted with cast alloy wheels at front and rear, early models being fitted with tubed tyres and all later models with tubeless tyres. All UK models are fitted with twin hydraulic disc front brakes, while all US models except the GPz versions are fitted with single hydraulic disc front brakes. At the rear, except for Z500 and all GPz models which are fitted with single hydraulic disc brakes, all models have a rod-operated single leading shoe drum brake.

2 Wheels: examination and renovation

1 Carefully check the complete wheel for cracks and chipping, particularly at the spoke roots and the edge of the rim. As a general rule a damaged wheel must be renewed as cracks will cause stress points which may lead to sudden failure under heavy load. Small nicks may be radiused carefully with a fine file and emery paper (No 600 – No 1000) to relieve the stress. If there is any doubt as to the condition of a wheel, advice should be sought from a reputable dealer or specialist repairer.

2 Each wheel is covered with a coating of lacquer, to prevent corrosion. If damage occurs to the wheel and the lacquer finish is penetrated, the bared aluminium alloy will soon start to corrode. A whitish grey oxide will form over the damaged area, which in itself is a protective coating. This deposit however, should be removed carefully as soon as possible and a new protective coating of lacquer applied.

3 Check the lateral runout at the rim by spinning the wheel and placing a fixed pointer close to the rim edge. If the maximum runout is greater than 0.5 mm (0.020 in) axially or 0.8 mm (0.032 in) radially, Kawasaki recommend that the wheel be renewed. This is, however, a counsel of perfection; a runout somewhat greater than this can probably be accommodated without noticeable effect on steering. No means is available for straightening a warped wheel without resorting to the expense of having the wheel skimmed on all faces. If warpage was caused by impact during an accident, the safest measure is to renew the wheel complete. Worn wheel bearings may cause rim runout. These should be renewed.

4 Note that impact damage or serious corrosion on models fitted with tubeless tyres has wider implications in that it could lead to a loss of pressure from the tubeless tyres. If in any doubt as to the wheel's condition, seek professional advice.

3 Front wheel: removal and refitting

1 Place the machine on its centre stand, leaving adequate working space around the wheel area. Slacken the knurled ring which retains the speedometer drive cable to the drive gearbox and pull the cable clear of the wheel. On machines with twin disc front brakes, remove the two caliper mounting bracket/fork lower leg mounting bolts and lift one caliper away from the fork; tie it to the frame to avoid straining the hydraulic hose. Place a wood wedge between the brake pads to prevent their being displaced should the brake be operated accidentally.

2 Remove the wheel spindle nut noting that a plain washer is fitted on KZ550 C2 models, then slacken the spindle clamp bolt. Place blocks beneath the crankcase so that the wheel is raised clear of the ground. Take the weight of the wheel and withdraw the spindle. The wheel can now be lifted away.

3 Note that on all models the wheel should not be placed on its side with weight resting on one of the brake discs. This can distort the disc. Place wooden blocks beneath the wheel rim or store the wheel against a wall to avoid this. Clean and grease the spindle to assist refitting.

4 On reassembly, refit the speedometer gearbox to the wheel left-hand side, ensuring that its drive engages correctly, then insert the spacer into the wheel right-hand side. Offer up the wheel, entering the disc between the brake pads, push through the spindle from right to left and refit the plain washer (KZ550 C2) and the spindle nut. Rotate the speedometer gearbox until its bottom edge is horizontal; a projection on the gearbox should be in contact with a lug on the fork lower leg. Check that the gearbox projection is in contact with the lug and not with the lower leg itself. On early KZ550 C1 models (up to frame number 002767) speedometer gearboxes were fitted with the projection cast wrongly; if such a machine is being worked on, ignore the projection and position the gearbox as described. Tighten firmly, but by hand only, the spindle nut.

5 Refit the removed brake caliper (where applicable) tightening its bolts to the recommended torque setting. Remove the support, push the machine off its stand, apply hard the front brake and pump the forks vigorously up and down to align both lower legs on the spindle. When normal fork movement is felt, tighten securely the clamp bolt, followed by the spindle nut, to their respective recommended torque settings.

6 Refit the speedometer cable ensuring that the inner engages correctly on its drive and that the knurled nut is securely fastened. Check that the front brake is working properly, pumping the lever to restore full pressure, before using the machine.

3.1a Disconnect speedometer drive and ...

3.1b ... where applicable, remove one brake caliper to provide clearance

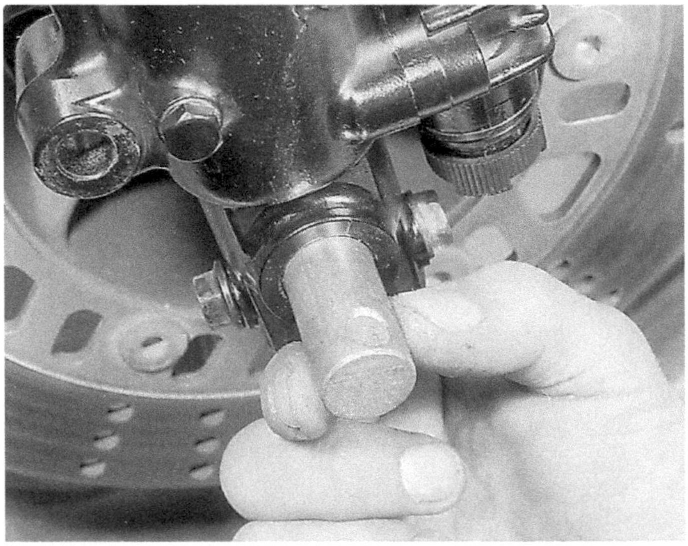

3.2 Release spindle nut and clamp bolt then withdraw spindle to release wheel

3.4a Ensure speedometer drive is correctly reassembled ...

3.4b ... and do not omit wheel right-hand side spacer

3.4c Ensure drive gearbox is positioned so that speedometer drive cable is correctly routed

3.5a Align fork legs on spindle before tightening clamp bolt

3.5b Where possible, tighten spindle nut to specified torque setting

4 Rear wheel: removal and refitting

All Z400J, Z500 and KZ/Z550 A, C and D models

1 Place the machine on its centre stand on level ground. On machines with drum rear brakes, unscrew the adjusting nut and disengage the brake rod from the operating arm, then remove the split-pin or R-clip, nut and lock washer and pull the torque arm off its mounting bolt. On Z500 models remove the torque arm retaining bolt from the caliper. On KZ/Z550 D1 models slacken the torque arm retaining bolt, remove the two axle bolts securing the caliper to its mounting bracket, lift the caliper up as far as possible without straining the brake hose and tie it to the frame, then remove the brake pads. Remove the chainguard.

2 Remove its securing split pin and slacken the spindle nut, then slacken their locknuts and unscrew both chain adjuster drawbolts by one or two full turns until the adjusters can be swung down clear of the swinging arm ends. Remove the single bolt securing each adjuster stop in the swinging arm fork ends and withdraw the stop. Push the wheel forward until the chain can be disengaged from the sprocket and hung over the swinging arm left-hand fork. Except for Z500 models, pull the wheel clear of the fork ends and manoeuvre it away from the machine. On Z500 models, take great care not to strain, kink or stretch the brake hose while pulling the wheel clear of the fork ends. As soon as possible, remove fully the spindle nut and pull the spindle out of the wheel, taking note of the chain adjusters and spacers, to release the caliper; replace the spindle through the caliper and swinging arm to hold it safely while the wheel is removed from the machine. Remove the spindle nut and spindle to separate sprocket carrier and, where applicable, the rear brake backplate from the wheel. Clean and grease the spindle to aid refitting.

3 On reassembly, check that all components are refitted in their correct places. The spindle is usually fitted from right to left, but on Z500 models it would ease future removal if the nut were on the right hand side. When working on a Z500 model, place the wheel as near as possible to the swinging arm ends, then refit the caliper on the spindle, followed by the right-hand chain adjuster and the nut. (Note that all chain adjusters should be fitted with their notched edges on the outside to align with the swinging arm marks.) Insert the rear wheel into the swinging arm and push it as far forwards as possible, refit the adjuster stops and tighten securely their bolts. Engage the chain on the sprocket and draw the wheel back until the adjuster can be swung upwards into place; the end of each adjuster drawbolt engages in a recess in its respective adjuster stop. Tighten the spindle nut by hand only at first, screw in the adjuster drawbolts to their original position and refit the chainguard. On Z500 models refit the torque arm retaining nut and bolt. On KZ/Z550 D1 models refit the pads on the mounting bracket, refit the caliper and tighten its axle bolts to a torque setting of 1.8 kgf m (13 lbf ft). On all models with drum rear brakes, refit the torque arm on its mounting bolt and refit the lock washer and retaining nut. Tighten lightly the retaining nut and refit the split pin or R-clip then connect the brake rod to the operating arm.

4 Working as described in Routine Maintenance, adjust the chain and rear brake (drum brakes only), tightening the spindle nut and torque arm nuts to their recommended torque settings and ensuring that all disturbed nuts and bolts are correctly fastened. Use a new split pin to secure the spindle nut. On models with disc rear brakes apply repeatedly the brake pedal until the pads are pushed back into contact with the disc and full pressure is restored.

All ZR400/550 models, KZ/Z550 H1, H2 and ZX550 A1, A1L models

5 Replace the machine on its centre stand on level ground. Remove the screw(s) or bolt(s) securing the chainguard and lift it clear, then remove the split pin from the spindle nut. On machines with drum rear brakes, remove the split pin (or R-clip), the nut and the lock washer securing the torque arm and pull the arm off its mounting bolt. Unscrew the adjusting nut and disengage the brake rod from the operating arm. On KZ/Z550 H1, H2 models slacken the torque arm retaining nut, remove the two caliper axle bolts and lift the caliper as far as possible without straining the brake hose. Tie it to the frame out of harm's way and withdraw the brake pads. Apply a similar procedure to ZX550 A1 and A1L models but removing instead the torque stay/mounting bracket bolts and tying the caliper assembly to the frame; if care is taken the torque stay can remain in place on the swinging arm. Place a wooden wedge between the pads and take care not to apply the brake.

6 Remove the wheel spindle nut and drive out the spindle, noting the chain adjusters and spacers which may fall clear; disengage the chain from the sprocket and hang it over the left-hand fork end, then manoeuvre the wheel clear of the machine. The sprocket carrier, and if applicable the brake backplate, can now be removed.

7 On reassembly, thoroughly clean and grease the spindle to aid refitting, check that the spacer is in place in the left-hand (sprocket carrier) oil seal, and that the chain adjusters are positioned correctly; each one is marked IN on one surface: that surface must face the wheel. Manoeuvre the wheel into place, engage the chain on the sprocket and lift the wheel so that the spindle can be fitted. Press the spindle through from right to left ensuring that it passes through the first chain adjuster (on each side of the fork end), the caliper mounting bracket/torque stay (if applicable), the first spacer, the wheel, the second spacer and the second chain adjuster. Refit and tighten by hand only the spindle nut, having ensured that the adjusters and caliper mounting bracket, if fitted, are correctly positioned. On KZ/Z550 H1, H2 models, refit the pads, place the caliper over them and tighten the axle bolts to a torque setting of 1.8 kgf m (13 lbf ft);

on ZX550 A1, A1L models lower the caliper assembly into place and tighten the mounting bolts to a setting of 3.0 kgf m (22 lbf ft). On machines with drum rear brakes, refit the torque arm on its mounting bolt and refit the lock washer and retaining nut. Tighten lightly the retaining nut and refit the split pin or R-clip then connect the brake rod to the operating arm. Refit the chainguard.

8 Working as described in Routine Maintenance, adjust the chain and rear brake (drum brakes only) tightening the spindle nut and torque arm nuts to their recommended torque settings and ensuring that all disturbed nuts and bolts are correctly fastened. Use a new split pin to secure the spindle nut. On models with disc rear brakes apply the pedal repeatedly until the pads are forced back into contact with the disc and full pressure is restored.

All shaft drive models

9 Place the machine on its centre stand on level ground and unscrew the brake adjusting nut, then disengage the rod from the operating arm. Remove the split pin or R-clip, the retaining nut and the lock washer, then pull the torque arm off its mounting bolt. Remove its split pin and unscrew the spindle nut, then drive out the spindle. It may be necessary to remove the suspension units from their bottom mountings on UK models so that this can be done. Displace the spacer, pull the wheel to the right, off its driving splines and manoeuvre it clear of the machine. The long spacer can then be removed.

10 On reassembly thoroughly clean and grease the spindle, placing the cover/seal against its head, and smear a small amount of grease on the driving splines and those of the cush drive hub. Insert the long spacer into the centre of the final gear case unit and partially push the spindle through to retain it. Lift the wheel into position, engaging it on its splines and pulling it fully to the left. Insert the shouldered spacer between the brake backplate and swinging arm fork end, then tap the spindle fully through and refit the spindle nut, tightening it by hand only. Refit the torque arm to its mounting bolt, followed by the lock washer and nut, then connect the brake rod to its operating arm. Apply the rear brake hard, to centralise the shoes and backplate on the drum and maintain pressure while the spindle nut is tightened to a torque setting of 7.5 kgf m (54 lbf ft). Secure the nut by fitting a new split pin. Tighten the torque arm retaining nut to a torque setting of 3.0 kgf m (22 lbf ft) and refit its split pin or R-clip. Adjust the rear brake, if necessary, as described in Routine Maintenance. If the suspension unit bottom mountings were released on UK models, lift the rear wheel up and refit each unit on its mounting stud, then tighten the mounting nut to a torque setting of 2.5 kgf m (18 lbf ft).

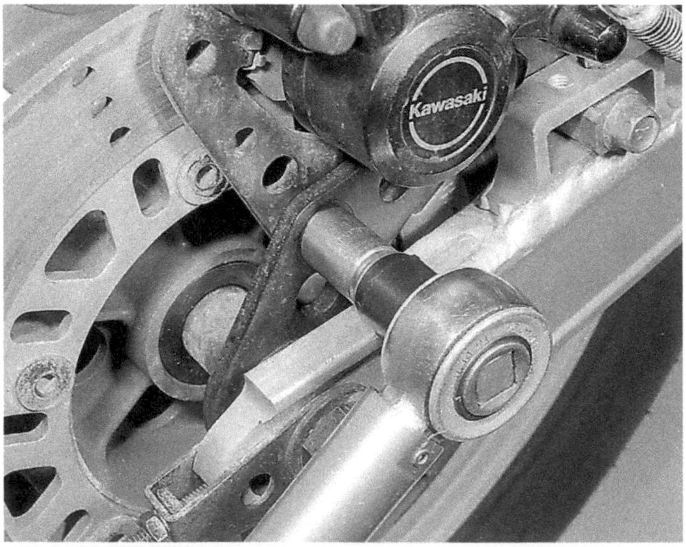

4.5 Remove caliper mounting bracket/torque stay bolts to release caliper – ZX550 A1

4.6 Withdraw sprocket carrier (chain drive models) to expose cush drive and wheel bearings

4.7a Do not forget to refit wheel left-hand ...

4.7b ... and right-hand spacers on ressembly

4.7c Chain adjusters must be refitted the correct way round

4.7d Tighten all retaining nuts and bolts to specified torque settings

5 Front wheel bearings: removal, examination and refitting

1 Remove the front wheel from the machine as described in Section 3 of this Chapter, then remove the speedometer gearbox from the wheel left-hand side and the spacer from the right-hand side. Although it is not necessary to remove the brake disc(s) for this task on any model, work is a great deal easier on Z400 J1, Z500 B1, B2 and KZ/Z550 A1 models if the cover plates are removed first, this meaning that the disc(s) will be released. On the KZ 550 A1 model, also the KZ 550 A2, A3, C2 and C3 models, a cover plate is secured by two screws to the wheel right-hand side; again while its removal is not strictly necessary, it should be removed to improve access and to make work much easier.

2 On all models except the Z400 J1, Z500 B1, B2 and KZ/Z550 A1 remove the circlip from the hub left-hand side and withdraw the speedometer drive dog, noting which way round it is fitted. Working from the right-hand side on all models, lever out the oil seal, taking care not to damage the casting, and remove the circlip behind it.

3 Arrange the wheel with the left-hand side uppermost and supported so that the hub is clear of the work surface. Pass a long drift through the hub, push the internal spacer to one side and drive out the right-hand bearing, tapping evenly all around its inner race. The spacer will drop out. Invert the wheel and drive out the remaining bearing.

4 Removing the bearings in this way will almost certainly damage them if they are a tight fit but there is no alternative. Wash each bearing thoroughly removing all old grease, then spin each one. If any signs of roughness can be heard or felt, if any free play can be felt or if any pitting can be seen on the balls or their tracks, the bearings must be renewed. The oil seal should be renewed as a matter of course.

5 On reassembly, pack the bearings with high melting-point grease and refit the left-hand bearing with its sealed surface facing outwards. Drive the bearing into place using a tubular drift such as a socket spanner which bears only on the bearing outer race. Turn the wheel over and refit the spacer with its shoulder on the left, then pack the central recess no more than $\frac{2}{3}$ full with high melting-point grease and refit the right hand bearing, sealed surface outwards, as described above. Refit the circlip, the oil seal and on the applicable models, the cover plate or right-hand disc and cover plate, if removed. Tighten the disc mounting bolts to a torque setting of 2.3 kgf m (16.5 lbf ft).

6 Turn the wheel over and refit the driving dog and circlip. Refit the disc and cover plate, if removed, as described above.

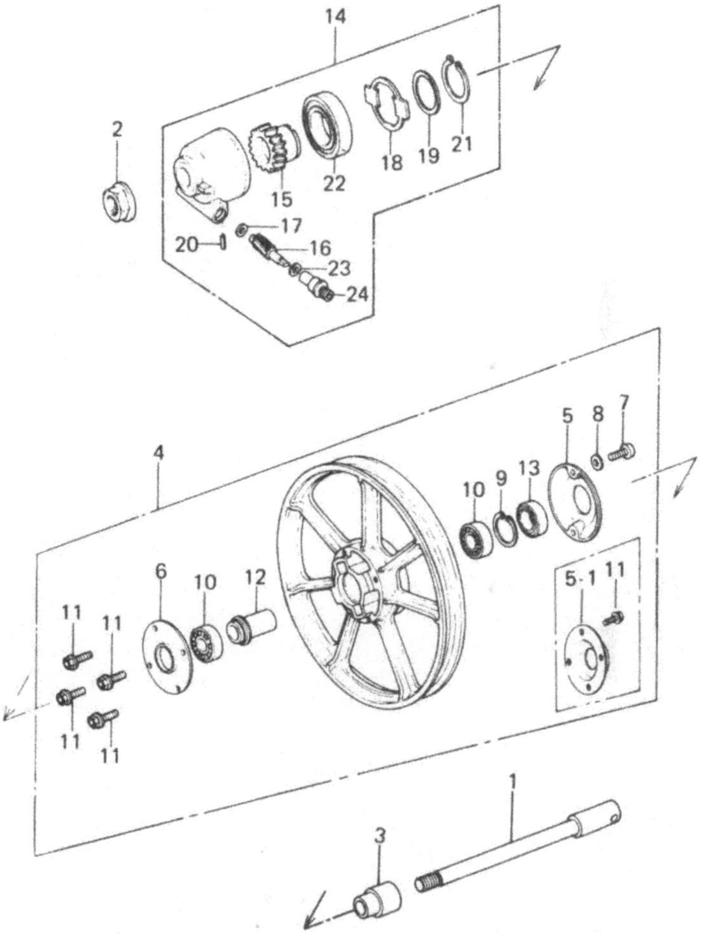

Fig. 5.1 Front wheel and speedometer drive – Z400 J1, Z500 B1, B2 and KZ/Z550 A1

1 Spindle	*12 Spacer*
2 Nut	*13 Oil seal*
3 Spacer	*14 Speedometer drive gearbox*
4 Wheel assembly	*15 Speedometer gear*
5 Dust cover – KZ550 A1	*16 Speedometer pinion*
5-1 Dust cover – Z400 J1, Z500 B1, B2, Z550 A1	*17 Thrust washer*
6 Dust cover	*18 Speedometer drive dog*
7 Screw – KZ550 A1	*19 Washer*
8 Washer – KZ550 A1	*20 Spring pin*
9 Circlip	*21 Circlip*
10 Bearing	*22 Oil seal*
11 Bolt – 4 off KZ550 A1, 8 off all other models	*23 Thrust washer*
	24 Bush

5.2a Speedometer drive dog (where fitted) is retained by a circlip

5.2b Lever out oil seal from hub right-hand side ...

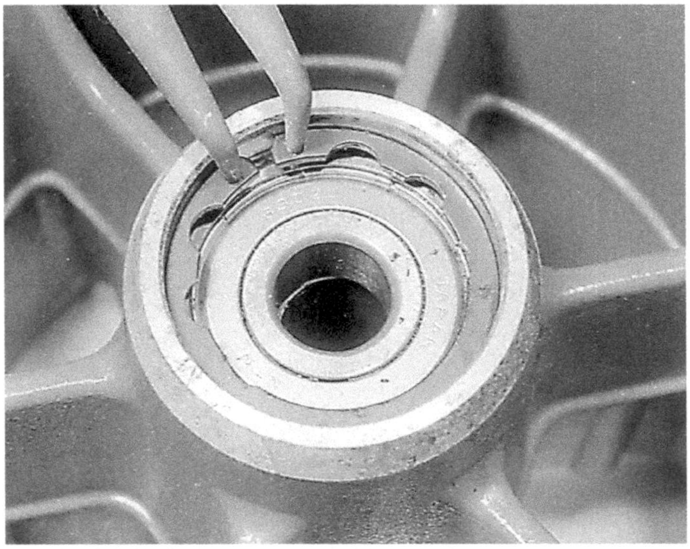

5.2c ... and withdraw bearing retaining circlip

5.3 Use long drift to tap out first bearing as shown

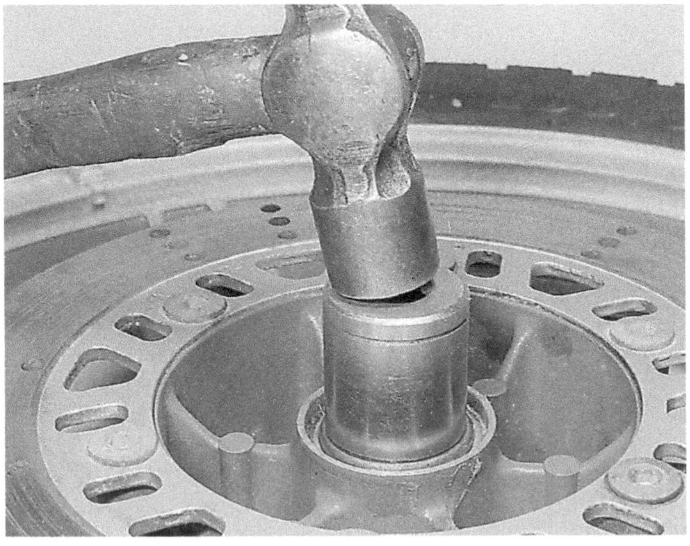

5.5a Refit first left-hand bearing, as shown ...

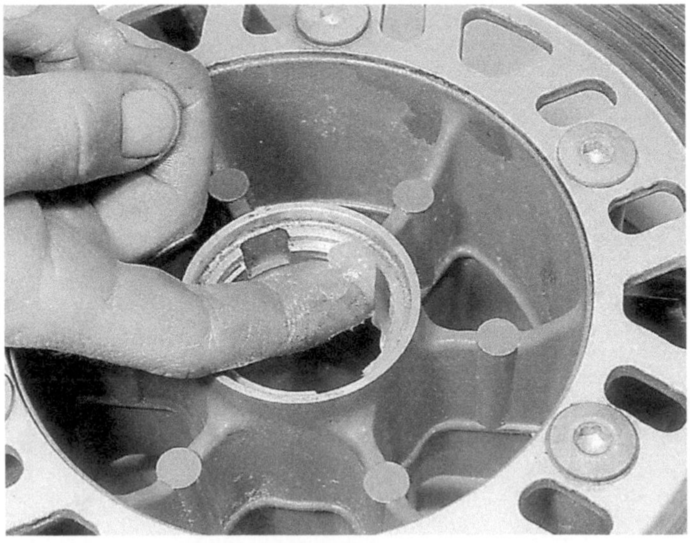

5.5b ... then pack central cavity with grease ...

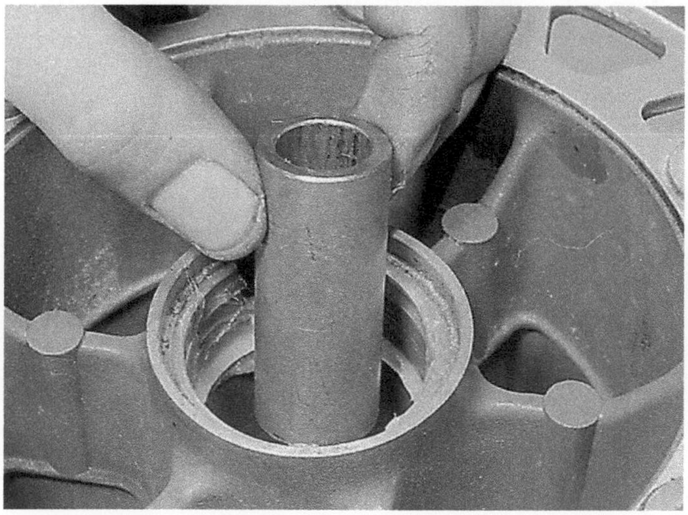

5.5c ... and refit central spacer

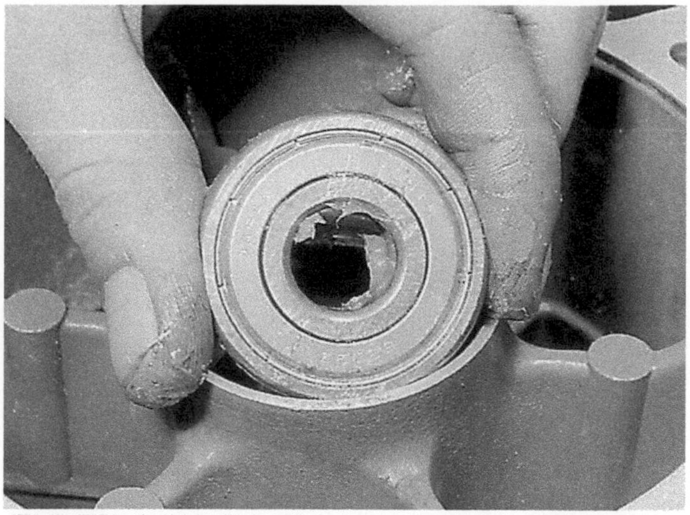

5.5d Bearing sealed surface must face outwards

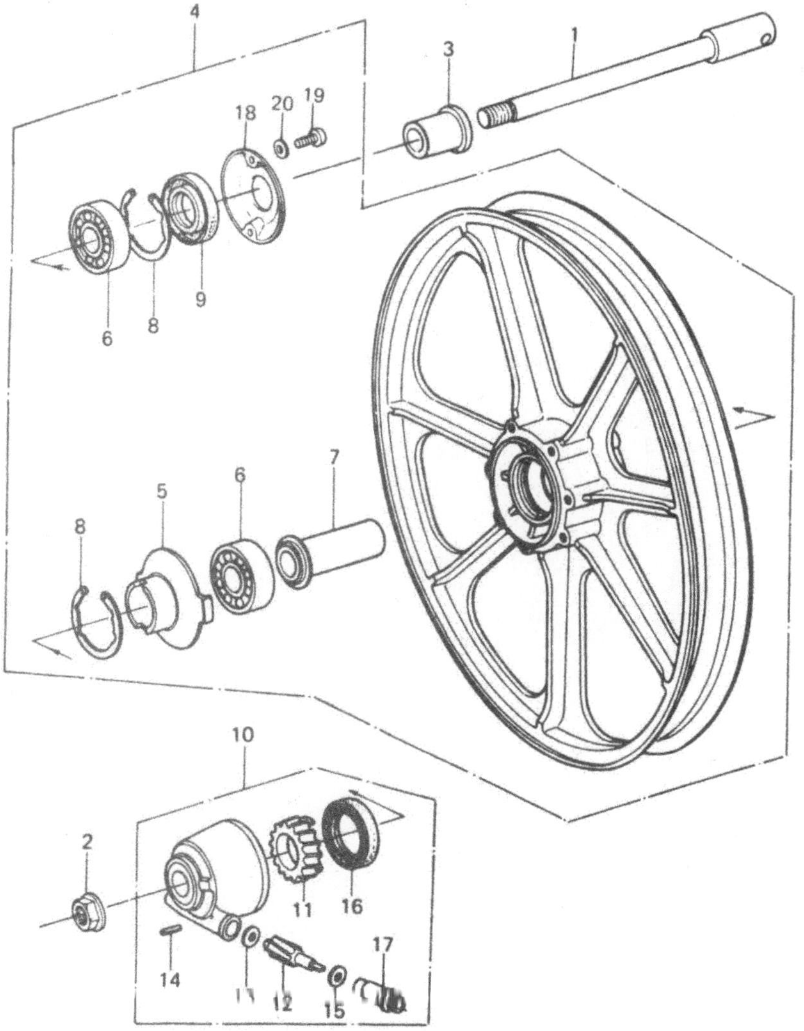

Fig. 5.2 Front wheel and speedometer drive – all other models (typical)

1 *Spindle*
2 *Nut*
3 *Spacer*
4 *Wheel assembly*
5 *Speedometer drive dog*
6 *Bearing*
7 *Spacer*
8 *Circlip*
9 *Oil seal*
10 *Speedometer drive gearbox*
11 *Speedometer gear*
12 *Speedometer pinion*
13 *Thrust washer*
14 *Spring pin*
15 *Thrust washer*
16 *Oil seal*
17 *Bush*
18 *Dust cover**
19 *Screw**
20 *Washer**
**KZ550 A2, A3, C2 and C3 models only*

6 Rear wheel bearings: removal, examination and refitting

1 Remove the rear wheel as described in Section 4 of this Chapter, then pull away the brake backplate, where applicable, and the sprocket carrier assembly on chain drive models. The cush drive assembly need not be disturbed on shaft drive models. Lever out the oil seal from the hub right-hand side (disc brake models only) and remove the circlip retaining the right-hand bearing (all models).

2 Remove, clean, examine, grease and refit the bearings exactly as described above for the front wheel. Note that some bearings may not be sealed; in this case they should be refitted with their numbered faces outwards.

6.1 Rear wheel components are removed in the same way as the front wheel

Fig. 5.3 Drum brake rear wheel – all chain drive models (typical)

1	*Split pin*	*15*	*Rubber cush drive block*
2	*Spindle nut*	*16*	*Spacer*
3	*Spacer*	*17*	*Wheel*
4	*Sprocket*	*18*	*Circlip*
5	*Drive chain*	*19*	*Brake shoe*
6	*Oil seal*	*20*	*Spring*
7	*Circlip*	*21*	*Brake backplate*
8	*Bearing*	*22*	*Dust seal*
9	*Sprocket carrier*	*23*	*Wear indicator*
10	*Shouldered spacer*	*24*	*Operating arm*
11	*Bolt*	*25*	*Brake camshaft*
12	*Nut*	*26*	*Spacer*
13	*Bearing*	*27*	*Pinch bolt*
14	*O-ring*	*28*	*Spindle*

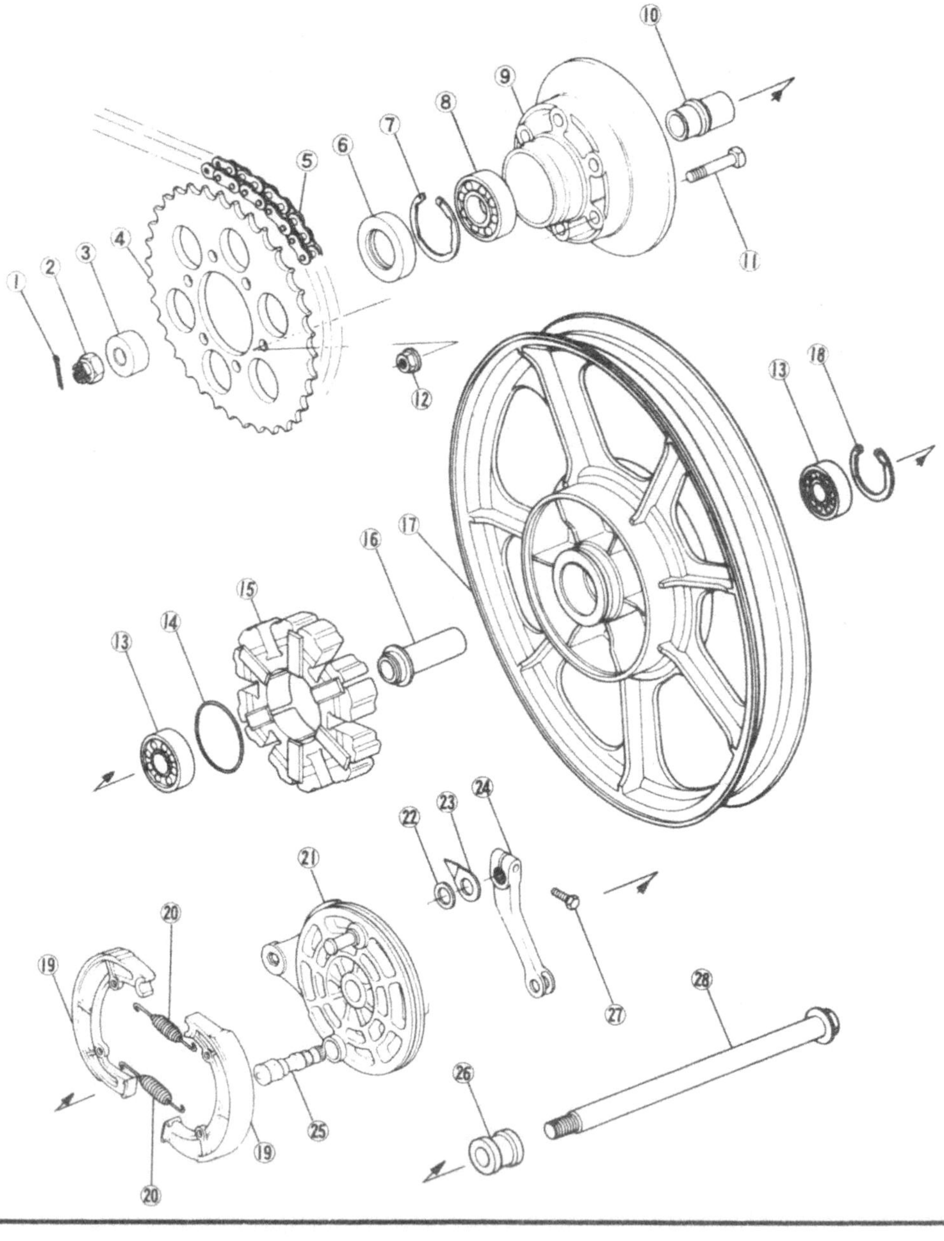

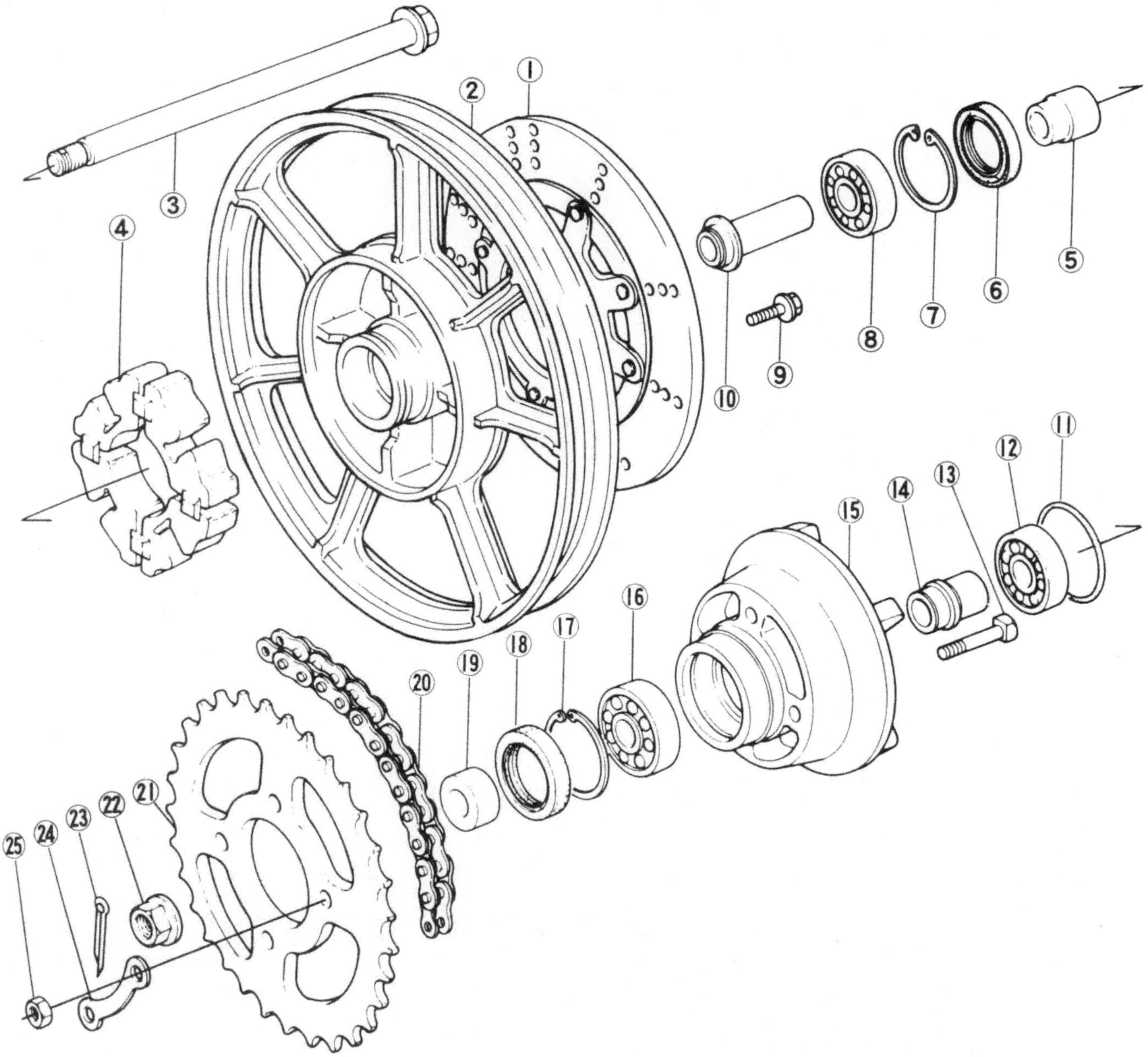

Fig. 5.4 Disc brake rear wheel – typical

1 Brake disc
2 Rear wheel
3 Spindle
4 Rubber cush drive block
5 Spacer
6 Oil seal
7 Circlip
8 Bearing
9 Bolt
10 Spacer
11 O-ring
12 Bearing
13 Bolt
14 Shouldered spacer
15 Sprocket carrier
16 Bearing
17 Circlip
18 Grease seal
19 Spacer
20 Drive chain
21 Rear sprocket
22 Spindle nut
23 Split pin
24 Tab washer
25 Nut

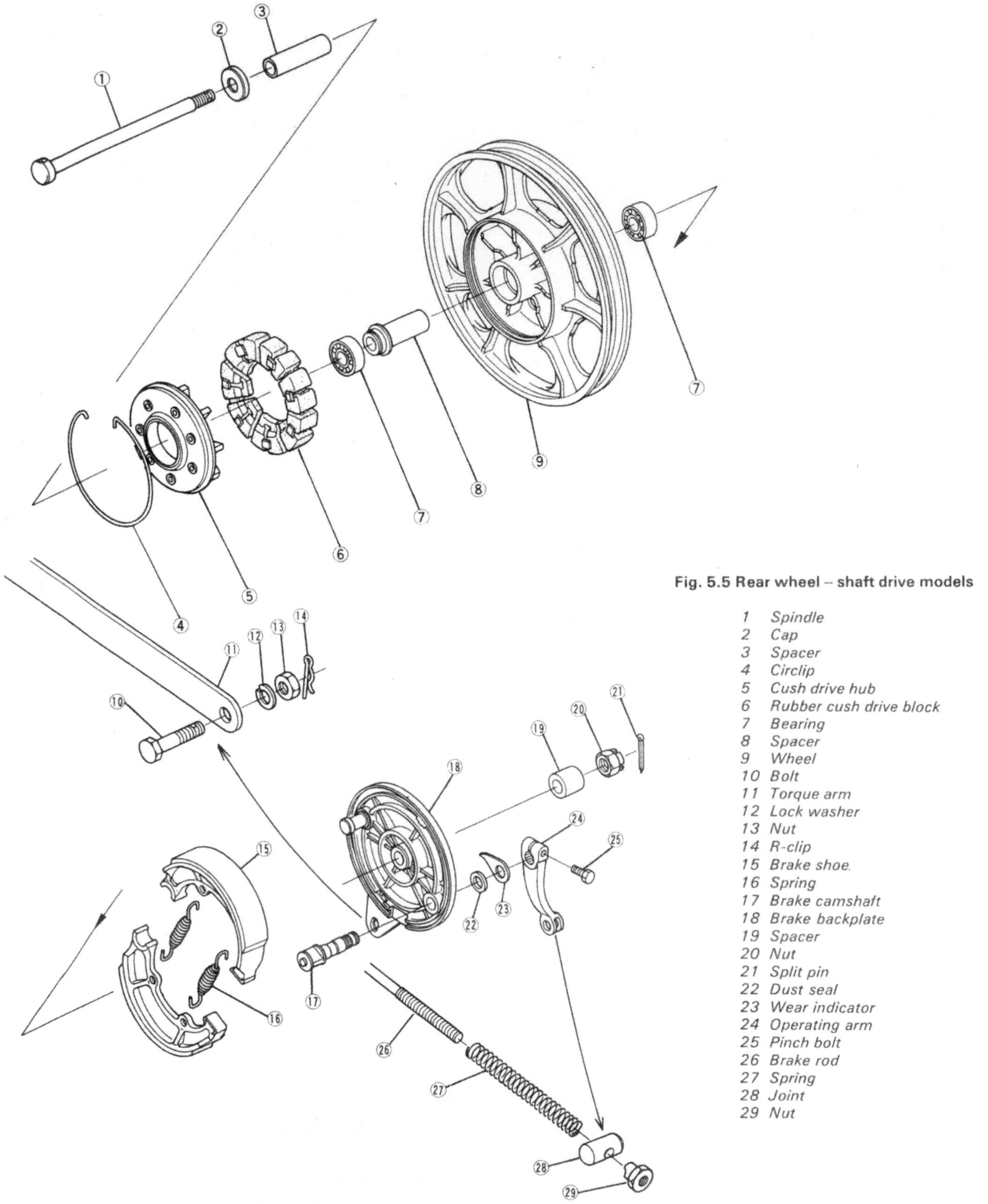

Fig. 5.5 Rear wheel – shaft drive models

1 *Spindle*
2 *Cap*
3 *Spacer*
4 *Circlip*
5 *Cush drive hub*
6 *Rubber cush drive block*
7 *Bearing*
8 *Spacer*
9 *Wheel*
10 *Bolt*
11 *Torque arm*
12 *Lock washer*
13 *Nut*
14 *R-clip*
15 *Brake shoe*
16 *Spring*
17 *Brake camshaft*
18 *Brake backplate*
19 *Spacer*
20 *Nut*
21 *Split pin*
22 *Dust seal*
23 *Wear indicator*
24 *Operating arm*
25 *Pinch bolt*
26 *Brake rod*
27 *Spring*
28 *Joint*
29 *Nut*

7 Rear wheel cush drive: examination and renovation

Chain drive models

1 Remove the rear wheel from the machine as described in Section 4 of this Chapter. Depending on the model, the sprocket is retained by six bolts or studs, or by four bolts; these are easiest to slacken while the assembly is on the wheel. Flatten the raised edges of the tab washers, where fitted, and slacken the retaining nuts, then remove the carrier assembly from the wheel. Remove the nuts (and tab washers) and withdraw the sprocket, pull out the spacers from both sides of the assembly, noting which way round the shouldered spacer is fitted,

then lever out the oil seal taking care not to scratch or damage the casting. On KZ/Z550 D1, H1 and H2 models the oil seal is retained by a circlip which must be displaced first, but on all Z500 and KZ/Z550 C models the circlip is behind the oil seal and must be removed to permit the bearing to be driven out.

2 Remove the bearing, wash it and check it for wear as described in Section 5 of this Chapter, then repack it with grease and refit it. Where applicable, secure it by refitting the circlip. The oil seal should be renewed whenever it is disturbed; do not forget the retaining circlip, where fitted.

3 If the rear sprocket teeth are hooked, chipped, missing or worn the sprocket must be renewed, but this should be done only in conjunction with a new gearbox sprocket and chain. Refit the sprocket on the carrier, followed by the tab washers (where fitted) and the nuts, applying thread locking compound to their threads. Tighten the nuts to the recommended torque setting and secure each nut by bending up an unused portion of the tab washer (where fitted). Note that the tab washers should be renewed as soon as all their locking tabs have been used once. Insert the spacers into both sides of the carrier assembly.

4 If the cush drive rubber block is perished, split, damaged or compressed to the extent that there is excessive movement between the sprocket carrier and wheel, it must be renewed. It can be pulled out of the hub by hand, but if a new one is a tight fit it should be lubricated using a very small amount of soapy water.

5 Some models have an O-ring fitted around the wheel left-hand bearing boss. This should be renewed if found to be worn or damaged to prevent grease leaking onto the cush drive rubbers. Lubricate it with a thin smear of grease whenever the carrier is refitted.

Shaft drive models

6 Remove the rear wheel as described in Section 4 of this Chapter. Use a pair of pliers to remove the large circlip retaining the cush drive hub, then pull the hub away from the rubber block. If its splines are worn or damaged, or if any other sign of damage is found, the hub should be renewed. Reassembly is the reverse of the removal procedure. The rubber block should be examined, and renewed if necessary, as described in paragraph 4 above.

8 Brakes: adjustment, examination and pad renewal

The full procedure for adjustment and for checks on pad or shoe wear and the braking system in general are given in Routine Maintenance.

9 Brake discs: examination and renovation

1 Examine the brake discs for scoring, particularly, where applicable, the rear unit which is more vulnerable to accumulations of road dirt. Damaged discs will cause poor braking and will wear pads quickly, and should therefore be renewed. The disc thickness can be measured with a micrometer and should not be less than the service limit specified.

2 Check for warpage with the relevant wheel raised clear of the ground, using a dial gauge probe running near the edge of the disc. Warpage must not exceed 0.3 mm (0.118 in) when the disc is rotated. A warped disc will often cause judder during braking and will necessitate renewal.

3 The discs can be removed after the appropriate wheel has been removed from the machine. Each disc is retained by a number of bolts. When refitting the disc, ensure that it and the hub are clean and that the chamfered hole side of the disc faces inwards or the marked surface outwards, as appropriate. Tighten the retaining bolts to 2.3 kgf m (16.5 lbf ft).

10 Brake master cylinder: removal, overhaul and refitting

Front

1 Connect a length of tubing from the caliper bleed nipple to a suitable container, unscrew the bleed nipple by 1 – 2 full turns and slowly pump the lever until all fluid is expelled. Tighten the bleed nipple and repeat the procedure on all remaining nipples in the system, where applicable. Remove the right hand mirror, then free the front brake light switch by depressing its locking pin using a small electrical screwdriver via the hole in the underside of the switch housing.

2 Pull back the banjo union dust cover and remove the union bolt and washers. Wipe up any residual hydraulic fluid. Slacken the two master cylinder clamp bolts, remove the clamp half and lift the master cylinder away.

3 Remove the two screws which retain the reservoir cover, remove the cover and diaphragm and empty out any remaining fluid. Remove the brake lever locknut and pivot bolt and remove the lever. Remove the white plastic liner by pushing in the locking tabs which retain it. Pull out the piston assembly, separating the spring, dust seal and piston stop from the piston.

4 Examine the piston surface and master cylinder bore for signs of wear or corrosion. Renew both components if damaged in any way; new seals will not compensate for scoring and will wear out quickly. Check the primary and secondary seals for damage or swelling, renewing them unless in perfect condition. The cups are sold as a kit together with the piston and spring. Renew the dust seal at the same time to preclude road dirt entering the caliper body. Ensure that the supply port and the smaller relief port between the cylinder and reservoir are clear, especially where swollen or damaged cups have been noted.

5 In most cases, the components can be measured for wear if required, and the readings checked against the figures given in Specifications.

Rear

6 Proceed as described above having removed the right-hand side panel and noting the following exceptions. Drain the system and disconnect the banjo union at the caliper. Disconnect the hose from the reservoir, where applicable, and wipe up any spilled fluid. Remove the split pin, washer and clevis pin which retain the brake arm to the push rod. Remove the two master cylinder mounting bolts and lift the cylinder away from the frame or footrest plate.

7 Pull off the dust seal and remove it together with the push rod. Displace the retaining clip and continue dismantling and overhaul as described above.

Front and rear

8 Reassemble the master cylinder by reversing the dismantling sequence, ensuring that all components are kept spotlessly clean. Lubricate the piston, cups and cylinder bore with new hydraulic fluid during installation. Use new sealing washers on the banjo unions and tighten the union bolts to 3.0 kgf m (22 lbf ft). Refill and bleed the hydraulic system and check brake operation before using the machine.

9.3 Remove mounting bolts to release brake discs – note minimum thickness marking

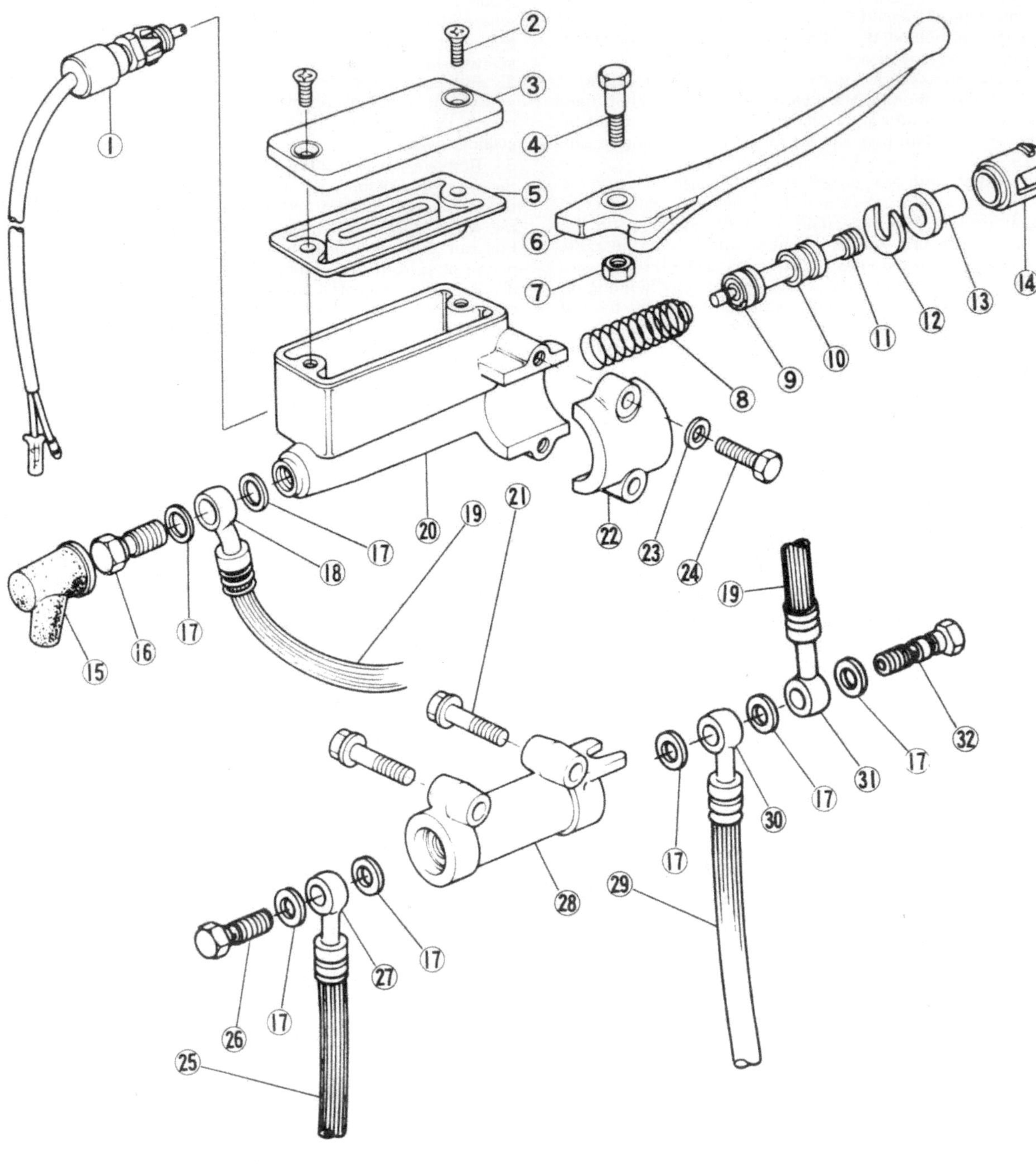

Fig. 5.6 Front brake master cylinder – typical

1 Front brake light switch
2 Screw
3 Master cylinder cover
4 Brake lever pivot bolt
5 Diaphragm
6 Brake lever
7 Locknut
8 Spring
9 Primary cup
10 Secondary cup
11 Piston
12 Piston stop
13 Dust seal
14 Liner
15 Dust cover
16 Banjo bolt
17 Flat washer
18 Hose fitting
19 Upper brake hose
20 Master cylinder body
21 Mounting bolt
22 Master cylinder clamp
23 Flat washer
24 Clamp bolt
25 Lower left-hand brake hose
26 Banjo bolt
27 Hose fitting
28 Two-way joint
29 Lower right-hand brake hose
30 Hose fitting
31 Hose fitting
32 Banjo bolt

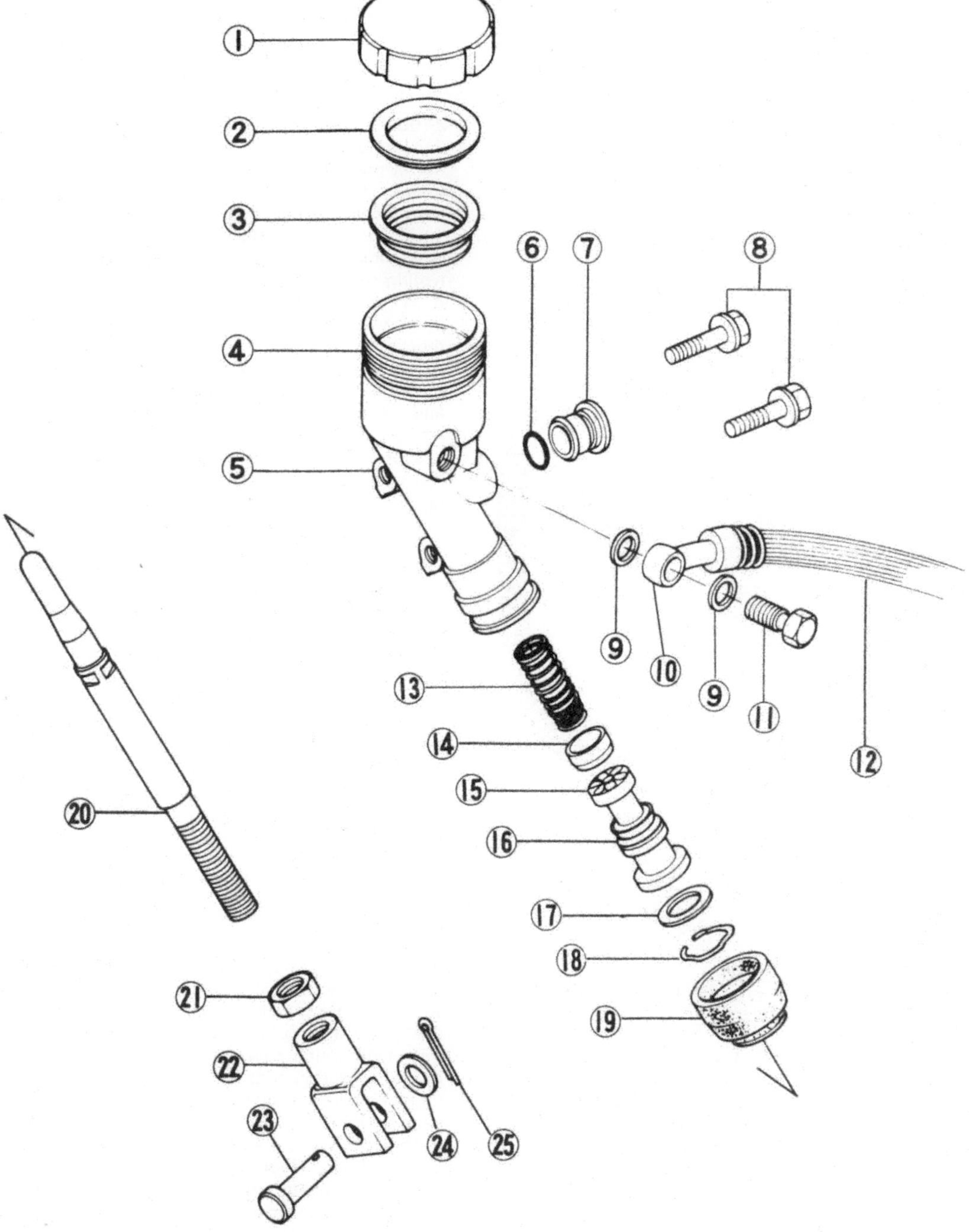

Fig. 5.7 Rear brake master cylinder – Z500 B1, B2 (KZ/Z550 D1 similar)

1 Cap
2 Retainer
3 Diaphragm
4 Reservoir
5 Master cylinder body
6 Gasket
7 Plug
8 Mounting bolts
9 Flat washer
10 Hose fitting
11 Banjo bolt
12 Brake hose
13 Return spring
14 Primary cup
15 Piston
16 Secondary cup
17 Piston stop
18 Retainer
19 Dust cover
20 Push rod
21 Locknut
22 Joint
23 Clevis pin
24 Flat washer
25 Split pin

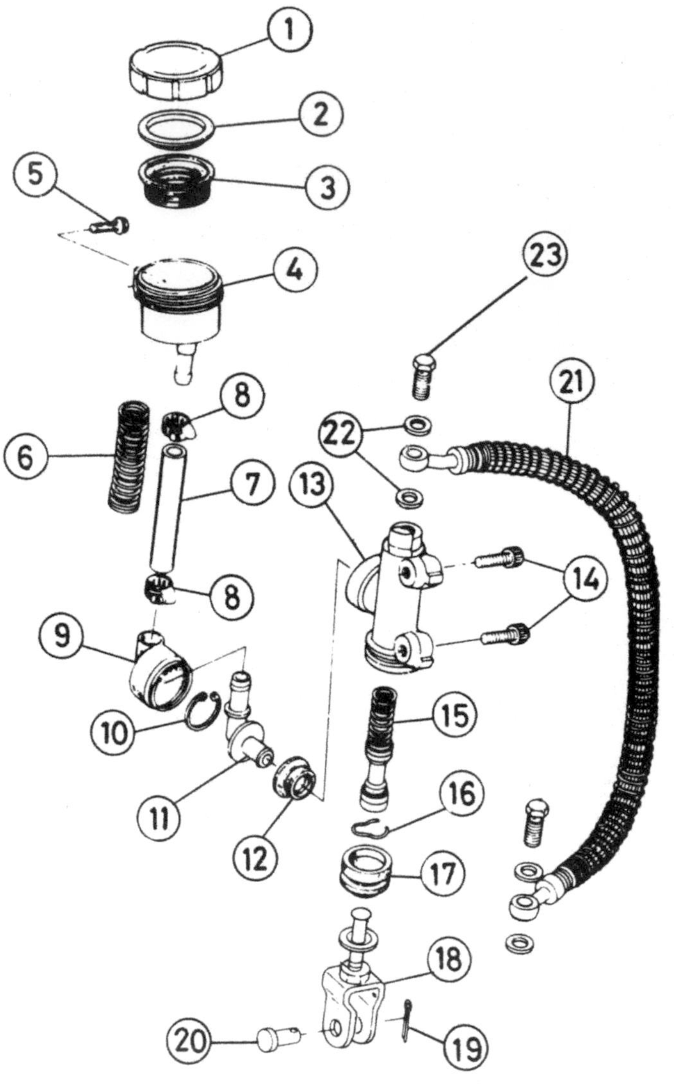

Fig. 5.8 Rear brake master cylinder – KZ/Z550 H1, H2, ZX550 A1 and A1L

1	*Cap*	*13*	*Master cylinder body*
2	*Retainer*	*14*	*Bolt*
3	*Diaphragm*	*15*	*Piston assembly*
4	*Reservoir*	*16*	*Spring clip*
5	*Bolt*	*17*	*Dust cover*
6	*Spring*	*18*	*Joint*
7	*Hose*	*19*	*Split pin*
8	*Clamp*	*20*	*Clevis pin*
9	*Boot*	*21*	*Brake hose*
10	*Circlip*	*22*	*Flat washer*
11	*Union*	*23*	*Banjo bolt*
12	*Seal*		

11 Brake caliper: removal, overhaul and refitting

1 The procedures required to dismantle and rebuild each type of caliper are given below under separate headings, with general notes applicable to all types given at the end of the Section. In all cases, commence work by removing the brake pads as described in Routine Maintenance. The simplest way of removing the piston is to place the caliper in a plastic bag, tying its neck around the brake hose to prevent a shower of fluid, then to pump the lever or pedal to expel the piston by hydraulic pressure. If both calipers of a twin-disc system are to be dismantled, perform the operation simultaneously on both calipers.

2 If the above method does not displace the piston, waste no further time on that caliper; it must be renewed as a complete assembly as it is too badly damaged or corroded to be of any further use. If the hydraulic system has been drained or disconnected, apply a jet of compressed air to the fluid passage or bleed nipple orifice (having removed the nipple) to displace the piston, but wrap the caliper first in a thick layer of rag to prevent the piston flying out.

Z400 J1, Z500 B1, B2 (front) and KZ/Z550 A1

3 Remove the two axle bolt retaining nuts and thick spacers, noting that it may be necessary to hold the caliper by refitting it temporarily to the machine, then pull out the axle bolts. Taking care not to damage the rubber dust covers, remove the mounting bracket from the caliper body. Displace the piston as described above, then disconnect the brake hose union and allow the fluid to drain into a suitable container. Noting the way in which each is fitted, remove the rubber seals.

4 On reassembly, refit the seals and piston as described below and fit new O-rings and dust covers to the mounting bracket, having thoroughly greased its internal bearing surfaces. Insert the bracket into the caliper and refit the axle bolts, rotating them as they pass through the mounting bracket to prevent damage to the seals. Refit the spacers and the nuts, tightening the nuts to a torque setting of 2.6 kgf m (19 lbf ft). Refit the pads and install the caliper on the machine as described in Routine Maintenance, then connect the brake hose to the caliper ensuring it is correctly routed and using new sealing washers; tighten the banjo bolt to a torque setting of 3.0 kgf m (22 lbf ft). Fill the system with fresh hydraulic fluid and bleed out any air.

Z500 B1, B2 – rear

5 Connect a clear plastic tube between the bleed nipple and a suitable container, unscrew the bleed nipple by 1-2 full turns and slowly pump the brake pedal until all fluid is drained from the system. Disconnect the brake hose from the caliper and place the hose end in a plastic bag to catch any surplus fluid. Remove the brake pads as described in Routine Maintenance, slacken the two caliper Allen bolts and remove the rear wheel from the machine as described in Section 4 of this Chapter. Disengage the caliper from the spindle, remove the two Allen bolts and separate the two caliper halves. Use a jet of compressed air to displace the pistons, taking note of the instructions given in paragraph 2 of this Section, then remove the piston seals from both halves, noting the way in which each is fitted. Keep separate the components from each caliper half.

6 On reassembly, refit the seals and pistons as described below, noting that the pistons, if reused, must be fitted in their original bores. Fit a new O-ring in the recess around the fluid passage of the right-hand caliper half, refit the left-hand half and tighten the two Allen bolts to a torque setting of 3.0 kgf m (22 lbf ft). Refit the caliper on the rear wheel spindle, install the rear wheel in the swinging arm and connect the torque arm to the caliper. Do not fully tighten any nuts or bolts at this stage. Using new sealing washers connect the brake hose to the caliper, ensuring that the hose is correctly routed, then tighten the banjo bolt to a torque setting of 3.0 kgf m (22 lbf ft). Refit the brake pads as described in Routine Maintenance, then fill the system with fresh hydraulic fluid and carry out the bleeding procedure to remove all traces of air. When the brake is working properly and full pedal pressure obtained, adjust the chain tension as described in Routine Maintenance and check that all disturbed fasteners are tightened to their recommended torque settings.

ZR400 A1, B1, ZR550 A1, A2, Z550 G1, G2 and ZX550 A1, A1L (front and rear)

7 Pushing the mounting bracket away from the piston, remove it from the caliper, then displace the two rubber dust covers and the anti-rattle spring. Displace the piston cap and expel the piston, as described above, then remove the piston seals and disconnect the brake hose.

8 On reassembly, refit the seals and piston as described below, press the rubber dust covers into place and refit the mounting bracket, followed by the anti-rattle spring. Using new sealing washers, refit the brake hose ensuring that it is correctly routed and that the banjo bolt is tightened to a torque setting of 3.0 kgf m (22 lbf ft). Refit the pads to the caliper and the caliper to the machine then fill the system with fresh hydraulic fluid and carry out the bleeding procedure to remove all traces of air.

All other models – front (and rear)

9 With the caliper separated from the machine, check that the mounting bracket is undamaged. Remove the pads and their guide shims, then withdraw from the mounting bracket the dust covers and the axle shafts noting that one shaft has a bush fitted around it. Remove the anti-rattle spring from the caliper body. Expel the piston, as described above, then displace the seals and disconnect the brake hose. Remove the mounting bracket from the machine only if necessary.

10 On reassembly, refit the bush to the smaller diameter axle shaft and press both shafts, and their dust covers, into place in the mounting bracket, followed by the guide shims and brake pads. Install the seals and refit the caliper piston as described below, then refit the brake hose, ensuring that it is correctly routed. Always use new sealing washers and tighten the banjo bolt to a torque setting of 3.0 kgf m (22 lbf ft). Refit the caliper to the mounting bracket, tightening the axle bolts to a torque setting of 1.8 kgf m (13 lbf ft), then fill the system with fresh hydraulic fluid and carry out the bleeding procedure to remove all traces of air.

Caliper overhaul – general

11 Before any repair work is undertaken, check with a local Kawasaki dealer to establish what is available for the caliper to be worked on. In some cases nearly all components are individually available, in other cases only complete assemblies can be obtained as replacement parts.

12 Clean all components carefully, removing all traces of road dirt, friction material and corrosion. Note that only clean hydraulic fluid (or ethyl or isopropyl alcohol) should be used to clean hydraulic components; petrol, paraffin or other normal cleaning solvents will attack the rubber seals. It is permissible to use a wire brush gently to remove dirt and corrosion except in the caliper bores and piston skirt.

13 Renew all piston seals (both fluid and dust seals) and all sealing O-rings as a matter of course. Never reuse a hydraulic seal after it has been disturbed and note that the piston seal must be in excellent condition as its secondary role is to return the piston when lever or pedal pressure is released, thus preventing brake drag. Carefully examine the axle bolt dust covers and any other rubber seals, renewing any that are perished, split or otherwise damaged. Similarly, discard the sealing washers fitted at the brake hose union; these should be renewed as a matter of course.

14 Examine the piston surface and caliper bore for signs of wear or scoring, normally caused by the presence of road dirt or corrosion. If wear is found, or deep scoring or scratches which might cause fluid leaks, the component concerned must be renewed. Where measurements are given in the Specifications Section of this Chapter, check that neither has worn to beyond the service limits.

15 Where applicable, check that there is no free play between the caliper body and its axle bolts, shafts or mounting bracket. Renew any component that is found to be worn. As mentioned in Routine Maintenance, it is essential that single-piston brake calipers can slide smoothly on their mountings. Make a final check that there are no signs of damage on any other part of the caliper assembly.

16 On reassembly, soak the new piston (fluid) seal in clean hydraulic fluid then refit it to the caliper bore, taking great care that it is seated correctly in its groove and that the bore is not scratched. Smear hydraulic fluid over the caliper bore and piston surface and refit the piston, rotating it slightly while keeping it square to the caliper bore so that it does not stick or displace the piston seal. Press the piston fully into the caliper, wipe off any surplus fluid, then refit the new dust seal ensuring that it locates correctly on the caliper lip.

17 Where applicable, apply PBC (Poly Butyl Cuprysil) grease to all sliding surfaces on the caliper body and mounting bracket or on the axle bolts or shafts, and pack grease into the recesses in the mounting bracket or caliper body. Be careful to wipe off all surplus grease once the caliper assembly is rebuilt. Check that the caliper body moves smoothly and fully from side to side before refitting the pads.

11.7a Remove anti-rattle spring ...

11.7b ... and push mounting bracket away from piston to withdraw from caliper

11.7c Renew rubber dust covers if worn or split

11.17 Use only recommended grease when lubricating caliper sliding components

Fig. 5.9 Front brake caliper – Z400 J1, Z500 B1, B2 and KZ/Z550 A1

1. *Dust cap*
2. *Bleed nipple*
3. *Caliper axle bolt*
4. *Caliper*
5. *Spacer*
6. *Nut*
7. *Piston seal*
8. *Piston*
9. *Dust seal*
10. *Caliper mounting bracket*
11. *O-ring*
12. *Dust cover*
13. *Fixed pad*
14. *Metal plate*
15. *Lock washer*
16. *Mounting screw*
17. *Moving pad*
18. *Mounting bolt*

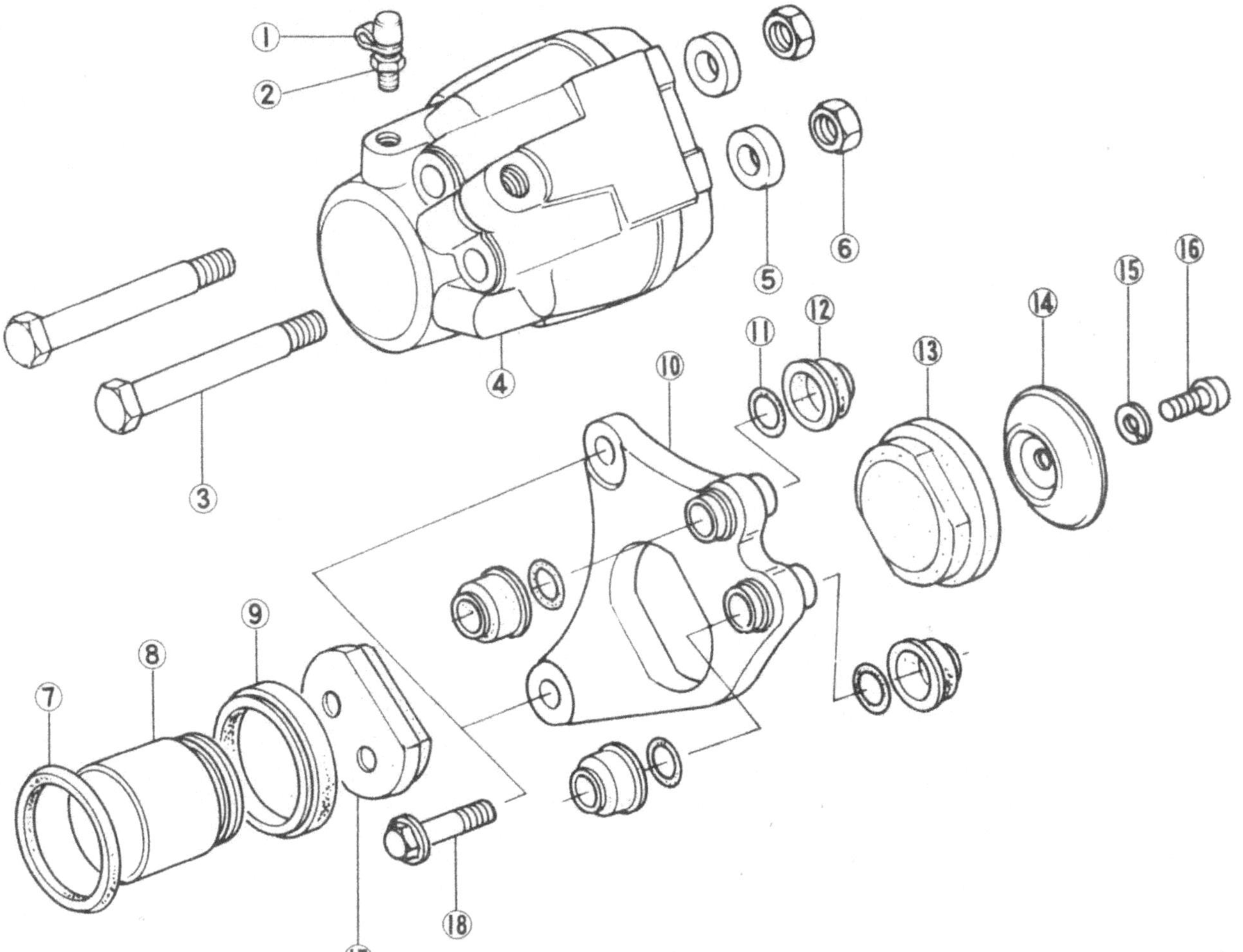

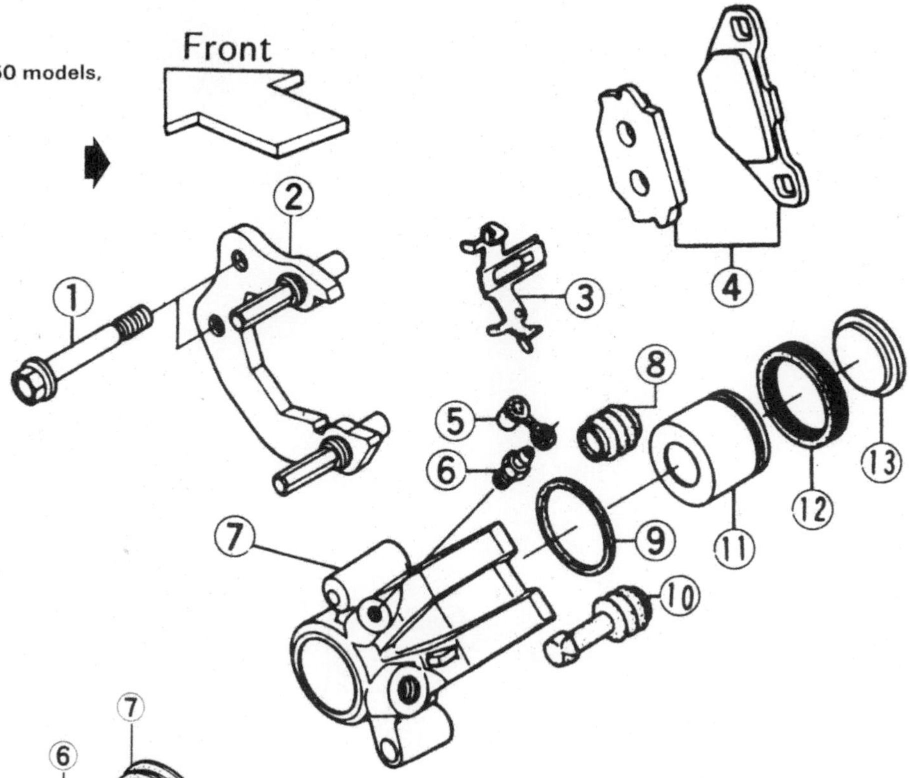

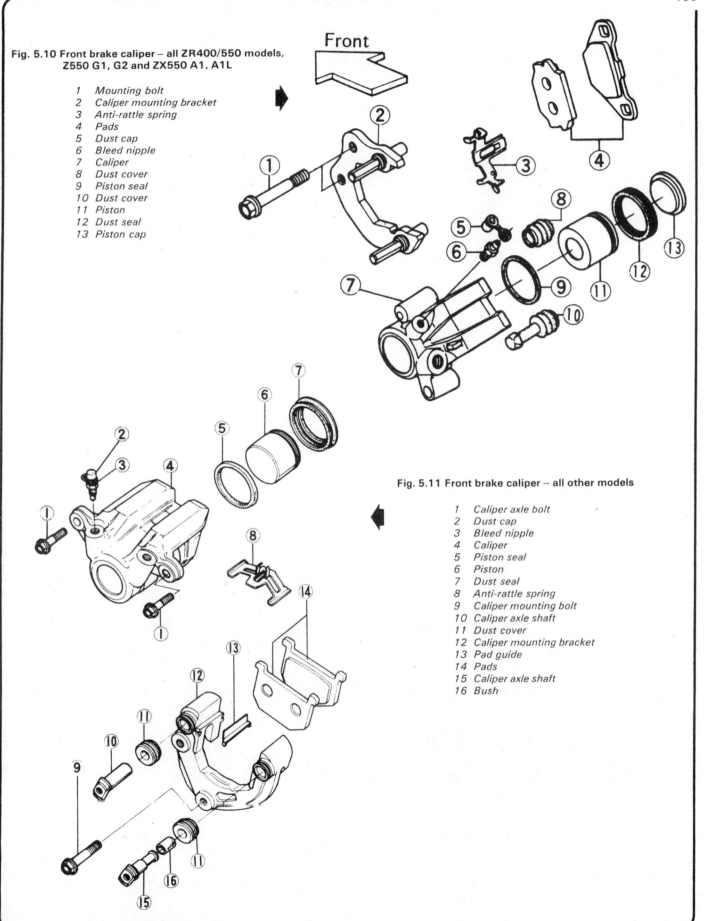

Fig. 5.10 Front brake caliper – all ZR400/550 models, Z550 G1, G2 and ZX550 A1, A1L

1 *Mounting bolt*
2 *Caliper mounting bracket*
3 *Anti-rattle spring*
4 *Pads*
5 *Dust cap*
6 *Bleed nipple*
7 *Caliper*
8 *Dust cover*
9 *Piston seal*
10 *Dust cover*
11 *Piston*
12 *Dust seal*
13 *Piston cap*

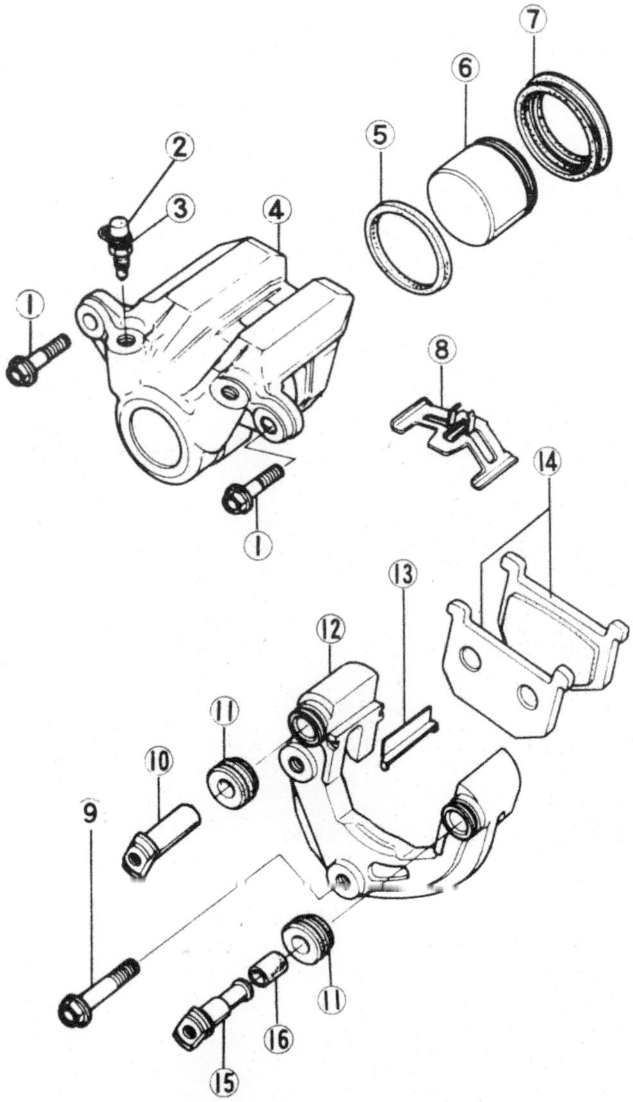

Fig. 5.11 Front brake caliper – all other models

1 *Caliper axle bolt*
2 *Dust cap*
3 *Bleed nipple*
4 *Caliper*
5 *Piston seal*
6 *Piston*
7 *Dust seal*
8 *Anti-rattle spring*
9 *Caliper mounting bolt*
10 *Caliper axle shaft*
11 *Dust cover*
12 *Caliper mounting bracket*
13 *Pad guide*
14 *Pads*
15 *Caliper axle shaft*
16 *Bush*

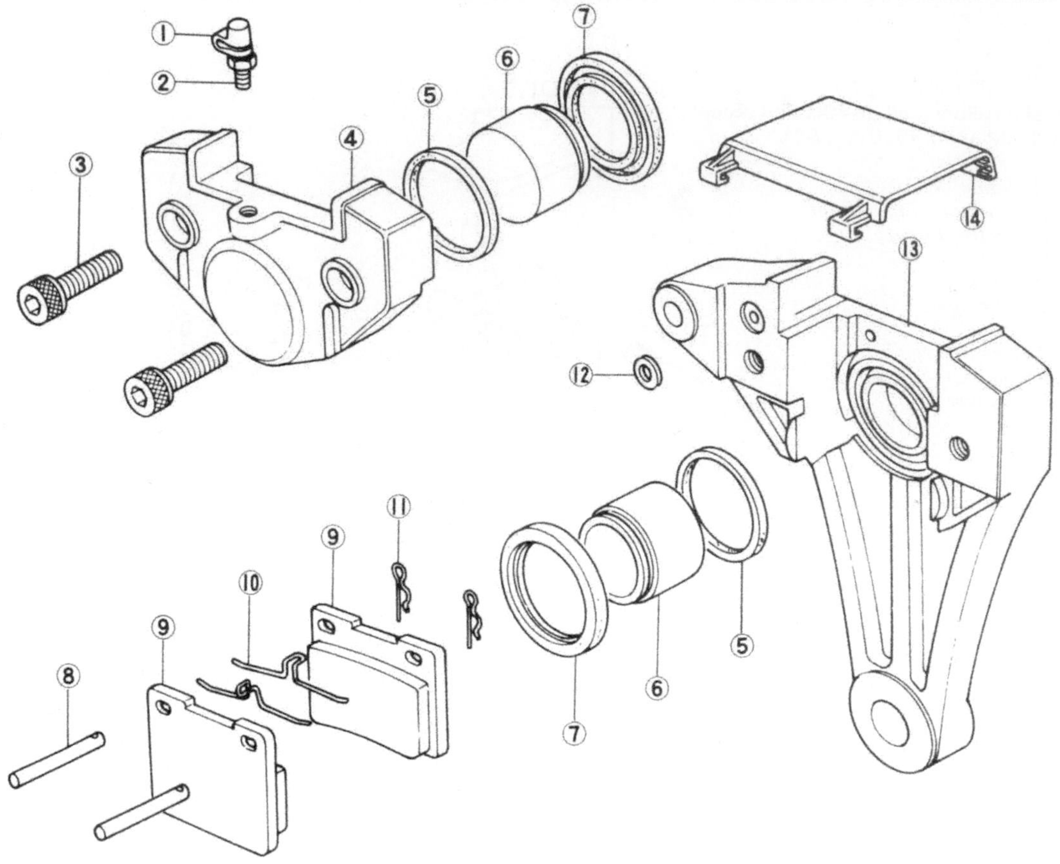

Fig. 5.12 Rear brake caliper – Z500 B1 and B2

1 *Dust cap*
2 *Bleed nipple*
3 *Allen bolt*
4 *Caliper half*
5 *Piston seal*
6 *Piston*
7 *Dust seal*
8 *Pin*
9 *Pad*
10 *Anti-rattle spring*
11 *R-clip*
12 *O-ring*
13 *Caliper half*
14 *Pad cover*

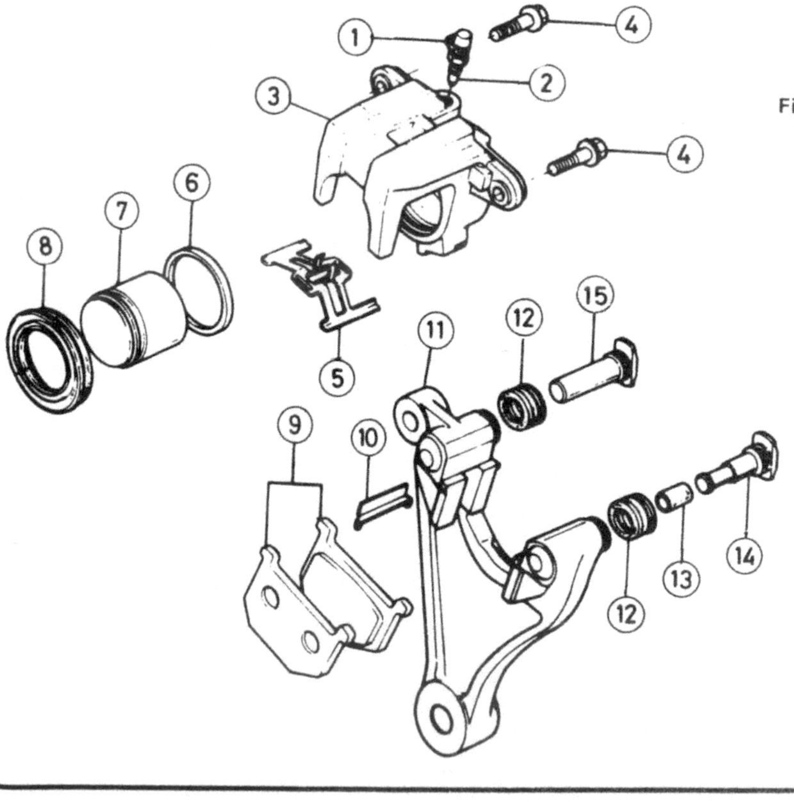

Fig. 5.13 Rear brake caliper – KZ/Z550 D1, H1 and H2

1 *Dust cap*
2 *Bleed nipple*
3 *Caliper*
4 *Axle bolt*
5 *Anti-rattle spring*
6 *Piston seal*
7 *Piston*
8 *Dust seal*
9 *Pads*
10 *Pad guide*
11 *Caliper mounting bracket*
12 *Dust cover*
13 *Bush*
14 *Axle shaft*
15 *Axle shaft*

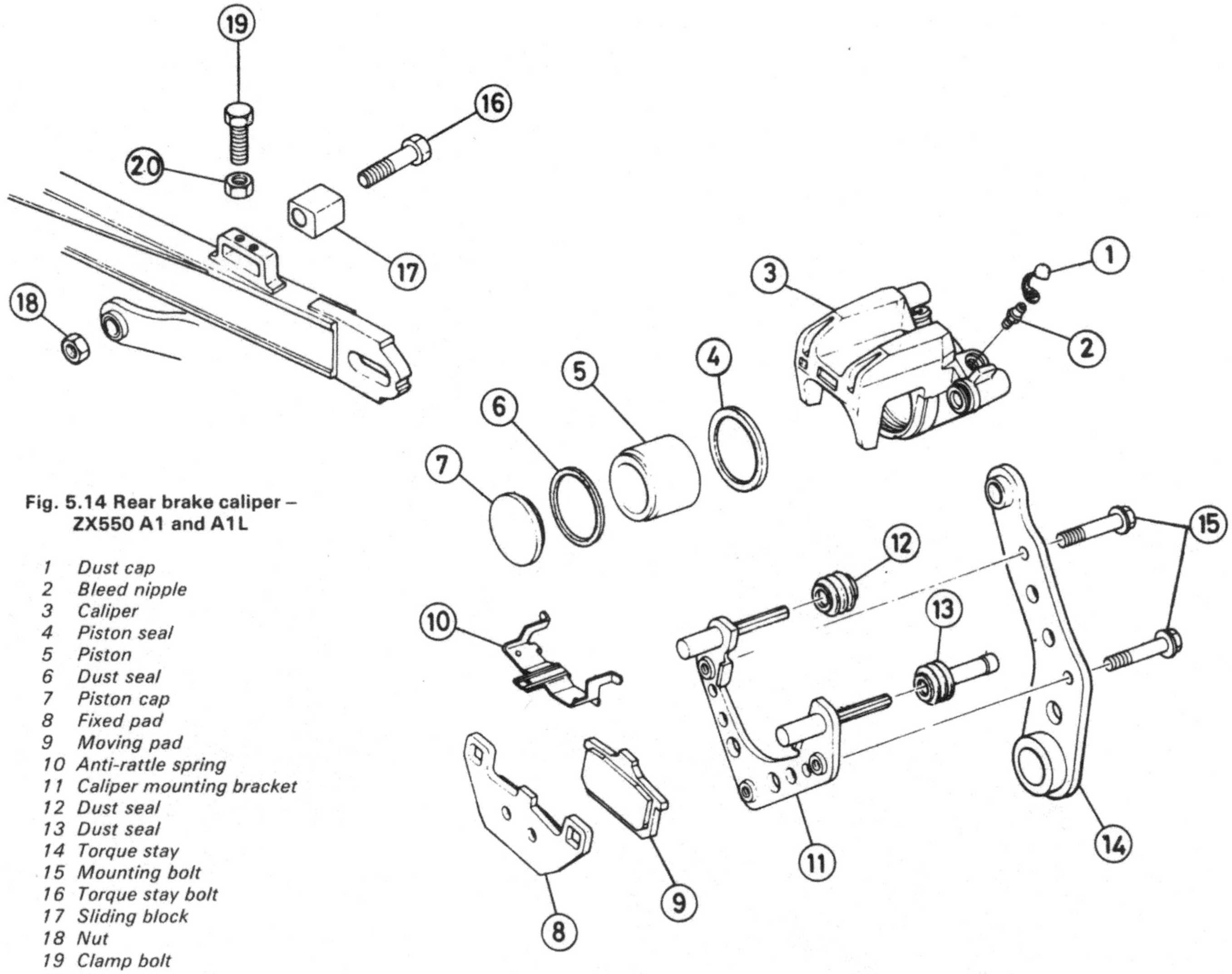

Fig. 5.14 Rear brake caliper – ZX550 A1 and A1L

1 Dust cap
2 Bleed nipple
3 Caliper
4 Piston seal
5 Piston
6 Dust seal
7 Piston cap
8 Fixed pad
9 Moving pad
10 Anti-rattle spring
11 Caliper mounting bracket
12 Dust seal
13 Dust seal
14 Torque stay
15 Mounting bolt
16 Torque stay bolt
17 Sliding block
18 Nut
19 Clamp bolt
20 Nut

12 Brake hoses and pipes: general

1 Drain the hydraulic system. Slacken the union bolts, noting the exact run of the faulty hose. Clean the unions, then refit the new hose using new sealing washers. Note in particular the notches in the three way union below the steering head, where fitted. These should locate the banjo unions when fitted properly. Tighten the union bolts to 3.0 kgf m (22 lbf ft) for ordinary hoses, or to 1.5 kgf m (11 lbf ft) for metal pipe gland nuts, if possible. Refill and bleed the system and check for leakage before using the machine.

2 Note that hoses deteriorate through age and must be renewed at a fixed interval for safety reasons. If any splits, kinks, leaks or any other damage, are found on a hose at any time, it must be renewed immediately.

13 Bleeding the hydraulic brake system

1 If the brake action becomes spongy, or if any part of the hydraulic system is dismantled (such as when a hose is replaced) it is necessary to bleed the system in order to remove all traces of air. The procedure for bleeding the hydraulic system is best carried out by two people.

2 Check the fluid level in the reservoir and top up with new fluid of the specified type if required. Keep the reservoir at least half full during the bleeding procedure; if the level is allowed to fall too far air will enter the system requiring that the procedure be started again from scratch. Refit the cap or cover securely to prevent the ingress of dust or the ejection of a spout of fluid.

3 Remove the dust cap from the caliper bleed nipple and clean the area with a rag. Place a clean glass jar below the caliper and connect a pipe from the bleed nipple to the jar. A clear plastic tube should be used so that air bubbles can be seen more easily. On twin disc systems start on one caliper, then repeat on the other. On ZX550 A1 and A1L models, start at the caliper and work up through the anti-dive plunger unit to the junction block nipple, then repeat on the other fork leg components. Place some clean hydraulic fluid in the glass jar so that the pipe is immersed below the fluid surface throughout the operation.

4 If parts of the system have been renewed, and thus the system must be filled, open the bleed nipple about one turn and pump the brake lever until fluid starts to issue from the clear tube. Tighten the bleed nipple and then continue the normal bleeding operation as described in the following paragraphs. Keep a close check on the reservoir level whilst the system is being filled.

5 Operate the brake lever as far as it will go and hold it in this position against the fluid pressure. If spongy brake operation has occurred, it may be necessary to pump rapidly the brake lever a number of times until pressure is achieved. With pressure applied, loosen the bleed nipple about half a turn. Tighten the nipple as soon as the lever has reached its full travel and then release the lever. Repeat this operation until no more air bubbles are expelled with the fluid into the glass jar. When this condition is reached, the air bleeding

operation should be complete, resulting in a firm feel to the brake operation. If sponginess is still evident, continue the bleeding operation; it may be that an air bubble trapped at the top of the system has yet to work down through the caliper. Repeat as necessary on other components in the system.
6 When all traces of air have been removed from the system, top up the reservoir and refit the diaphragm and cap or cover. Check the entire system for leaks, and check also that the brake system in general is functioning efficiently before using the machine on the road.
7 Brake fluid drained from the system will almost certainly be contaminated, either by foreign matter or by the absorption of water from the air. All hydraulic fluids are capable of drawing water from the atmosphere, and thereby degrading their specifications. In view of this, and the relative cheapness of the fluid, old fluid should always be discarded.
8 Great care should be taken not to spill hydraulic fluid on any painted cycle parts; it is a very effective paint stripper. Also, the plastic glasses in the instrument heads, and most other plastic parts, will be damaged by contact with this fluid.

14 Anti-dive system: testing and renewal – ZX550 A1, A1L

1 The anti-dive system is activated by hydraulic pressure whenever the brake is applied, pressure being transmitted via metal brake pipes from the junction block at the top of each fork leg to the anti-dive unit.
2 To test the system, place the machine on its centre stand, unbolt the junction block from each fork lower leg and remove the two Allen screws securing the plunger assembly to the top of each anti-dive valve unit, then withdraw the plunger assemblies, taking care not to distort the brake pipe.
3 Lightly apply the front brake with a finger over each plunger in turn. The plunger should move out by 2 mm when pressure is applied at the lever and should return easily under finger pressure when the lever is released.
4 If this is not the case, or if any signs of hydraulic fluid leakage are discovered, the plunger assembly must be renewed; no component parts are available and no repairs are possible. If it is sticking, it is possible to remove the plunger for cleaning once the large hexagon-headed top plug has been removed. Withdraw the brake pipe before the plug is disturbed and take great care not to damage the plunger sealing O-rings. On reassembly, tighten the top plug securely. The two Allen screws must not be overtightened; the recommended torque setting is 0.4 kgf m (3 lbf ft).
5 Note that the plunger assemblies must be renewed at fixed intervals for safety reasons alone, regardless of their apparent condition.
6 **Note**: Do not try to cure fluid leaks by overtightening the gland nuts of the metal brake pipes. The recommended torque setting is 1.5 kgf m (11 lbf ft).

13.3 Follow carefully specified procedure when bleeding brake system – ZX550 A1

14.4a Remove mounting screws to separate ...

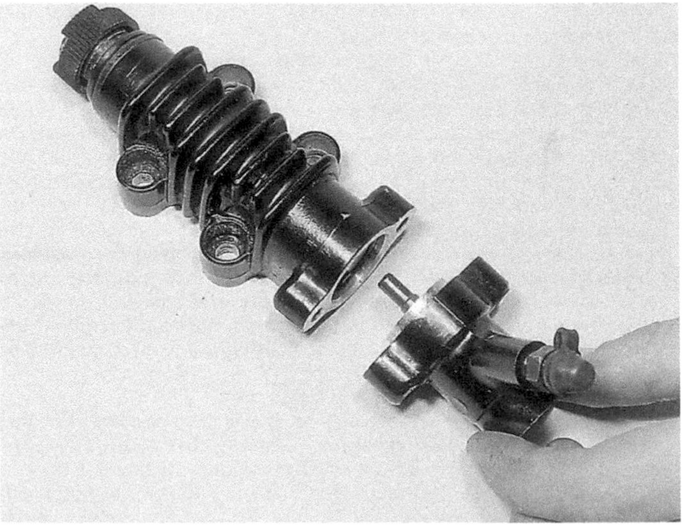

14.4b ... plunger assembly from anti-dive valve body

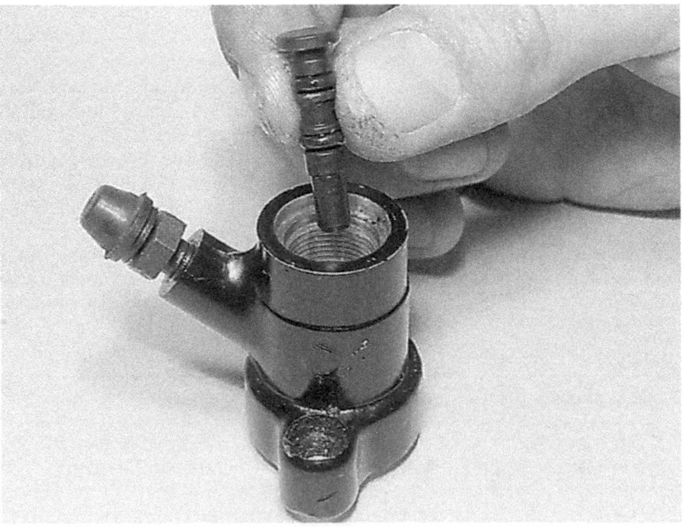

14.4c Plunger can be removed for cleaning – complete assembly must be renewed if wear or damage is found

Tyre changing sequence - tubed tyres

A Deflate tyre. After pushing tyre beads away from rim flanges push tyre bead into well of rim at point opposite valve. Insert tyre lever adjacent to valve and work bead over edge of rim.

B Use two levers to work bead over edge of rim. Note use of rim protectors

C Remove inner tube from tyre

D When first bead is clear, remove tyre as shown

E When fitting, partially inflate inner tube and insert in tyre

F Work first bead over rim and feed valve through hole in rim. Partially screw on retaining nut to hold valve in place.

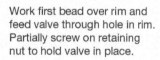

G Check that inner tube is positioned correctly and work second bead over rim using tyre levers. Start at a point opposite valve.

H Work final area of bead over rim whilst pushing valve inwards to ensure that inner tube is not trapped

15 Rear drum brake: examination and renovation

1 Remove the rear wheel as described in Section 4 of this Chapter, then withdraw the brake backplate.
2 Use a punch or felt marker to mark the original position of the brake camshaft and the brake operating arm. Remove the pinch bolt and arm. Remove the wear indicator pointer and dust seal, then the brake shoes by prying them up evenly and removing them followed by the brake camshaft.
3 Take off the brake shoe return springs. Inspect the brake drum for scoring, wear or ovality. If the drum is scored or warped slightly, it is possible to have it skimmed on a lathe by a specialist repairer but if the scoring is too deep or the warpage too great, a new rear wheel is necessary.
4 Inspect the brake shoes for excessive or uneven wear, or for oil or grease on the linings. If the friction material of either shoe is worn at any point to the service limit given, both shoes must be renewed. If they are fouled with oil or grease they must be renewed regardless of the amount of wear remaining.
5 Check that the return springs are not corroded, marked or damaged in any way, also that they have not stretched to the service limits given or beyond. If damage or excessive wear is found, both springs must be renewed.
6 Where measurements are given in the Specifications Section of this Chapter, check that the camshaft and backplate are not worn. On all other models, check that the camshaft is a reasonably tight fit with no excessive signs of sloppiness when installed. Renew any components found to be excessively worn or damaged.
7 On reassembly, apply a thin smear of high-melting point grease to the brake camshaft bearing surfaces and to the brake backplate pivot pins and camshaft passage. Refit the camshaft and dust seal, ensuring that they are correctly aligned, then assemble the shoes and return springs as a single unit before pressing them into place. If the original shoes are being reused, the wear indicator pointer and operating arm should be refitted in their original positions. If new shoes are to be fitted, install the pointer so that it aligns with the extreme right-hand (or forward) end of the 'Usable Range' arc; the position of the operating arm may require adjustment so that the angle between it and the brake rod is 80 – 90° with the brake firmly applied, as described in Routine Maintenance. Ensure that the arm is tapped firmly into place on the camshaft splines and that the pinch bolt is securely fastened.

16 Tyres: removal, repair and refitting – tubed tyres

1 To remove the tyre from either wheel, first detach the wheel from the machine. Deflate the tyre by removing the valve core, and when the tyre is fully deflated, push the bead away from the wheel rim on both sides so that the bead enters the centre well of the rim. Remove the locking ring and push the tyre valve into the tyre itself.
2 Insert a tyre lever close to the valve and lever the edge of the tyre over the outside of the rim. Very little force should be necessary; if resistance is encountered it is probably due to the fact that the tyre beads have not entered the well of the rim all the way round. Prevent damage to the soft alloy by tyre levers by the use of plastic rim protectors.
3 Once the tyre has been edged over the wheel rim, it is easy to work round the wheel rim so that the tyre is completely free from one side. At this stage the inner tube can be removed.
4 Now working from the other side of the wheel, ease the other edge of the tyre over the outside of the wheel rim that is furthest away. Continue to work around the rim until the tyre is completely free from the rim.
5 If a puncture has necessitated the removal of the tyre, reinflate the inner tube and immerse it in a bowl of water to trace the source of the leak. Mark the position of the leak, and deflate the tube. Dry the tube, and clean the area around the puncture with a petrol soaked rag. When the surface has dried, apply rubber solution and allow this to dry before removing the backing from the patch, and applying the patch to the surface.
6 It is best to use a patch of self vulcanizing type, which will form a permanent repair. Note that it may be necessary to remove a protective covering from the top surface of the patch after it has sealed into position. Inner tubes made from a special synthetic rubber may require a special type of patch and adhesive, if a satisfactory bond is to be achieved.
7 Before replacing the tyre, check the inside to make sure that the article that caused the puncture is not still trapped inside the tyre. Check the outside of the tyre, particularly the tread area, to make sure nothing is trapped that may cause a further puncture.
8 If the inner tube has been patched on a number of past occasions, or if there is a tear or large hole, it is preferable to discard it and fit a replacement. Sudden deflation may cause an accident.
9 To replace the tyre, inflate the inner tube for it just to assume a circular shape but only to that amount, and then push the tube into the tyre so that it is enclosed completely. Lay the tyre on the wheel at an angle, and insert the valve through the hole in the wheel rim. Attach the locking ring on the first few threads, sufficient to hold the valve captive in its correct location.
10 Starting at the point furthest from the valve, push the tyre bead over the edge of the wheel rim until it is located in the central well. Continue to work around the tyre in this fashion until the whole of one side of the tyre is on the rim. It may be necessary to use a tyre lever during the final stages.
11 Make sure there is no pull on the tyre valve and again commencing with the area furthest from the valve, ease the other bead of the tyre over the edge of the rim. Finish with the area close to the valve, pushing the valve up into the tyre until the locking ring touches the rim. This will ensure that the inner tube is not trapped when the last section of bead is edged over the rim with a tyre lever.
12 Check that the inner tube is not trapped at any point. Reinflate the inner tube, and check that the tyre is seating correctly around the wheel rim. There should be a thin rib moulded around the wall of the tyre on both sides, which should be an equal distance from the wheel rim at all points. If the tyre is unevenly located on the rim, try bouncing the wheel when the tyre is at the recommended pressure. It is probable that one of the beads has not pulled clear of the centre well.
13 Always run the tyres at the recommended pressures and never under or over inflate. The correct pressures are given in the Specifications Section of this Chapter.
14 Tyre replacement is aided by dusting the side walls, particularly in the vicinity of the beads, with a liberal coating of french chalk. Washing up liquid can also be used to good effect.
15 Never fit a tyre that has a damaged tread or sidewalls. Apart from legal aspects, there is a very great risk of a blowout, which can have very serious consequences on a two wheeled vehicle.
16 Tyre valves rarely give trouble, but it is always advisable to check whether the valve itself is leaking before removing the tyre. Do not forget to fit the dust cap, which forms an effective extra seal.

17 Valve cores and caps: tubed tyres

1 Valve cores seldom give trouble, but do not last indefinitely. Dirt under the seating will cause a puzzling 'slow-puncture'. Check that they are not leaking by applying spittle to the end of the valve and watching for air bubbles.
2 A valve cap is a safety device, and should always be fitted. Apart from keeping dirt out of the valve, it provides a second seal in case of valve failure, and may prevent an accident resulting from sudden deflation.

18 Tyres: removal and refitting – tubeless tyres

1 It is strongly recommended that should a repair to a tubeless tyre be necessary, the wheel is removed from the machine and taken to a tyre fitting specialist who is willing to do the job or taken to an official dealer. This is because the force required to break the seal between the wheel rim and tyre bead is considerable and beyond the capabilities of an individual working with normal tyre removing tools. Any abortive attempt to break the rim to bead seal may also cause damage to the wheel rim, resulting in an expensive wheel replacement. If, however, a suitable bead releasing tool is available, and experience has already been gained in its use, tyre removal and refitting can be accomplished as follows.
2 Remove the wheel from the machine by following the instructions for wheel removal as described in the relevant Section of this Chapter. Deflate the tyre by removing the valve insert and when it is fully deflated, push the bead of the tyre away from the wheel rim on both

TYRE CHANGING SEQUENCE - TUBELESS TYRES

Deflate tyre. After releasing beads, push tyre bead into well of rim at point opposite valve. Insert lever next to valve and work bead over edge of rim.

Use two levers to work bead over edge of rim. Note use of rim protectors.

When first bead is clear, remove tyre as shown.

Before installing, ensure that tyre is suitable for wheel. Take note of any sidewall markings such as direction of rotation arrows.

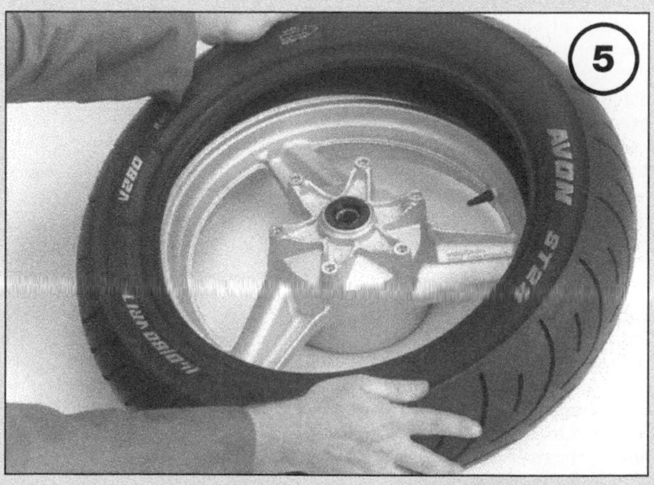

Work first bead over the rim flange.

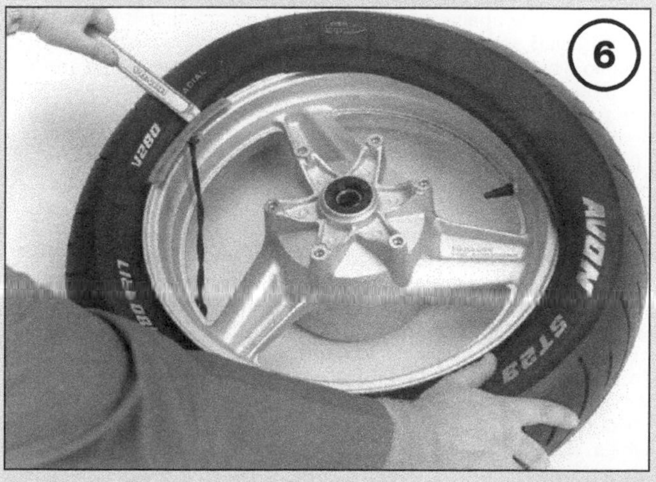

Use a tyre lever to work the second bead over rim flange.

sides so that the bead enters the centre well of the rim. As noted, this operation will almost certainly require the use of a bead releasing tool.
3 Insert a tyre lever close to the valve and lever the edge of the tyre over the outside of the wheel rim. Very little force should be necessary; if resistance is encountered it is probably due to the fact that the tyre beads have not entered the well of the wheel rim all the way round the tyre. Should the initial problem persist, lubrication of the tyre bead and the inside edge and lip of the rim will facilitate removal. Use a recommended lubricant, a diluted solution of washing-up liquid, or french chalk. Lubrication is usually recommended as an aid to tyre fitting but its use is equally desirable during removal. The risk of lever damage to wheel rims can be minimised by the use of proprietary plastic rim protectors placed over the rim flange at the point where the tyre levers are inserted. Suitable rim projectors may be fabricated very easily from short lengths (4-6 inches) of thick-walled nylon petrol pipe which have been split down one side using a sharp knife. The use of rim protectors should be adopted whenever levers are used and, therefore, when the risk of damage is likely.
4 Once the tyre has been edged over the wheel rim, it is easy to work around the wheel rim so that the tyre is completely free on one side.
5 Working from the other side of the wheel, ease the other edge of the tyre over the outside of the wheel rim, which is furthest away. Continue to work around the rim until the tyre is freed completely from the rim.
6 Refer to the following Section for details relating to puncture repair and the renewal of tyres. See also the remarks relating to the tyre valves in Section 20.
7 Refitting of the tyre is virtually a reversal of removal procedure. If the tyre has a balance mark (usually a spot of coloured paint), as on the tyres fitted as original equipment, this must be positioned alongside the valve. Similarly, any arrow indicating direction of rotation must face the right way.
8 Starting at the point furthest from the valve, push the tyre bead over the edge of the wheel rim until it is located in the central well. Continue to work around the tyre in this fashion until the whole of one side of the tyre is on the rim. It may be necessary to use a tyre lever during the final stages. Here again, the use of a lubricant will aid fitting. It is recommended strongly that when refitting the tyre only a recommended lubricant is used because such lubricants also have sealing properties. Do not be over generous in the application of lubricant or tyre creep may occur.
9 Fitting the upper bead is similar to fitting the lower bead. Start by pushing the bead over the rim and into the well at a point diametrically opposite the tyre valve. Continue working round the tyre, each side of the starting point, ensuring that the bead opposite the working area is always in the well. Apply lubricant as necessary. Avoid using tyre levers unless absolutely essential, to help reduce damage to the soft wheel rim. The use of the levers should be required only when the final portion of bead is to be pushed over the rim.
10 Lubricate the tyre beads again prior to inflating the tyre, and check that the wheel rim is evenly positioned in relation to the tyre beads. Inflation of the tyre may well prove impossible without the use of a high pressure air hose. The tyre will retain air completely only when the beads are firmly against the rim edges at all points and it may be found when using a foot pump that air escapes at the same rate as it is pumped in. This problem may also be encountered when using an air hose on new tyres which have been compressed in storage and by virtue of their profile hold the beads away from the rim edges. To overcome this difficulty, a tourniquet may be placed around the circumference of the tyre, over the central area of the tread. The compression of the tread in this area will cause the beads to be pushed outwards in the desired direction. The type of tourniquet most widely used consists of a length of hose closed at both ends with a suitable clamp fitted to enable both ends to be connected. An ordinary tyre valve is fitted at one end of the tube so that after the hose has been secured around the tyre it may be inflated, giving a constricting effect. Another possible method of seating beads to obtain initial inflation is to press the tyre into the angle between a wall and the floor. With the airline attached to the valve additional pressure is then applied to the tyre by the hand and shin, as shown in the accompanying illustration. The application of pressure at four points around the tyre's circumference whilst simultaneously applying the airhose will often effect an initial seal between the tyre beads and wheel rim, thus allowing inflation to occur.
11 Having successfully accomplished inflation, increase the pressure to 40 psi and check that the tyre is evenly disposed on the wheel rim. This may be judged by checking that the thin positioning line found on each tyre wall is equidistant from the rim around the total circumference of the tyre. If this is not the case, deflate the tyre, apply additional lubrication and reinflate. Minor adjustments to the tyre position may be made by bouncing the wheel on the ground.
12 Always run the tyre at the recommended pressures and never under or over-inflate. The correct pressures for various weights and configurations are given in the Specifications Section of this Chapter.

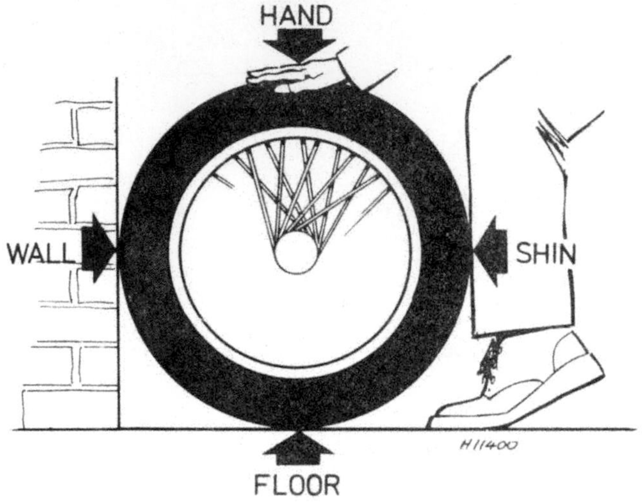

Fig. 5.15 Method of seating the beads on tubeless tyres

19 Puncture repair and tyre renewal: tubeless tyres

1 The primary advantage of the tubeless tyre is its ability to accept penetration by sharp objects such as nails etc without loss of air. Even if loss of air is experienced, because there is no inner tube to rupture, in normal conditions a sudden blow-out is avoided.
2 If a puncture of the tyre occurs, the tyre should be removed for inspection for damage before any attempt is made at remedial action. The temporary repair of a punctured tyre by inserting a plug from the outside should not be attempted. Although this type of temporary repair is used widely on cars, the manufacturers strongly recommend that no such repair is carried out on a motorcycle tyre. Not only does the tyre have a thinner carcass, which does not give sufficient support to the plug, the consequences of a sudden deflation are often sufficiently serious that the risk of such an occurrence should be avoided at all costs.
3 The tyre should be inspected both inside and out for damage to the carcass. Unfortunately the inner lining of the tyre – which takes the place of the inner tube – may easily obscure any damage and some experience is required in making a correct assessment of the tyre condition.
4 There are two main types of tyre repair which are considered safe for adoption in repairing tubeless motorcycle tyres. The first type of repair consists of inserting a mushroom-headed plug into the hole from the inside of the tyre. The hole is prepared for insertion of the plug by reaming and the applications of an adhesive. The second repair is carried out by buffing the inner lining in the damaged area and applying a cold or vulcanised patch. Because both inspection and repair, if they are to be carried out safely, require experience in this type of work, it is recommended that the tyre be placed in the hands of a repairer with the necessary skills, rather than repaired in the home workshop.
5 In the event of an emergency, the only recommended 'get-you-home' repair is to fit a standard inner tube of the correct size. If this course of action is adopted, care should be taken to ensure that the cause of the puncture has been removed before the inner tube is fitted. It will be found that the valve hole in the rim is considerably larger

than the diameter of the inner tube valve stem. To prevent the ingress of road dirt, and to help support the valve, a spacer should be fitted over the valve.

6 In the event of the unavailability of tubeless tyres, ordinary tubed tyres fitted with inner tubes of the correct size may be fitted. Refer to the manufacturer or a tyre fitting specialist to ensure that only a tyre and tube of equivalent type and suitability is fitted, and also to advise on the fitting of a valve nut to the rim hole.

20 Tyre valves: description and renewal – tubeless tyres

1 It will be appreciated from the preceding Sections that the adoption of tubeless tyres has made it necessary to modify the valve arrangement, as there is no longer an inner tube which can carry the valve core. The problem has been overcome by fitting a separate tyre valve which passes through a close-fitting hole in the rim, and which is secured by a nut and locknut. The valve is fitted from the rim well, and it follows that the valve can be removed and replaced only when the tyre has been removed from the rim. Leakage of air from around the valve body is likely to occur only if the sealing seat fails or if the nut and locknut become loose. Check both are fastened to 0.15 kgf m (1 lbf ft).

2 The valve core is of the same type as that used with tubed tyres, and screws into the valve body. The core can be removed with a small slotted tool which is normally incorporated in plunger type pressure gauges. Some valve dust caps incorporate a projection for removing valve cores. Although tubeless tyre valves seldom give trouble, it is possible for a leak to develop if a small particle of grit lodges on the sealing face. Occasionally, an elusive slow puncture can be traced to a leaking valve core, and this should be checked before a genuine puncture is suspected.

3 The valve dust caps are a significant part of the tyre valve assembly. Not only do they prevent the ingress of road dirt in the valve, but also act as a secondary seal which will reduce the risk of sudden deflation if a valve core should fail.

21 Wheel balancing

1 It is customary on all high performance machines to balance the wheels complete with tyre and tube. The out of balance forces which exist are eliminated and the handling of the machine is improved in consequence. A wheel which is badly out of balance produces through the steering a most unpleasant hammering effect at high speeds.

2 Some tyres have a balance mark on the sidewall, usually in the form of a coloured spot. This mark must be in line with the tyre valve, next to the valve or opposite it, depending on the tyre manufacture. Even then the wheel may require the addition of balance weights, to offset the weight of the tyre valve itself.

3 If the wheel is raised clear of the ground and is spun, it will probably come to rest with the tyre valve or the heaviest part downward and will always come to rest in the same position. Balance weights must be added to a point diametrically opposite this heavy spot until the wheel will come to rest in ANY position after it is spun.

4 Balance weights are available from Kawasaki dealers in 10, 20 and 30 gram sizes. These are recommended since they are designed to fit Kawasaki rims and will not cause deflation problems when used with tubeless tyres.

5 To fit the weights, clip the hooked end over the bead and tap the weight home, deflating the tyre slightly to allow this, if necessary.

6 While the rear wheel is much more tolerant of out-of-balance forces, it should be balanced also whenever a new tyre is fitted or whenever any serious vibration problems are encountered. Remove the wheel from the machine and place it on a stand to test it; the drag of the chain or shaft would make any check ineffective.

Chapter 6 Electrical system

For information relating to the 1984 on models, see Chapter 7

Contents

Specifications

Electrical system

Voltage	12
Earth (ground)	Negative
Alternator type:	
Z400 J1, J2, Z500 B1, B2, KZ/Z550 A1, C1	Single-phase AC generator
All other models	3-phase AC generator
Alternator output:	
Z400 J1, J2, Z500 B1, B2, KZ/Z550 A1, C1	210W @ 10 000 rpm
All other models	259W @ 10 000 rpm

Battery

Make	Furukawa (FB)
Type:	
ZR400 A1, B1	FB12A-AK
All other models	FB12A-A
Capacity:	
ZR400 A1, B1	10 Ah
All other models	12 Ah

Alternator

Rated output:	
Z400 J1, J2, Z500 B1, B2, KZ/Z550 A1, C1	15.0A @ 10 000 rpm, 14 volts
All other models	18.5A @ 10 000 rpm, 14 volts
No-load voltage – approximate:	
Z400 J1, J2, Z500 B1, B2, KZ/Z550 A1, C1	75 volts @ 4000 rpm
All other models	50 volts @ 4000 rpm
Stator coil resistance – cold:	
Z400 J1, J2, Z500 B1, B2, KZ/Z550 A1, C1	0.32 – 0.48 ohm
All other models	0.36 – 0.54 ohm
Charging output – across battery terminals:	
Z400 J1, J2, Z500 B1, B2, KZ/Z550 A1, C1	0.5 – 1.5A, 14 – 15.5 volts @ 4000 rpm
All other models	4.0A, 14 – 15.5 volts @ 4000 rpm

Bulbs	UK models	US models
Headlamp Z400 J1, Z500 B1, B2	12V,45/40W	N/App
Headlamp – KZ/Z550 A1, C1	12V,45/40W	12V,50/35W
Headlamp – all other models	12V,60/55W	12V,60/55W
Stop/tail lamp	12V,21/5W	12V,27/8W
Parking lamp	12V,4W	N/App
Turn signal lamps – front all UK models and KZ550 A1, C1 – rear all models	12V, 21W	12V,23W
Front turn signal/running lamps – all US models except KZ550 A1, C1	N/App	12V,23/8W
Instrument illuminating and warning lamps – except ZX550 A1, A1L	12V,3.4W	12V,3.4W
Instrument illuminating and warning lamps – ZX550 A1, A1L	12V,3W	12V,3W
Fuel tank console illuminating lamps – ZX550 A1, A1L	12V,1.4W	12V,1.4W

Fuse rating	
Main	20A
Head	10A
Tail	10A
Accessory wires:	
Z550 G1, G2, all US models except KZ550 A1, C1, ZX550 A1, A1L	10A x 2
ZX550 A1 (US), A1L	10A x 1

Torque wrench settings

Component	kgf m	lbf ft
Alternator rotor bolt	7.0	50.5
Alternator stator Allen screws	1.0	7.0
Neutral switch	1.5	11.0
Oil pressure switch – where fitted	1.5	11.0
Starter motor terminal nut	1.1	8.0
Turn signal assembly mounting nuts:		
Standard	1.3	9.5
Maximum permissible	1.5	11.0

1 General description

The electrical system is based on a single- or three-phase alternator mounted on the left-hand end of the crankshaft, the output from which is rectified and controlled by an electronic regulator/rectifier unit before being passed to the battery and main electrical circuit.

Before starting work on any part of the electrical system, refer to the relevant wiring diagram at the back of this Manual for details of all components in any particular circuit, then refer to the relevant Section of this Chapter for information on testing.

2 Electrical system: general information and preliminary checks

1 In the event of an electrical system fault, always check the physical condition of the wiring and connectors before attempting any of the test procedures described here and in subsequent Sections. Look for chafed, trapped or broken electrical leads and repair or renew these as necessary. Leads which have broken internally are not easily spotted, but may be checked using a multimeter or a simple battery and bulb circuit as a continuity tester. This arrangement is shown in the accompanying illustration. The various multi-pin connectors are generally trouble-free but may corrode if exposed to water. Clean them carefully, scraping off any surface deposits, and pack with silicone grease during assembly to avoid recurrent problems. The same technique can be applied to the handlebar switches.

2 A sound, fully charged battery is essential to the normal operation of the system. There is no point in attempting to locate a fault if the battery is partly discharged or worn out. Check battery condition and recharge or renew the battery before proceeding further.

3 Many of the test procedures described in this Chapter require that voltages or resistances be checked. This requires the use of some form of test equipment such as a simple and inexpensive multimeter of the type sold by electronics or motor accessory shops.

4 If you doubt your ability to check safely the electrical system entrust the work to a Kawasaki dealer. In any event have your findings double checked before consigning expensive components to the scrap bin.

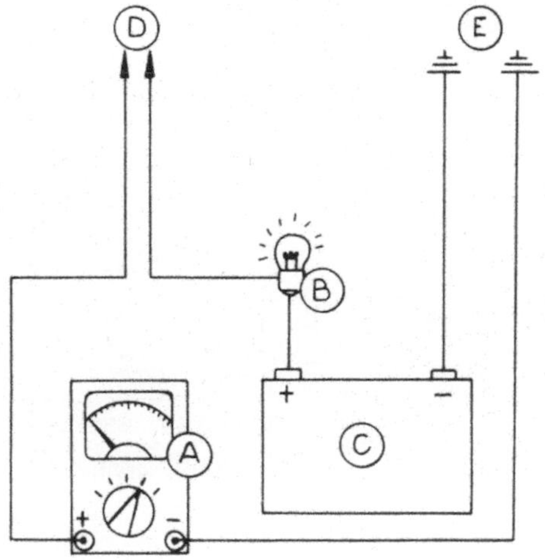

Fig. 6.1 Simple testing arrangement for checking the wiring

A Multimeter
B Bulb
C Battery
D Positive probe
E Negative probe

3 Battery: examination and maintenance

1 To check the battery thoroughly it is best to disconnect it and remove it from the machine; always disconnect the negative (-) terminal first to prevent the risk of short circuits damaging any component, particularly the IC ignitor unit, where applicable. It is housed in a recess below the dualseat, except on KZ/Z550 H1, H2 and all ZR400/550 models, where it is behind the right-hand sidepanel and is retained by a single strap. Batteries can be dangerous if mishandled. See 'Safety First' and note the precautions described for handling

them. Wear overalls or old clothing in case of accidental acid spillage. Clean the outside of the battery carefully, and remove any deposits from the terminals, which should be coated with petroleum jelly prior to installation. Connect the negative (-) terminal last.

2 When new, the battery is filled with an electrolyte of dilute sulphuric acid having a specific gravity of 1.280 at 20°C (68°F). Subsequent evaporation, which occurs in normal use, can be compensated for by topping up with distilled or demineralised water only. Never use tap water as a substitute and do not add fresh electrolyte unless spillage has occurred.

3 The state of charge of a battery can be checked using a hydrometer.

4 The normal charge rate for a battery is 1/10 of its rated capacity, thus for a 12 ampere hour unit charging should take place at 1.2 amp. Exceeding this figure could cause the battery to overheat, buckling the plates and rendering it useless. Few owners will have access to an expensive current controlled charger, so if a normal domestic charger is used check that after a possible initial peak, the charge rate falls to a safe level. If the battery becomes hot during charging **stop**. Further charging will cause damage. Note that cell caps should be loosened and vents unobstructed during charging to avoid a build-up of pressure and risk of explosion.

5 After charging, top up with distilled water as required, then check the specific gravity and battery voltage. Specific gravity should be above 1.250 and a sound, fully charged battery should produce 15 – 16 volts. If the recharged battery discharges rapidly if left disconnected it is likely that an internal short caused by physical damage or sulphation has occurred. A new battery will be required. A sound item will tend to lose its charge at about 1% per day.

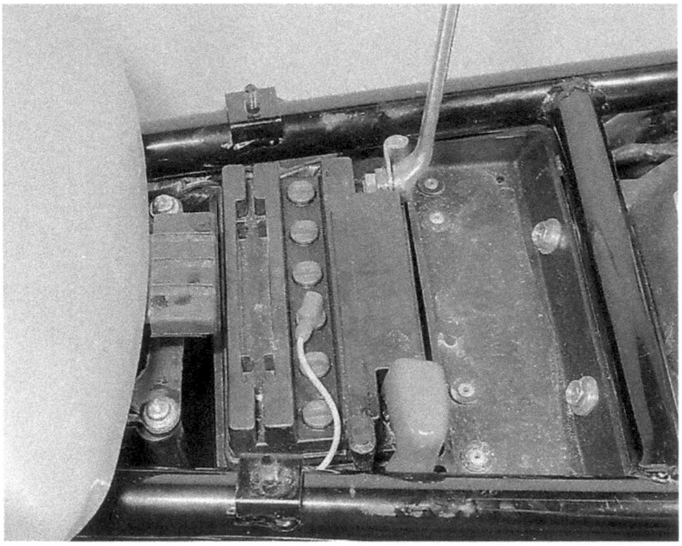

3.1 Always disconnect negative terminal first when removing battery

4 Checking the charging system

1 Before attempting to test the output of the charging system, check the wiring connections and battery condition (Sections 2 and 3). Open or remove the dualseat to gain access to the battery terminals. Set the multimeter on the 0–20 volts dc scale, and attach the negative (-) probe to the negative terminal and the positive (+) probe to the positive terminal. Start the engine, and note the meter reading at about 4000 rpm. If all is well a reading of about 14.5 volts should be indicated.

2 A reading significantly in excess of 14.5 volts indicates a possible defective regulator/rectifier unit or loose or broken wiring connections. Check these and repeat the test to find out if the problem has been resolved. At idle, battery voltage will be shown on the meter. If this does not increase as the engine speed rises the alternator or regulator/rectifier may be at fault or disconnected.

5 Checking the alternator

1 Trace and disconnect the three yellow alternator output leads after the engine has been warmed up to normal operating temperature. Set the meter to the 250 volts ac scale and connect the probes to any two of the output leads. Start the engine and measure the voltage at about 4000 rpm. Note the reading, then repeat the test until all combinations of leads have been checked (three tests in all). A reading of about 50 (or 75, as appropriate) volts should be obtained in each case, in which case the fault must lie with the regulator/rectifier unit.

2 A reading significantly lower than that shown above indicates a fault in the alternator itself, and the alternator winding resistances should be measured to discover the nature of the fault. With the engine off, measure the resistance between each pair of leads, making three tests as described above. The multimeter should be set on the ohm x 1 scale. A sound winding will give a reading similar to that given in the Specifications Section. If infinite resistance is shown, the windings are open (broken), whilst a much lower reading or zero resistance indicates a short. In both instances the alternator stator must be renewed.

3 Set the meter on its highest resistance range, normally ohm x 1000 or kilo ohms, and check for insulation between each alternator lead and earth (ground). Anything less than infinity is indicative of a short between the stator core and its windings, again requiring renewal.

6 Checking the regulator/rectifier unit

1 Check that the ignition switch is off and remove the seat, on KZ/Z550 H1, H2 and all ZR400/550 models, or the left-hand side panel on all other models. Release the electrical panel cover, where fitted, to gain access to the regulator/rectifier. Trace and disconnect the red/white lead, where separate, and the multi-pin connector from the unit.

2 Set the multimeter on the ohm x 10 or ohm x 100 scale, and measure the resistance between the red/white lead and each of the three yellow leads. Note the reading, then reverse the meter probes and repeat. Once these six (four, on models with single-phase alternators) tests have been completed, repeat the sequence using the black lead in the connector in place of the red/white lead, another six or four tests. In each test, a very high resistance should be shown in one direction, with a very low reading if the meter probes are reversed. The actual resistance figures are not important, but a large difference between the two readings indicates that the particular diode is functioning normally. If with any pair of leads a similar reading is shown in both directions, a diode has failed and the unit must be renewed.

3 To check the regulator unit, three 12 volt car or motorcycle batteries and a 12 volt bulb rated at 3-6W will be required. Using the accompanying illustration for reference, connect the topmost battery with the test lamp as shown, that is, with the positive (+) battery terminal to one of the yellow leads and the negative (-) terminal to the black lead via the test bulb. The bulb should remain off at this stage.

4 Connect the remaining two batteries in series as shown to produce a 24 volt source, connecting the positive (+) terminal to the brown lead. Read the notes below before connecting the negative (-) lead. **Important note**: Do not use a meter or bulb of a different wattage in place of the test lamp specified. It acts both as an indicator and as a current limiter. On no account apply more than 24 volts to the unit and do not apply even this voltage for more than a few seconds. The unit may be destroyed if these precautions are not observed.

5 Touch the negative (-) lead from the 24 volt source **briefly** against the black terminal of the connector. If the regulator stage is functioning normally, the test lamp should light. Repeat the test with the 12 volt positive (+) lead connected to each of the remaining yellow leads in turn. It should be noted that whilst the above tests will usually reveal a regulator fault, the sequence is not infallible. If the tests indicate a sound unit, no other charging system faults can be found but the problem persists, it will be necessary to check the unit by substitution.

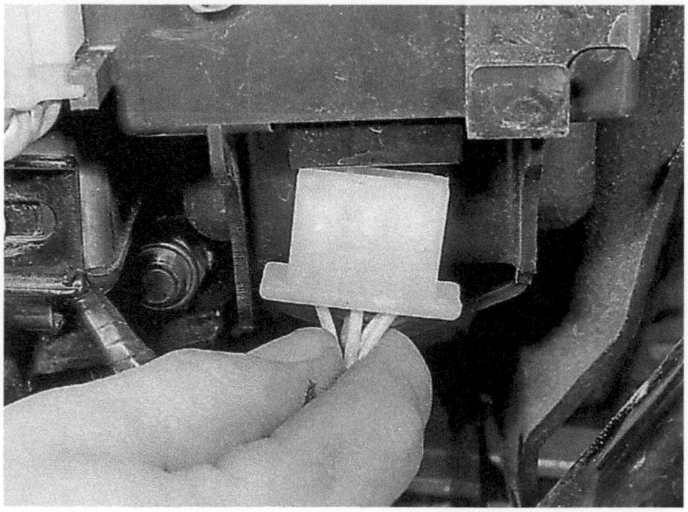

6.1 Location of regulator/rectifier unit – ZX550 A1

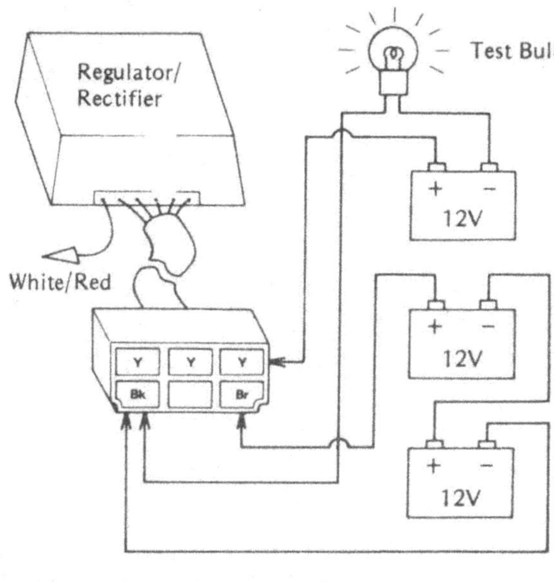

Fig. 6.2 Regulator testing circuit (3-phase alternator shown)

7 Fuses: location and renewal

1 The electrical system is protected by three fuses, with one or two fuses being fitted to protect the accessory circuit. Refer to the wiring diagrams at the end of the Chapter for details of specific arrangements. The fuses are housed in a plastic holder behind the left-hand side panel, and electrical panel cover where fitted, except on KZ/Z550 H1, H2 and all ZR400/550 models where they are under the seat. The accessory fuse(s) are in a separate holder mounted next to the main fuse holder or under the seat except on all US shaft drive models where they are fitted in the main fuse holder.

2 The function of a fuse is to introduce an intentional weak link in a circuit so that it will 'blow' before the expensive electrical components are damaged. Fuses do age and may occasionally fail through this or vibration, but the system must be checked for possible faults if the replacement fails soon afterwards. Avoid using fuses of the wrong rating unless circumstances make this unavoidable. Remember that using the wrong fuse or wrapping the blown fuse in metal foil will **not** protect the electrical system. Fit the correct fuse as soon as possible, and replace any spare fuse used at the same time.

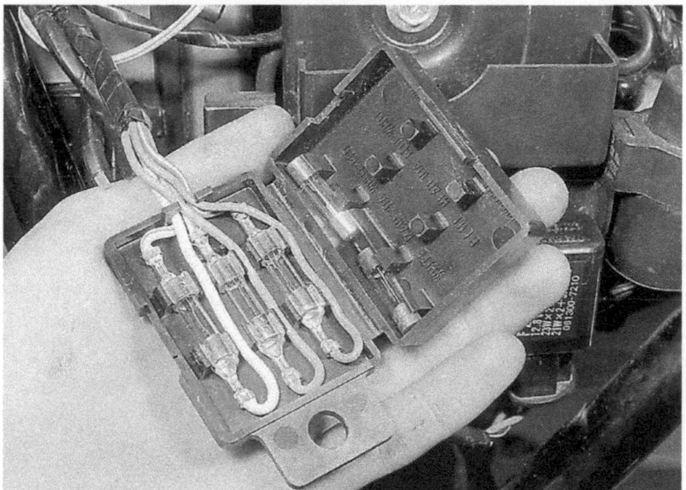

7.1 Fuse rating and circuit protected is marked clearly on fuse holder

8 Switches: general

1 While the switches should give little trouble, they can be tested using a multimeter set to the resistance function or a battery and bulb test circuit. Using the information given in the wiring diagrams at the end of this Manual, check that full continuity exists in all switch positions and between the relevant pairs of wires. When checking a particular circuit follow a logical sequence to eliminate the switch concerned.

2 As a simple precaution always disconnect the battery before removing any of the switches, to prevent the possibility of a short circuit. Most troubles are caused by dirty contacts, which can be cleaned, but in the event of the breakage of some internal part, it will be necessary to renew the complete switch.

3 Note that handlebar switches are secured by clamping screws which may not be the same length; ensure each screw is refitted correctly on reassembly, and note that in many cases the switch is located by a lug projecting into a hole drilled in the handlebar.

9 Starter circuit tests

1 The starter system is of robust construction and will rarely malfunction. In the event of a fault, always check that the battery is in good condition, noting that a failing battery may operate the general electrical system adequately whilst being unable to produce the heavy starting current. Look also for broken, chafed or corroded wiring before proceeding further.

Starter solenoid (relay)

2 Remove the left-hand side panel and electrical panel cover to gain access to the relay, then detach the heavy starter motor lead from the relay terminal. Set the multimeter to the ohm x 1 scale and connect one lead to the starter motor lead terminal and the other to earth (ground). Switch on the ignition, pull in the clutch lever and press the starter button. An audible click from the relay should be accompanied by a zero ohm reading if the unit is sound. **Note:** On machines equipped with a side-stand interlock switch, make sure that the stand is retracted during the check.

3 If the relay clicks normally but the meter still indicates that the contacts are open, the unit must be considered defective and renewed. If there is no sign of activity from the relay, check that the starter switch circuit is operating correctly.

Starter switch circuit

4 Disconnect the black lead and the yellow/red lead from the relay. Set the multimeter on the 0-20 volts dc scale, and connect the

negative (-) probe to the yellow/red lead and the positive (+) probe to the black lead. Switch on the ignition, pull in the clutch lever and press the starter button. If battery voltage is shown, but the relay would not work when tested as described above, it should be considered defective and renewed. It is a sealed unit and cannot be repaired. If battery voltage is not shown, check the wiring, connections and interlock switches to locate the fault.

Starter button test

5 Remove the fuel tank (Chapter 2) and disconnect the 4-pin connector and the separate black lead from the right-hand switch cluster. Set the meter on the ohm x 1 scale and connect the probes to the leads. Press the starter button and check that zero resistance is indicated. If not, the starter button contacts are faulty and should be cleaned by spraying aerosol contact cleaner into the switch housing. If this fails, renew the switch cluster.

Interlock switch test

6 Remove the fuel tank (Chapter 2) and disconnect the leads from the clutch interlock switch. Set the multimeter to the ohm x 1 scale and test for continuity when the clutch lever is pulled in. If a reading of zero ohms is not obtained, the switch is defective and should be renewed. The switch can be freed by depressing its locating pin with a small screwdriver and pulling it out of the clutch lever housing. In an emergency, bypass the switch by joining the leads together. Renew the switch as soon as possible. Similarly the side stand switch, where fitted, should have continuity when the stand is raised and can be bypassed in an emergency. The neutral indicator switch should have continuity between its terminal and earth when neutral gear is selected. Note that an override switch is fitted on some models to prevent the engine from cutting out when the side stand is lowered.

Starter control (circuit) relay

7 Disconnect the multi-pin block connector and connect a fully-charged 12 volt battery and an ohmmeter (set to the x 1 ohm range) to the relay as shown in the accompanying illustration. With the battery connected, a reading of zero ohms should be indicated across the switch terminals, but with it disconnected a reading of infinite resistance should be obtained.

Diode pack

8 Where fitted, the diode pack is employed to prevent feedback between the various circuits and may affect the starter motor circuit, if faulty. Refer to Chapter 3 for details.

10 Starter motor: removal, overhaul and refitting

1 In the event of a starter fault, check first that the battery is fully charged and that the solenoid is operating correctly (Section 9). Remove the engine sprocket cover or front gear case cover, as appropriate. Remove the two starter motor mounting bolts and pull the motor clear of the crankcase. Once it is partly clear, turn the motor body to gain access to the terminal. Slide back the rubber boot and remove the nut to free the starter cable. Clean the mounting lugs to ensure a good earth and renew the O-ring if worn or damaged.
2 Place the motor on a clean workbench and dismantle it, following the photographic sequence which accompanies this Section. Lay out each part in sequence as a guide during reassembly. Clean the motor components using a non-greasy high flash-point solvent. It will be noted that the motor shown in the photographs is of the later, four brush, type. Earlier models made use of a two brush motor of similar construction, and this may be dismantled as follows.
3 Remove the two long retaining screws and lock washers and remove the starter motor end covers. Remove the toothed washer, thrust washers and armature from the right-hand end of the motor. Moving to the left-hand end, release the screw which retains the field coil lead to the brush plate and remove the plate and brushes. **Do not** attempt to remove the field coil windings from the motor body.

Two brush motors

4 Pull back and displace the brush springs to allow the carbon brushes to be measured. Renew them if they are 6 mm (0.236 in) or less in length. The brush springs should exert a pressure of 560 – 580 grams. In practice, the brush springs can be considered serviceable if they press firmly on the brushes. Clean the commutator surface with fine abrasive paper to restore a smooth, polished surface. Clean out the grooves between the commutator segments, using a hacksaw blade ground to the correct width. Each segment should be straight sided, with an undercut of 0.5 – 0.8 mm (0.020 – 0.032 in). If the depth of undercut is less than 0.2 mm (0.008 in) the armature should preferably be renewed. It is possible to re-cut the grooves using the modified hacksaw blade mentioned above, but this requires care and patience. Do not cut into the segment material, or leave the groove anything other than square sided.
5 Set the multimeter on the ohm x 1 scale, then check the resistance between each commutator segment and its neighbour. A very high or infinite resistance indicates an open circuit and the armature must be renewed. Next, set the multimeter on its highest resistance scale, normally ohm x 1000 (kilo ohms) and check the resistance between the armature core (shaft) and each of the commutator segments. There should be no conductivity shown in this test, any reading indicating a partial or complete short circuit, again necessitating renewal.
6 Place the motor body on the workbench with the brushes towards you and the starter motor lead terminal to the right-hand side. With the meter on the ohm x 1 range, check for continuity between the positive (+) brush (located on the left-hand side, opposite the terminal) and the terminal. If a reading close to zero ohms is not obtained, the field coil windings are open (broken) and the assembly should be renewed. Now set the meter on its highest range and check for resistance between the positive (+) brush and the motor body. Anything other than infinite resistance indicates a short circuit, again requiring renewal.

Four brush motors

7 These should be dealt with as described above, except for the following points. Brush spring tension has been decreased to 340 – 460 grams, but again in practice the springs can be considered acceptable if they bear firmly upon the brushes. Measure the resistance between each positive (+) brush (attached to the motor body) and the body, with the meter set on the ohm x 1000 scale. No reading should be shown. Next, set the meter to ohm x 1 and measure the resistance between the two positive (+) brushes. Unless the reading is at or close to zero ohms, renew the brushes and leads to correct the open (broken) circuit.
8 Moving to the brush plate and the negative (-) brushes, set the meter on the ohm x 1 scale and check for resistance between the two negative brushes. If a high or infinite resistance is shown, the brush plate assembly should be renewed. Set the meter on the ohm x 1000 scale and measure the resistance between each brush holder and the brush plate. There should be no conductivity between the two, any reading indicating the need for renewal.

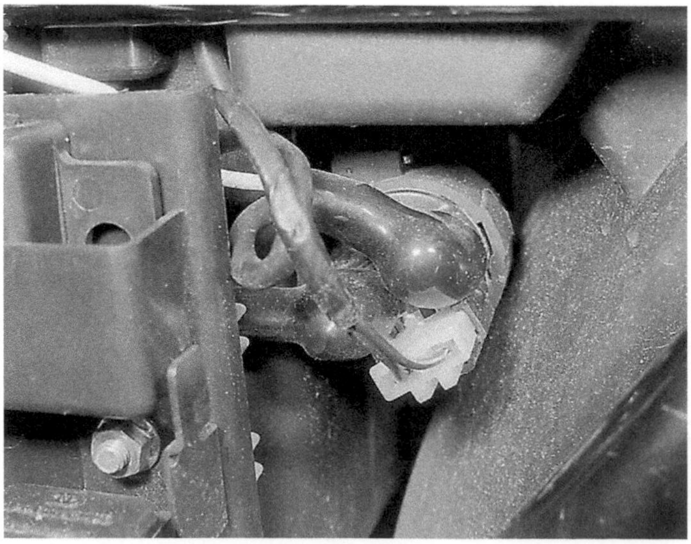
9.2 Starter relay is identified by black, heavy-gauge starter motor lead connected to it

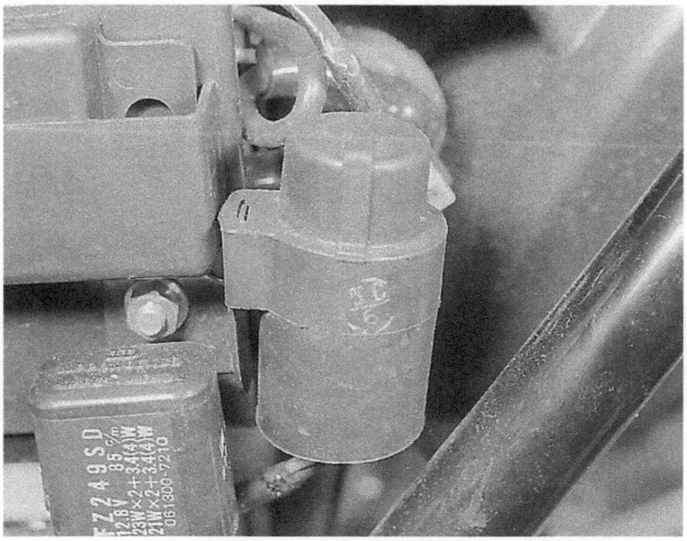

9.7 Starter control (circuit) relay – ZX550 A1

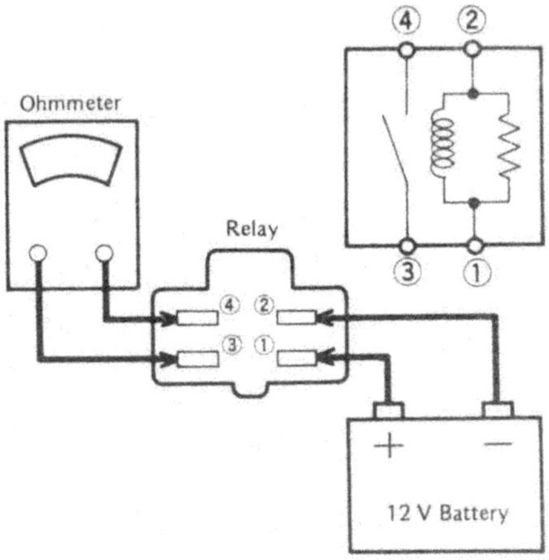

Fig. 6.3 Testing the starter control (circuit) relay

1 *Coil terminal – Black/white wire*
2 *Coil terminal – Yellow/green wire*
3 *Switch terminal – Black wire*
4 *Switch terminal – Black/white wire*

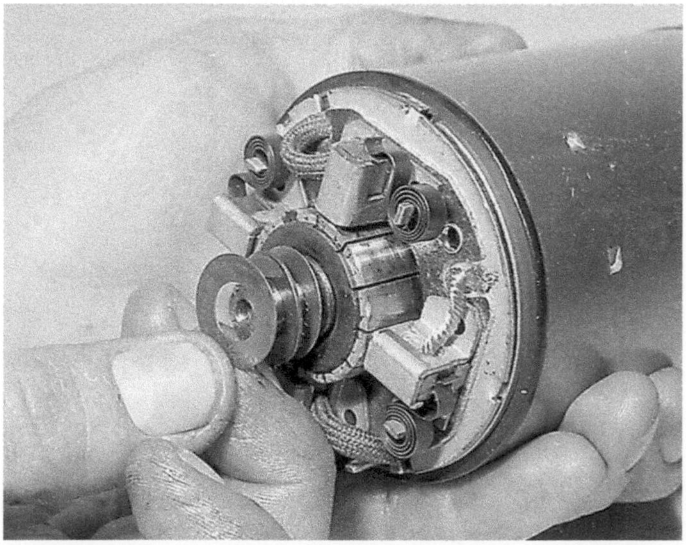

10.2a Note carefully position and number of shims on each end of armature when dismantling

10.2b Always refit shims exactly as they were removed

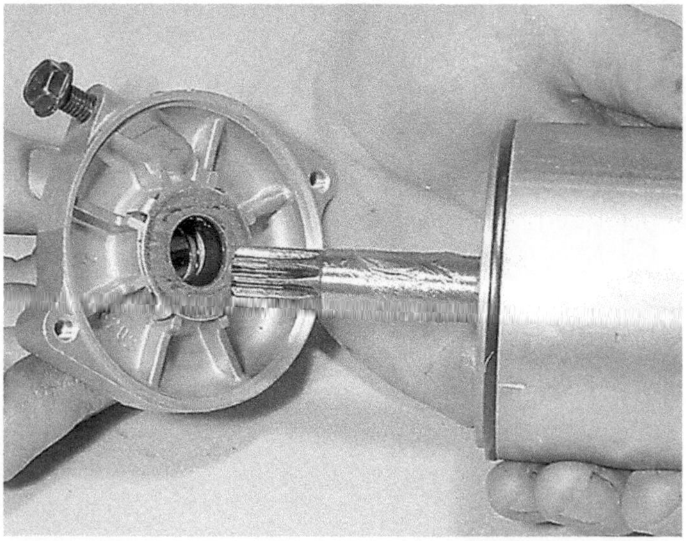

10.2c On refitting end covers ...

10.2d ... align scribed lines with those on motor body, as shown

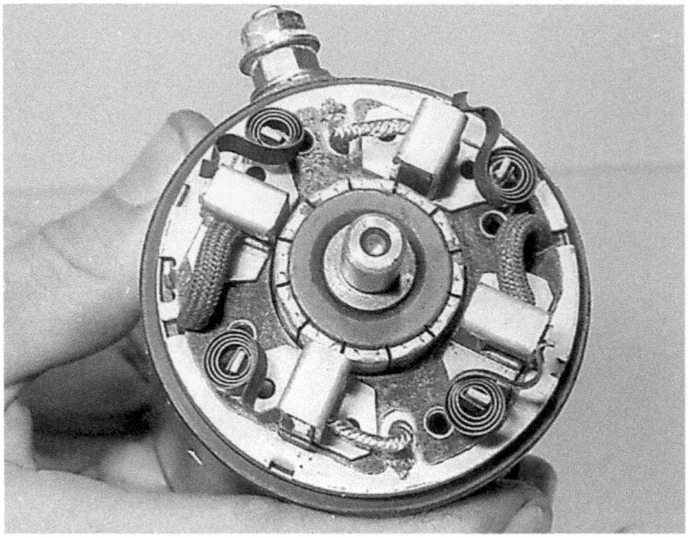

10.2e Disengage brush springs as shown on refitting armature

10.7 Renew brushes if excessively worn – check that they move easily in holders

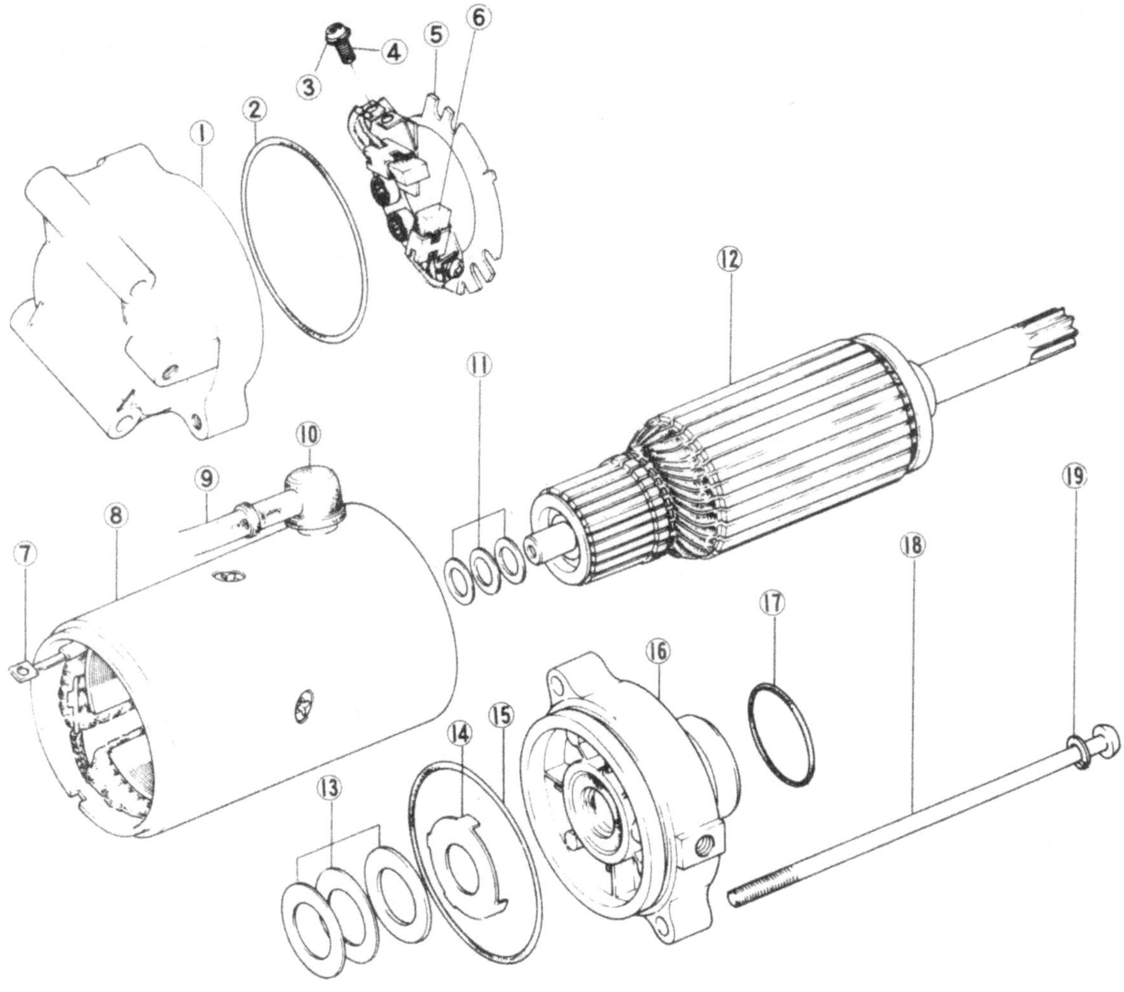

Fig. 6.4 Starter motor – early models

1 *End cover*
2 *O-ring*
3 *Lock washer*
4 *Screw*
5 *Brush plate*
6 *Brush*
7 *Field coil lead*
8 *Motor body*
9 *Starter motor lead*
10 *Terminal cover*
11 *Shims*
12 *Armature*
13 *Shims*
14 *Toothed washer*
15 *O-ring*
16 *End cover*
17 *O-ring*
18 *Screw*
19 *Lock washer*

11 Headlamp and reserve lighting system: testing

1 All later US models employ a reserve lighting system which automatically switches in the remaining bulb filament in the event of headlamp failure. For details of the headlamp and lighting system of all other models, see Section 13.
2 To check the operation of the reserve lighting system, first release the headlamp unit by removing the retaining screws. Disconnect the headlamp connector and make up three insulated test leads which should be connected between the headlamp and the wiring connector as shown in the accompanying illustration. Set the dipswitch to the low beam position and turn on the ignition switch to operate the lights. Disconnect the test lead from the red/yellow low beam lead and check the main beam comes on, though more dimly than normal, and that the headlamp failure warning lamp is lit. Reconnect the test lead and switch to main beam. Disconnect the test lead from the red/black main beam lead and check that low beam comes on together with the failure warning lamp.
3 To test the dipstick, remove the fuel tank and disconnect the red/black, red/yellow, blue/yellow and blue leads from the dipswitch. Use the multimeter as a continuity tester to check that the appropriate switch terminals are connected in the two switch positions, as shown in the relevant wiring diagram. If the switch proves faulty, dismantle and clean the contacts using fine abrasive paper and aerosol contact cleaner. If this fails, renew the switch unit. For details on testing the reserve lighting unit, see Section 12.

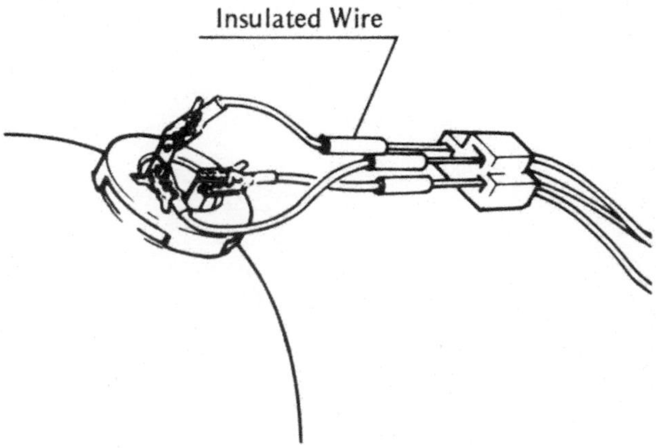

Fig. 6.5 Reserve lighting operation test

12 Reserve lighting unit: testing

1 Unlock and open the seat and remove the right-hand side panel. Separate the 6-pin connector from the reserve lighting unit. Set the meter on the 20 volts dc range, connecting the negative (-) probe to earth (ground). Use the positive (+) probe to test the wiring to the unit as described below.

System fails to select remaining filament after one has failed
2 Check the voltage on the blue/orange lead. If about 12 volts is shown, both bulb filaments have failed or the black/yellow lead is broken or disconnected. If less than 12 volts is shown, check the voltage on the blue lead. If about 12 volts is shown the reserve lighting unit is defective. If 0 volts is shown the ignition switch or wiring is faulty.

Both bulb filaments are selected
3 Check the voltage on the blue/orange lead. If 12 volts, the reserve lighting unit is defective. If 0 volts, the dipswitch or wiring is at fault.

Main beam is not dimmed when selected after low beam failure
4 Check the voltage on the red/black lead. If 12 volts, the reserve lighting unit has failed. If 0 volts, the wiring harness is broken or disconnected.

Failure warning lamp inoperative
5 Check the voltage on the light green/red lead. If 12 volts, the bulb has failed or is disconnected. If zero volts, the reserve lighting unit has failed.

13 Headlamp and lighting system: testing

1 All models except those covered by Section 11 are equipped with a conventional manual lighting system. In the event of a failure, check the operation of the lighting switch and dipswitch as follows. On UK models only, remove the fuel tank and disconnect the brown/white lead, brown lead and blue lead from the lighting switch. Using a multimeter set on the resistance scale (ohms) as a continuity tester, check that the appropriate switch terminals are connected at the various switch positions, referring to the relevant wiring diagram.
2 The dipswitch can be checked in the same way, this time tracing the red/black, blue and red/yellow leads and checking the switch contacts. A fault in the headlamp flasher, or pass, switch can be traced by checking the red/black and brown leads and the switch for continuity.
3 Z400 J1, Z500 B1, B2 and Z550 A1 and C1 models are fitted with a resistor in the (blue) feed wire from the fuse to the dipswitch; the resistor being mounted on a bracket bolted to the fork bottom yoke. In many cases this will have been bypassed and removed, especially where a more powerful headlamp bulb has been fitted. Its function is to prevent the headlamp bulb from blowing by soaking up excess power surges; to test it measure the resistance across its wire terminals. A reading of 5 ohms should be retained; if the measured resistance is significantly above or below this, the resistor must be renewed.

14 Headlamp bulb/sealed beam unit: renewal and adjustment

1 The headlamp unit comprises a lens and reflector unit, either round or rectangular, secured in the headlamp shell by two screws. On all GPz models, remove the fairing as described in Chapter 4 to gain access to the headlamp; on ZX550 A1 and A1L models the headlamp shell must be removed from the fairing sub-frame to permit the bulb to be changed. On all models, to free the unit from the shell, remove the two short screws which pass through the lower edge of the rim, noting that the alignment screw(s) should not be disturbed. Lift and disengage the unit and pull off the wiring connector, and on European models, the front parking lamp or 'city lamp' lead.
2 Pull off the rubber boot from the back of the bulb and release the bulb retainer. Depending on the type of bulb used a ring-type retainer may be fitted, this being a bayonet fitting, or a wire spring retainer may be used. Free the retainer and remove the bulb. **Note:** If a quartz halogen bulb is fitted, never touch the quartz glass envelope with the hands; it will be damaged by oil or skin acids. When fitting a new bulb, check that it is positioned correctly and secure the retainer. Refit the rubber boot, ensuring that the moisture drain channel faces downward and that it seats fully against the back of the reflector.
3 On US models fitted with sealed beam headlamp units, unscrew fully the beam horizontal alignment screw and remove the two mounting screws to separate the unit from the rim. Ensure that the 'Top' mark is uppermost on reassembly.
4 Refit the headlamp in the shell and check that the horizontal alignment screw in the front edge of the rim is set so that the beam shines straight ahead. If vertical adjustment is necessary, slacken the headlamp mounting nuts inside the shell and the vertical adjustment nut below the unit. Angle the headlamp to comply with local legislation, noting that the rider should be seated normally, then secure the nuts.
5 UK models are equipped with a low wattage parking (city) lamp incorporated in the headlamp unit. This can be renewed when the headlamp unit has been detached as described above. Pull out the bulbholder from the rubber grommet which retains it to gain access to the bayonet fitting bulb.

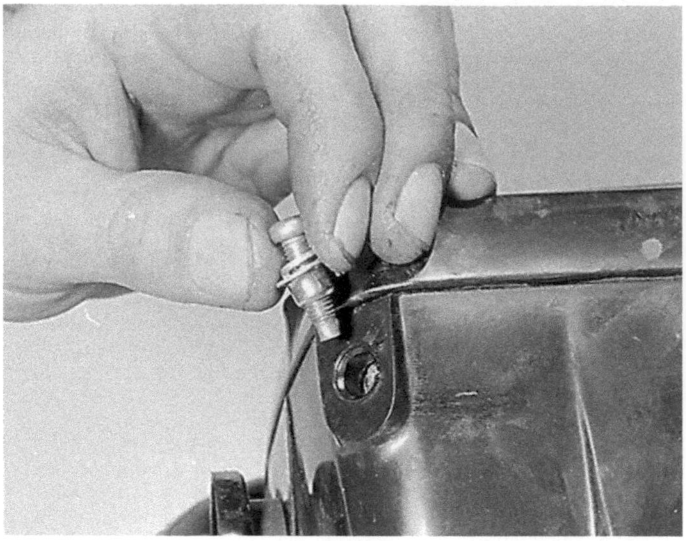

14.1a Remove retaining screws to release headlamp unit from shell ...

14.1b ... and withdraw unit to expose connectors

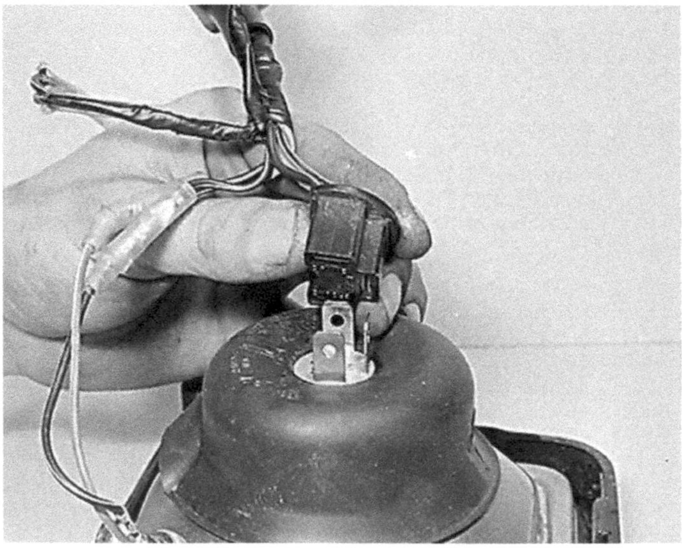

14.1c Headlamp connector is of three-pin spade type

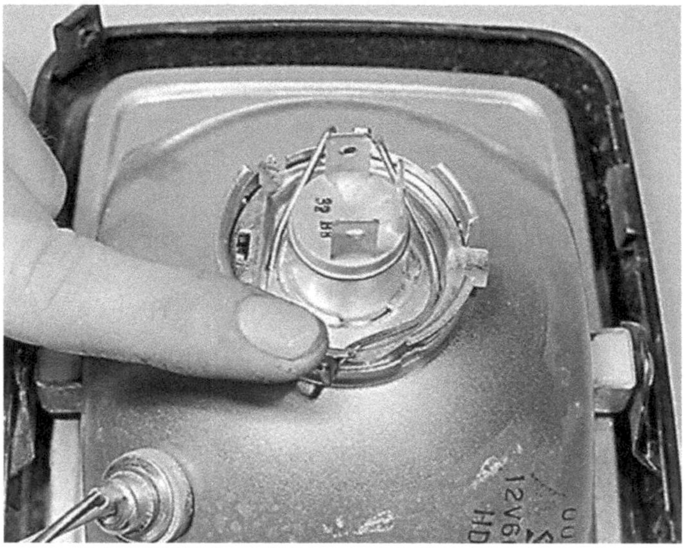

14.2a Release spring retainer ...

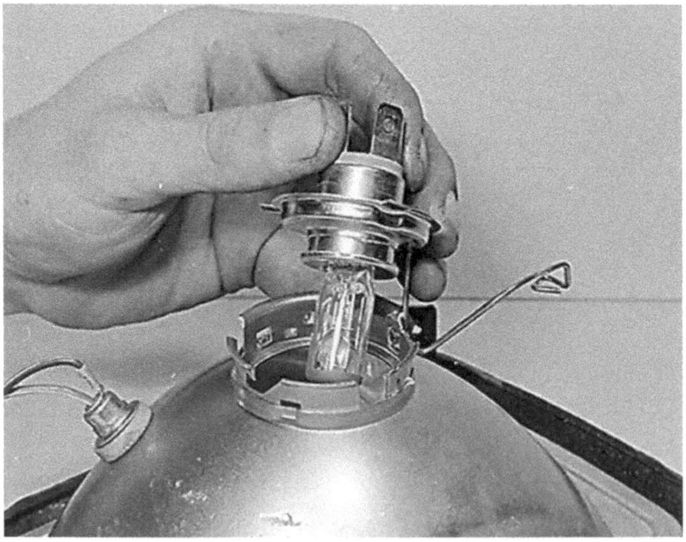

14.2b ... and withdraw headlamp bulb – do not touch glass envelope

14.4 Horizontal alignment is adjusted by rotating screw as shown

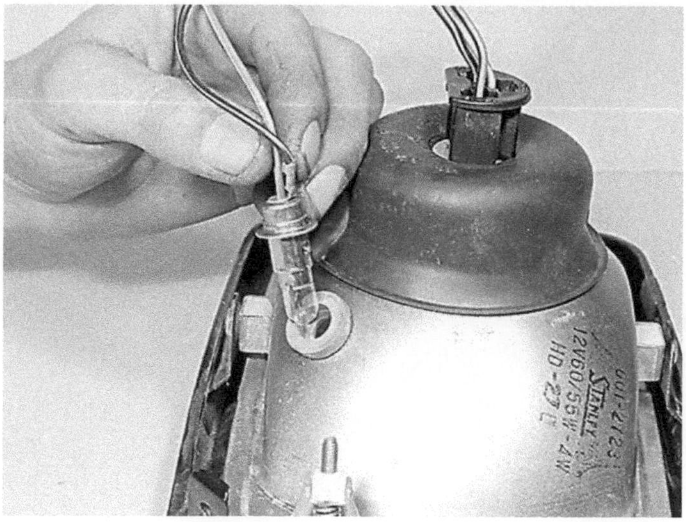

14.5 UK models are fitted with separate parking lamp bulb

15 Turn signal and stop/tail lamps: bulb renewal

1 All stop/tail and turn signal lamp lenses are retained by two screws. Remove these screws and withdraw the lens, taking care not to tear the sealing gasket.
2 All bulbs are of the conventional bayonet fitting, ie press in and twist anti-clockwise to release. Refitting is the reverse of the above, but note that stop/tail lamp bulbs have offset pins so that the bulb can be fitted one way only. The same is true of the front turn signal/running lamp bulbs fitted to some later US models. Do not overtighten lens retaining screws or the lens may crack.
3 Remove all traces of moisture and corrosion from the lamp interior and renew the sealing gasket if it is found to be torn or compressed. Turn signal lamps with black interiors can be improved by lining with cooking foil; this will greatly improve their visibility in bright sunlight.

16 Tail/brake lamp circuit: general description and testing

1 All early models were fitted with a brake light failure warning circuit which illuminates a warning lamp to indicate failure of the bulb filament or a broken connection in the brake lamp circuit. All later models employ twin bulbs in the tail lamp to obviate the need for the warning circuit, the bulbs being connected in parallel so that one will still work if its partner has blown. In the event of failure of the brake lamp in either system, first check that the bulb has not blown (see Section 15) then proceed as follows.
2 To check that the front brake switch is functioning normally, remove the headlamp unit from the shell and disconnect the brown/blue leads from the switch. Connect the probes from a multimeter set on the ohm x 1 scale to the switch leads and check that no resistance is shown when the lever is operated. If the switch is faulty it must be renewed, the sealed construction precluding repair or cleaning. To free the switch use a small screwdriver to push in the locating tab via the hole in the underside of the lever assembly.
3 The rear brake switch is located inboard of the right-hand footrest plate and is operated via a spring by the brake pedal. The operation of the switch can be checked as described above, noting that it is possible to adjust the position of the switch, and thus the point at which it comes on, by moving it up or down in relation to the mounting bracket.
4 The brake light failure warning circuit (where fitted) is controlled by a switch unit housed behind the left-hand side panel. If the system is working normally, the warning lamp should come on when the brake is applied and go off when it is released. If the brake lamp bulb is blown or disconnected, the warning lamp should still come on when the brake is applied, but will flash when released.
5 If a fault occurs, it is preferable to check the switch unit by substituting a new item. If this cannot be arranged, check the system wiring and switch unit with a multimeter as follows. Disconnect the switch at its block connector, check that the ignition is switched off and use an ohmmeter set to the x 1 ohm range to check that there is no measurable resistance between the black/yellow wire terminal and a good earth point on the frame.
6 Using a multimeter set to the 20 volts dc range, connect the meter negative (-) probe securely to a good earth point on the frame and switch on the ignition. Check that full battery voltage is available at the green/white wire terminal; if not, check that the warning lamp bulb is now blown. Connecting the meter positive (+) probe to the blue wire terminal, check that full battery voltage is obtainable when either the brake lever or pedal is applied, and that no reading is obtainable with the brake released. If any results are not as stated, check carefully for broken or damaged wires and connectors.
7 To test the switch itself, connect it again to the wiring loom, switch on the ignition and connect the meter as described in paragraph 6 above. Make the tests by inserting the meter positive (+) probe into the back of the connector block to contact the terminal of the wire concerned, while the brake lever or pedal is applied and released. Check that a reading of full battery voltage is available at the yellow wire when the brake is applied, and that no reading is obtainable when it is released. On the green/white wire, no reading should be obtainable when the brake is applied, but full battery voltage should be available when it is released. If either test fails to produce the stated result, the switch is faulty and must be renewed.

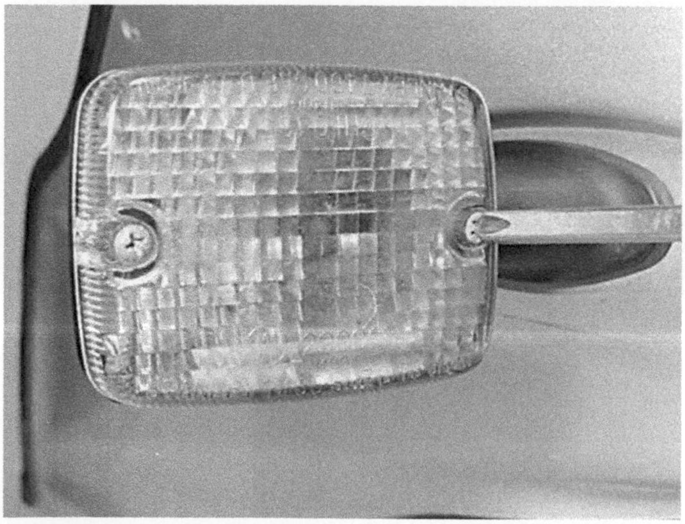

15.1 Lens is retained by two screws – do not overtighten on reassembly

15.2 Bulb is bayonet fitting – some may have offset pins to ensure correct installation

17 Turn signal and hazard warning circuit: testing

1 As a general guide, turn signal problems affecting the whole system are usually attributable to the relay, wiring or switch, whilst if the fault is confined to one side, the relay and its supply can be considered sound. In this case examine the bulbs, lamp wiring and switch for faults. If the system fails totally, check first that the battery is in good condition and fully charged, and that all wiring and connections are sound.

2 Remove the left-hand side panel and open the electrical panel cover. Pull off the brown lead and orange/green lead from the relay, and check the resistance between the two relay terminals. This should be close to zero ohms if the relay is sound. If a higher resistance is found, renew the relay.

3 If the relay is sound, set the multimeter to the 20 volt dc range and connect the positive (+) probe to the brown lead and the negative (-) probe to the orange/green lead. Switch on the ignition and note the meter reading when the turn signal switch is moved to the 'L' and 'R' positions. If battery voltage is not shown in both positions check the switch contacts and wiring. If no voltage is shown in either position, check the fuse, ignition switch and wiring.

4 If both turn signal lamps come on when selected, but do not flash, or flash very slowly, check that the battery is charged and that all wiring connections are secure. Next, check that the bulbs are of the correct wattage. If the above checks fail to reveal the cause of the fault, renew the relay.

5 If only one of the two lamps comes on, and fails to flash or flashes weakly, check that the inoperative bulb is not blown and is of the correct wattage. Other possible causes are broken or disconnected wiring or a poor earth connection. If neither lamp comes on, check that the switch and switch wiring is sound.

6 The lamps should flash at 60 – 120 flashes per minute. If this speeds up excessively, the regulator/rectifier may be faulty, the relay may be faulty or the bulb wattages incorrect.

Hazard warning system

7 A hazard warning system is fitted to all later US models. In the event of a suspected fault, start by testing the turn signal system as described above.

8 The hazard warning switch can be checked by tracing the wiring from the left-hand switch cluster to the 6-pin or 9-pin connector beneath the fuel tank. Separate the connector and check the switch connections using a multimeter as a continuity tester. See the wiring diagram for details of switch position.

9 The hazard warning relay is located on the electrical panel, just forward of the turn signal relay. To check the relay, disconnect the leads and measure the resistance across the two terminals. If this is significantly lesser or greater than 60 ohms, renew the relay.

10 If the switch and relay prove to be serviceable, check for battery voltage between the grey lead from the main harness and the grey lead from the handlebar switch with the ignition switch at the 'Park' and 'On' positions and the hazard switch on. Repeat the test with the meter connected to the green lead from the switch unit as described above. If battery voltage is not shown in both positions, check the fuse, hazard warning switch and wiring.

17.2 Location of turn signal relay – ZX550 A1

18 Automatic turn signal cancelling system: testing

Distance sensor

1 Remove the headlamp unit and trace the distance sensor wiring from the speedometer, separating it at the four-pin connector. Release the lower end of the speedometer drive cable by unscrewing the knurled retaining ring. Connect multimeter probes to the red lead and light green sensor lead and select one of the ohms scales to check continuity. Slowly rotate the speedometer inner cable and note how many times the sensor switches on and off during each revolution. If the sensor is working correctly, there should be four on/off pulses per revolution. Failing this, it will be necessary to obtain a new speedometer or to remember to switch the turn signals off manually. No repair is possible.

Turn signal and combination switches

2 Remove the fuel tank and trace the wiring from the left-hand switch cluster back to the 9-pin connectors. Refer to the switch diagrams shown in the main wiring diagrams at the end of this Chapter and check the continuity of the turn signal switch contacts, and also those of the combination switch (where fitted). If an open or short circuit is discovered, check the switch wiring for damage and dismantle and clean the switch contacts. If the problem persists, renew the left-hand switch cluster. Note that individual parts are not available for the switch clusters, making repair impracticable.

3 To check the operation of the solenoid which resets the turn signal switch, connect a test lead to the battery positive (+) terminal. Set the turn signal switch to the left or right, then touch momentarily the test

Meter Range	Connections*	Ignition Switch	Selector Switch Position	Turn Signal Switch Position	Reading
25V DC	Meter (+) → Yellow, Blue/White	ON	A	Any (R, L, Neu.)	Battery voltage
		OFF	M	Any	0 V
	Meter (+) → White/Red	ON	A	R or L	Battery voltage
		OFF	M	Neutral	0 V

*Connect the meter negative (–) lead to ground.

Fig. 6.6 Automatic turn signal cancelling system – switch wiring test

lead on the white/green lead to the solenoid. If the solenoid fails to reset the switch from both the left and right positions, renew the switch assembly. Note that battery voltage should not be applied to the solenoid for more than a second or so, or the solenoid windings may be burned out.

System wiring checks

4 Remove the right-hand side panel and locate the 6-pin connector which will be found below the battery tray and the regulator/rectifier unit. Using the accompanying table, check the voltage on the various leads with the switches set as described. Note that this test sequence requires the system to be intact, with all connections made normally. If a discrepancy is noted, make a careful check of all connectors and leads. If this fails to resolve the fault, renew the control unit. This is located to the front of the battery tray or on the electrical panel.

19 Low fuel and oil level warning system: testing – all models without computer warning system

1 As previously stated, refer to the wiring diagram appropriate to the machine being worked on before starting work. On some models the same warning lamp is used for brake lamp failure as well as for low fuel, while on others the same lamp is used for low oil level warning and low fuel. Check the components of all related circuits before condemning any component in particular.

2 Later models are fitted with a self-checker. When the ignition is first switched on the warning lamp will light for about 3 seconds, then switch off. After a delay of up to three minutes, the self-checker switches in the level sensor which will light the appropriate bulb if either level is low.

3 Before checking any fault, ensure that the battery is fully charged and that the fuel tank level is well above the half-way mark. The engine oil level should also be checked and topped up if necessary. If either bulb fails to come on, check that it has not blown before moving on to the tests described below.

System with low fuel level warning lamp only

4 First check that the brake light failure warning circuit is functioning normally; the two circuits use the same warning lamp. If the fuel level is low but the warning lamp fails to come on, trace and disconnect the sender leads at the 2-pin connector below the fuel tank. Check the voltage on the sender leads on the harness side of the connector. With the multimeter set to the 20 volts dc range, connect the positive (+) probe to the green/white lead and the negative (-) probe to the black/yellow lead. If, with the ignition switched on, battery voltage is indicated, the fault lies with the sensor. If no reading is shown, check the wiring for breaks or loose connections.

5 If the warning lamp stays on irrespective of the amount of fuel in the tank, and the brake light failure circuit is sound, the sender unit can be assumed to be faulty. Intermittent flashing of the warning lamp is usually due to fuel surge in the tank during braking or acceleration. Other possible causes are damaged wiring earthing against the frame or a regulator/rectifier fault.

6 If the sender unit is to be renewed, remove and drain the fuel tank (see Chapter 2), then release the two screws which retain it to the underside of the tank. Use a new O-ring when fitting the new unit and check for leaks before refitting the tank.

System with low fuel level warning lamp and self-checker unit

7 Make all preliminary checks as described in paragraph 3 above. The self-checker unit is mounted behind the right-hand side panel and is connected via a three-pin block connector. If the warning lamp fails to light when the ignition is first switched on, check that the bulb is sound, then disconnect the self-checker and use a meter set to the 20 volts DC range to check that full battery voltage is available when the ignition is switched on, connecting the meter positive (+) probe to the terminal of the brown wire leading into the main loom and its negative (-) probe to the black/yellow wire. If full battery voltage is not obtainable, the wiring is at fault. If it is, connect the self-checker again, ensuring that the connectors are clean and securely fastened, and check whether the lamp still fails to light. If it does not light, the self-checker is faulty and must be renewed.

8 If the fuel level is low but the warning lamp fails to come on, first check that it lights when the ignition is first switched on, then test the sender unit and wiring as described in paragraph 4 above. If the warning lamp stays on irrespective of the amount of fuel in the tank, or if it flashes intermittently, see paragraph 5 above, but check that the fault does not persist with the self-checker disconnected; if it does the sender unit or other items are at fault, but if the problem is cured then the self-checker is faulty.

System with low fuel and oil warning lamp

9 The self-checker unit is mounted either behind the right-hand side panel, or on shaft drive models, in the frame top tube at the rear of the fuel tank. Disconnect its three-pin connector, then trace and disconnect the fuel level sender leads at the connector below the fuel tank. Make up a test lead from a length of insulated wire with both ends bared for about $\frac{1}{2}$ inch. Switch on the ignition and use the test lead to connect the terminals described below, noting the response from the warning lamps. The object of these tests is to simulate the operation of the sensors by bridging the appropriate terminals with the test lead.

10 Starting with the 6-pin connector from the self-checker unit, bridge the male terminals of the green/white lead and black/yellow lead. If all is well, the warning lamp should come on. Now repeat the test on the same leads at the fuel level sensor. If the lamp still comes on, the fault lies with the fuel level sensor unit.

11 To check the oil level sensor circuit, connect the test lead between the male terminals of the yellow/blue lead and blue/red lead at the 6-pin connector. If the circuit is sound but the oil level switch is at fault, the warning lamp will come on. If in the above tests the warning lamp(s) fail to operate, check the wiring, connectors, bulbs and sensors.

12 The power supply to the self-checker unit can be checked using a multimeter on the 20 volts dc scale. Connect the positive (+) probe to the brown lead and the negative (-) probe to the black/yellow lead on the male terminal side of the 6-pin connector. Switch on the ignition and check that battery voltage is shown. If this is not the case, check the leads for damage.

13 To check the fuel level sensor, remove and drain the fuel tank (Chapter 2) and release the two screws which retain the sensor to the underside of the tank. Fill a jar with fuel so that the cylindrical thermistor can be submerged in fuel for tests purposes, taking normal precautions to avoid fire risks. Connect the sensor to the main harness via its two-pin connector and place the thermistor in the jar of fuel. Switch on the ignition and observe the warning lamp, which should come on for about 3 seconds, and then go off. Remove the jar of fuel to expose the sensor to air. After an interval of between 20 seconds and 3 minutes, the warning lamp should come on. If the sensor fails to operate normally and the rest of the system appears to function correctly, renew the sensor.

14 To check the oil level switch, drain the engine oil and remove the switch from the underside of the sump by releasing the two retaining bolts. Using a multimeter on the ohm x 1 scale, connect one probe to the switch lead and the other to the switch body. When held upright, a reading of infinite resistance should be shown, whilst when the switch is inverted, less than 0.5 ohm should be indicated. If the switch does not conform to the above, or works erratically, it must be renewed.

System with fuel level gauge

15 This system incorporates a variable resistance (rheostat) operated by a float on a pivoting arm, the whole sender unit being mounted by five screws on the underside of the fuel tank. A simple car-type gauge unit is mounted on the instrument panel. To decide which is defective if a fault occurs, disconnect the two-pin connector from the sender unit under the tank and switch on the ignition; the gauge should read 'E'. Use a length of spare wire to connect the black/yellow and white/yellow wires leading into the main loom from the sender unit connector; the gauge should read 'F'. If the above tests produce the expected results, the sender unit is at fault; if not the fault is in the gauge or wiring.

16 To check the gauge, disconnect the large multi-pin block connector joining the instrument panel wires to the main loom, then set a multimeter to the 20 volts dc range and connect its positive (+) probe to the terminal of the brown wire coming from the main loom and its negative (-) probe to the black/yellow wire. Full battery voltage should be obtainable when the ignition is switched on. Having established that the gauge is receiving the correct power supply, set the multimeter to the relevant resistance scale and measure the resistance across the terminals of the brown and black/yellow wires; a reading of 60 – 80 ohms should be obtained. If the measured

resistance is significantly above or below the set figure, the gauge is faulty and must be renewed.
17 To check the sender unit, remove and drain the tank, then release the sender unit by releasing the securing screws. Take care not to damage the sealing gasket or bend the float arm. Check that the float assembly moves smoothly and evenly with no signs of sticking. Set a multimeter on the appropriate resistance range and check the operation of the variable resistor. At the full (highest) position a reading of 1 – 5 ohm should be indicated. This should increase smoothly to a value of 103 – 117 ohm in the empty (lowest) position. If the sender does not conform to this, or is erratic in its operation, it should be renewed. On reassembly, ensure that the arrow marking on the mounting flange faces forwards.

20 Computer warning system: description and initial check

1 KZ/Z550 H1, H2, ZX550 A1, A1L and Z550 G1, G2 models are fitted with a microprocessor-controlled monitor system. A series of red LCD (liquid crystal display) segments warns the rider if the side stand is down, the oil level is low or the battery electrolyte level is low. A nine-segment display monitors fuel level, the bottom segment flashing to indicate that the fuel level is low. In addition to the above, a separate LED (light emitting diode) flashes to attract the rider's attention when any of the warning segments comes on.
2 The microprocessor monitors the system via sensors, and switches on the appropriate segments. It includes an automatic checking sequence which runs through the various system functions whenever the ignition is switched on.
3 If a fault develops in the system, always check the more obvious possible causes before dismantling anything. Start by switching on the ignition and watching the display panel during the automatic self-checking procedure. If nothing at all happens it can be assumed that the power supply to the system has failed. Refer to Section 21 for further details. If the self-checking sequence is completed normally, the fault can be assumed to lie in one of the sensors or the associated wiring. Refer to Section 23 for further details. If one or more of the LCD segments fail to come on, or work erratically the display panel microprocessor may be at fault. Refer to Section 22 for further details.
4 **Note**: If all tests fail to locate a fault, try the system with and without the engine running; engine vibration may be upsetting the delicate components or causing an intermittent wiring fault.

21 Computer warning system: power supply tests

1 Turn the ignition switch off and trace the red 6-pin connector joining the display panel to the wiring loom. Separate the connector, set the multimeter to the 20 volts scale, and connect the meter probes to the supply side of the connector, the positive (+) probe going to the brown wire terminal and the negative (-) probe to the black/yellow wire terminal. Switch on the ignition and check that full battery voltage is measured, then switch off and check that no reading is obtainable. If the results obtained are as stated, the power supply is working; if the display still fails to operate in the self-checking mode, refer to Section 22. If the power supply is not working, trace the brown lead and black/yellow lead back until the fault is located. If the fault appears to be confined to one sensor, refer to Section 23.

22 Computer warning system: microprocessor and display tests

1 Prepare a 12 volt power source (the machine's battery is ideal for this and may be left in place if long test leads are used) and six test leads. Connect the test leads to the display side of the red six-pin connector as described below. On ZX550 A1 and A1L models, connect an additional test wire between the terminals of the green/yellow wires in the connector male and female halves so that the separate warning lamp functions normally during the tests.

Power supply. Connect a test lead from the battery positive (+) terminal to the brown lead. Connect a test lead from the battery negative (-) terminal to the black/yellow lead.

Side stand switch. Connect a test lead from the battery positive (+) terminal (negative terminal on ZX550 A1, A1L models) to the green/white lead.

Oil level sensor. Connect a test lead from the battery negative (-) terminal to the blue/red lead.

Battery level sensor. Connect a test lead from the battery positive (+) terminal to the pink lead.

Fuel level sensor. Connect a test lead from the battery negative (-) terminal to the white/yellow lead.

2 Briefly disconnect and reconnect one of the power supply leads. The system should then go into the self-checking mode, after which all warning segments should go off leaving the fuel gauge display showing a 'full tank'. Now disconnect each of the sensor test leads in turn and check that the appropriate warning segments and the red LED begin to flash. Note that in the case of the fuel level system a delay circuit is incorporated, and a pause of up to 12 seconds will be noted before the microprocessor acknowledges a change.
3 If any of the sensor circuits fails to provoke the appropriate display, or if the self-checking mode does not start, the unit must be considered faulty and renewed. If all is well in the display panel proceed to the tests of the wiring and sensors described in the next Section.

23 Computer warning system: wiring and sensor tests

Wiring test

1 With the red 6-pin connector separated as described in Section 21, check that the kill switch is set to the 'Run' position. The following tests should be made on the female terminal (main harness) half of the connector. Note that the ignition switch must be **on** for the side stand switch test only. If the readings obtained conflict with those shown in the accompanying table, trace back through the wiring, checking for damaged wires or loose connectors.
2 Note that the table refers to chain drive models but there is only one difference when working on Z550 G models; when testing the electrolyte level wires, connect the outer positive (+) probe to the green/blue wire, not the pink as indicated. Results must be as shown.
3 If any of the tests reveal a wiring fault, check carefully all wires and connectors, repairing or renewing all damaged components. If all is well, test the individual sensors as described below.

Sensor tests

4 Test the side stand switch as described in Section 9, paragraph 6, the oil level switch as described in Section 19, paragraph 14, and the fuel level sensor unit as described in Section 19, paragraph 17. Renew any faulty component.
5 The battery is equipped with a sensor which consists of a small electrode fitted in place of one of the cell caps. If the electrolyte falls below a specific level, the electrode is exposed to the air and the lack of current triggers the warning panel. Check that the electrolyte is above the minimum mark and that the sensor is fitted next to the arrow mark on the battery casing. Connect the positive (+) probe of the multimeter to the sensor lead and set the meter to the 10 volts dc range. Touch the negative (-) probe to earth (ground) and check that at least 6 volts is shown. If a low reading is obtained, remove the sensor and wash it with copious quantities of water. When dry, clean the electrode section with a wire brush, then refit it and perform the voltage check again. Note that if that particular cell of the battery fails, it may not be possible to persuade the sensor to operate correctly. Check the battery as described in Section 3, and renew it if necessary.

24 Electronic tachometer and voltmeter: testing

Tachometer

1 If a fault occurs first check that the instrument is mounted securely, with all nuts and bolts correctly fastened, and that its mounting rubbers are in good condition so that it is isolated from vibration. Check this, tightening or renewing components if necessary. If this makes no difference to the fault, proceed as follows.
2 With a multimeter set to the 20 volts dc scale connect the positive (+) probe to the brown wire in the back of the connector block and the negative (-) probe to the black/yellow wire. Switch on the ignition and check that full battery voltage is obtainable. If not check the brown and black/yellow wires for damage.
3 If the power supply is working, transfer the meter negative (-) probe to the black wire terminal. The meter should show zero volts when the engine is stopped and 2 – 4 volts when it is running. If no reading is obtained, trace back and check the two leads.

4 If all the above tests show the wiring to be sound, the instrument must be at fault, but this can be checked only by substituting a new component.

Combined tachometer/voltmeter unit

5 On some models a secondary scale and circuit allows the tachometer to function as a voltmeter when a change-over switch is pressed down. Before performing any other tests in the event of a suspected fault, check that the instrument panel is secure and that all damper rubbers are in place and in good condition. The battery must be fully charged and the ignition system working normally.
6 Test the tachometer unit as described above. If the voltmeter is receiving the correct power supply, the only other component likely to be at fault is the switch itself. Disconnect the green, red and yellow leads from the changeover switch. Using a multimeter as a continuity tester, check that the green and red leads are connected when the switch is released and that the yellow and red leads are connected when it is pressed. If the switch is faulty, renew it, but if it is in good condition the instrument must be at fault and should be renewed.

25 Instrument panel and warning lamps: bulb renewal

1 Remove the fairing on GPz models. Slacken the drive cable retaining ring (where appropriate) to free the cable from the underside of the instrument to be removed. Remove the two domed nuts or the self tapping screws and free the bottom cover. Pull the instrument head upwards and clear of the rubber mounts. Unplug the bulbholders and lift the instrument clear.
2 The warning lamp panel can be removed after its two securing screws have been released. The bulbs are all rated at 12 volts 3.4W except on ZX550 A1, A1L models which have 3W items, and are a bayonet fitting; use the wiring colours to identify the suspect bulb.
3 On ZX550 A1 and A1L models, some warning lamps are mounted in the ignition switch/top yoke cover. This is secured by two screws and can be lifted sufficiently for the bulbs to be displaced.

26 Horn: location and adjustment

1 A single or twin horn arrangement is fitted according to the model. Each horn is mounted on a resilient steel bracket on the frame front downtubes. If the horn fails to operate, or works feebly, it can be adjusted by slackening the locknut and turning the adjuster screw in or out by a small amount until the best sound is obtained.
2 If the horn fails to work at all, first check that power is reaching it by disconnecting the wires. Substitute a 12 volt bulb, switch on the ignition and press the horn button. If the bulb lights, the circuit is proved good and the horn is at fault; if the bulb does not light, there is a fault in the circuit which must be found and rectified.
3 To test the horn itself, connect a fully-charged 12 volt battery directly to the horn. If it does not sound, a gentle tap on the outside may free the internal contacts. If this fails, the horn must be renewed as repairs are not possible.

Wire	Meter Range	Connections	Meter Reading (Criteria)
Side stand warner	25V DC	oMeter (+) → Green/white wire oMeter (–) → Black/yellow wire	oBattery voltage when side stand is up. o0 V when side stand is down.
Oil level warner	x 10 Ω	oOne meter lead → Blue/red wire oOther meter lead → Black/yellow wire	oLess than 0.5 Ω when engine oil level is higher than "lower level line" next to the oil level gauge. o∞ Ω when engine oil level is much lower than the "lower level line".
Battery electrolyte level warner	10V DC	oMeter (+) → Pink wire oMeter (–) → Black/yellow wire	oMore than 6 V when electrolyte level is higher than "lower level line." o0 V when electrolyte level is lower than "lower level line."
Fuel gauge and low fuel warner	x 10 Ω	oOne meter lead → White/yellow wire oOther meter lead → Black/yellow wire	o1 – 117 Ω

Fig. 6.7 Computer warning system – wiring test

Right-hand view of the ZX400 C2

Right-hand view of the ZX550 A4

Chapter 7 The 1984 on models

Contents

Specifications – ZX400 C2 model

Note: Information is given only where different from that shown in the Specifications Sections of Chapters 1 to 6. Refer back to the previous model in such cases, except for the ZX400 C2 model where reference should be made to the UK ZX550 A1 specifications.

Model dimensions and weight – ZX400 C2

Overall length	2180 mm (85.8 in)
Overall height	1145 mm (45.1 in)
Wheelbase	1435 mm (56.5 in)
Dry weight	179.0 kg (395 lb)

Routine Maintenance

Rear suspension unit damper standard setting – all ZX550 models and ZX400 C2	Number 1

Specifications relating to Chapter 1 – ZX400 C2

Engine

Capacity	399 cc (24.3 cu in)
Bore	55.0 mm (2.17 in)
Stroke	42.0 mm (1.65 in)
Compression ratio	9.7:1
Maximum power	54 bhp @ 11 500 rpm
Maximum torque	3.5 kgf m (25.3 lbf ft) @ 9500 rpm
Compression pressure	7.7 – 12.0 kg/cm² (109 – 171 psi)

Cylinder block

Cylinder bore ID	55.000 – 55.012 mm (2.1654 – 2.1658 in)
Service limit	55.100 mm (2.1693 in)

Pistons

Piston OD	54.965 – 54.980 mm (2.1640 – 2.1646 in)
Service limit	54.830 mm (2.1587 in)
Piston ring groove width:	
Top	1.02 – 1.04 mm (0.0402 – 0.0410 in)
Service limit	1.12 mm (0.0441 in)
Second	1.22 – 1.24 mm (0.0480 – 0.0488 in)
Service limit	1.32 mm (0.0520 in)
Ring/groove clearance – top and second compression rings:	
Standard	0.030 – 0.070 mm (0.0012 – 0.0028 in)
Service limit	0.170 mm (0.0067 in)

Piston rings

Thickness:	
Top compression ring	0.970 – 0.990 mm (0.0382 – 0.0390 in)
Service limit	0.900 mm (0.0354 in)
Second compression ring	1.170 – 1.190 mm (0.0461 – 0.0469 in)
Service limit	1.100 mm (0.0433 in)
Compression rings end gap – installed	0.15 – 0.35 mm (0.0059 – 0.0138 in)
Service limit	0.70 mm (0.0276 in)

Primary drive

Reduction ratio	3.277:1 (27/23 x 67/24T)

Clutch

Friction plate thickness service limit	2.8 mm (0.1102 in)
Spring free length	30.8 mm (1.2126 in)
Service limit	30.0 mm (1.1811 in)

Final drive

Reduction ratio	2.625:1 (42/16T)
Chain size	520 (5/8 x 1/4) x 106 links

Specifications relating to Chapter 2

Fuel tank capacity – ZX400 C2

Reserve	4.0 lit (0.88 Imp gal)

Carburettors

	ZX400 C2	Z550 G4	ZX550 A4
Make	Keihin	Keihin	Keihin
Model	CV30	CV30	CV30
Choke size	30 mm	30 mm	30 mm
Main jet	112	110	110
Main air jet:			
Cylinders 1 and 4	180	130	N/Av
Cylinders 2 and 3	110	130	N/Av
Jet needle:			
Cylinders 1 and 4	N16B	N16E	N16E
Cylinders 2 and 3	N16C	N16E	N16E
Needle jet	6	6	6
Pilot jet	35	35	35
Pilot air jet	130	145	N/Av
Pilot screw – turns out from fully closed	2 1/2	2 1/2	2 1/2
Starter jet	N/Av	52	N/Av
Starter air jet	N/Av	2.1 mm	N/Av
Fuel level	2.0 ± 1.0 mm (0.078 ± 0.039 in) **above** bottom edge of carburettor body		
Float height	16.5 mm (0.64 in)	16.5 mm (0.64 in)	N/Av
Idle speed	1200 ± 50 rpm	1050 ± 50 rpm	1050 ± 50 rpm

Specifications relating to Chapter 3 – ZX400 C2

Ignition timing

Initial	15° BTDC @ 1200 rpm
Full advance	40° BTDC @ 7000 rpm

Specifications relating to Chapter 4 – ZX400 C2

Front forks

Oil capacity – on reassembly:	
Left-hand leg	250 ± 2.5 cc (8.80 ± 0.09 Imp fl oz)
Right-hand leg	220 ± 2.5 cc (7.74 ± 0.09 Imp fl oz)
Oil level – both legs	440.0 ± 2.0 mm (17.32 ± 0.08 in)

Specifications relating to Chapter 5 – ZX400 C2

Brake disc

Minimum thickness – front brake	3.5 mm (0.1378 in)

Tyres

	Front	Rear
Size	100/90 – 18 56H	110/90 – 18 61H

Tyre pressures

	Front	Rear – at specified load
0 – 97.5 kg (0 – 215 lb) load	2.00 kg/cm² (28 psi)	2.25 kg/cm² (32 psi)
97.5 – 185 kg (215 – 408 lb) load	2.00 kg/cm² (28 psi)	2.50 kg/cm² (36 psi)

Torque wrench settings

	kgf m	lbf ft
Front brake disc mounting bolts	2.8	20.0

Specifications relating to Chapter 6 – ZX400 C2

Battery

Type	YB12A-AK
Capacity	10 Ah

1 Introduction

1 The first six Chapters of this Manual cover all models up to 1984. This Chapter covers those models introduced from late 1984 onwards, describing only those modifications which require an alteration in specifications or working procedure. If working on a later model, refer to this Chapter to note any relevant changes in specifications or working procedure, before referring to the relevant Section of Chapters 1 to 6.

2 To ensure that there is no confusion over model coverage, all models available in 1984 or later are listed below, grouped according to their coverage in this manual:

Models covered in Chapters 1 to 6:
- ZR400 B1
- ZR550 A2
- ZX550 A1/A1L (UK and US)
- Z550 G2
- KZ550 F2/F2L

Models covered in this Chapter:
- ZX400 C2
- ZX550 A2/A2L
- ZX550 A3
- ZX550 A4
- Z550 G3
- Z550 G4

3 For a specific introduction to the six models covered in this Chapter, refer to the appropriate sub-section below.

ZX400 C2 (Z400F-II)

4 Introduced to the UK in October 1984, this model is best described as having a more powerful version of the ZR400 B1 engine/gearbox unit (with ZX550 A1 camshafts and 30 mm Keihin carburettors) housed in the cycle parts of the ZX550 A1 model. The only differences in the cycle parts are that this model uses different tyres, has an anti-dive unit on the left-hand fork leg only, is not fitted with a fairing, the fuel tank warning console is replaced by a storage compartment, and that it is finished in Ebony. For all other information, refer to that given in Chapters 1 to 6 for the ZX550 A1 (UK) model. Its engine and frame numbers are as follows:

Engine number	ZX400AE038001 on
Frame number	ZX400C – 000001 on

ZX550 A2/A2L (GPz 550)

5 Introduced in 1985, these models are identical to the ZX550 A1/A1L models apart from the clutch release modification described in Section 2 of this Chapter, except that the US models are finished in Ebony. For all other information, refer to that given for the ZX550 A1/A1L models in Chapters 1 to 6. Their engine and frame numbers are as follows:

UK model:	
Engine number	KZ550DE052001 on
Frame number	ZX550A – 009401 on
US models:	
Engine number	KZ550DE052001 – 056800
Frame number	JKAZXFA1*FA009401 – 013800

ZX550 A3 (GPz 550)

6 This model was introduced to the UK only in 1986 (US models having been discontinued in favour of the liquid-cooled GPZ models). Apart from the modifications to the clutch release and selector fork shaft described in this Chapter it is mechanically identical to the ZX550 A1 (UK) model described in Chapters 1 to 6. It can be distinguished by its paintwork which is now in Ebony, with yellow and orange graphics, and by its grey-painted wheels. Engine and frame numbers as follows:

Engine number	KZ550DE056801 on
Frame number	ZX550A-013801 on

ZX550 A4 (GPz 550)

7 Introduced to the UK in late 1987/early 1988, this model is identical to the ZX550 A1 (UK) model described in Chapters 1 to 6 apart from the modifications to the clutch release and selector fork shaft, the modified air filter cleaning procedure and the Keihin carburettors, as described in this Chapter. Its finish is identical to the ZX550 A3 model described above, the only identifying feature being the different make of carburettor. Engine and frame numbers not available at the time of writing.

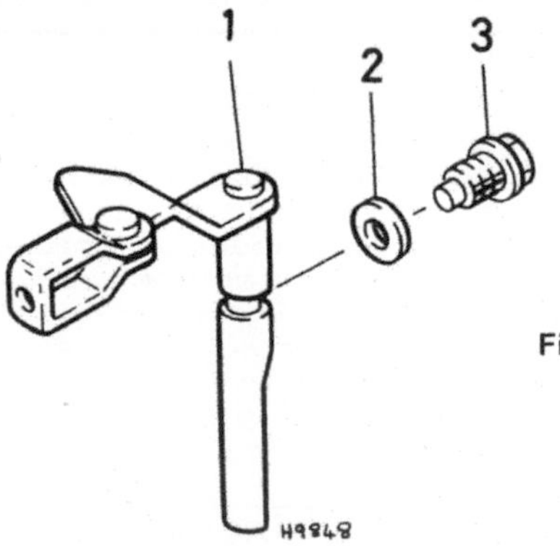

Fig. 7.1 Clutch release lever

1 Release lever
2 Sealing washer
3 Special bolt

Z550 G3 (GT550)

8 Introduced to the UK in 1986, this model is finished in Luminous Polaris Blue and, apart from the selector fork shaft modification described in this Chapter, is identical to the Z550 G2 model described in Chapters 1 to 6. For all other information, refer to that given for the Z550 G2 in Chapters 1 to 6. Engine and frame numbers are as follows:

Z550 G4 (GT550)

9 Introduced to the UK in 1987, this model is identical to the Z550 G2 model described in Chapters 1 to 6, apart from the modification to the selector fork shaft, the modified air filter cleaning procedure and the Keihin carburettors, as described in this Chapter. Its finish is identical to that of the Z550 G3 model described above, the only identifying feature being the different make of carburettor. Engine and frame numbers as follows:

Engine number	KZ550FE010601 on
Frame number	KZ550G-004501 on

2 Routine maintenance: cleaning the air filter element – ZX550 A4 and Z550 G4

1 These models are fitted with a different air filter element which must be cleaned by tapping gently to dislodge any large particles of dirt and by using compressed air to blow clear the element from the inside outwards. If the element or its sealing strips are torn or damaged, or if the element is too badly clogged with dirt to be of any further use, it must be renewed.

2 The removal and refitting procedure, and also the service interval, remain as described in Routine Maintenance.

3 Clutch release: modification – ZX400 C2 and ZX550 A2/A2L, A3, A4

1 These models are fitted with a modified clutch release mechanism. The release lever is altered at the point where it bears on the pull rod and is located in the clutch cover by a special bolt, see the accompanying illustration.

2 Removal, examination and refitting is as described in Chapter 1, but note that the locating bolt must be removed before the release lever can be pulled out of the cover. On refitting, renew the sealing washer if worn or damaged and tighten the bolt to a torque setting of 0.7 kgf m (5 lbf ft). Check that the release lever is correctly located and free to rotate before refitting the cover.

4 Selector fork shaft: modification – ZX550 A3, A4 and Z550 G3, G4

With reference to Fig. 1.7, the selector fork shaft has been modified to accept a 10 mm E-clip instead of the 12 mm external circlip previously used.

5 Unleaded fuel: general – all models

1 Owners should note that all Kawasaki motorcycles since 1974 have been designed and built to use unleaded fuel. All models covered in this Manual will therefore run better on unleaded fuel than on leaded or low-lead fuel; owners are advised to use unleaded fuel wherever possible to prevent spark plug fouling and to gain smoother running.
2 Note that the fuel used, whether unleaded or leaded, must have an octane rating of at least 91 (Research Method/RON). UK owners should appreciate that this means using three-star (where available) or four-star; UK two-star petrol has *too low* a rating to be suitable.

6 Carburettors: general – ZX400 C2, ZX550 A4 and Z550 G4

1 These models are fitted with Keihin carburettors which are essentially similar in design and construction to the Teikei units described in Chapter 2. Refer to Routine Maintenance and/or Chapter 2 for details of regular adjustments, noting the different settings given in the Specifications Section of this Chapter.
2 All other working procedures are basically the same as those given in Chapter 2, but refer closely to the accompanying illustration and note any differences before starting work.

7 Front forks: ZX400 C2

1 Refer to Routine Maintenance for details of changing the fork oil, following the instructions given for the ZX550 A1 model, but note the different fork oil level; no oil capacity is specified for oil changes.
2 For information on removal and refitting refer to Chapter 4, following the instructions given for the ZX550 A1 model, but note that the air union is no longer retained by a circlip.
3 If the left-hand fork leg, or any part of the anti-dive mechanism, is to be serviced, refer to the instructions given for the ZX550 A1 model. If the right-hand fork leg is to be serviced, refer to the instructions given for all other models. Refer to the accompanying illustration before starting work.

8 Front brake disc: mountings – ZX400 C2

Note the different torque wrench setting specified for the front brake disc mounting bolts, also that a shim is fitted between each disc and the wheel hub; do not omit the shim on refitting.

9 Computer warning system: testing – ZX400 C2

With reference to Chapter 6, Section 22.1, there is no green/yellow wire (no separate warning lamp) to be connected on this model. All other test procedures are as given for the ZX550 A1 model.

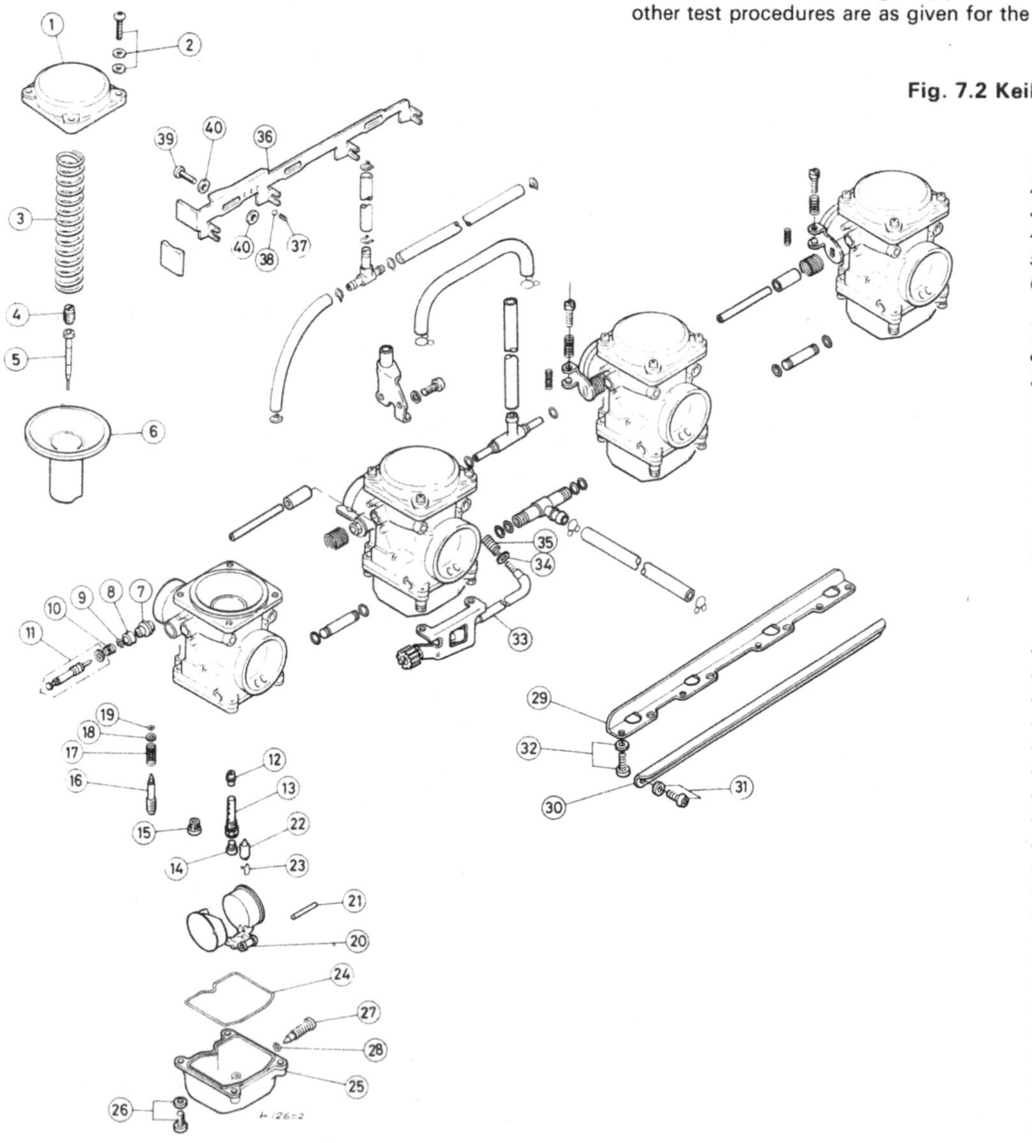

Fig. 7.2 Keihin carburettors – typical

1 *Cover*
2 *Screw and washers*
3 *Spring*
4 *Stop*
5 *Jet needle*
6 *Vacuum piston and diaphragm*
7 *Choke plunger cap*
8 *Cover*
9 *Circlip*
10 *Spring*
11 *Choke plunger*
12 *Needle jet*
13 *Needle jet holder*
14 *Main jet*
15 *Pilot jet*
16 *Pilot screw*
17 *Spring*
18 *Washer*
19 *O-ring*
20 *Float*
21 *Pivot pin*
22 *Valve needle*
23 *Clip*
24 *O-ring*
25 *Float bowl*
26 *Screw and washer*
27 *Drain screw*
28 *O-ring*
29 *Lower mounting bracket*
30 *Upper mounting bracket*
31 *Screw and washer*
32 *Screw and washer*
33 *Idle adjusting screw*
34 *Washer*
35 *Spring*
36 *Choke shaft*
37 *Spring*
38 *Steel ball*
39 *Screw*
40 *Washer*

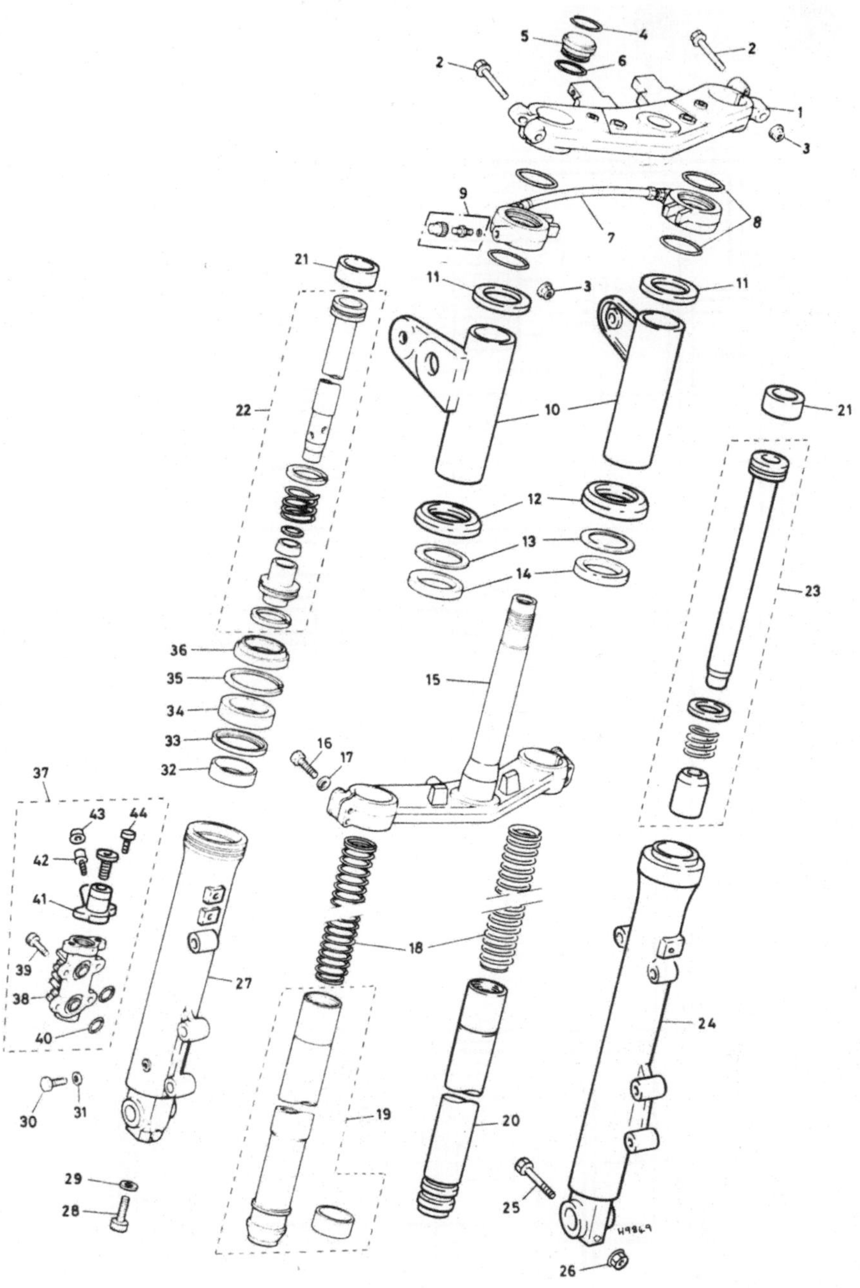

Fig. 7.3 Front forks – ZX400 C2

1 Top yoke
2 Pinch bolt
3 Nut
4 Circlip
5 Top plug
6 O-ring
7 Air union
8 O-ring
9 Air valve
10 Headlamp bracket
11 Guide
12 Guide
13 Washer
14 Rubber spacer
15 Bottom yoke
16 Pinch bolt
17 Spring washer
18 Spring
19 Left-hand stanchion
20 Right-hand stanchion
21 Bottom bush
22 Damper rod assembly – left-hand leg
23 Damper rod assembly – right-hand leg
24 Right-hand lower leg
25 Pinch bolt
26 Nut
27 Left-hand lower leg
28 Allen bolt
29 Gasket
30 Drain plug
31 Gasket
32 Top bush
33 Seal seat
34 Oil seal
35 Circlip
36 Dust seal
37 Anti-dive unit assembly
38 Anti-dive housing
39 Allen screw
40 O-ring
41 Top cover
42 Bleed valve
43 Cap
44 Allen screw

Wiring diagram component key

1 Battery
2 Alternator
3 Regulator/rectifier
4 Starter motor
5 Starter relay
6 Starter lockout switch
7 Starter button
8 Fuses
9 Ignition coils
10 IC ignitor
11 Spark plugs
12 Pickup coils
13 Neutral switch
14 Neutral indicator light
15 Oil pressure switch
16 Oil pressure warning light
17 Oil level sensor
18 Oil level warning light
19 Headlamp
20 Headlamp pass switch
21 Headlamp dip switch
22 High beam indicator light
23 Headlamp failure warning light
24 Parking lamp
25 Lighting switch
26 Reserve lighting unit
27 Horn
28 Horn switch
29 Turn signal switch
30 Front left-hand turn signal
31 Front left-hand turn signal/running light
32 Front right-hand turn signal
33 Front right-hand turn signal/running light
34 Rear left-hand turn signal
35 Rear right-hand turn signal
36 Turn signal relay
37 Left-hand turn signal warning light
38 Right-hand turn signal warning light
39 Turn signal control unit
40 Distance sensor
41 Automatic turn signal cancelling switch
42 Hazard switch
43 Hazard relay
44 Fuel gauge
45 Fuel level sender unit
46 Fuel gauge light
47 LCD fuel gauge
48 Low fuel level warning light
49 Low fuel level/brake light failure warning light
50 Instrument illuminating light
51 Tachometer
52 Tachometer/voltmeter
53 Tachometer/voltmeter switch
54 Speedometer
55 LCD oil level warning unit
56 Side-stand switch
57 Side-stand relay
58 Side-stand release switch
59 LCD side-stand warning unit
60 Warning light self-checker
61 Ignition switch
62 Engine stop switch
63 Front brake light switch
64 Rear brake light switch
65 Brake light failure switch
66 Brake light failure warning light
67 Tail/stop lamp
68 Electrical accessory terminals
69 Fuel gauge/low fuel level warning indicator
70 Battery electrolyte level warning indicator
71 Battery electrolyte level sensor
72 Diode pack
73 Starter control relay
74 General warning light
75 Combination switch

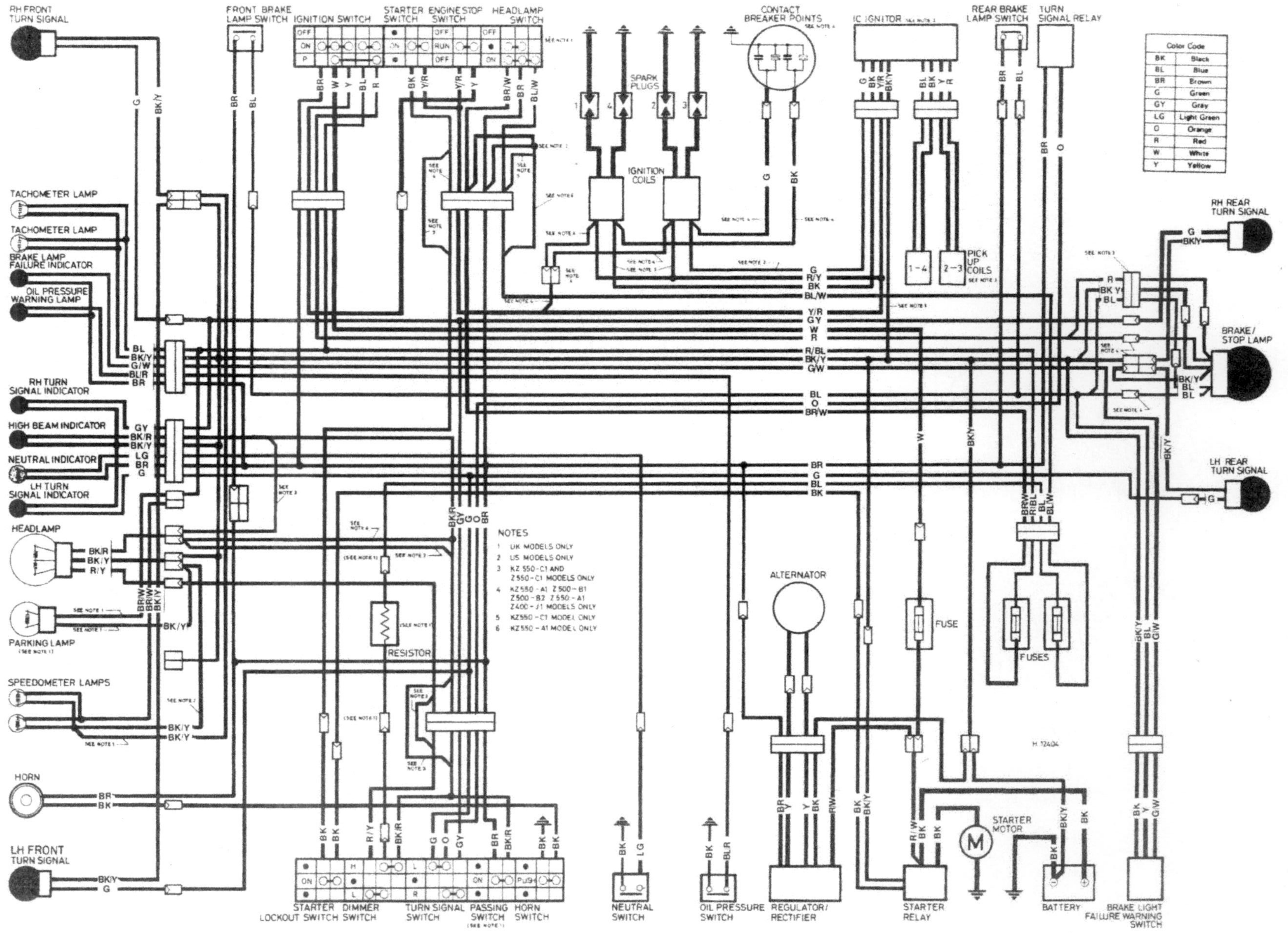

RH FRONT TURN SIGNAL
FRONT BRAKE LAMP SWITCH
IGNITION SWITCH
STARTER SWITCH
ENGINE STOP SWITCH
HEADLAMP SWITCH
SPARK PLUGS
IGNITION COILS
CONTACT BREAKER POINTS
IC IGNITOR
PICK UP COILS
REAR BRAKE LAMP SWITCH
TURN SIGNAL RELAY
Color Code
BK Black
BL Blue
BR Brown
G Green
GY Gray
LG Light Green
O Orange
R Red
W White
Y Yellow
RH REAR TURN SIGNAL
BRAKE/STOP LAMP
LH REAR TURN SIGNAL
TACHOMETER LAMP
TACHOMETER LAMP
BRAKE LAMP FAILURE INDICATOR
OIL PRESSURE WARNING LAMP
RH TURN SIGNAL INDICATOR
HIGH BEAM INDICATOR
NEUTRAL INDICATOR
LH TURN SIGNAL INDICATOR
HEADLAMP
PARKING LAMP
SPEEDOMETER LAMPS
HORN
LH FRONT TURN SIGNAL
NOTES
1 UK MODELS ONLY
2 US MODELS ONLY
3 KZ 550-C1 AND Z 550-C1 MODELS ONLY
4 KZ 550 - A1, Z 500 - B1 Z 500 - B2, Z 550 - A1 Z 400 - J1 MODELS ONLY
5 KZ550 - C1 MODEL ONLY
6 KZ550 - A1 MODEL ONLY
RESISTOR
ALTERNATOR
FUSE
FUSES
STARTER MOTOR
STARTER LOCKOUT SWITCH
DIMMER SWITCH
TURN SIGNAL SWITCH
PASSING SWITCH
HORN SWITCH
NEUTRAL SWITCH
OIL PRESSURE SWITCH
REGULATOR/RECTIFIER
STARTER RELAY
BATTERY
BRAKE LIGHT FAILURE WARNING SWITCH
H 12404

Wiring diagram – Z400 J1, Z500 B1/B2, KZ/Z550 A1 and KZ/Z550 C1

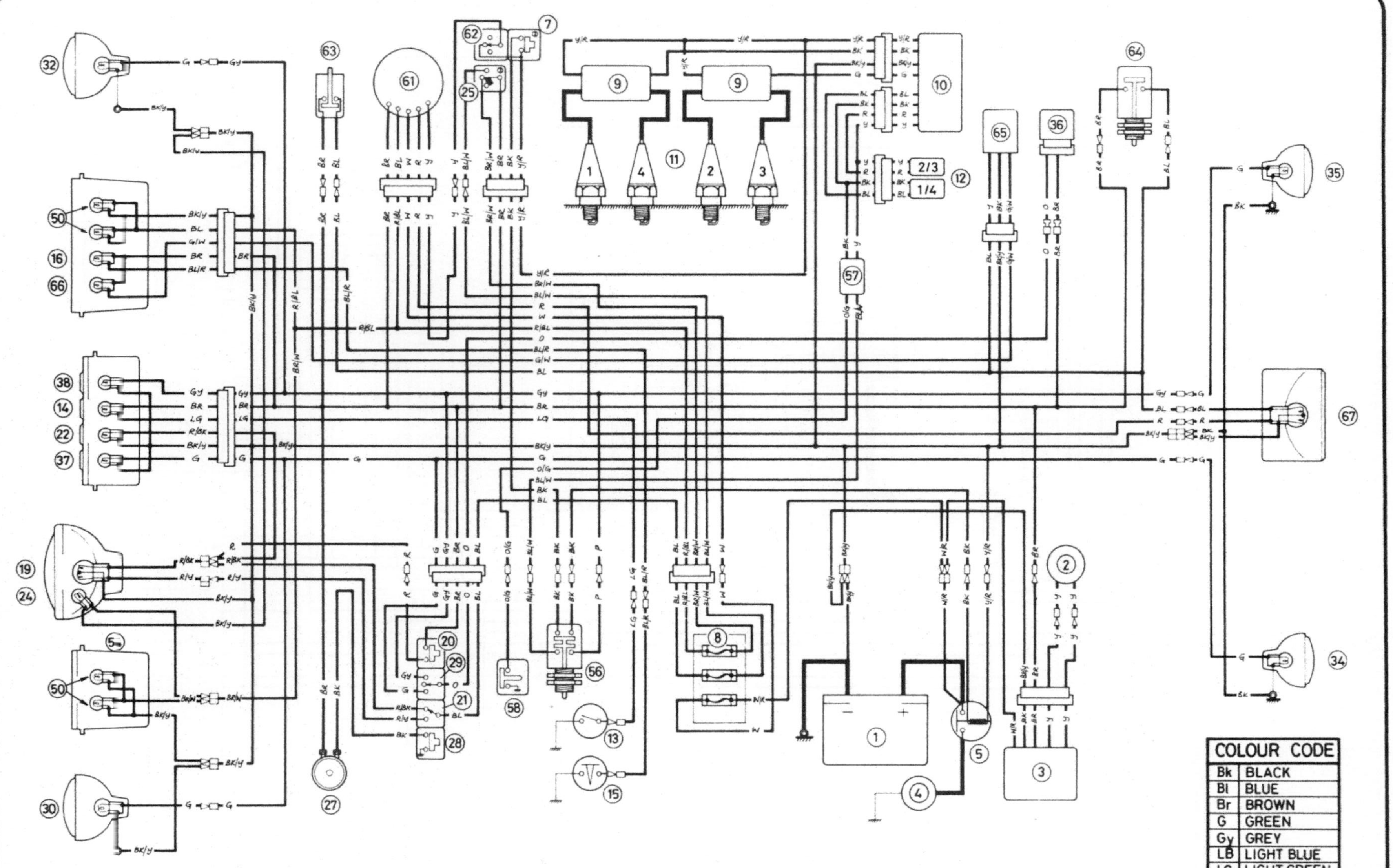

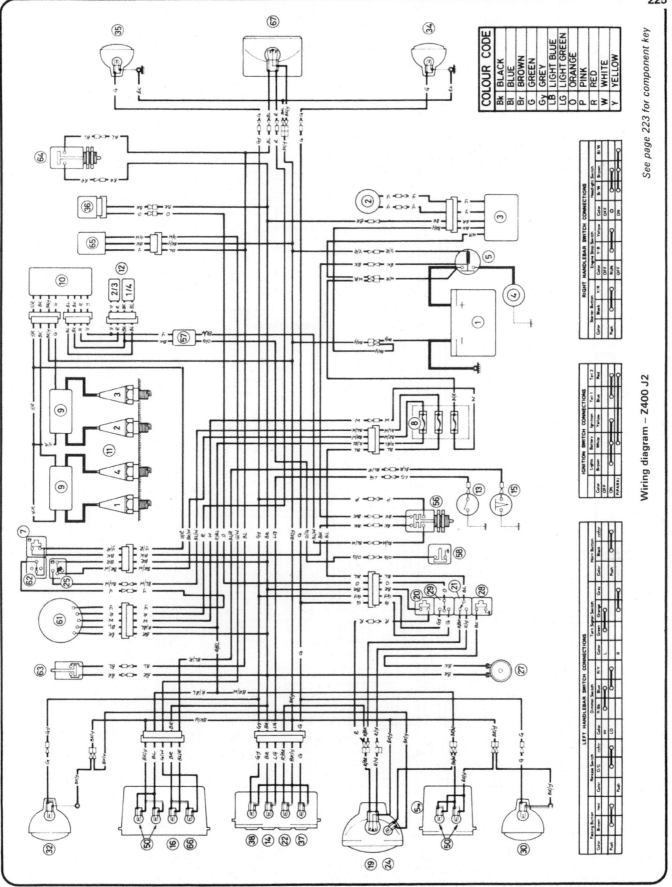

COLOUR CODE	
Bk	BLACK
Bl	BLUE
Br	BROWN
G	GREEN
Gy	GREY
LB	LIGHT BLUE
LG	LIGHT GREEN
O	ORANGE
P	PINK
R	RED
W	WHITE
Y	YELLOW

LEFT HANDLEBAR SWITCH CONNECTIONS

Passing Button			Release Switch			Dimmer Switch				Turn Signal Switch				Horn Button		
Color	Brown	Red	Color	O/G	⏚	Color	R/Bk	Blue	R/Y	Color	Green	Orange	Gray	Color	Black	⏚
						HI	O	O		L	O	O				
Push	O	O		O	O	LO		O	O					Push	O	O
			Push							R		O	O			

IGNITION SWITCH CONNECTIONS

	Lights	Battery	Ignition	Tail 1	Tail 2
Color	Brown	White	Yellow	Blue	Red
OFF					
ON	O	O	O	O	O
P(PARK)		O			O

RIGHT HANDLEBAR SWITCH CONNECTIONS

Starter Button			Engine Stop Switch			Headlight Switch			
Color	Black	Y/R	Color	Y/R	Yellow	Color	Br/W	Brown	Bl/W
			OFF			OFF			
Push	O	O	RUN	O	O	O	O	O	
			OFF			ON	O	O	O

Wiring diagram – Z400 J2

See page 223 for component key

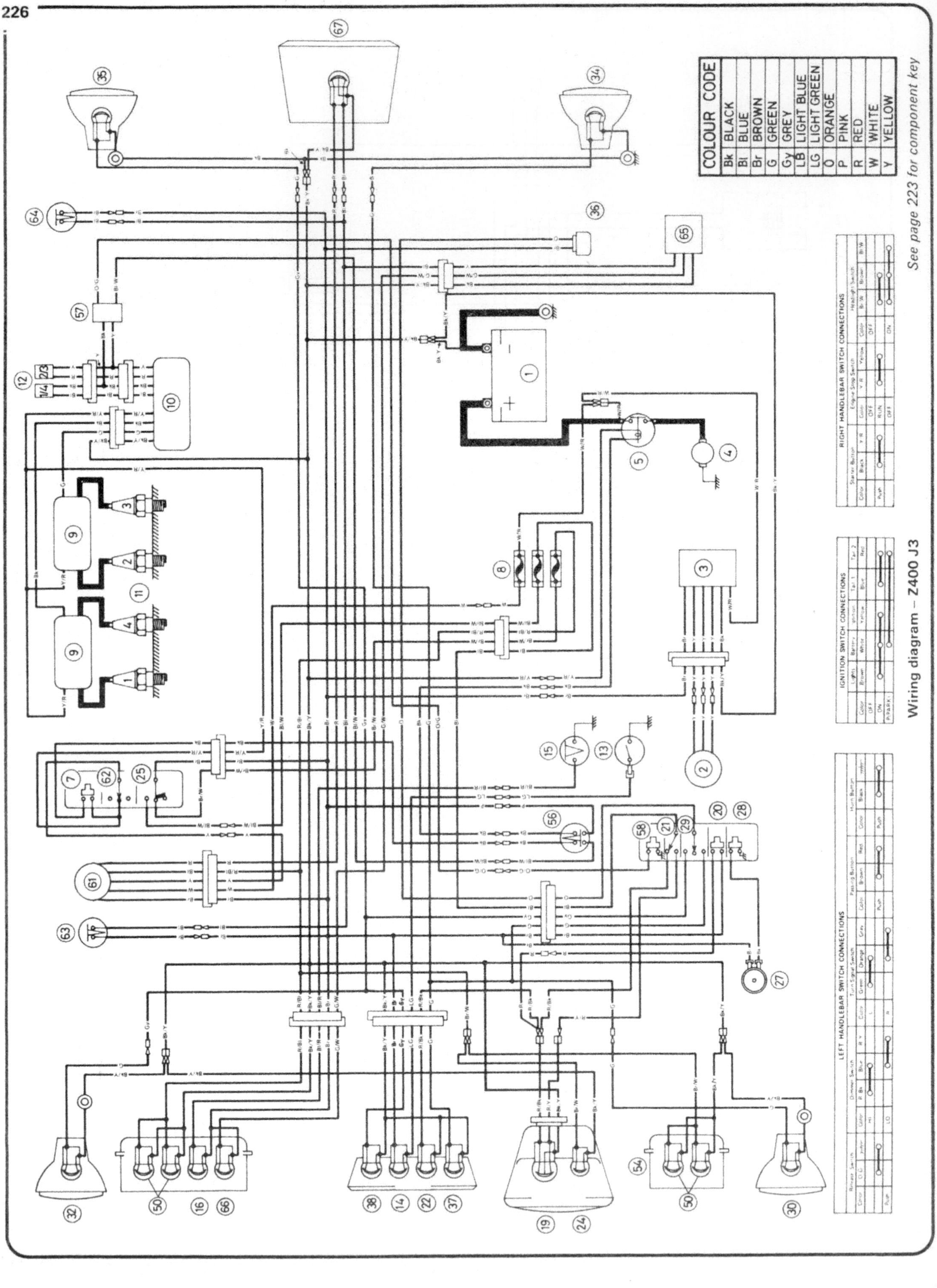

Wiring diagram – Z400 J3

See page 223 for component key

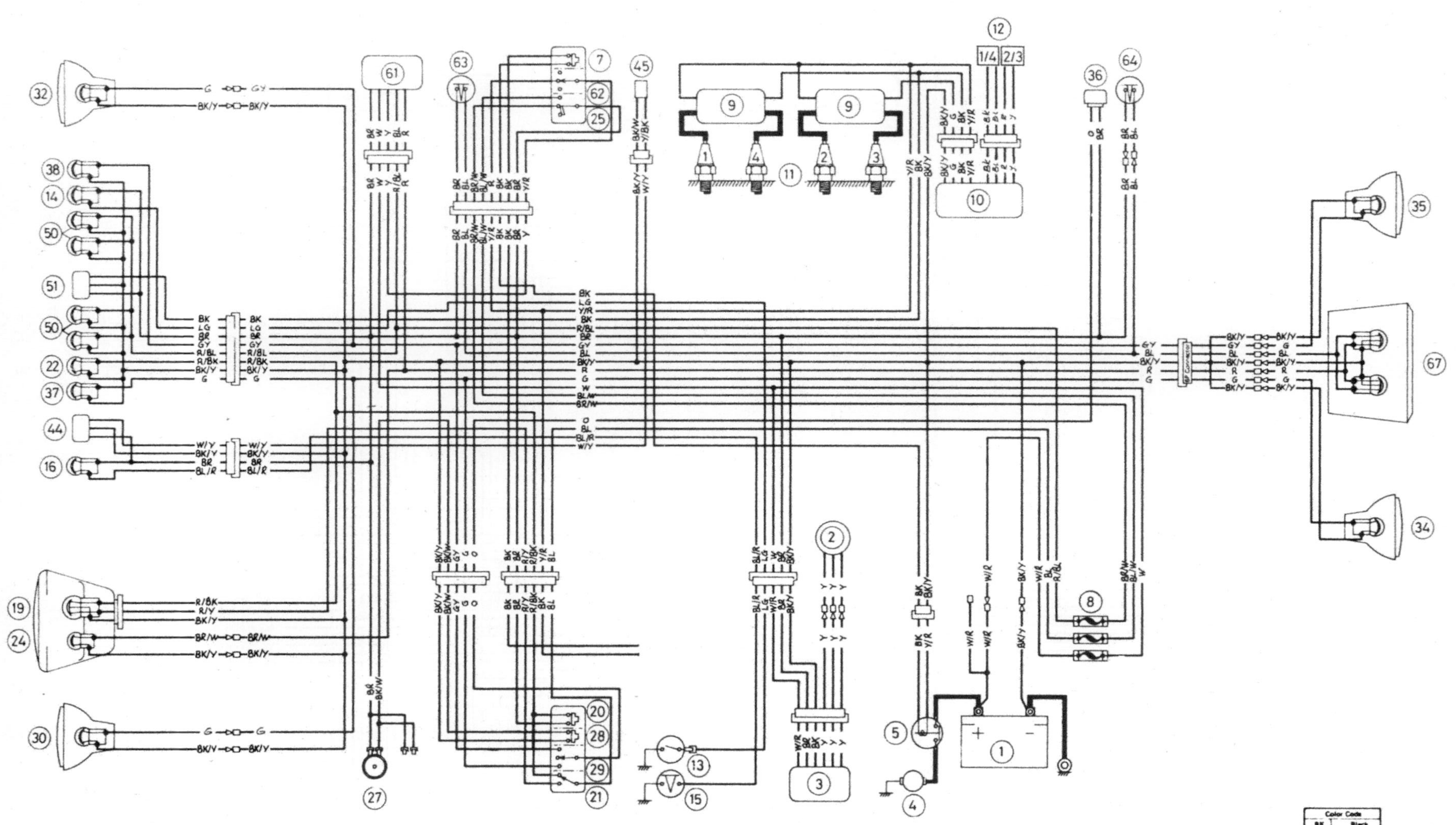

LEFT HANDLEBAR SWITCH CONNECTIONS												
Turn Signal Switch				Dimmer Switch				Horn Switch			Passing Switch	
Color	Green	Orange	Gray	Color	Blue	R/BK	R/Y	Color	BK/W	BK/Y	Color	R/BK	Brown
L	o	o		HI	o	o							
N								ON	o	o	ON	o	o
R		o	o	LO	o	o	o						

IGNITION SWITCH CONNECTIONS					
	Battery	Light	Ignition	Tail 1	Tail 2
Color	White	Brown	Yellow	Blue	Red
OFF (and LOCK)					
ON	o	o	o	o	o
P(PARK)	o	o	o	o	o

RIGHT HANDLEBAR SWITCH CONNECTIONS								
Engine Stop/Starter Switch					Headlight Switch			
Color	Black	Black	Y/R	Red	Color	BR/W	Brown	BL/W
OFF					OFF			
RUN			o	o	PO	o	o	
PUSH with RUN	o	o	o	o	ON	o	o	o

Color Code	
BK	Black
BL	Blue
BR	Brown
G	Green
GY	Gray
LB	Light Blue
LG	Light Green
O	Orange
P	Pink
R	Red
W	White
Y	Yellow

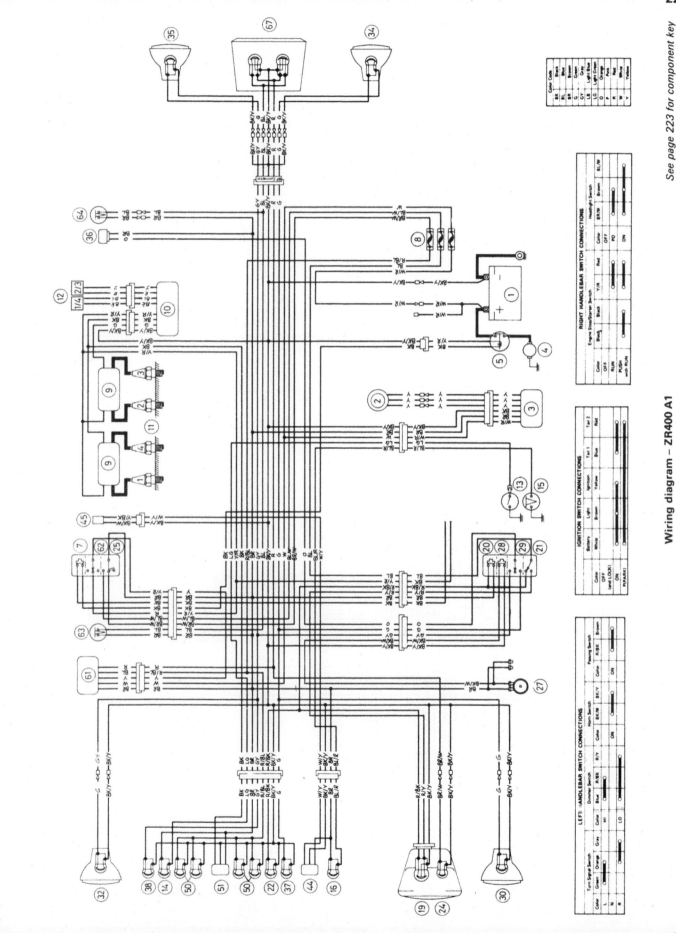

Wiring diagram – ZR400 A1

See page 223 for component key

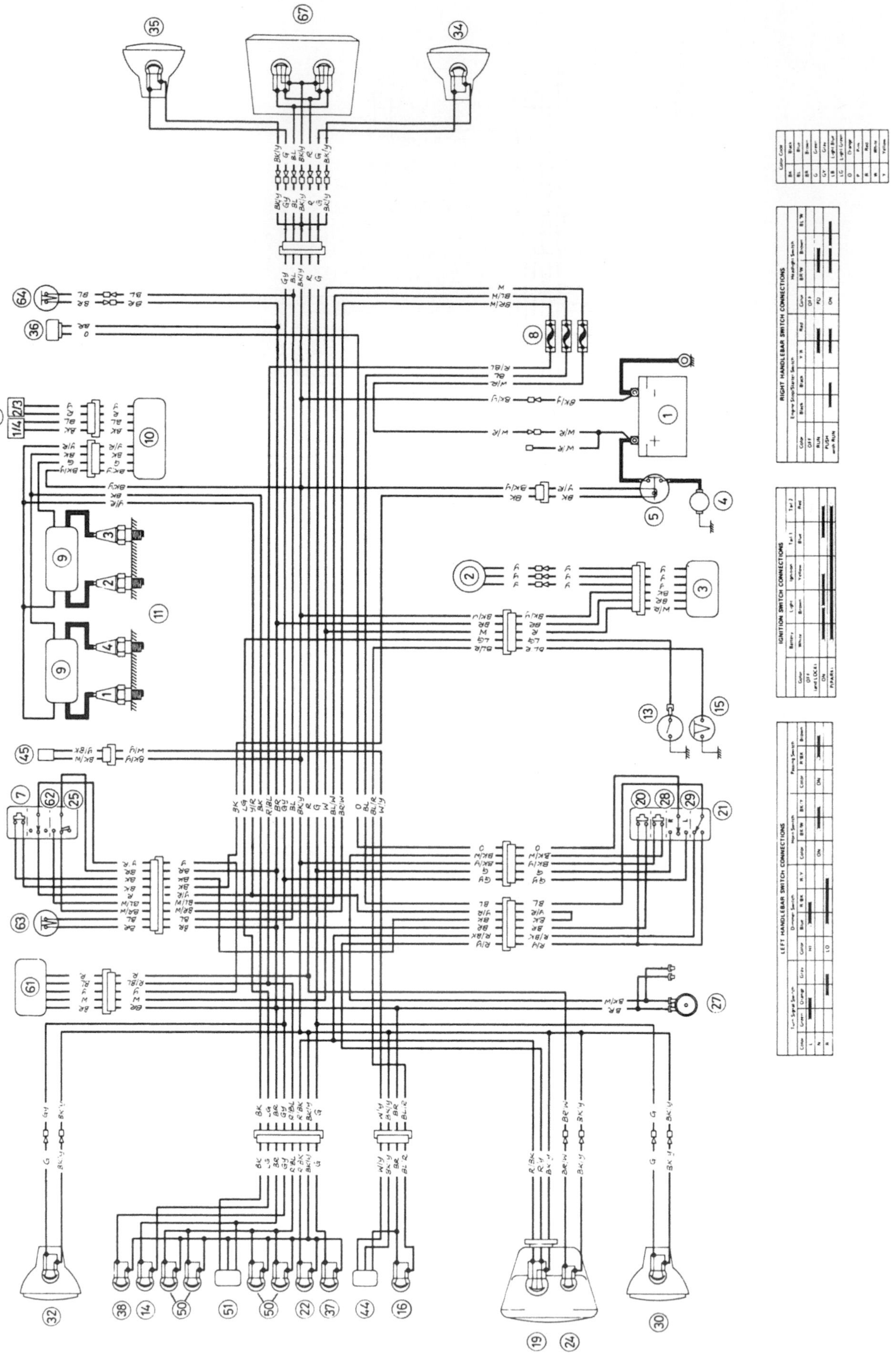

Wiring diagram – ZR400 B1

See page 223 for component key

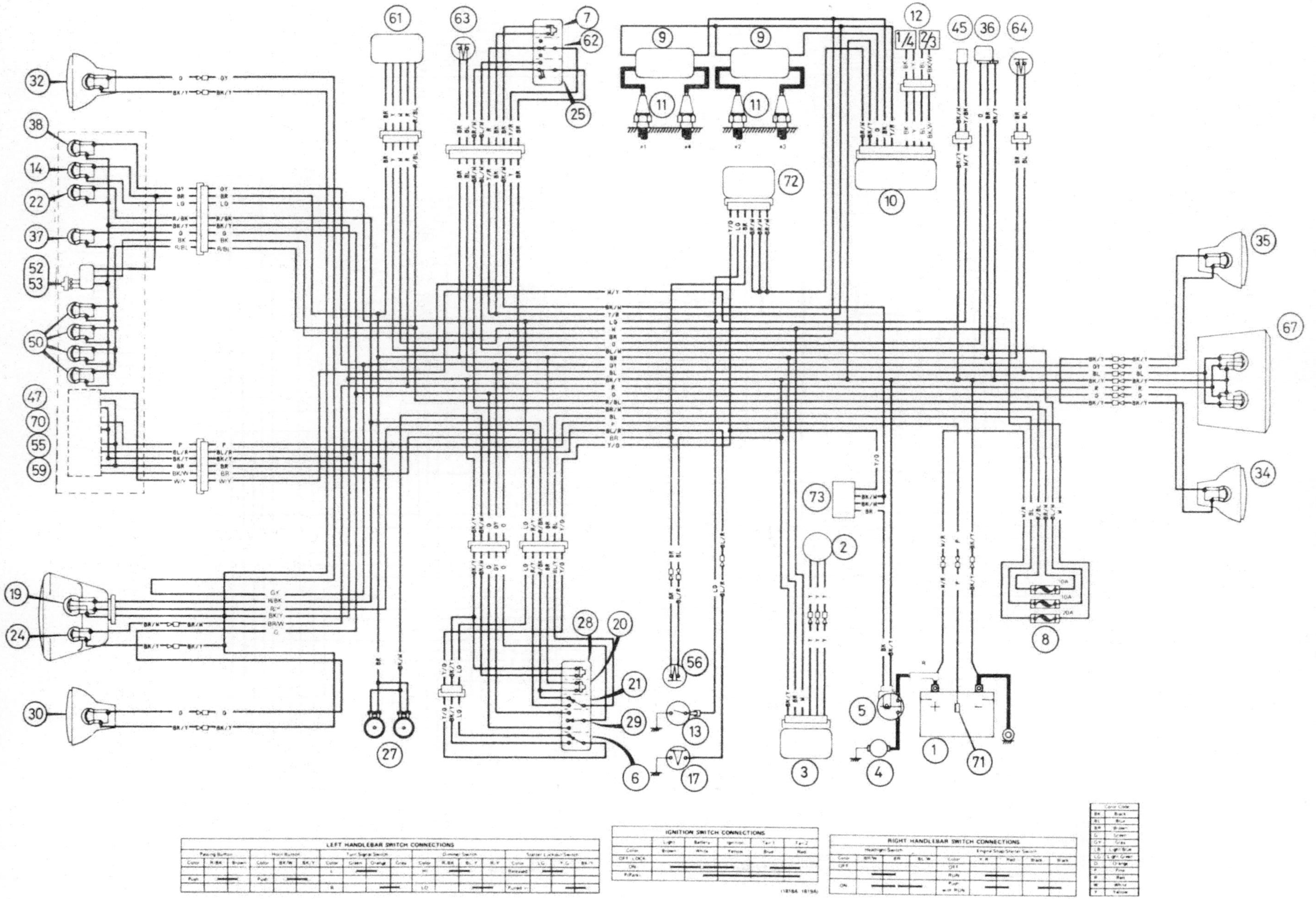

Wiring diagram – ZX400 C2

See page 223 for component key

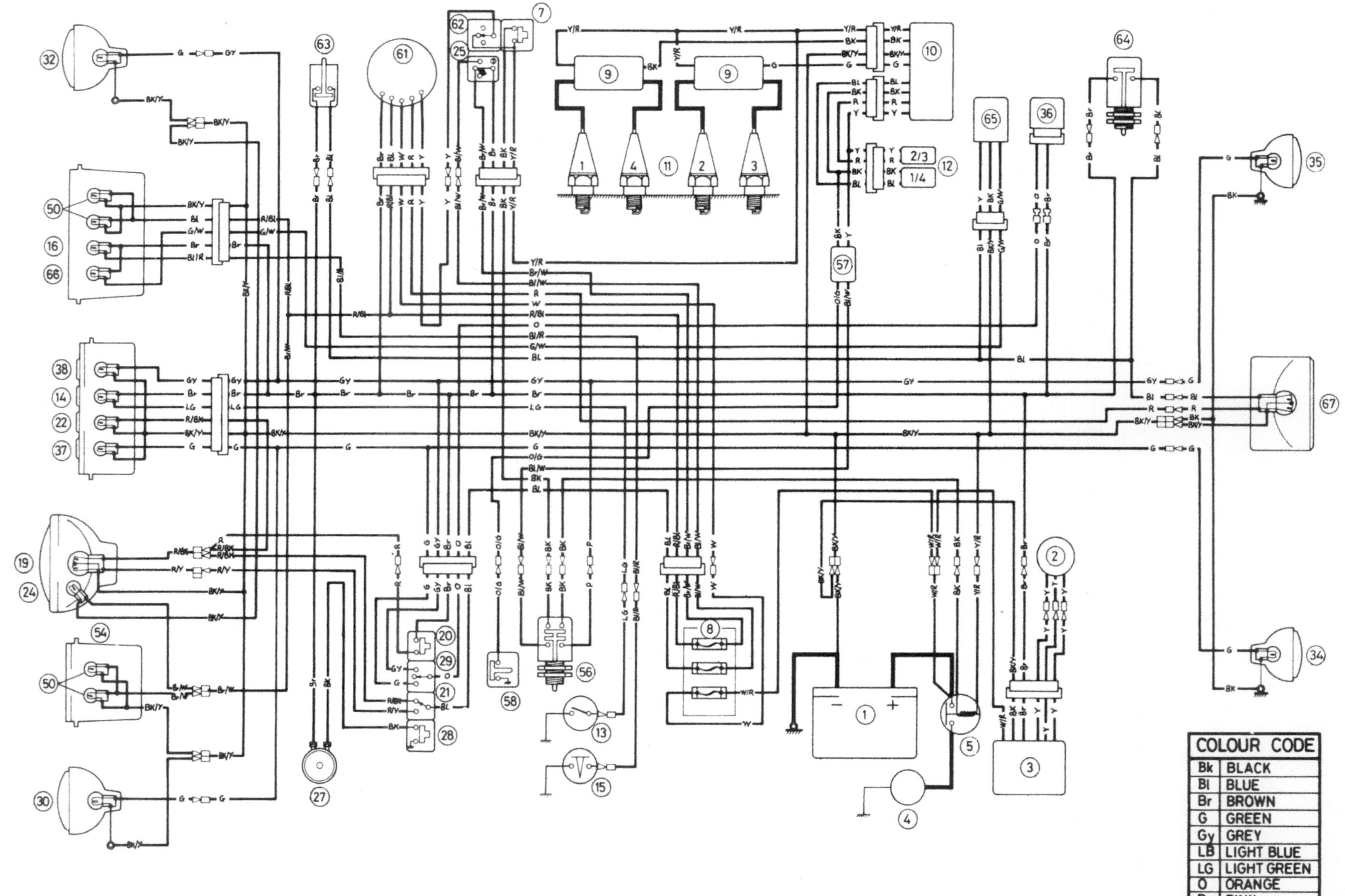

COLOUR CODE	
Bk	BLACK
Bl	BLUE
Br	BROWN
G	GREEN
Gy	GREY
LB	LIGHT BLUE
LG	LIGHT GREEN
O	ORANGE
P	PINK
R	RED
W	WHITE
Y	YELLOW

LEFT HANDLEBAR SWITCH CONNECTIONS

Passing Button			Release Switch			Dimmer Switch				Turn Signal Switch				Horn Button		
Color	Brown	Red	Color	O/G	earth	Color	R/Bk	Blue	R/Y	Color	Green	Orange	Gray	Color	Black	earth
						HI	O	O		L	O	O				
Push	O	O		O	O	LO		O	O					Push	O	O
			Push							R		O	O			

IGNITION SWITCH CONNECTIONS

	Lights	Battery	Ignition	Tail 1	Tail 2
Color	Brown	White	Yellow	Blue	Red
OFF					
ON	O	O	O	O	O
P(PARK)		O			O

RIGHT HANDLEBAR SWITCH CONNECTIONS

Starter Button			Engine Stop Switch			Headlight Switch			
Color	Black	Y/R	Color	Y/R	Yellow	Color	Br/W	Brown	Bl/W
			OFF			OFF			
Push	O	O	RUN	O	O	O	O	O	
			OFF			ON	O	O	O

Wiring diagram – Z550 A2

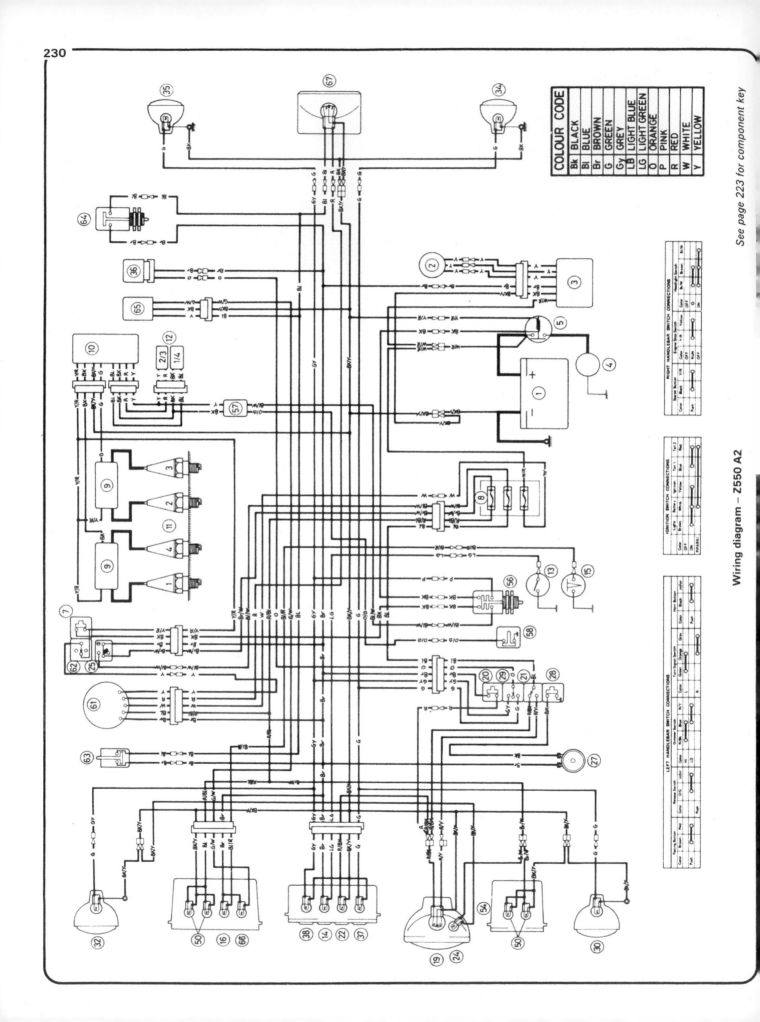

See page 223 for component key

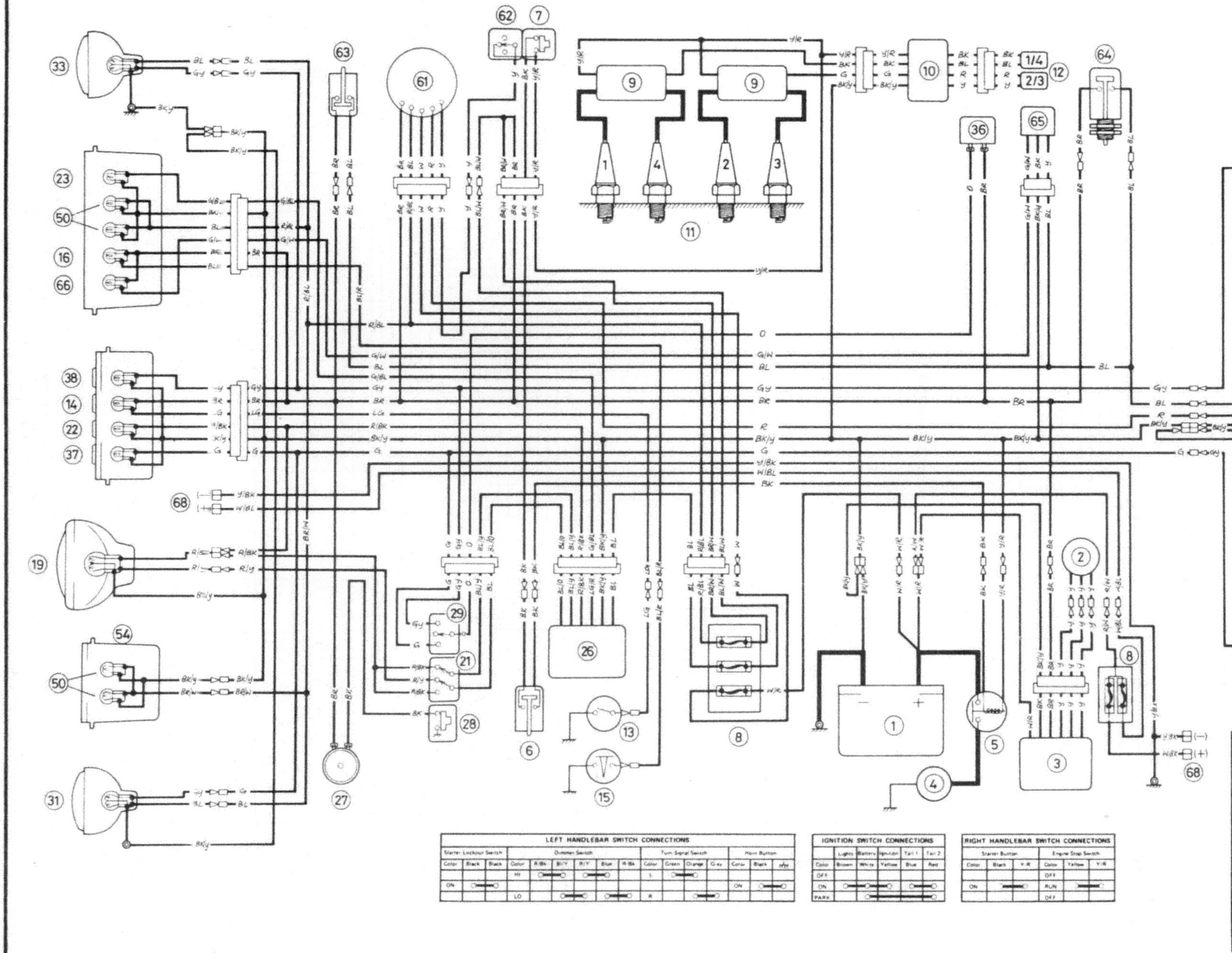

COLOUR CODE	
Bk	BLACK
Bl	BLUE
Br	BROWN
G	GREEN
Gy	GREY
LB	LIGHT BLUE
LG	LIGHT GREEN
O	ORANGE
P	PINK
R	RED
W	WHITE
Y	YELLOW

Wiring diagram – KZ550 A2

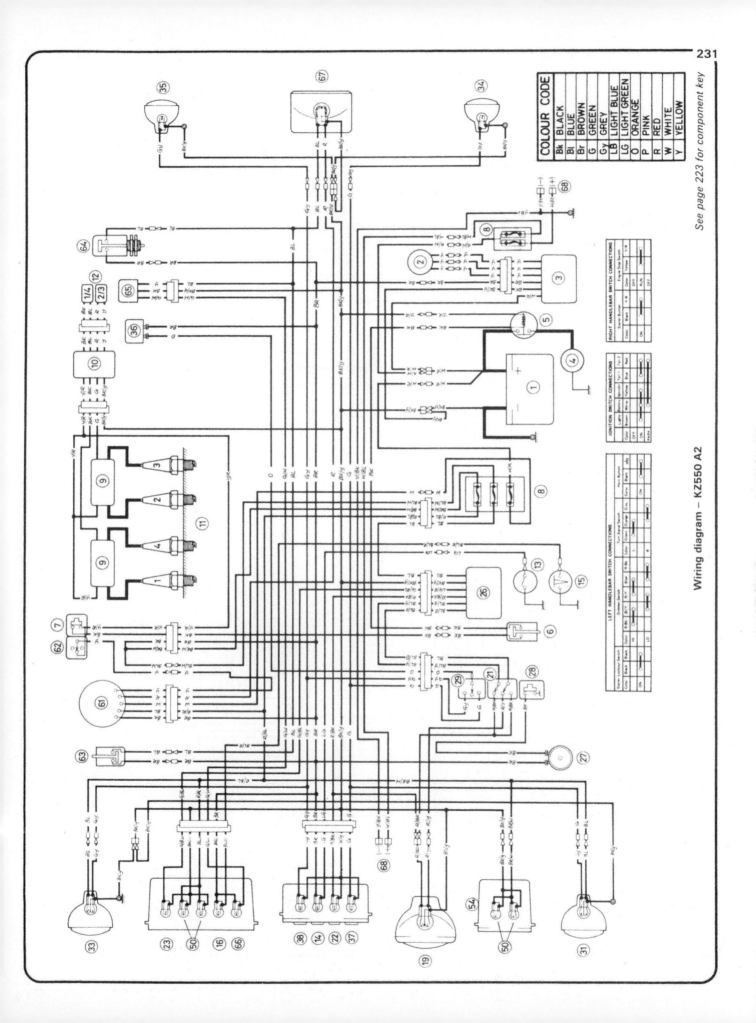

See page 223 for component key

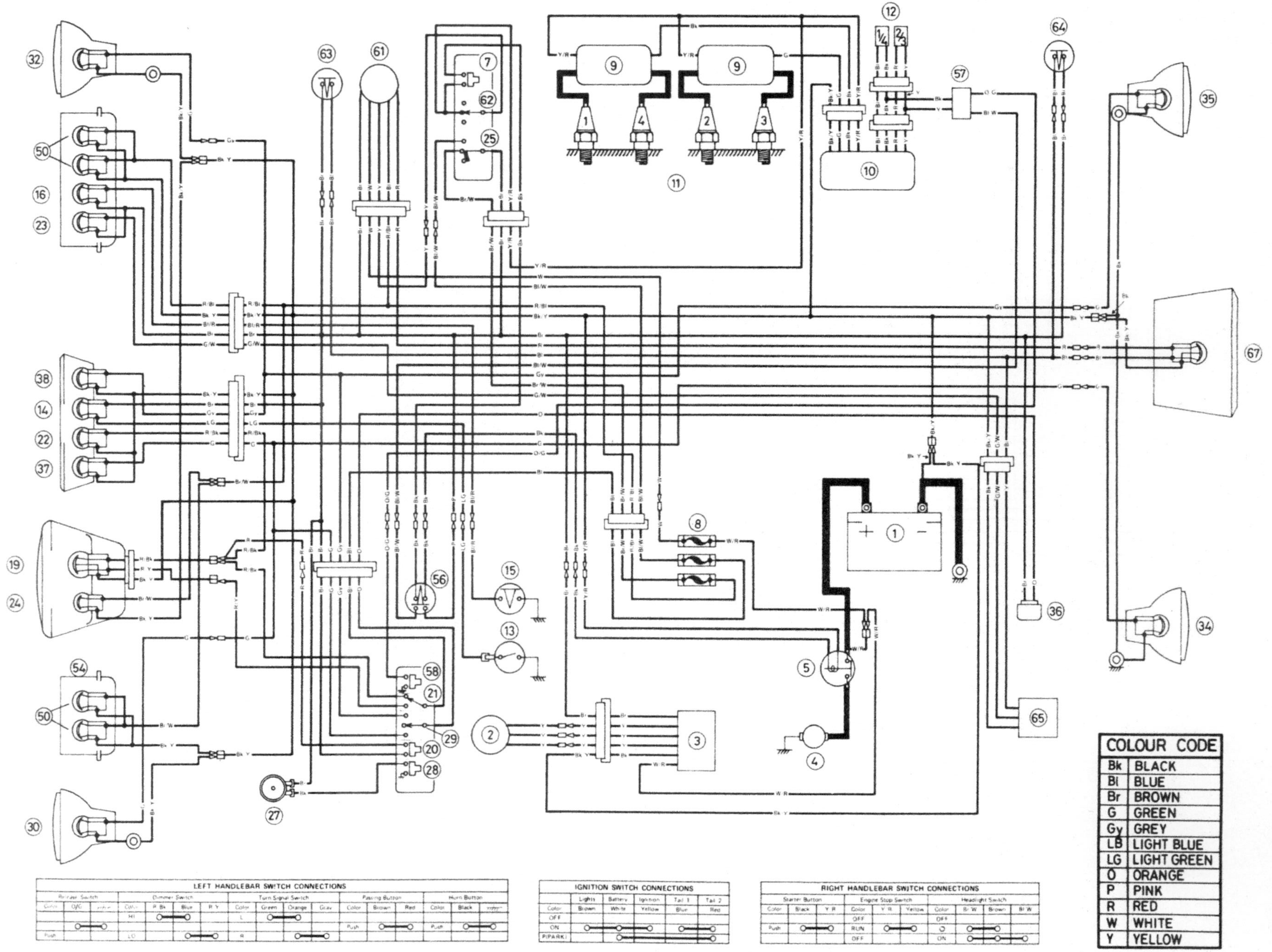

COLOUR CODE	
Bk	BLACK
Bl	BLUE
Br	BROWN
G	GREEN
Gy	GREY
LB	LIGHT BLUE
LG	LIGHT GREEN
O	ORANGE
P	PINK
R	RED
W	WHITE
Y	YELLOW

Wiring diagram – Z550 A3

See page 223 for component key

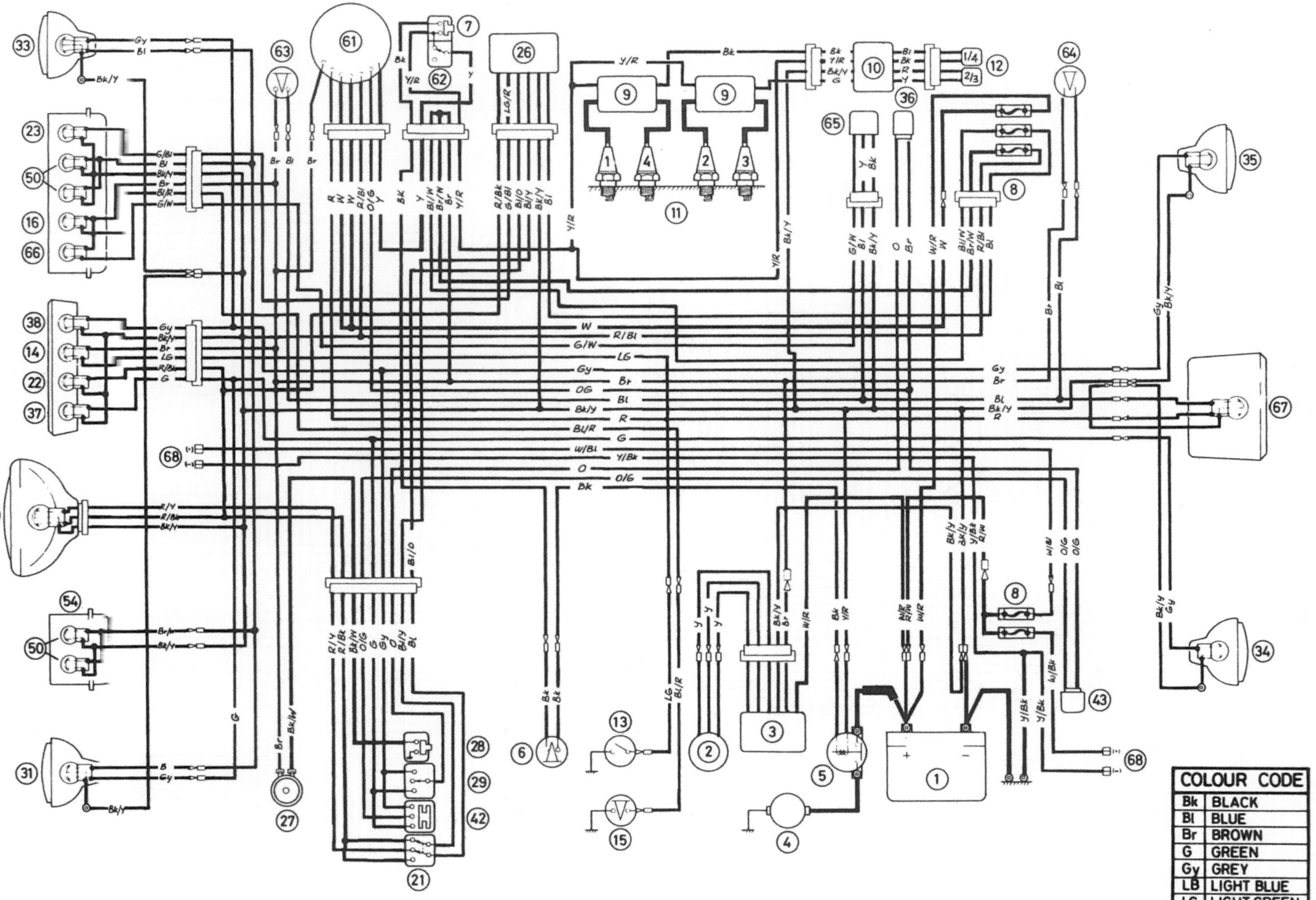

COLOUR CODE	
Bk	BLACK
Bl	BLUE
Br	BROWN
G	GREEN
Gy	GREY
LB	LIGHT BLUE
LG	LIGHT GREEN
O	ORANGE
P	PINK
R	RED
W	WHITE
Y	YELLOW

LEFT HANDLEBAR SWITCH CONNECTIONS

Horn Button			Turn Signal Switch				Hazard Switch				Dimmer Switch						Starter Lockout Switch		
Color	Bk	(earth)	Color	G	O	Gy	Color	O/G	Gy	G	Color	R/Bk	Bl/Y	R/Y	Bl	R/Bk	Color	Bk	Bk
			L	○	○						HI	○	○	○	○				
Push	○	○					ON	○	○	○							ON	○	○
			R		○	○					LO		○	○	○	○			

IGNITION SWITCH CONNECTIONS

	Lights	Bat 1	Ignition	Tail 1	Tail 2	Bat 2	Tail 3
Color	Br	W	Y	Bl	R	W	O/G
OFF							
ON	○	○	○	○	○	○	○
P (Park)		○			○	○	○

RIGHT HANDLEBAR SWITCH CONNECTIONS

Starter Button			Engine Stop Switch		
Color	Bk	Y/R	Color	Y/R	Y
			OFF		
Push	○	○	RUN	○	○
			OFF		

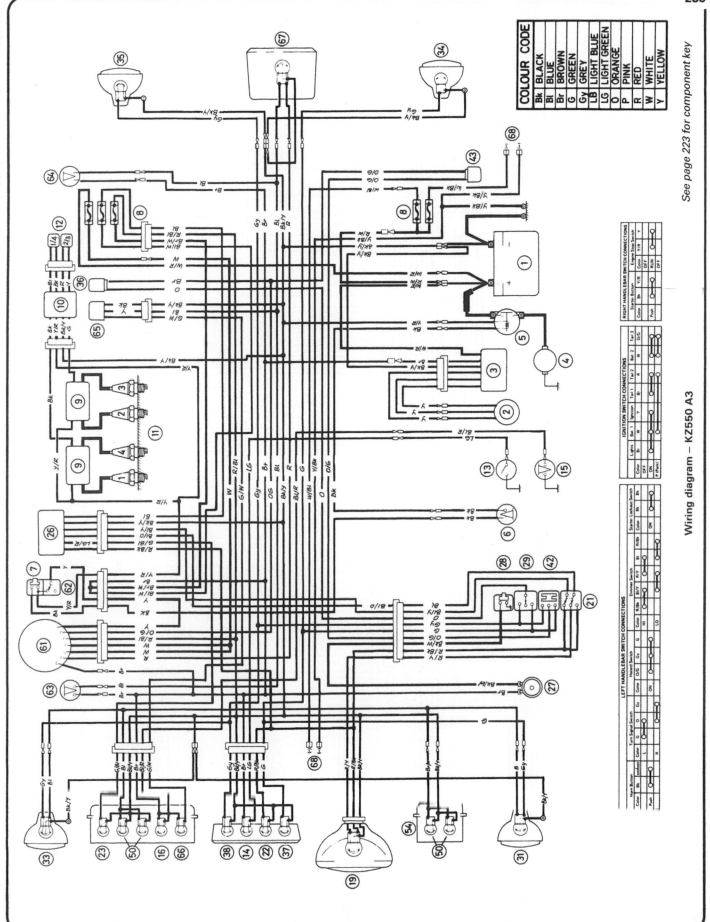

Wiring diagram – KZ550 A3

See page 223 for component key

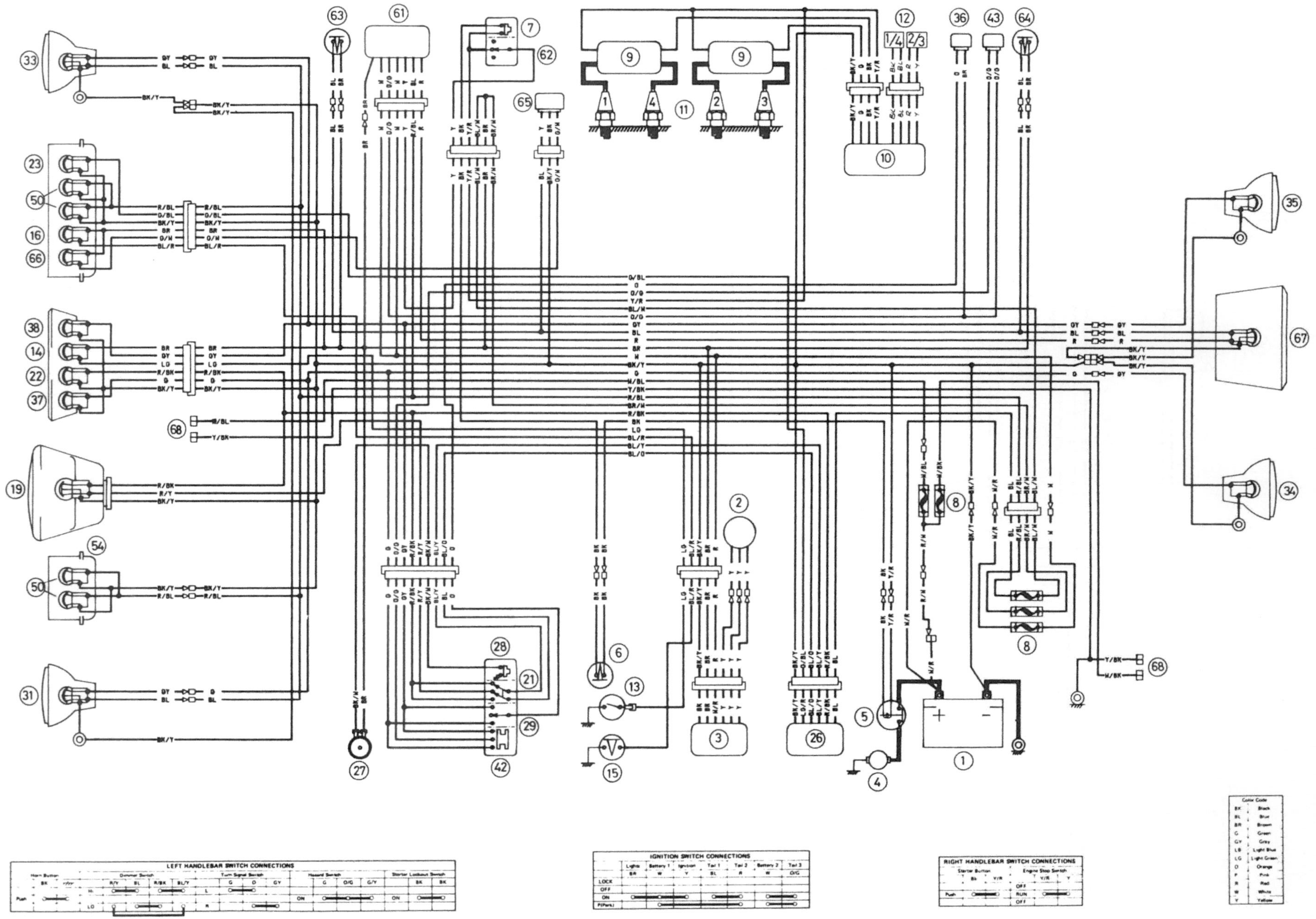

Wiring diagram – KZ550 A4

See page 223 for component key

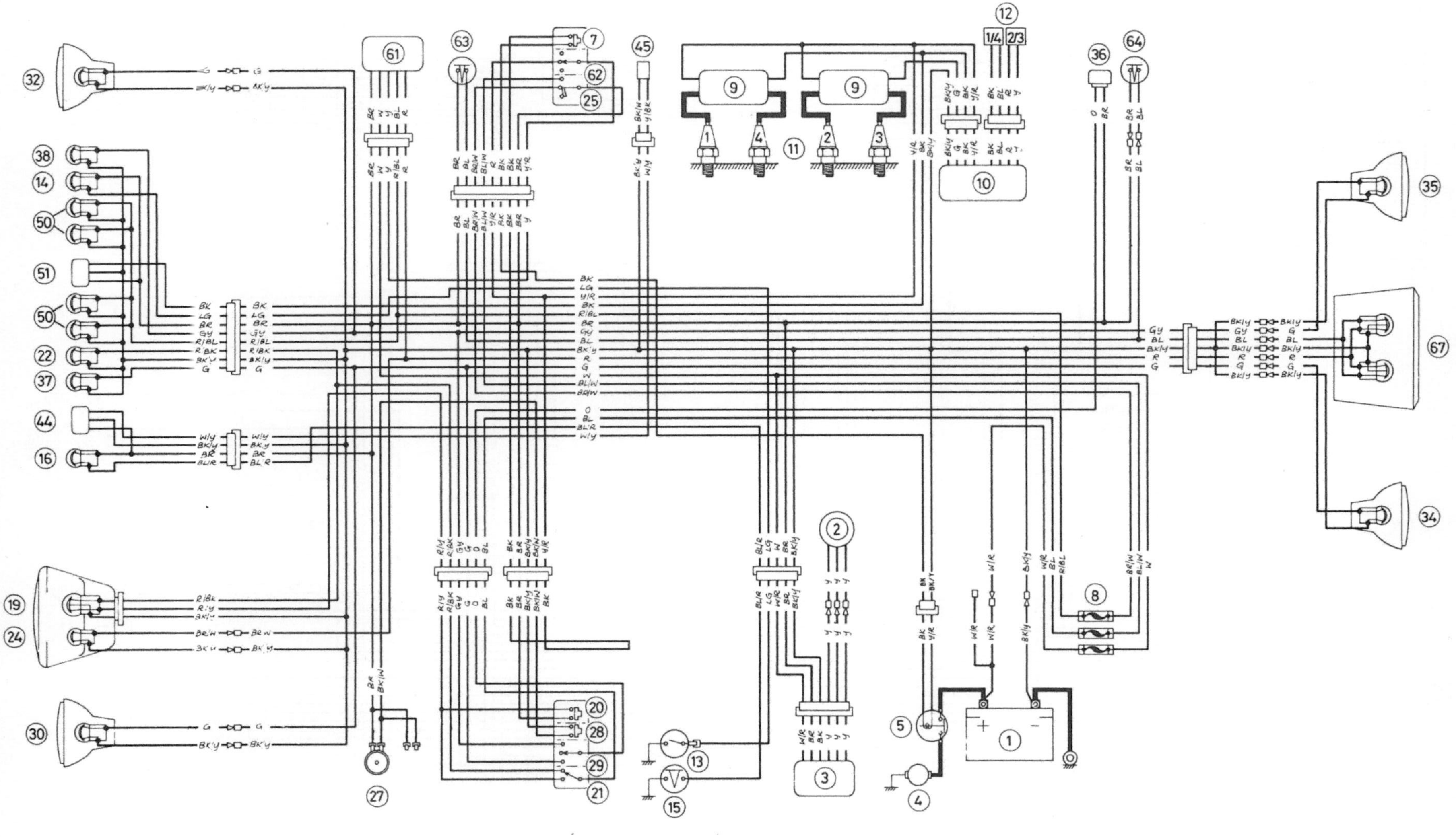

LEFT HANDLEBAR SWITCH CONNECTIONS															
Turn Signal Switch				Dimmer Switch				Horn Switch			Passing Switch		
Color	Green	Orange	Gray	Color	Blue	R/BK	R/Y	Color	BK/W	BK/Y	Color	R/BK	Brown
L	●	●		HI	●	●							
N								ON	●	●	ON	●	●
R		●	●	LO	●	●	●						

IGNITION SWITCH CONNECTIONS					
	Battery	Light	Ignition	Tail 1	Tail 2
Color	White	Brown	Yellow	Blue	Red
OFF (and LOCK)					
ON	●	●	●	●	●
P/PARK	●	●	●	●	●

RIGHT HANDLEBAR SWITCH CONNECTIONS								
Engine Stop/Starter Switch					Headlight Switch			
Color	Black	Black	Y/R	Red	Color	BR/W	Brown	BL/W
OFF					OFF			
RUN			●	●	PO	●	●	
PUSH with RUN	●	●	●	●	ON	●	●	●

Color Code	
BK	Black
BL	Blue
BR	Brown
G	Green
GY	Gray
LB	Light Blue
LG	Light Green
O	Orange
P	Pink
R	Red
W	White
Y	Yellow

Wiring diagram – ZR550 A1

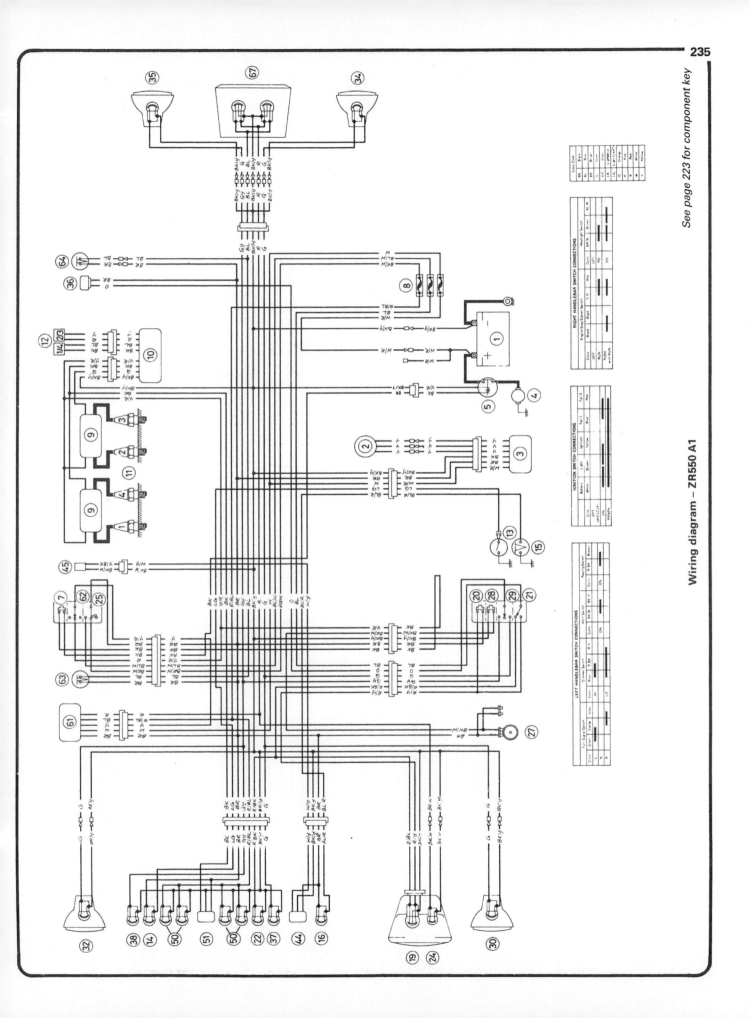

See page 223 for component key

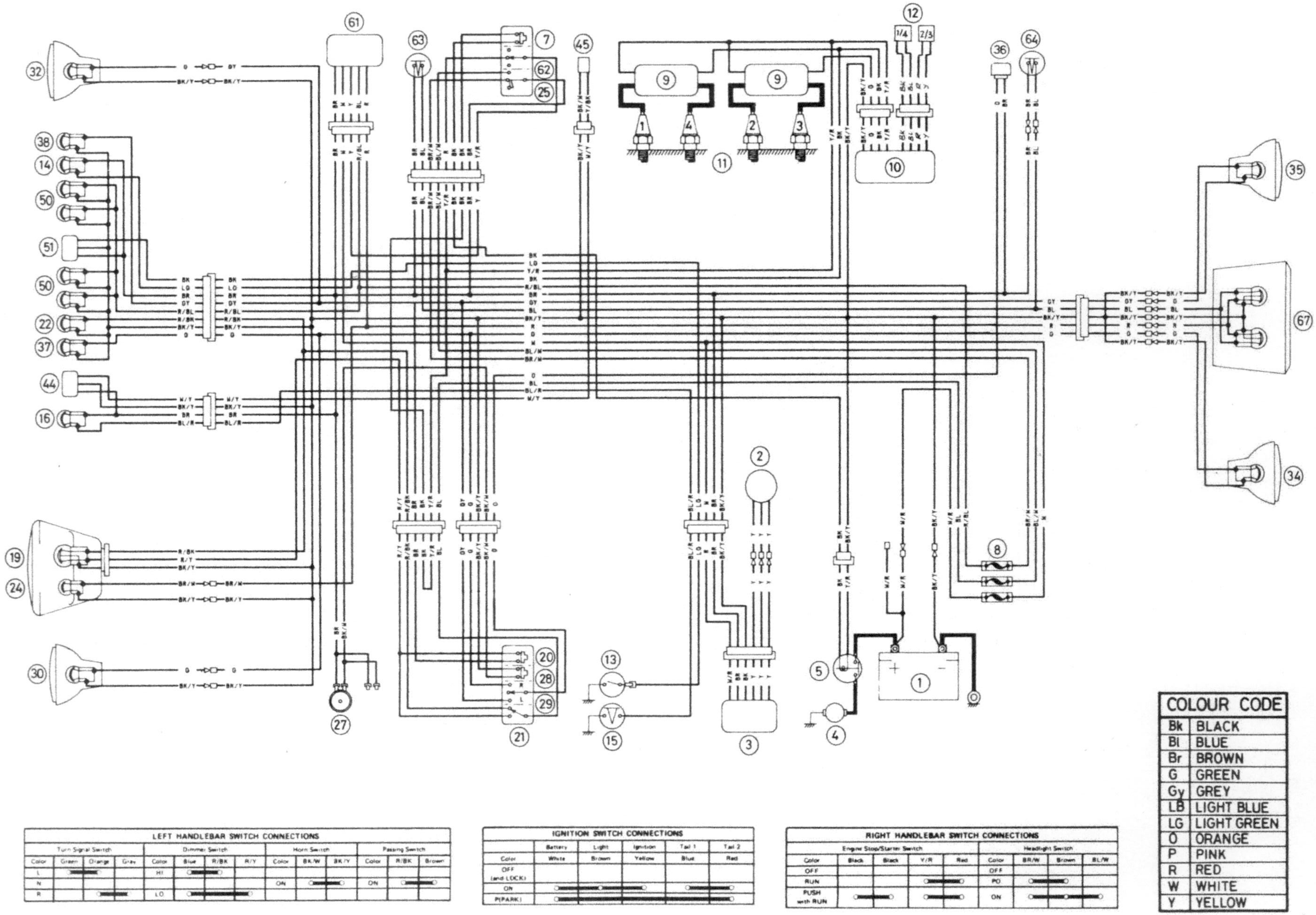

COLOUR CODE	
Bk	BLACK
Bl	BLUE
Br	BROWN
G	GREEN
Gy	GREY
LB	LIGHT BLUE
LG	LIGHT GREEN
O	ORANGE
P	PINK
R	RED
W	WHITE
Y	YELLOW

Wiring diagram – ZR550 A2

See page 223 for component key

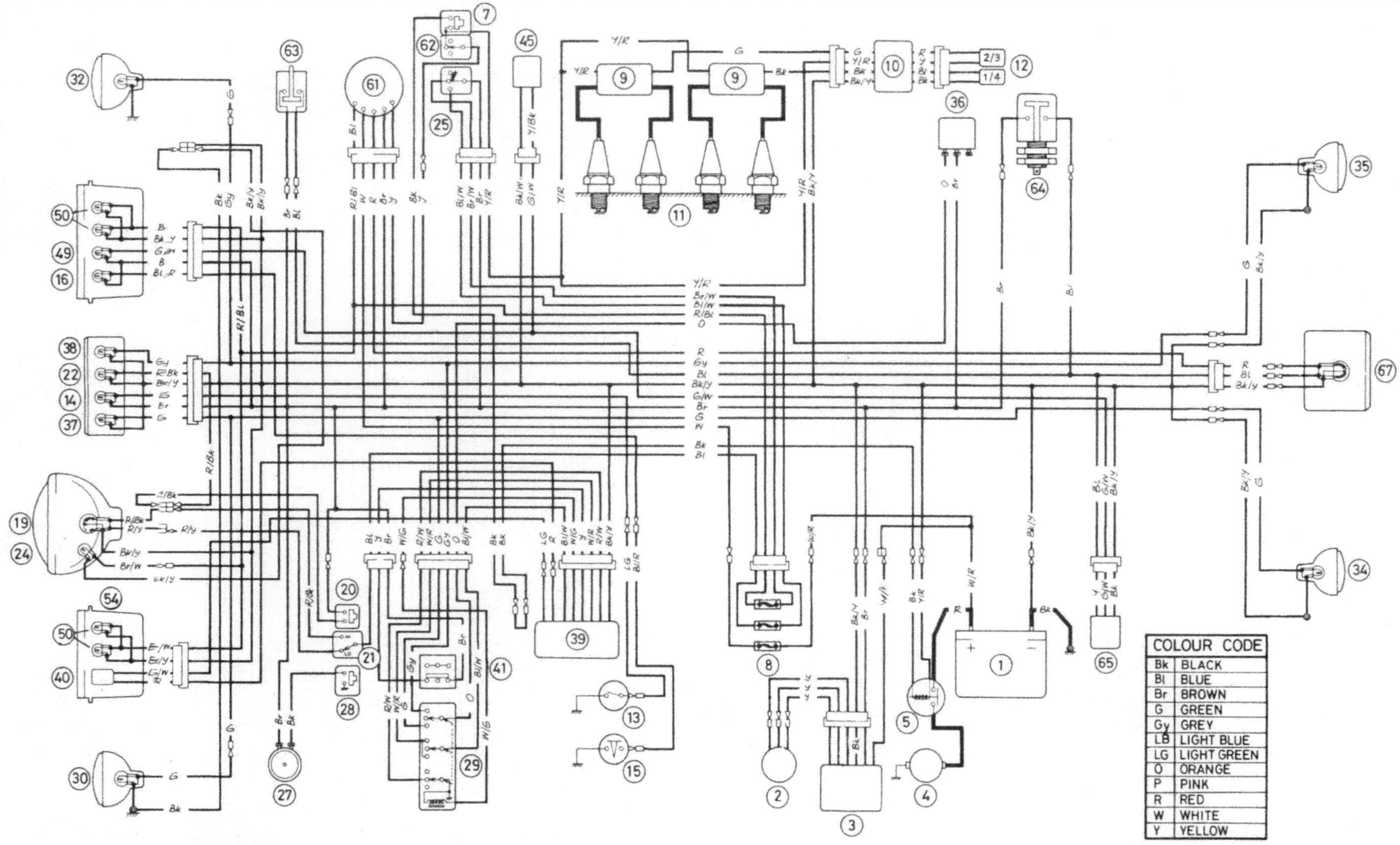

COLOUR CODE	
Bk	BLACK
Bl	BLUE
Br	BROWN
G	GREEN
Gy	GREY
LB	LIGHT BLUE
LG	LIGHT GREEN
O	ORANGE
P	PINK
R	RED
W	WHITE
Y	YELLOW

LEFT HANDLEBAR SWITCH CONNECTIONS

STARTER LOCKOUT SW			DIMMER SWITCH				PASSING SWITCH			HORN BUTTON			TURN SIGNAL SWITCH								CANCEL SWITCH		
	Bk	Bk		R/Bk	Bl	R/Y		Br	Bk/R		Bk	⏚		Gy	O	G	R/W	⏚	W/R	Bl/W		Br	Y
			Hi	O—	—O								L	O—	—O				O—	—O	M		
ON	O—	—O					PUSH	O—	—O		O—	—O					O—	—O					
			Lo		O—	—O							R		O—	—O			O—	—O	A	O—	—O

IGNITION SWITCH

	L	Bat	Ign	T1	T2
	Br	W	Y	Bl	R
OFF					
ON	O—	—O—	—O		
P		O—	—	—	—O

RH HANDLEBAR SWITCH CONNECTIONS

STARTER SWITCH			ENGINE STOP SW			HEADLIGHT SWITCH			
	Bk	Y/R		Y/R	Y		Br/W	Br	Bl/W
			OFF			OFF			
PUSH	O—	—O	RUN	O—	—O	0	O—	—O	
			OFF			ON	O—	—O—	—O

Wiring diagram – Z550 C2

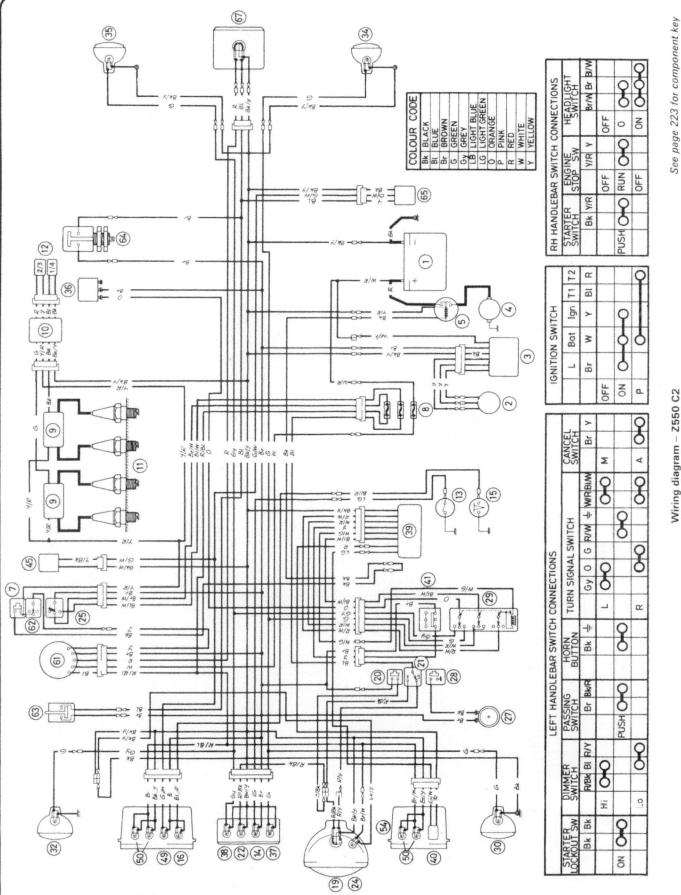

See page 223 for component key

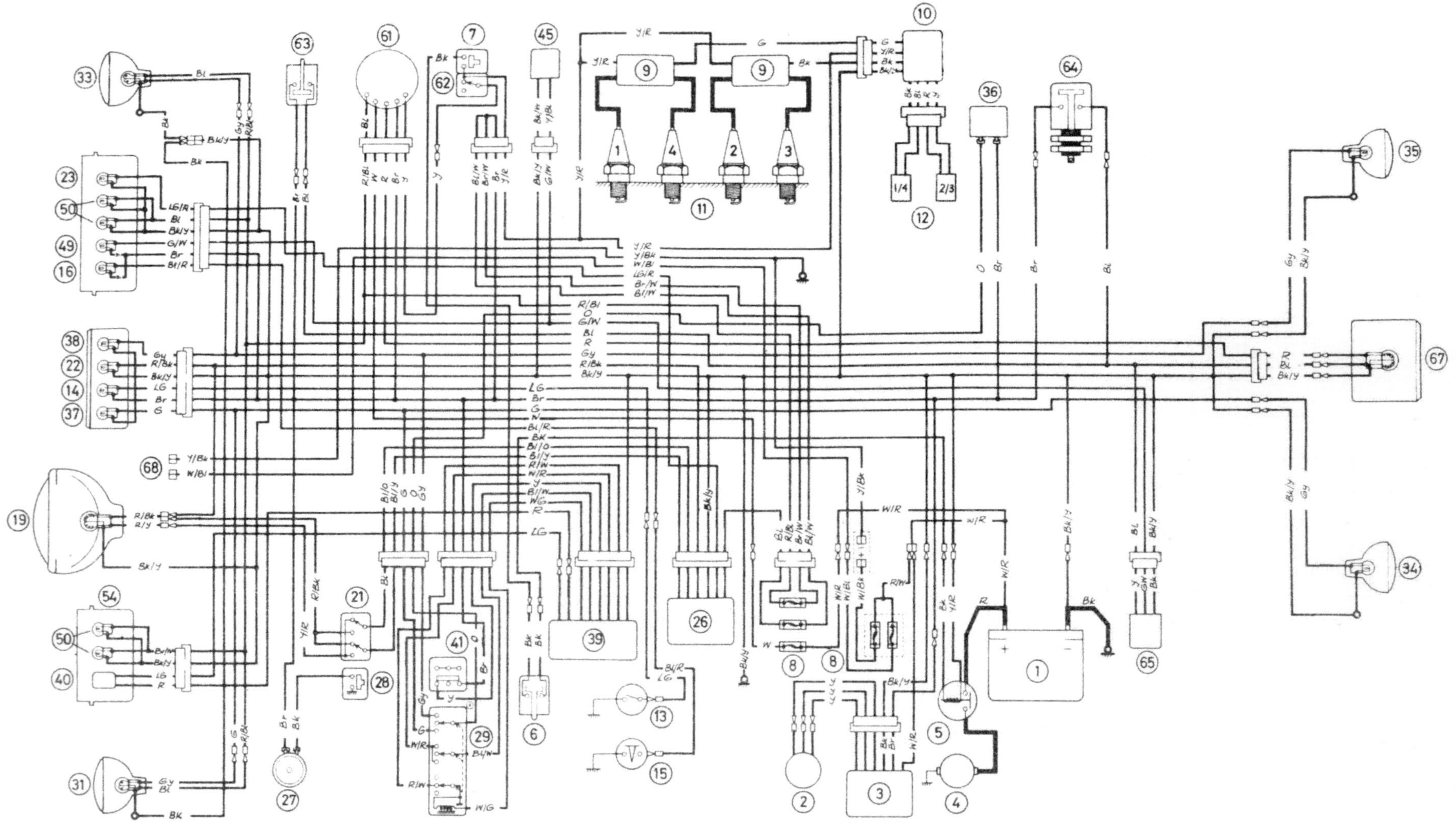

COLOUR CODE	
Bk	BLACK
Bl	BLUE
Br	BROWN
G	GREEN
Gy	GREY
LB	LIGHT BLUE
LG	LIGHT GREEN
O	ORANGE
P	PINK
R	RED
W	WHITE
Y	YELLOW

IGNITION SWITCH

	L	Bat	Ig	TL1	TL2
	Br	W	Y	Bl	R
OFF					
ON	●	●	●	●	●
P		●			●

LEFT HANDLEBAR SWITCH

STARTER LOCKOUT SW

	Bk	Bk
ON	●	●

DIMMER SWITCH

	R/Bk	Bl/Y	R/Y	Bl	R/Bk
Hi	●	●	●	●	
Lo		●	●	●	●

TURN SIGNAL SWITCH

	Gy	O	G	R/W	⏚	W/R	Bl/W
L	●	●				●	●
				●	●		
R		●	●			●	●

CANCEL SWITCH

	Br	Y
M		
	-	
A	●	●

HORN SWITCH

	Bk	⏚
PUSH	●	●

RIGHT HANDLEBAR SWITCH

STARTER BUTTON

	Bk	Y/R
PUSH	●	●

ENGINE STOP SW

	Y/R	Y
OFF		
RUN	●	●
OFF		

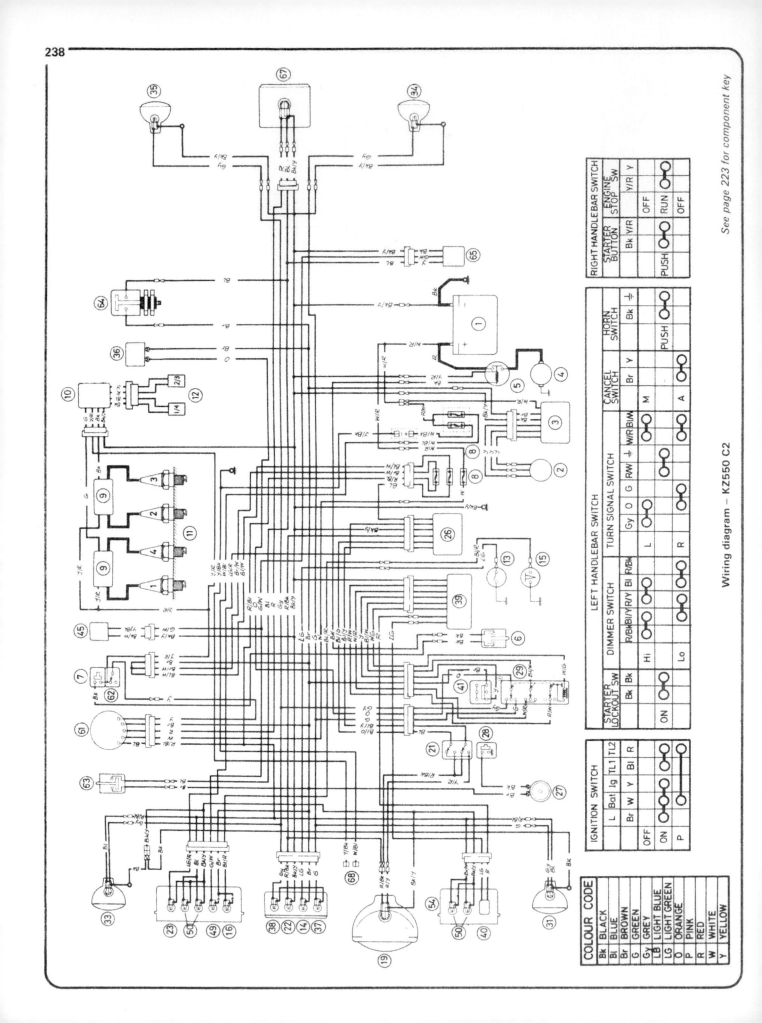

Wiring diagram – KZ550 C2

See page 223 for component key

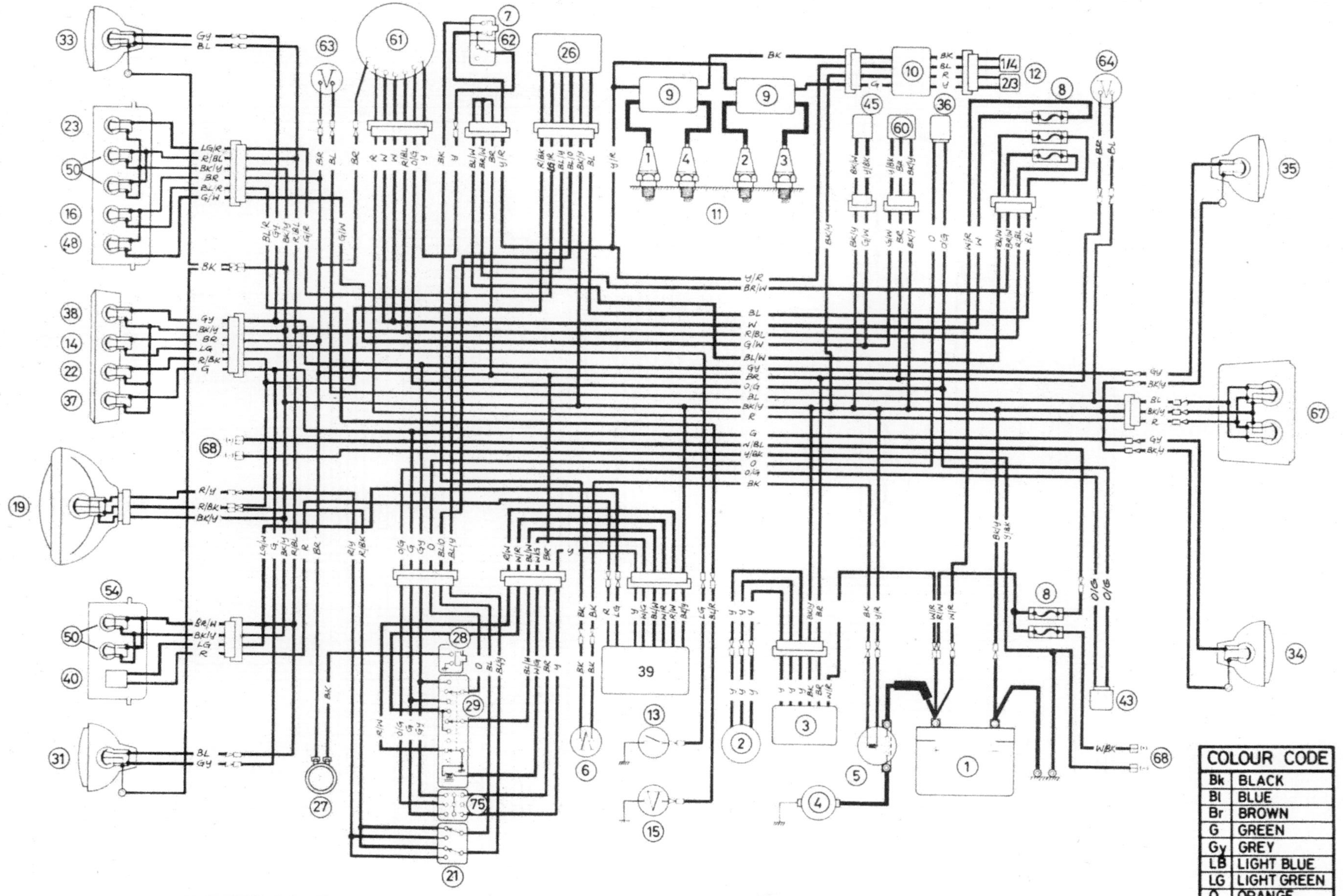

LEFT HANDLEBAR SWITCH CONNECTIONS

Horn Button	Black	Earth
Push	O	O

Turn Signal Switch	Gray	O	Green	R/W	Earth	W/R	Bl/W
L	O	O				O	O
N				O	O		
R		O	O			O	O

Combination Switch	Brown	Y	Grey	O/G	Green
Δ			O	O	O
M					
A	O	O			

Dimmer Switch	R/Bk	Bl/Y	R/Y	Blue	R/Bk
Hi	O	O	O	O	
Lo		O	O	O	O

Starter Lockout Switch	Black	Black
On	O	O

IGNITION SWITCH CONNECTIONS

	Lights	Bat. 1	Ignition	Tail 1	Tail 2	Bat. 2	Tail 3
	Brown	White	Yellow	Blue	Red	White	O/G
OFF							
ON	O	O	O	O	O	O	O
P(Park)		O			O	O	O

RIGHT HANDLEBAR SWITCH

Starter Button	Black	Y/R
Push	O	O

Engine Stop Switch	Y/R	Yellow
OFF		
RUN	O	O
OFF		

COLOUR CODE	
Bk	BLACK
Bl	BLUE
Br	BROWN
G	GREEN
Gy	GREY
LB	LIGHT BLUE
LG	LIGHT GREEN
O	ORANGE
P	PINK
R	RED
W	WHITE
Y	YELLOW

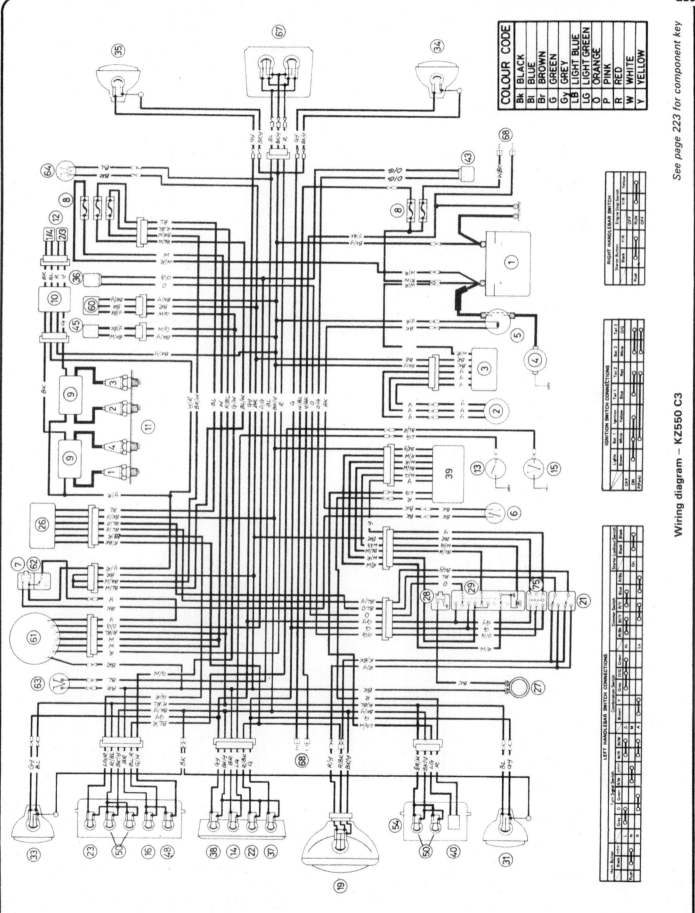

Wiring diagram – KZ550 C3

See page 223 for component key

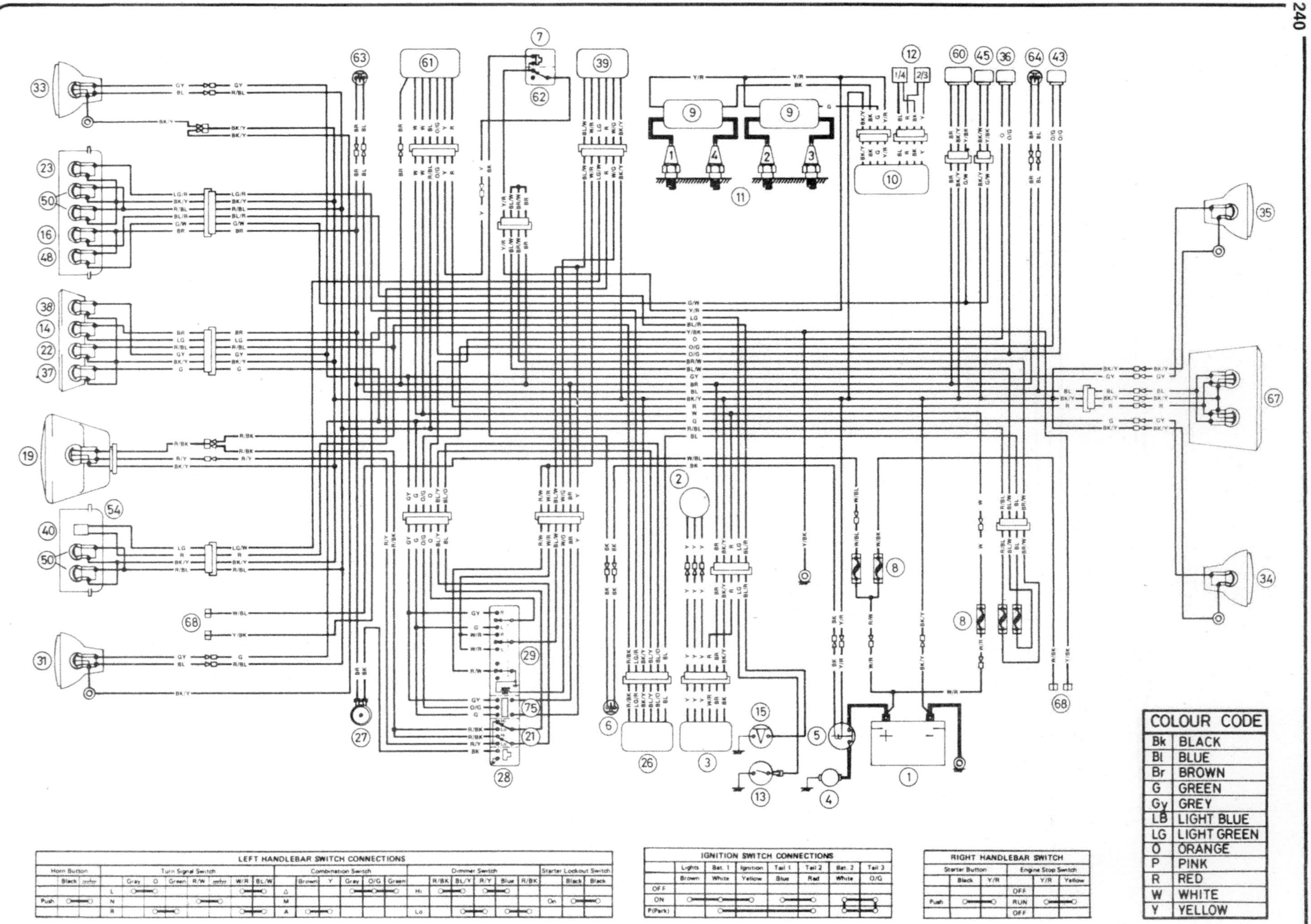

COLOUR CODE	
Bk	BLACK
Bl	BLUE
Br	BROWN
G	GREEN
Gy	GREY
LB	LIGHT BLUE
LG	LIGHT GREEN
O	ORANGE
P	PINK
R	RED
W	WHITE
Y	YELLOW

Wiring diagram – KZ550 C4

See page 223 for component key

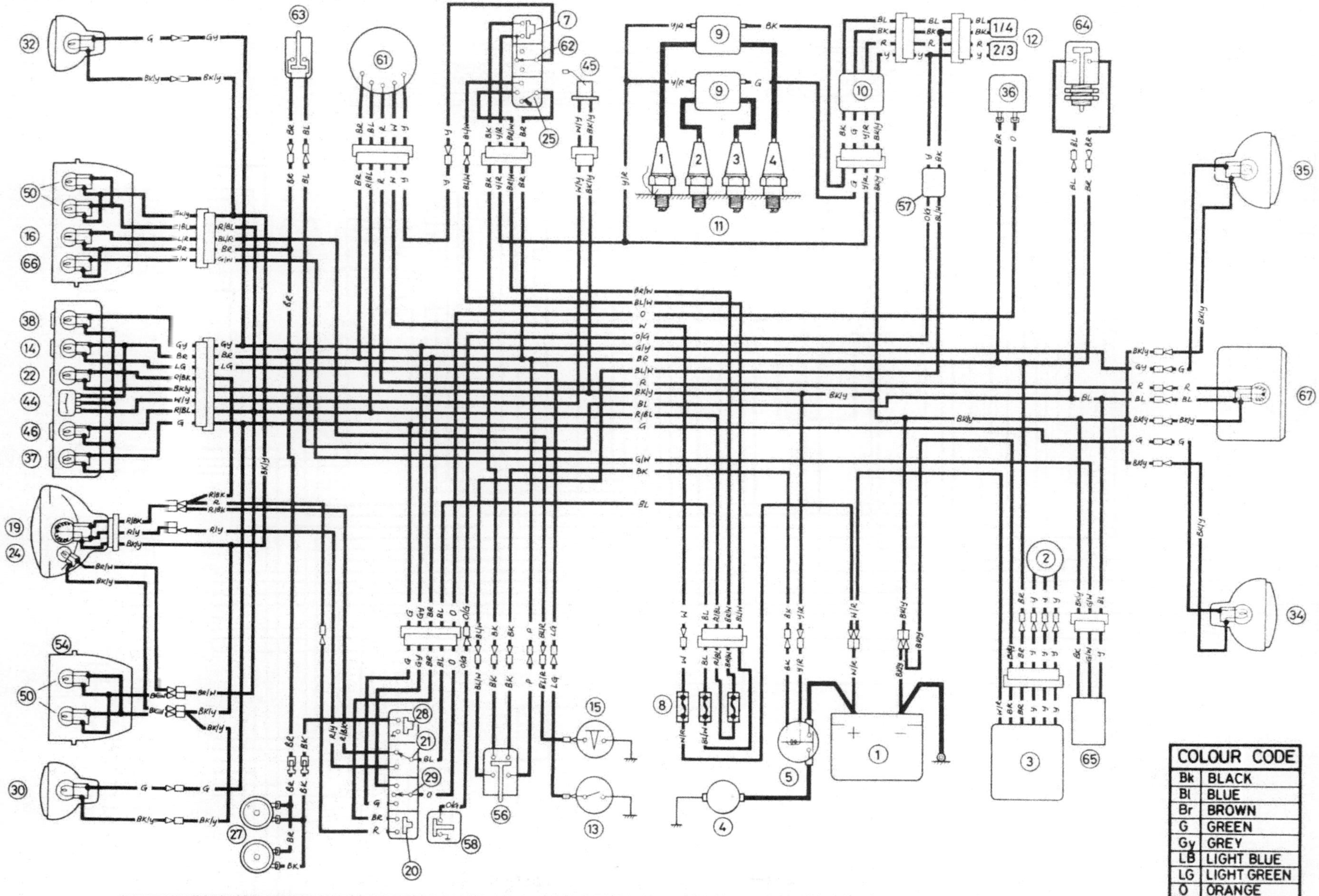

LEFT HANDLEBAR SWITCH CONNECTIONS

Dimmer Switch				Turn Signal Switch				Passing Button			Horn Button			Release Switch		
	R/Bk	Blue	R/Y		Green	Orange	Gray		Brown	Red		Bk/W	Earth		O/G	Earth
HI	O	O		L	O	O									O	O
								Push	O	O	Push	O	O	Push		
LO		O	O	R		O	O									

IGNITION SWITCH CONNECTIONS

	Lights	Battery	Ignition	Tail 1	Tail 2
	Brown	White	Yellow	Blue	Red
OFF					
ON	O	O	O	O	O
PARK		O			O

RIGHT HANDLEBAR SWITCH CONNECTIONS

Engine Stop Switch			Headlight Switch				Starter Button		
	Yellow	Y/R		Brown	Bl/W	Br/W		Black	Y/P
OFF			OFF						
RUN	O	O	O	O		O	ON	O	O
OFF			ON	O	O	O			

COLOUR CODE	
Bk	BLACK
Bl	BLUE
Br	BROWN
G	GREEN
Gy	GREY
LB	LIGHT BLUE
LG	LIGHT GREEN
O	ORANGE
P	PINK
R	RED
W	WHITE
Y	YELLOW

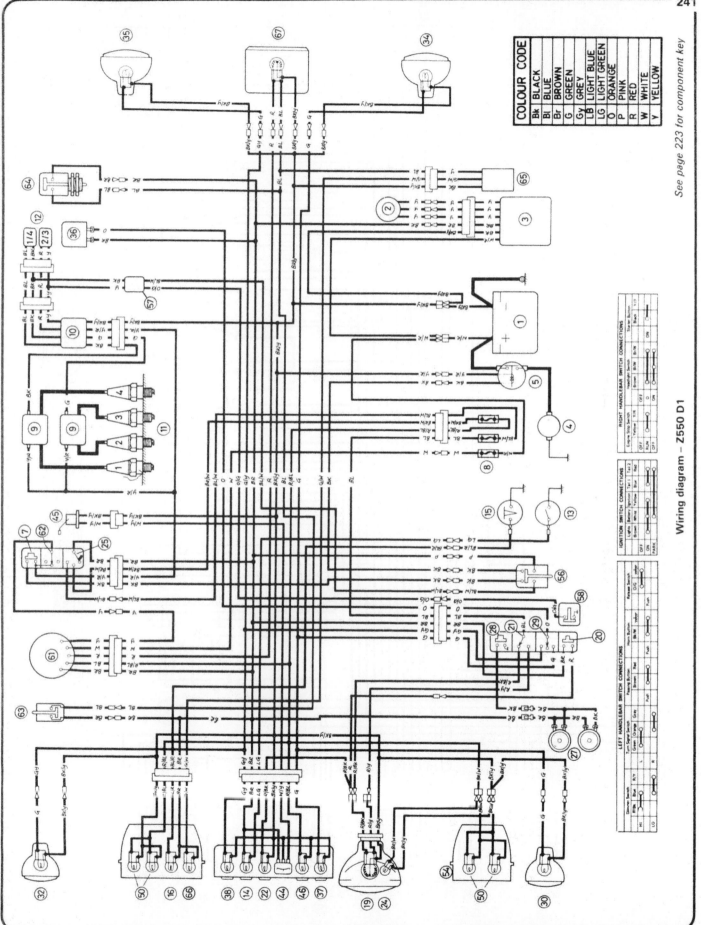

Wiring diagram – Z550 D1

See page 223 for component key

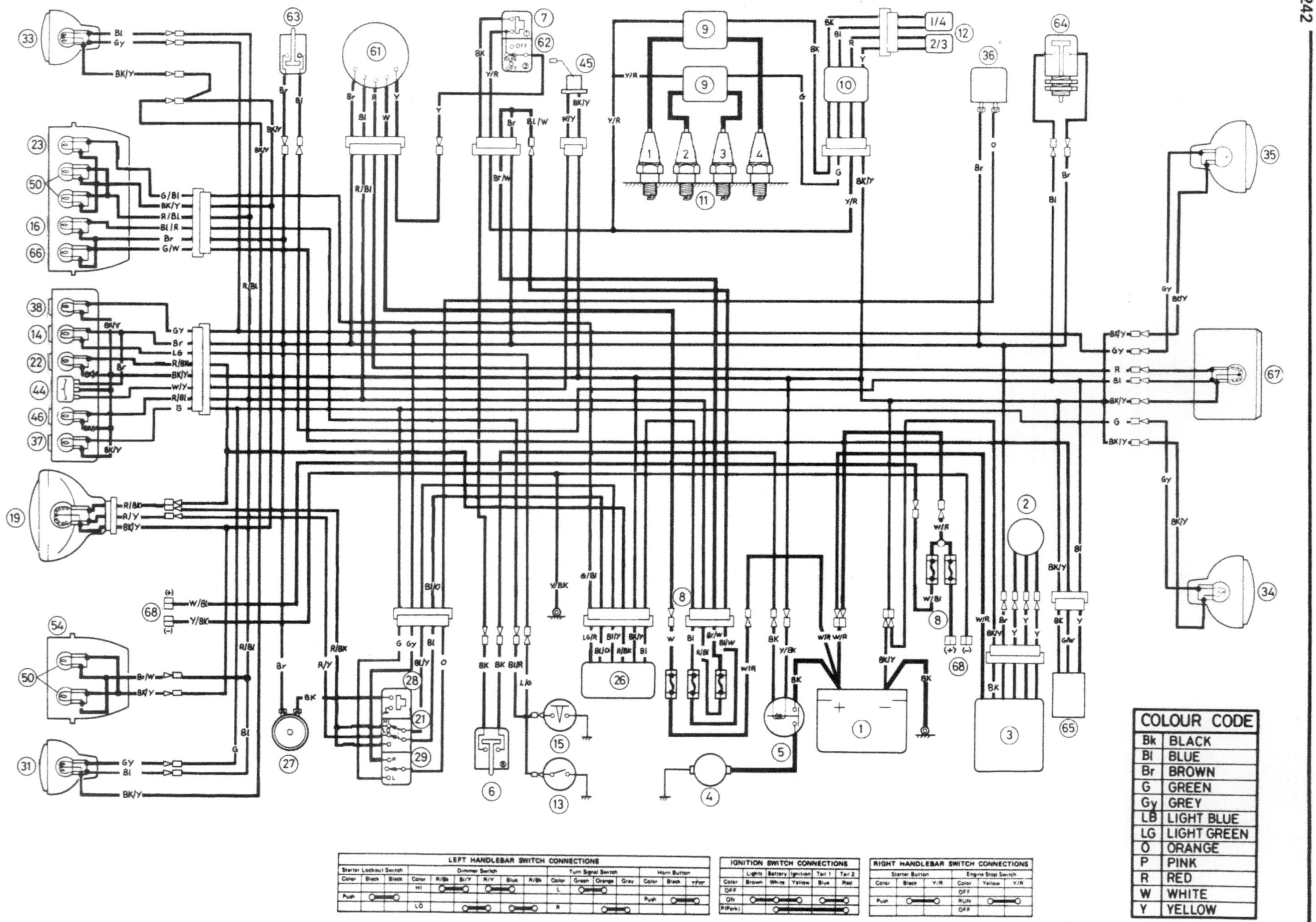

COLOUR CODE	
Bk	BLACK
Bl	BLUE
Br	BROWN
G	GREEN
Gy	GREY
LB	LIGHT BLUE
LG	LIGHT GREEN
O	ORANGE
P	PINK
R	RED
W	WHITE
Y	YELLOW

Wiring diagram – KZ550 D1

See page 223 for component key

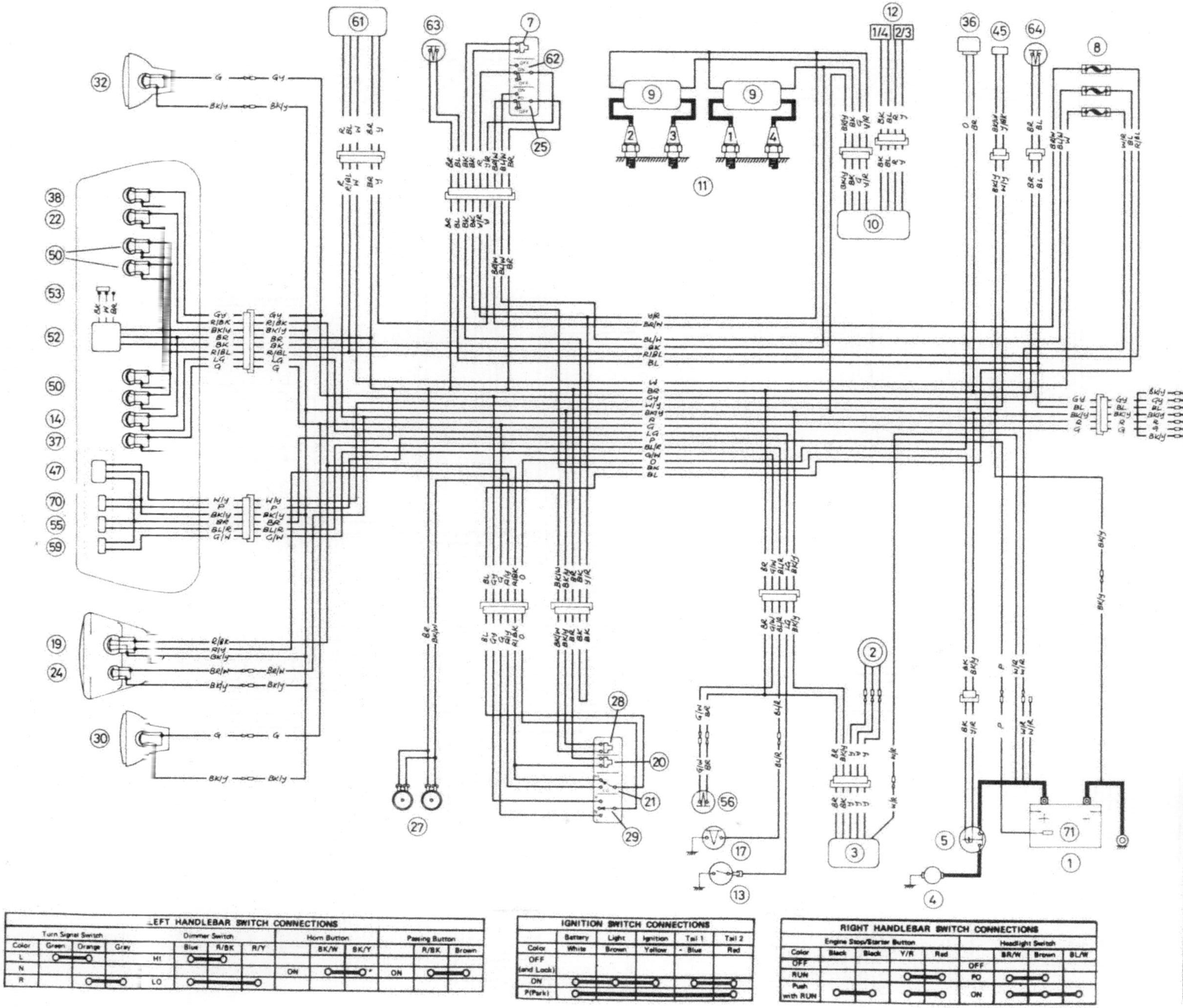

COLOUR CODE	
Bk	BLACK
Bl	BLUE
Br	BROWN
G	GREEN
Gy	GREY
LB	LIGHT BLUE
LG	LIGHT GREEN
O	ORANGE
P	PINK
R	RED
W	WHITE
Y	YELLOW

Wiring diagram – Z550 H1

See page 223 for component key

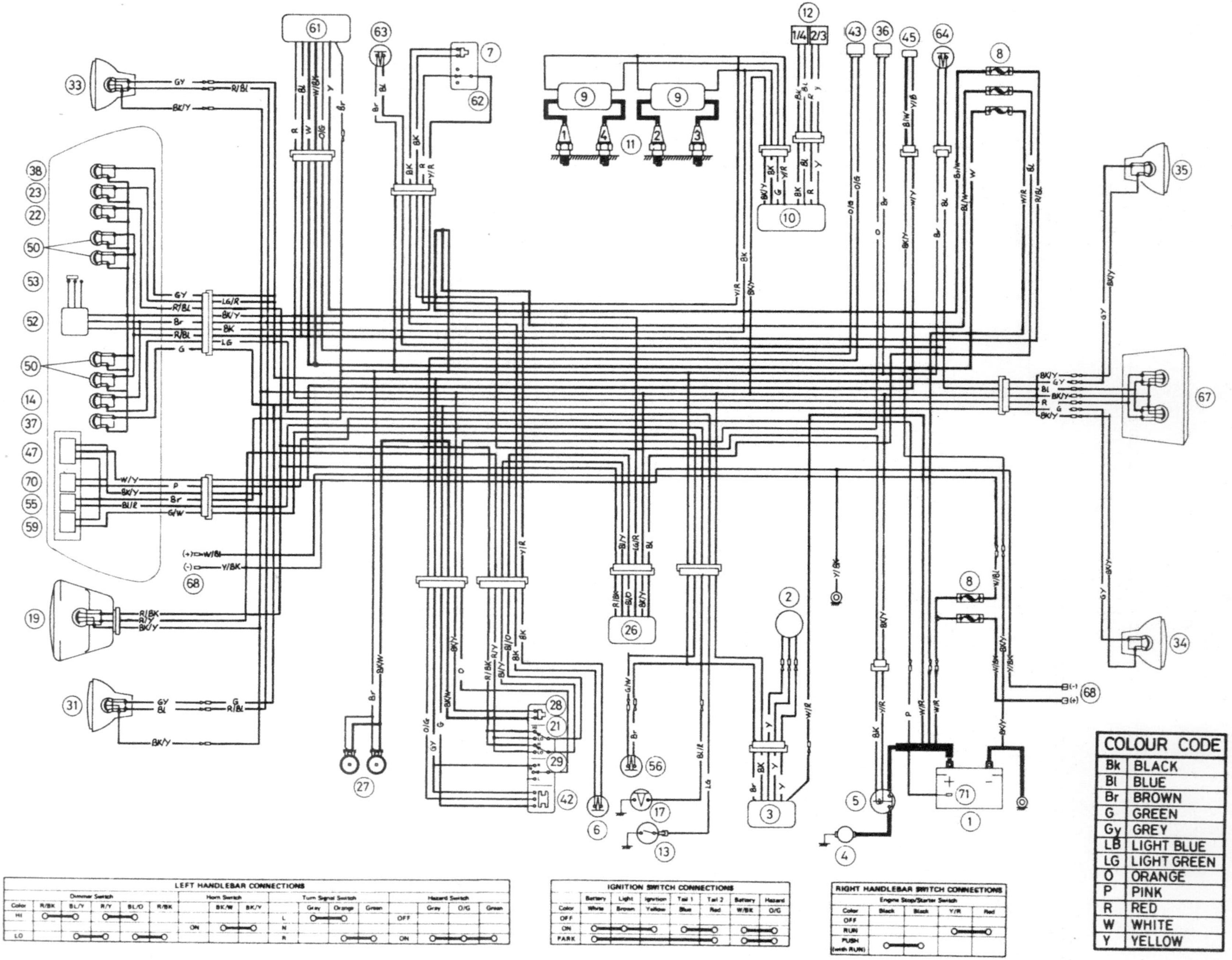

Wiring diagram – KZ550 H1

See page 223 for component key

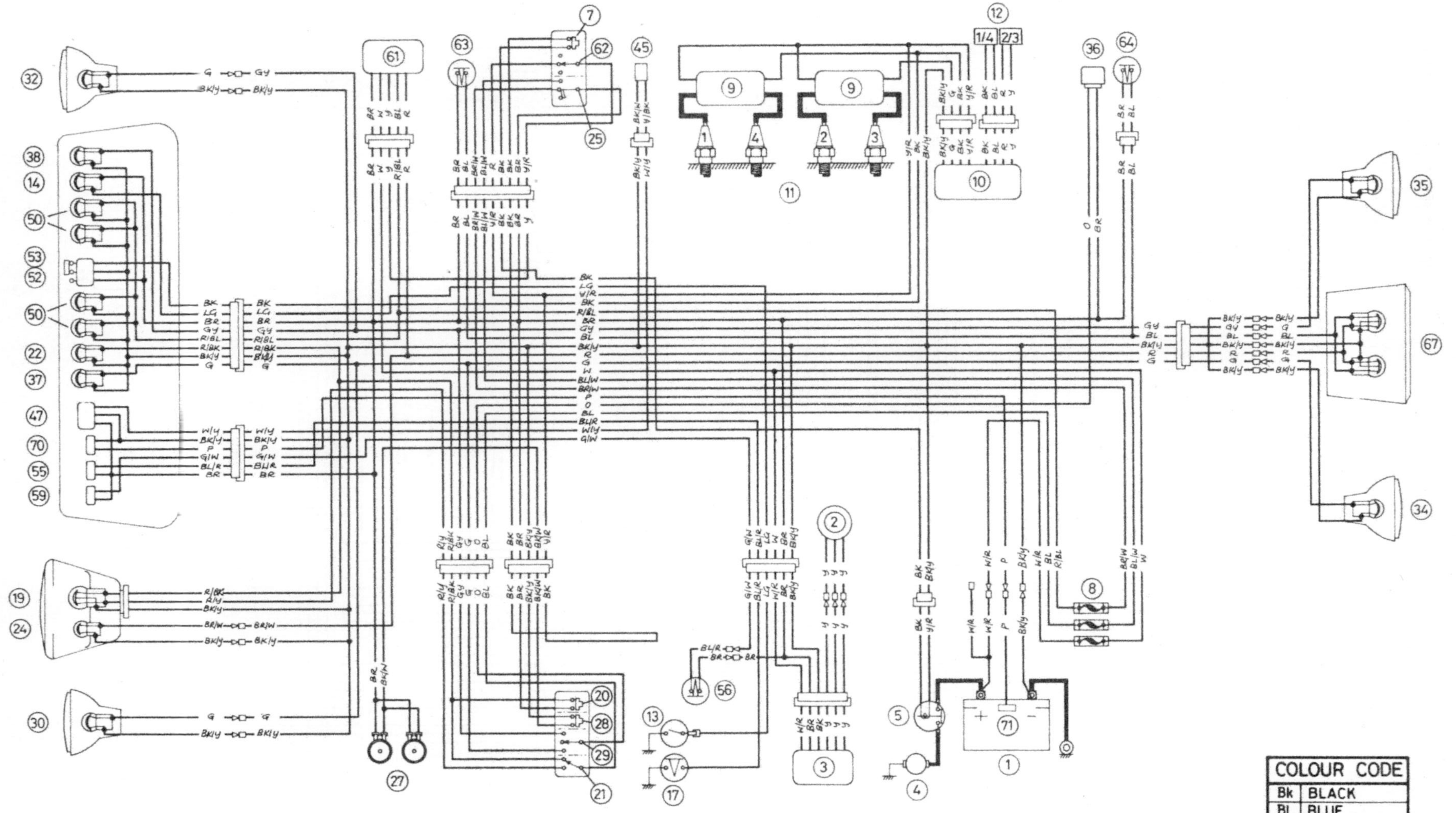

LEFT HANDLEBAR SWITCH CONNECTIONS														
Turn Signal Switch				Dimmer Switch				Horn Switch			Passing Button			
Color	Green	Orange	Gray		Blue	R/BK	R/Y		BK/W	BK/Y		R/BK	Brown	
L	○—	—○		H	○—	—○								
N								ON	○—	—○	ON	○—	—○	
R		○—	—○	LC	○—	—	—○							

IGNITION SWITCH CONNECTIONS					
	Battery	Light	Ignition	Tail 1	Tail 2
Color	White	Brown	Yellow	Blue	Red
OFF (and LOCK)					
ON	○—	—○—	—○	○—	—○
P(PARK)	○—	—	—	—	—○

RIGHT HANDLEBAR SWITCH CONNECTIONS								
Engine Stop/Starter Switch					Headlight Switch			
Color	Black	Black	Y/R	Red		BR/W	Brown	BL/W
OFF					OFF			
RUN			○—	—○	PO	○—	—○	
Push with RUN	○—	—○	○—	—○	ON	○—	—○—	—○

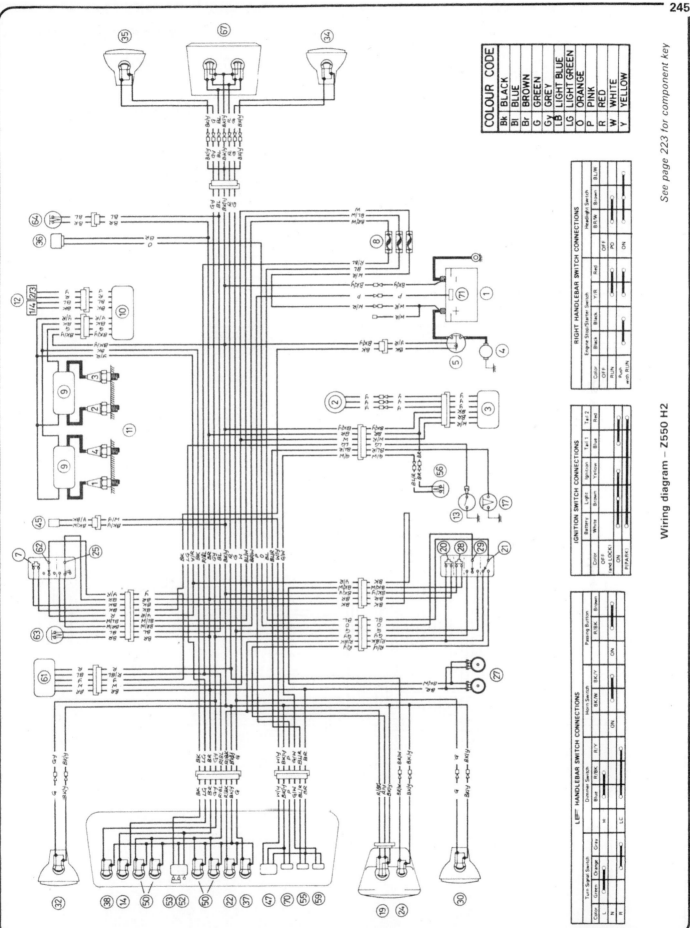

COLOUR CODE	
Bk	BLACK
Bl	BLUE
Br	BROWN
G	GREEN
Gy	GREY
LB	LIGHT BLUE
LG	LIGHT GREEN
O	ORANGE
P	PINK
R	RED
W	WHITE
Y	YELLOW

Wiring diagram – Z550 H2

See page 223 for component key

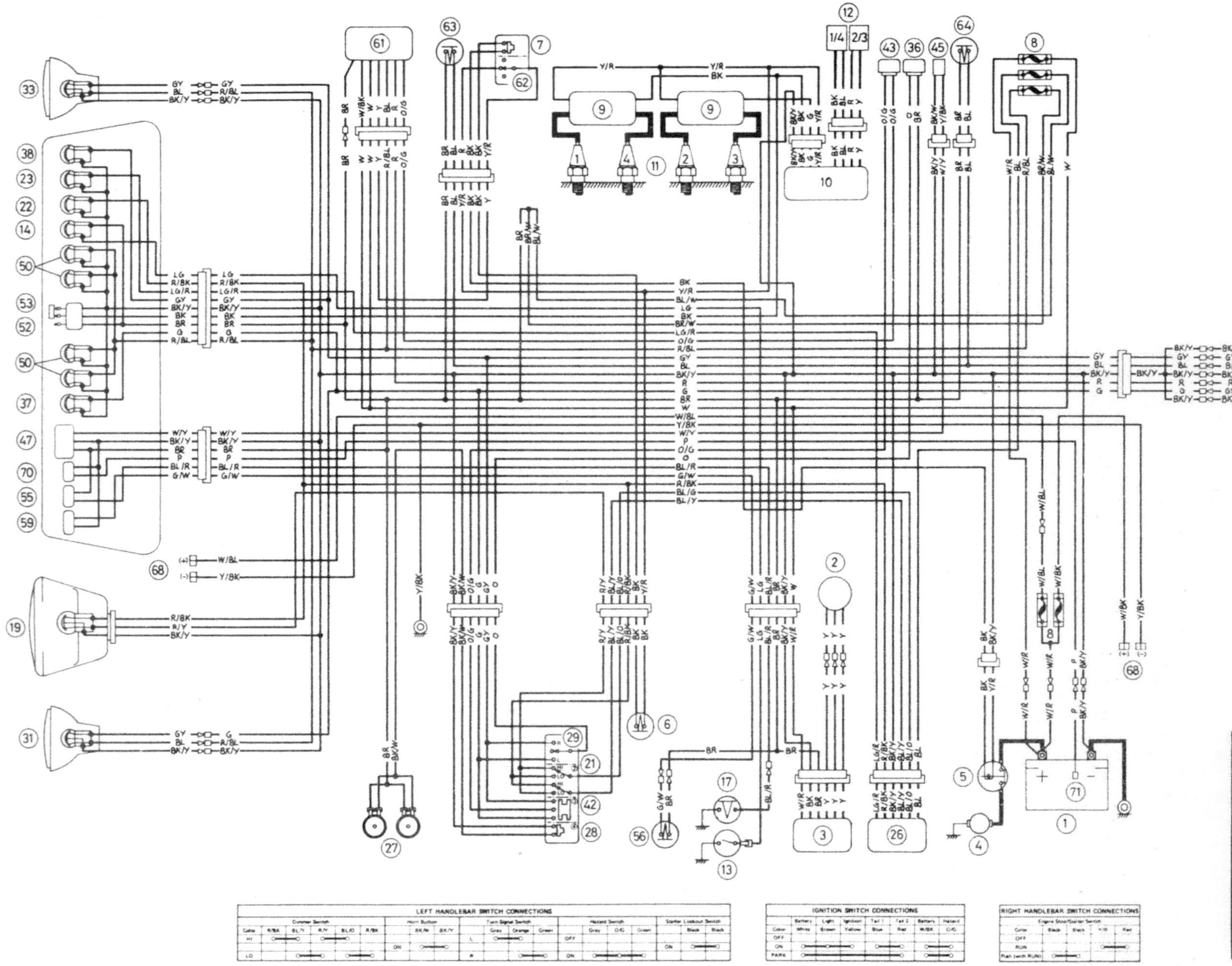

COLOUR CODE	
Bk	BLACK
Bl	BLUE
Br	BROWN
G	GREEN
Gy	GREY
LB	LIGHT BLUE
LG	LIGHT GREEN
O	ORANGE
P	PINK
R	RED
W	WHITE
Y	YELLOW

Wiring diagram – KZ550 H2

See page 223 for component key

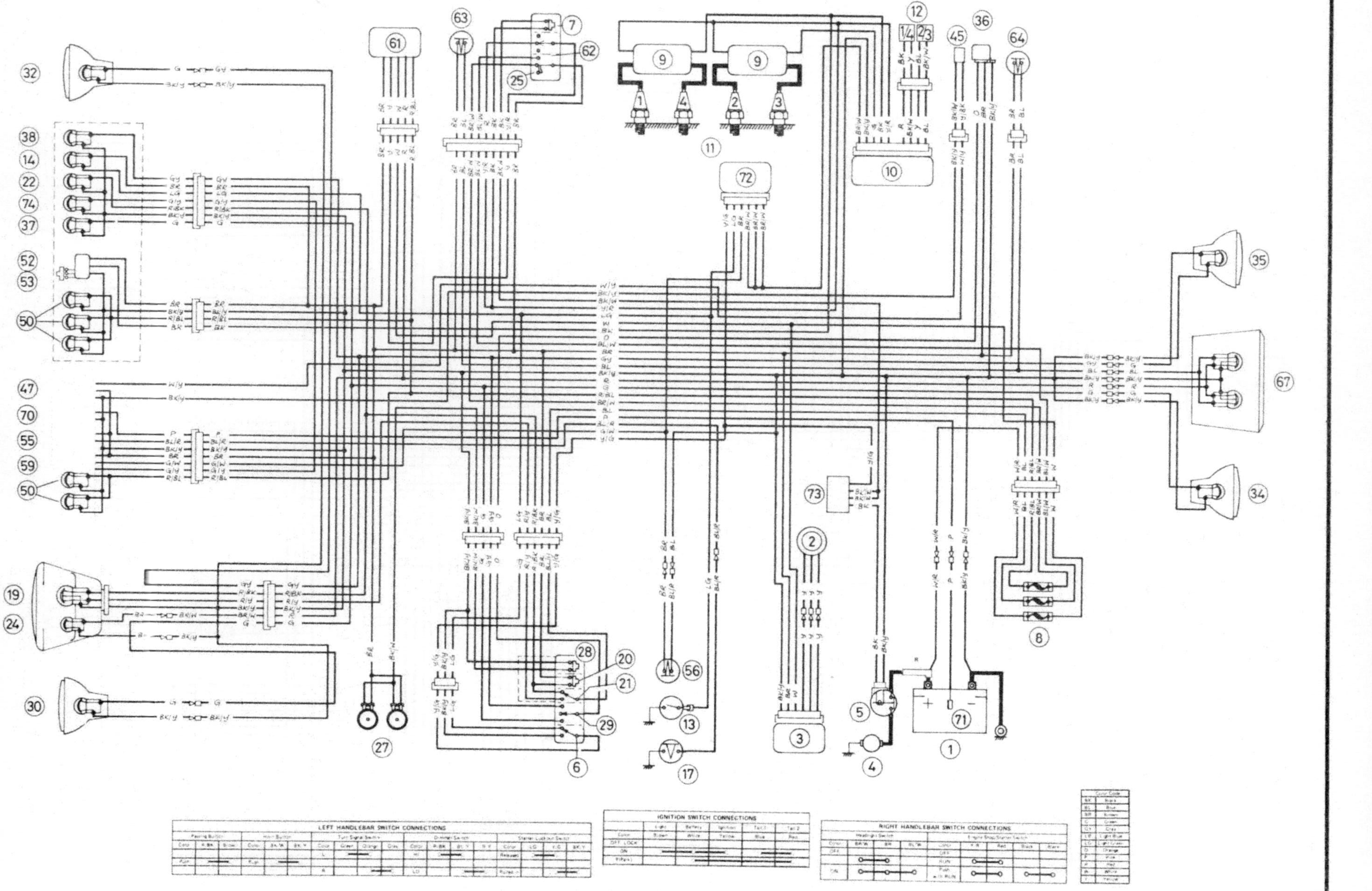

Wiring diagram – ZX550 A1, A2, A3 and A4 (UK)

See page 223 for component key

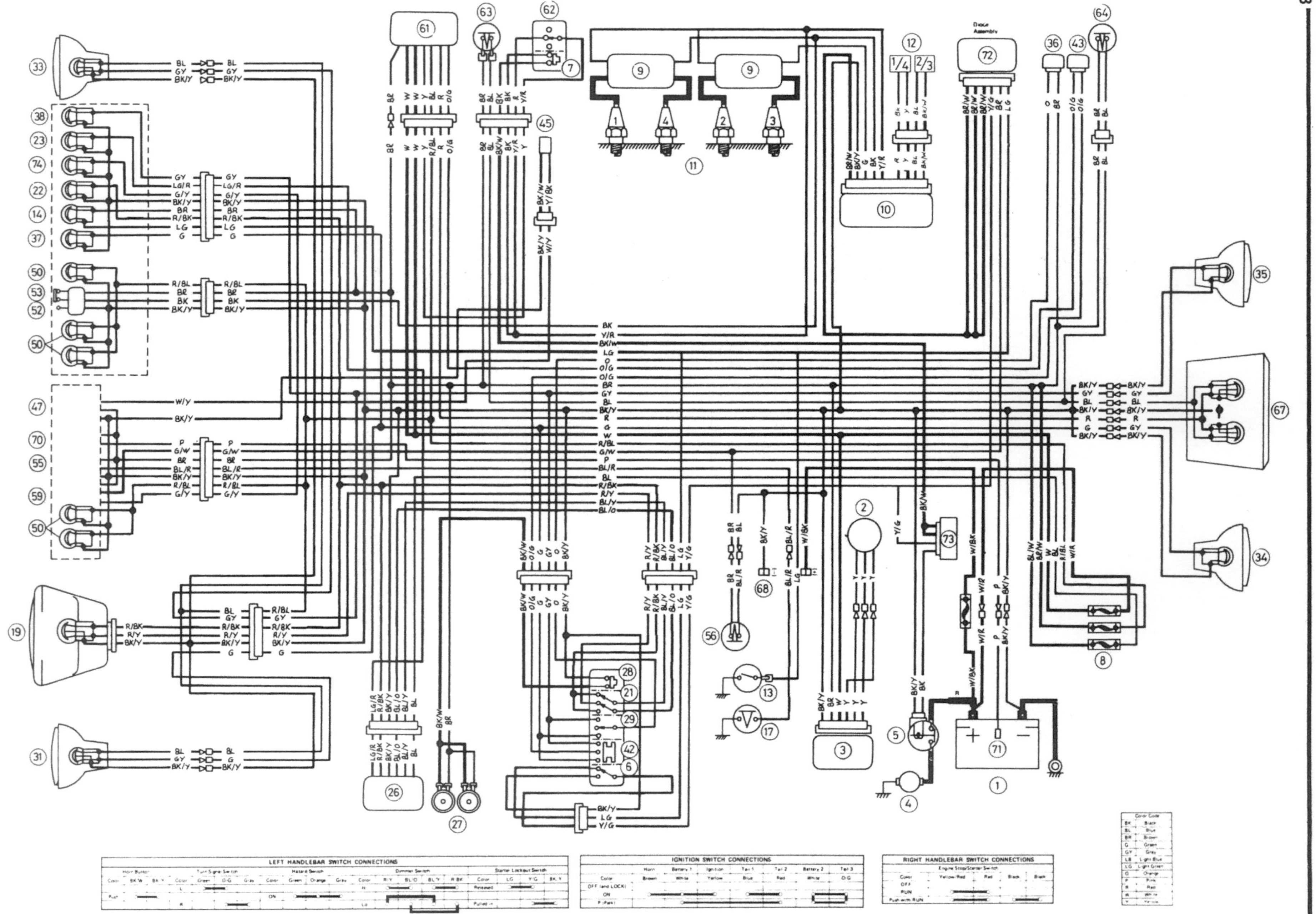

See page 223 for component key

Wiring diagram – ZX550 A1, A1L, A2 and A2L (US)

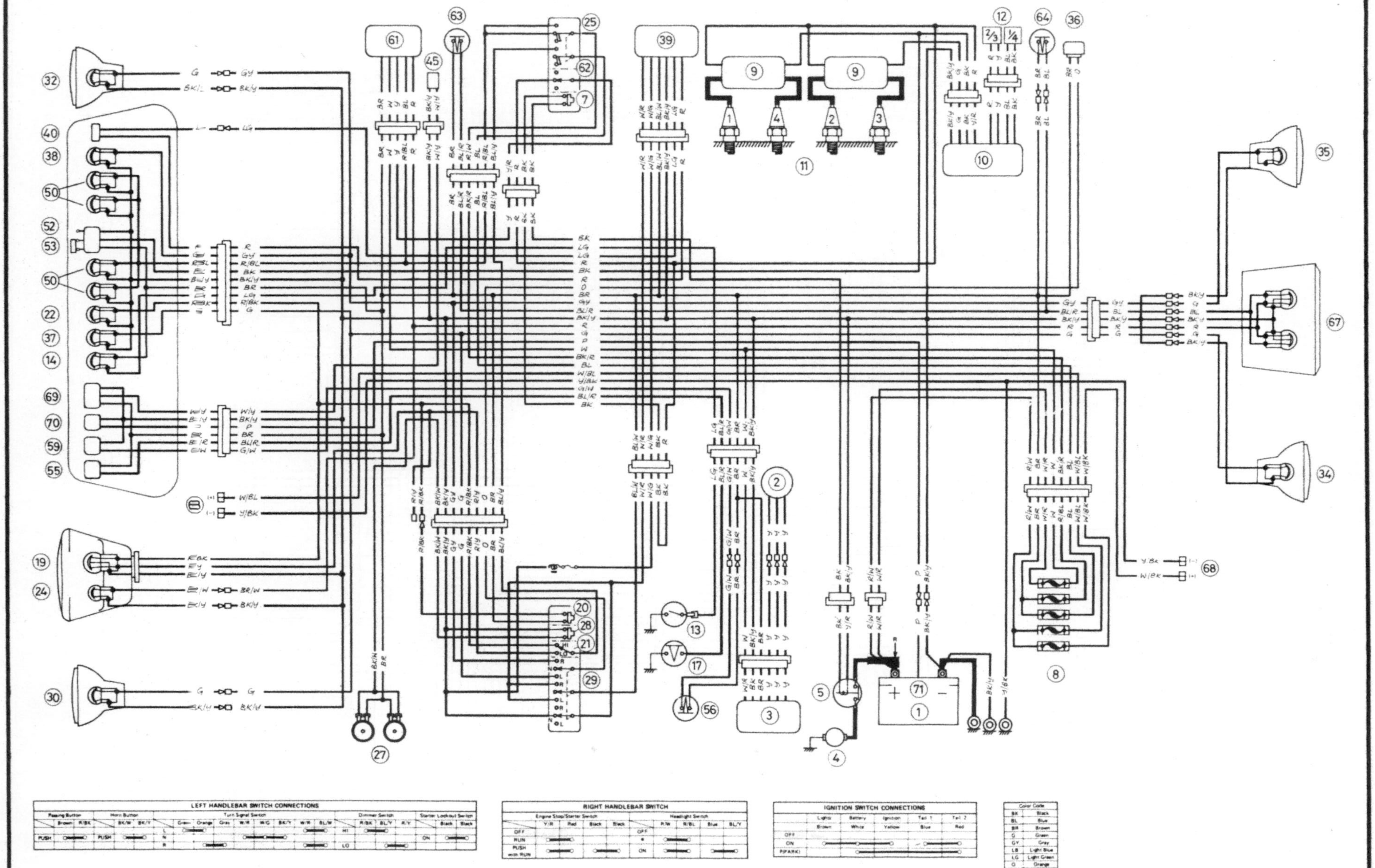

Color Code	
BK	Black
BL	Blue
BR	Brown
G	Green
GY	Gray
LB	Light Blue
LG	Light Green
O	Orange
P	Pink
R	Red
W	White
Y	Yellow

Wiring diagram – Z550 G1

See page 223 for component key

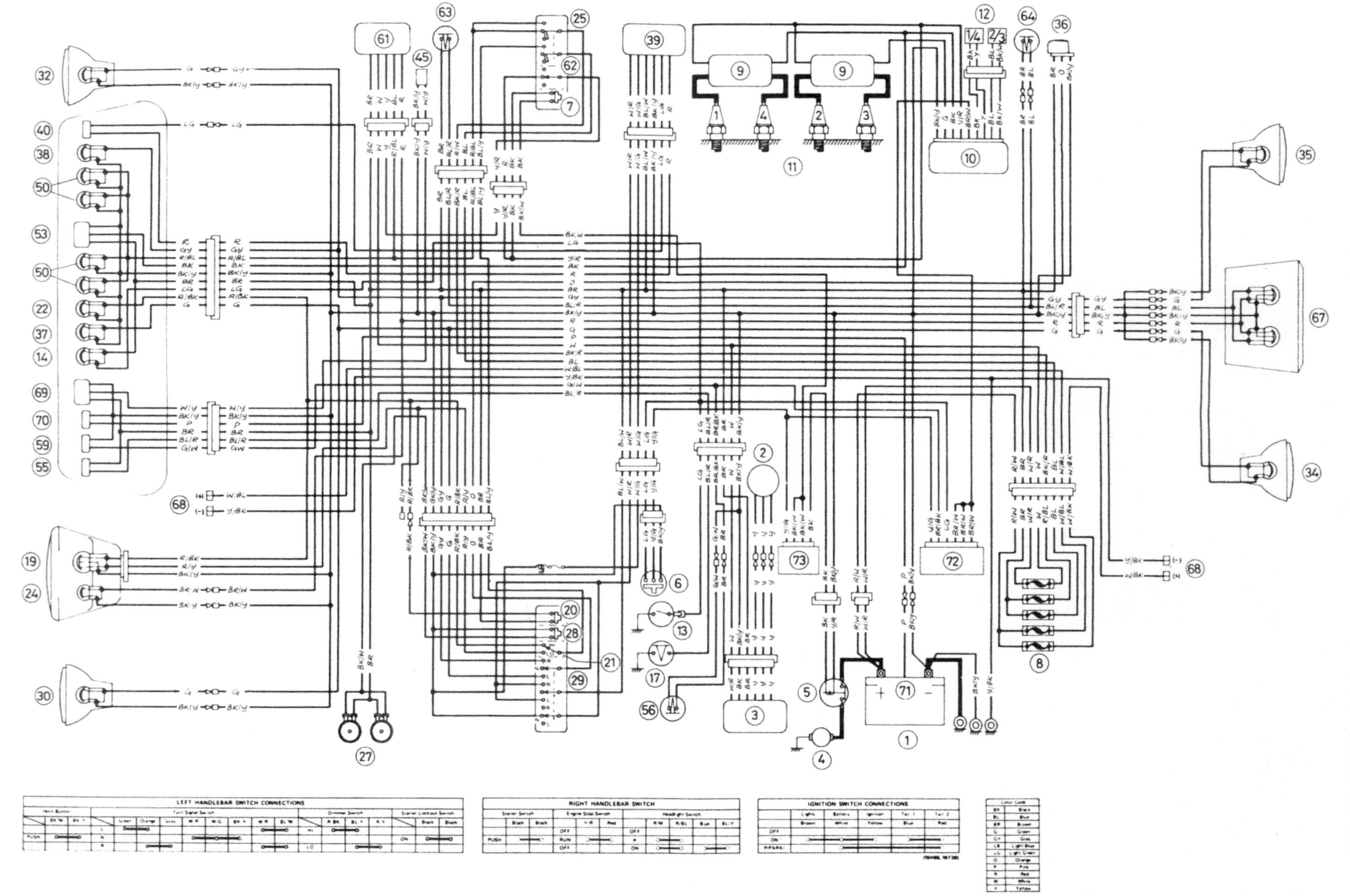

Wiring diagram – Z550 G2, G3 and G4

See page 223 for component key

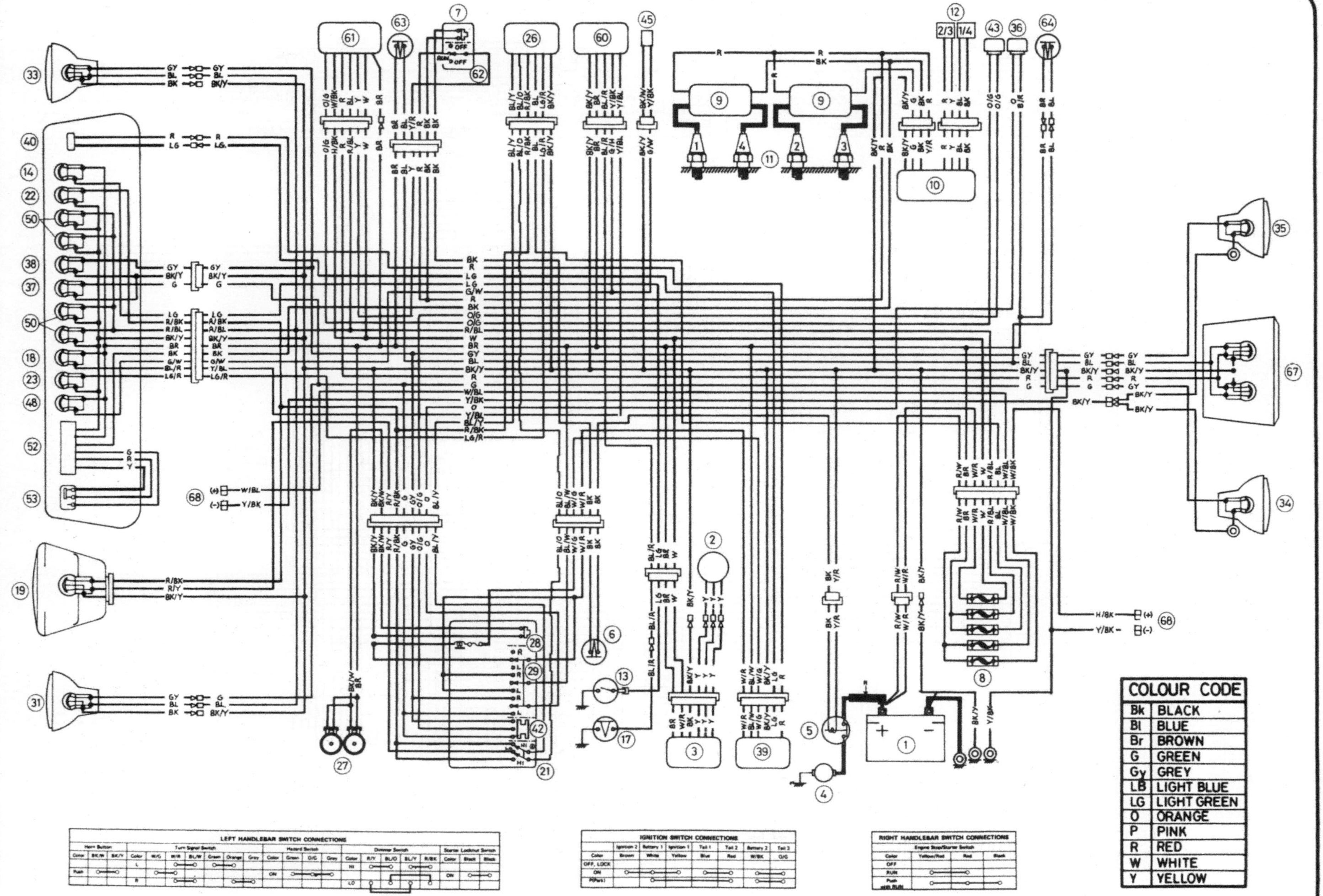

COLOUR CODE	
Bk	BLACK
Bl	BLUE
Br	BROWN
G	GREEN
Gy	GREY
LB	LIGHT BLUE
LG	LIGHT GREEN
O	ORANGE
P	PINK
R	RED
W	WHITE
Y	YELLOW

Wiring diagram – KZ550 F1

See page 223 for component key

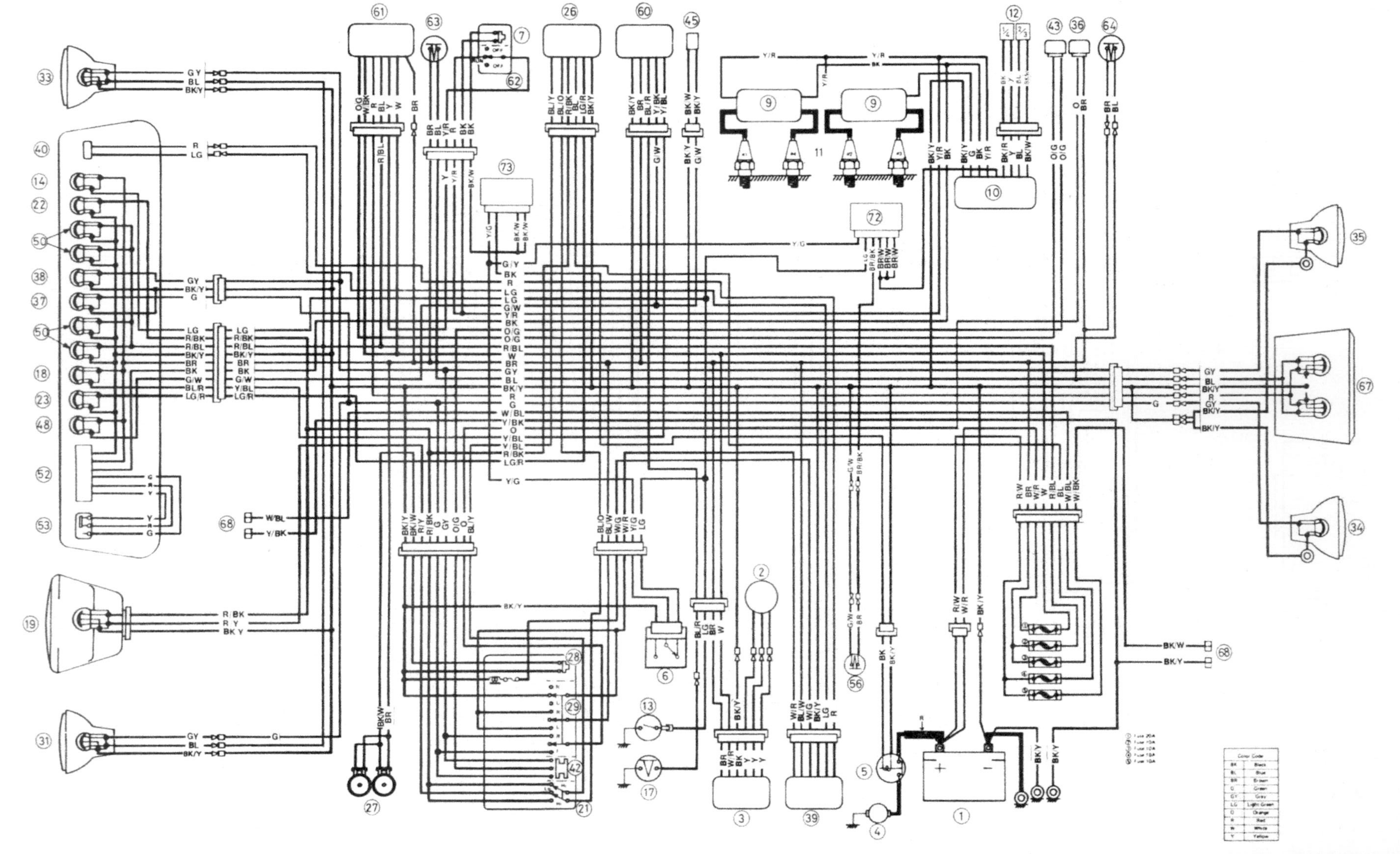

LEFT HANDLEBAR SWITCH CONNECTIONS

Horn Button			Turn Signal Switch							Hazard Switch				Dimmer Switch					Starter Lockout Switch			
Color	BK/W	BK/Y	Color	W/G	W/R	BL/W	Green	Orange	Gray	Color	Green	O/G	Gray	Color	R/Y	BL/O	BL/Y	R/BK	Color	BK/Y	Y/G	LG
			L		o	o	o	o						HI	o	o	o	o				
Push	o	o		o	o					ON	o	o	o						ON	o	o	
			R		o	o		o	o					LO	o	o	o	o	Released		o	o

IGNITION SWITCH CONNECTIONS

	Ignition 2	Battery 1	Ignition 1	Tail 1	Tail 2	Battery 2	Tail 3
Color	Brown	White	Yellow	Blue	Red	W/BK	O/G
OFF, LOCK							
ON	o	o	o	o	o	o	o
P(Park)		o			o	o	o

(1962A)

RIGHT HANDLEBAR SWITCH CONNECTIONS

Engine Stop/Starter Switch			
Color	Yellow/Red	Red	Black
OFF			
RUN	o	o	
Push with RUN	o	o	o

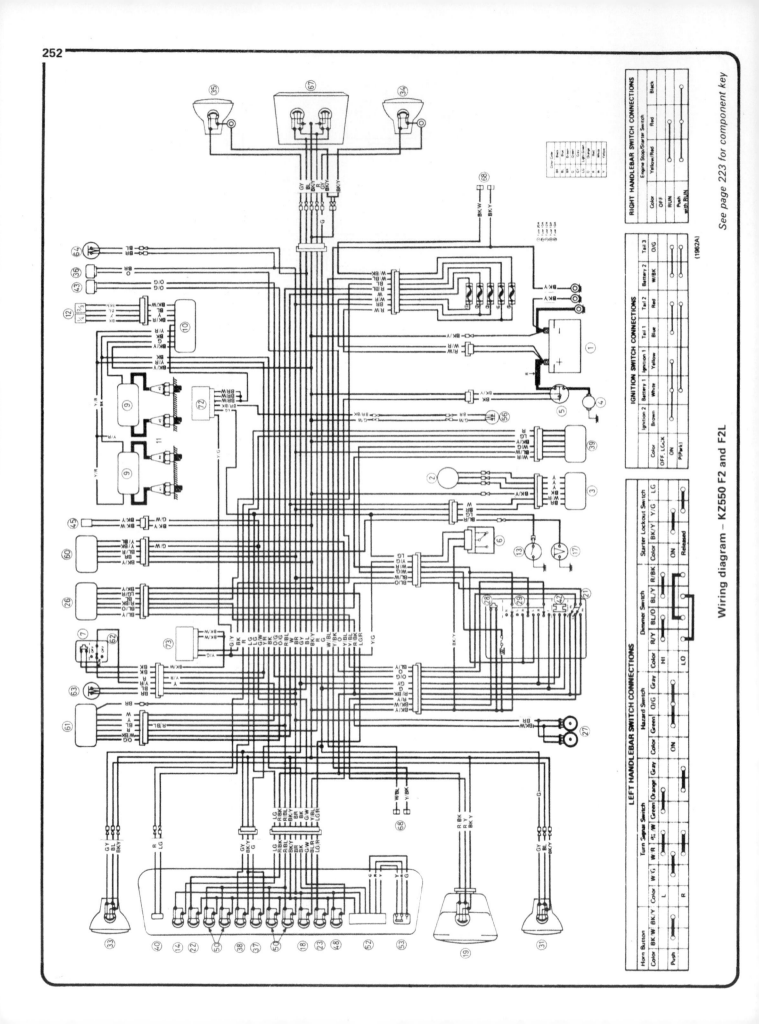

Wiring diagram – KZ550 F2 and F2L

See page 223 for component key

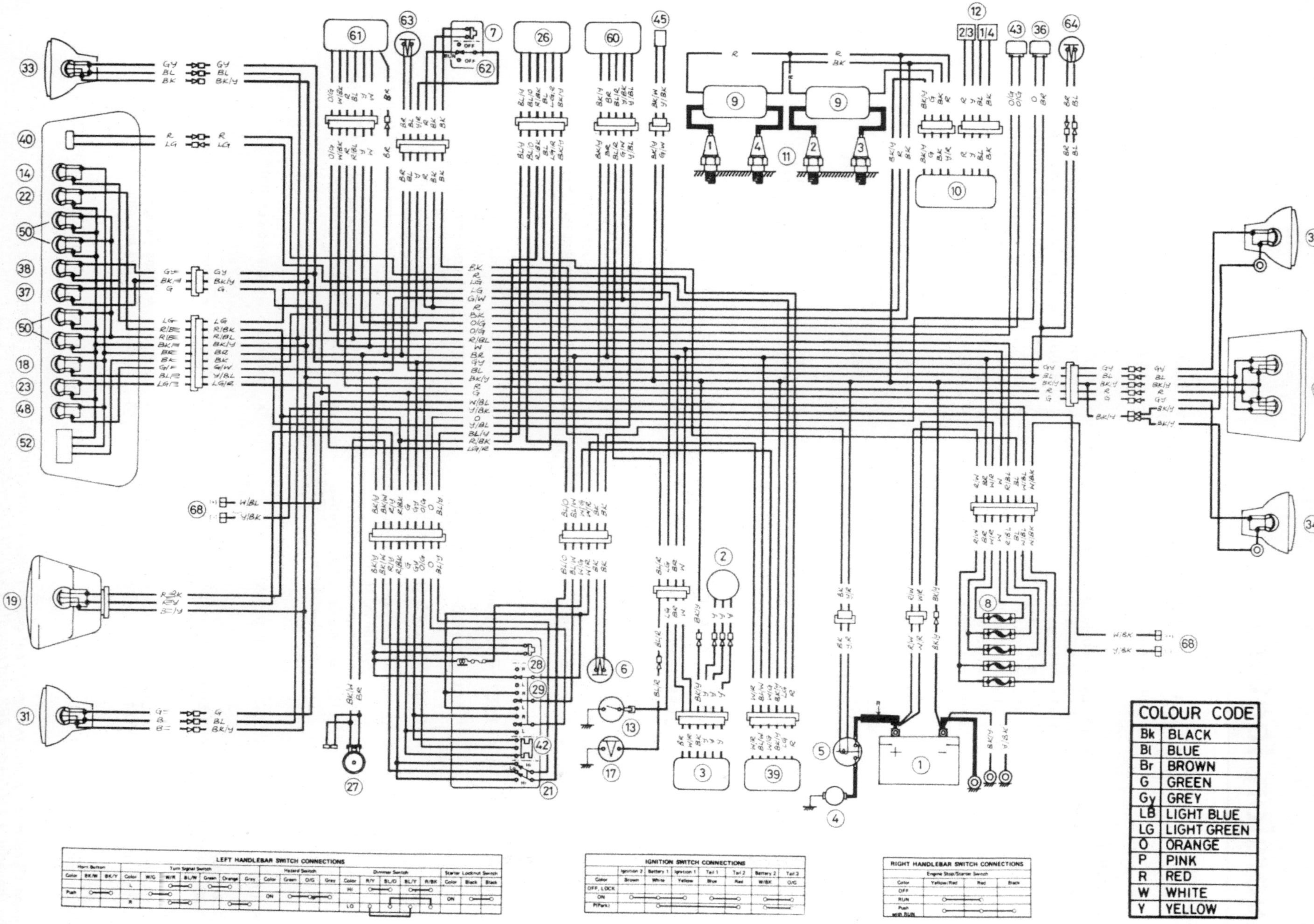

COLOUR CODE	
Bk	BLACK
Bl	BLUE
Br	BROWN
G	GREEN
Gy	GREY
LB	LIGHT BLUE
LG	LIGHT GREEN
O	ORANGE
P	PINK
R	RED
W	WHITE
Y	YELLOW

Wiring diagram – KZ550 M1

See page 223 for component key

Conversion factors

Length (distance)

Inches (in)	X 25.4	= Millimetres (mm)	X 0.0394	= Inches (in)
Feet (ft)	X 0.305	= Metres (m)	X 3.281	= Feet (ft)
Miles	X 1.609	= Kilometres (km)	X 0.621	= Miles

Volume (capacity)

Cubic inches (cu in; in^3)	X 16.387	= Cubic centimetres (cc; cm^3)	X 0.061	= Cubic inches (cu in; in^3)
Imperial pints (Imp pt)	X 0.568	= Litres (l)	X 1.76	= Imperial pints (Imp pt)
Imperial quarts (Imp qt)	X 1.137	= Litres (l)	X 0.88	= Imperial quarts (Imp qt)
Imperial quarts (Imp qt)	X 1.201	= US quarts (US qt)	X 0.833	= Imperial quarts (Imp qt)
US quarts (US qt)	X 0.946	= Litres (l)	X 1.057	= US quarts (US qt)
Imperial gallons (Imp gal)	X 4.546	= Litres (l)	X 0.22	= Imperial gallons (Imp gal)
Imperial gallons (Imp gal)	X 1.201	= US gallons (US gal)	X 0.833	= Imperial gallons (Imp gal)
US gallons (US gal)	X 3.785	= Litres (l)	X 0.264	= US gallons (US gal)

Mass (weight)

Ounces (oz)	X 28.35	= Grams (g)	X 0.035	= Ounces (oz)
Pounds (lb)	X 0.454	= Kilograms (kg)	X 2.205	= Pounds (lb)

Force

Ounces-force (ozf; oz)	X 0.278	= Newtons (N)	X 3.6	= Ounces-force (ozf; oz)
Pounds-force (lbf; lb)	X 4.448	= Newtons (N)	X 0.225	= Pounds-force (lbf; lb)
Newtons (N)	X 0.1	= Kilograms-force (kgf; kg)	X 9.81	= Newtons (N)

Pressure

Pounds-force per square inch (psi; lbf/in^2; lb/in^2)	X 0.070	= Kilograms-force per square centimetre (kgf/cm^2; kg/cm^2)	X 14.223	= Pounds-force per square inch (psi; lbf/in^2; lb/in^2)
Pounds-force per square inch (psi; lbf/in^2; lb/in^2)	X 0.068	= Atmospheres (atm)	X 14.696	= Pounds-force per square inch (psi; lbf/in^2; lb/in^2)
Pounds-force per square inch (psi; lbf/in^2; lb/in^2)	X 0.069	= Bars	X 14.5	= Pounds-force per square inch (psi; lbf/in^2; lb/in^2)
Pounds-force per square inch (psi; lbf/in^2; lb/in^2)	X 6.895	= Kilopascals (kPa)	X 0.145	= Pounds-force per square inch (psi; lbf/in^2; lb/in^2)
Kilopascals (kPa)	X 0.01	= Kilograms-force per square centimetre (kgf/cm^2; kg/cm^2)	X 98.1	= Kilopascals (kPa)

Torque (moment of force)

Pounds-force inches (lbf in; lb in)	X 1.152	= Kilograms-force centimetre (kgf cm; kg cm)	X 0.868	= Pounds-force inches (lbf in; lb in)
Pounds-force inches (lbf in; lb in)	X 0.113	= Newton metres (Nm)	X 8.85	= Pounds-force inches (lbf in; lb in)
Pounds-force inches (lbf in; lb in)	X 0.083	= Pounds-force feet (lbf ft; lb ft)	X 12	= Pounds-force inches (lbf in; lb in)
Pounds-force feet (lbf ft; lb ft)	X 0.138	= Kilograms-force metres (kgf m; kg m)	X 7.233	= Pounds-force feet (lbf ft; lb ft)
Pounds-force feet (lbf ft; lb ft)	X 1.356	= Newton metres (Nm)	X 0.738	= Pounds-force feet (lbf ft; lb ft)
Newton metres (Nm)	X 0.102	= Kilograms-force metres (kgf m; kg m)	X 9.804	= Newton metres (Nm)

Power

Horsepower (hp)	X 745.7	= Watts (W)	X 0.0013	= Horsepower (hp)

Velocity (speed)

Miles per hour (miles/hr; mph)	X 1.609	= Kilometres per hour (km/hr; kph)	X 0.621	= Miles per hour (miles/hr; mph)

*Fuel consumption**

Miles per gallon, Imperial (mpg)	X 0.354	= Kilometres per litre (km/l)	X 2.825	= Miles per gallon, Imperial (mpg)
Miles per gallon, US (mpg)	X 0.425	= Kilometres per litre (km/l)	X 2.352	= Miles per gallon, US (mpg)

Temperature

Degrees Fahrenheit = (°C x 1.8) + 32

Degrees Celsius (Degrees Centigrade; °C) = (°F - 32) x 0.56

**It is common practice to convert from miles per gallon (mpg) to litres/100 kilometres (l/100km), where mpg (Imperial) x l/100 km = 282 and mpg (US) x l/100 km = 235*

English/American terminology

Because this book has been written in England, British English component names, phrases and spellings have been used throughout. American English usage is quite often different and whereas normally no confusion should occur, a list of equivalent terminology is given below.

English	American
Air filter	Air cleaner
Alignment (headlamp)	Aim
Allen screw/key	Socket screw/wrench
Anticlockwise	Counterclockwise
Bottom/top gear	Low/high gear
Bottom/top yoke	Bottom/top triple clamp
Bush	Bushing
Carburettor	Carburetor
Catch	Latch
Circlip	Snap ring
Clutch drum	Clutch housing
Dip switch	Dimmer switch
Disulphide	Disulfide
Dynamo	DC generator
Earth	Ground
End float	End play
Engineer's blue	Machinist's dye
Exhaust pipe	Header
Fault diagnosis	Trouble shooting
Float chamber	Float bowl
Footrest	Footpeg
Fuel/petrol tap	Petcock
Gaiter	Boot
Gearbox	Transmission
Gearchange	Shift
Gudgeon pin	Wrist/piston pin
Indicator	Turn signal
Inlet	Intake
Input shaft or mainshaft	Mainshaft
Kickstart	Kickstarter
Lower leg	Slider
Mudguard	Fender
Number plate	License plate
Output or layshaft	Countershaft
Panniers	Side cases
Paraffin	Kerosene
Petrol	Gasoline
Petrol/fuel tank	Gas tank
Pinking	Pinging
Rear suspension unit	Rear shock absorber
Rocker cover	Valve cover
Selector	Shifter
Self-locking pliers	Vise-grips
Side or parking lamp	Parking or auxiliary light
Side or prop stand	Kick stand
Silencer	Muffler
Spanner	Wrench
Split pin	Cotter pin
Stanchion	Tube
Sulphuric	Sulfuric
Sump	Oil pan
Swinging arm	Swingarm
Tab washer	Lock washer
Top box	Trunk
Torch	Flashlight
Two/four stroke	Two/four cycle
Tyre	Tire
Valve collar	Valve retainer
Valve collets	Valve cotters
Vice	Vise
Wheel spindle	Axle
White spirit	Stoddard solvent
Windscreen	Windshield

Index

D

E

F